W9-DGE-799

2009 IBC® HANDBOOK
Fire- and Life-Safety Provisions

Douglas W. Thornburg, AIA

CD-ROM INCLUDED

THE AMERICAN INSTITUTE OF ARCHITECTS

ICC
INTERNATIONAL
CODE COUNCIL®

2009 IBC Handbook
Fire- and Life-Safety Provisions

ISBN: 978-1-58001-878-4

Cover Design:	Carmel Gieson
Illustrator/Interior Design:	Beverly Ledbetter
Manager of Development:	Doug Thornburg
Project Head:	Doug Thornburg
Project Editor:	Mary Lou Luif
Publications Manager:	Mary Lou Luif
Typesetting:	Beverly Ledbetter

INTERNATIONAL
CODE COUNCIL®

First Printing: October 2009

PRINTED IN THE U.S.A.

Preface

Internationally, code officials recognize the need for a modern, up-to-date building code addressing the design and installation of building systems through requirements emphasizing performance. The *International Building Code®* (IBC®) meets those needs by providing model code regulations that safeguard the public health and safety in all communities, large and small.

The IBC is one of a family of codes published by the International Code Council® (ICC®) that establishes comprehensive minimum regulations for building systems using prescriptive and performance-related provisions. It is founded on broad-based principles that use new materials and new building designs. Additionally, the IBC is compatible with the entire family of *International Codes®* published by the ICC.

There are three major subdivisions to the IBC:

1. The text of the IBC

2. The referenced standards listed in Chapter 35

3. The appendices

The first 34 chapters of the IBC contain both prescriptive and performance provisions that are to be applied. Chapter 35 contains those referenced standards that, although promulgated and published by separate organizations, are considered as a part of the IBC as applicable. The provisions of the appendix do not apply unless specifically included in the adoption ordinance of the jurisdiction enforcing the code.

The 2009 *IBC Handbook: Fire- and Life-Safety Provisions* is designed to present commentary only for those portions of the code for which commentary is helpful in furthering the understanding of the provision and its intent. This handbook uses many drawings and figures to help clarify the application and intent of many code provisions.

This handbook examines many design provisions for the administrative, fire- and life-safety, and inspection components of the IBC. It addresses in detail: means of egress, occupancy classification, allowable height and area, type of construction, fire protection systems and fire-resistance-rated construction.

As a benefit to the reader, ICC has included a CD-ROM containing the complete handbook. The CD-ROM also includes the complete ICC A117.1-2003 standard. This CD allows you to navigate easily through the handbook and ICC A117.1-2003, search for specific text, copy images from figures and tables, and copy and paste code provisions into correspondence or reports.

Questions or comments concerning this handbook are encouraged. Please direct any correspondence to *handbook@iccsafe.org*.

Dedication

This book is dedicated to the memory of the late Robert B. Feldner, Sr., former Superintendent of Central Inspection for the City of Wichita, Kansas. Bob, a registered architect and professional engineer, was the ultimate professional and leader, respected by all with whom he interacted. His dedication and contribution to the development and enforcement of construction codes was significant.

Foreword

How often have you heard these questions when discussing building codes: "What is the intent of this section?" Or, "How do I apply this provision?" This publication offers the code user a resource that addresses much of the intent and application principles of the major design provisions of the 2009 *International Building Code*® (IBC®).

It is impossible for building codes and similar regulatory documents to contain enough information, both prescriptive and explanatory narrative, to remove all doubt as to the intent of the various provisions. If such a document were possible, it would be so voluminous that it would be virtually useless.

Because the *International Building Code* must be reasonably brief and concise in its provisions, the user—and particularly the enforcement official—must have knowledge of the intent and background of these provisions to apply their intent appropriately. The *International Building Code* places great reliance on the judgment of the designer and building official for the specific application of its provisions. Where the designer and building official have knowledge of the rationale behind the provisions, the design and enforcement of the code will be based on informed judgment rather than arbitrariness or rote procedure.

The information that this handbook provides, coupled with the designer's and enforcement official's experience and education, will result in better use of the IBC and more uniformity in its application. As lengthy as this document may seem, it still cannot provide all of the answers to questions of code intent; that is why the background, training and experience of the reader must also be called on to properly apply, interpret and enforce the code provisions.

The preparation of a document of this nature requires consulting a large number of publications, organizations and individuals. Even so, the intent of many code provisions is not completely documented. Sometimes the discussion is subjective; therefore, individuals may disagree with the conclusions presented. It is, however, important to note that the explanatory narratives are based on many decades of experience by the authors and the other contributors to the manuscript.

Acknowledgements

The publication of this handbook is based on many decades of experience by the authors and other contributors. Since its initial publication, the handbook has become a living document subject to changes and refinements as newer code editions are released. This latest edition reflects extensive modifications based on the requirements found in the 2009 *International Building Code*.

The initial handbook, on which this document is based, was published in 1988. It was authored by Vincent R. Bush. In developing the discussions of intent, Mr. Bush drew heavily on his 25 years of experience in building regulation. Mr. Bush, a structural engineer, was intimately involved in code development work for many years.

In addition to the expertise of Mr. Bush, major contributions were made by John F. Behrens. Mr. Behrens' qualifications were as impressive as the original author's. He had vast experience as a building official, code consultant and seminar instructor. Mr. Behrens provided the original manuscript of the egress chapter and assisted in the preparation of many other chapters.

Revisions to the handbook occurred regularly over the years, with content based on the provisions of the *International Building Code* authored by Doug Thornburg, AIA, C.B.O. Mr. Thornburg, a certified building official and registered architect, has over 28 years of experience in the building regulatory profession. Previously a building inspector, plan reviewer, building code administrator, seminar instructor and code consultant, he is currently Technical Director of Product Development for the International Code Council. In his present role, Mr. Thornburg develops and reviews technical publications, reference books and resource materials relating to construction codes. He continues to present building code seminars nationally and has developed numerous educational texts. Mr. Thornburg was presented with ICC's inaugural Educator of the Year Award in 2008, recognizing his outstanding contributions in education and training.

The information and opinions expressed in this handbook are those of the present and past authors, as well as the many contributors, and do not necessarily represent the official position of the International Code Council. Additionally, the opinions may not represent the viewpoint of any enforcing agency. Opinions expressed in this handbook are only intended to be a resource in the application of the IBC, and the building official is not obligated to accept such opinions. The building official is the final authority in rendering interpretations of the code.

Contents

SCOPE AND ADMINISTRATION

Section 101 *General*

In addition to title and scope, Chapter 1 covers general subjects such as the purpose of the code, the duties and powers of the building official, performance provisions relating to alternate methods and materials of construction, applicability of the provisions, and creation of the department of building safety. This chapter also contains requirements for the issuance of permits, subsequent inspections and certificates of occupancy. The provisions in Chapter 1 are of such a general nature as to apply to the entire *International Building Code*® (IBC®).

101.2   **Scope.** The intent of the code as outlined in this section is that the IBC applies to virtually anything that is built or constructed. The definitions of "Building" and "Structure" in Chapter 2 are so inclusive that the code intends that any work of any kind that is accomplished on any building or structure comes within the scope of the code. Thus, the code would apply to a major high-rise office building as well as to a small wooden fence that might enclose a portion of the rear yard of a person's property. However, certain types of work are exempt from the permit process as indicated in the discussion of required permits in this chapter.

Whereas initially the IBC appears to address all construction-related activities, the design and construction of detached one- and two-family dwellings and townhouses, as well as their accompanying accessory structures, are intended to be regulated under the *International Residential Code*® (IRC®). However, in order for such structures to fall under the authority of the IRC, two limiting factors have been established. First, each such building is limited to a maximum height of three stories above grade plane as established by the definition of "Story above grade plane" in IRC Section R202. Where a basement space is located predominantly above the adjoining exterior ground level, it would be considered in the total number of stories above grade plane for evaluation of its regulation by the IRC. It is quite possible that a residential unit with four floor levels will be regulated by the IRC, provided that the bottom floor level is established far enough below the exterior grade that it would not qualify as a story above grade plane. For further discussion on the determination of a story above grade plane as similarly regulated in the IBC, see the commentary on Section 202. Secondly, each dwelling unit of a two-family dwelling or townhouse must be provided with a separate egress system. Although the definition of an IBC means of egress would require travel extending to the public way, for the purpose of this requirement it is acceptable to provide individual and isolated egress only until reaching the exterior of the dwelling at the required egress door. Once reaching the exterior, the building occupants could conceivably share a stairway, sidewalk or similar pathway to the public way. The IRC does not regulate egress beyond the structure itself; thus, any exit discharge conditions would only be applicable to IBC structures.

Townhouse design and construction is also regulated by the IRC. Section 202 defines a townhouse as a grouping of three or more single-family dwelling units in the same structure. The units must each extend individually from the ground to the sky, with open space provided on at least two sides of each dwelling unit. The effect of such limitations maintains the concept of "multiple single-family dwellings."

The requirement for open space on a minimum of two sides of each townhouse unit allows for interpretation regarding the degree of openness. Although not specific in language, the provision intends that each townhouse be provided with a moderate degree of exterior wall, thus allowing for adequate fire department access to each individual unit.

Structures such as garages, carports and storage sheds are also regulated by the IRC where they are considered accessory to the residential buildings previously mentioned.

Such accessory buildings are limited to 3,000 square feet (279 m²) in floor area and two stories in height.

Even though the IRC may use the IBC as a reference for certain design procedures, the intent is to solely utilize the IRC for the design and construction of one- and two-family dwellings, multiple single-family dwellings (townhouses), and their accessory structures. This does not preclude the use of the IBC by a design professional for the design of the types of residential buildings specified. However, unless specifically directed to the IBC by provisions of the IRC, it is not the intent of the IRC to utilize the IBC for provisions not specifically addressed. For example, the maximum allowable floor area of a residence based on the building's type of construction is not addressed in the IRC. Therefore, there is no limit to the floor area permitted in the dwelling unit. It would not be appropriate to utilize the IBC to limit the residence's floor area based upon construction type.

Appendices. A number of subjects are addressed in Appendices A through K. The topics **101.2.1** range from detailed information on the creation of a board of appeals to more general provisions for grading, excavation and earthwork construction. Although the code clearly indicates that the appendices are not considered a part of the IBC unless they are specifically adopted by the jurisdiction, this does not mean they are of any less worth than those set forth in the body of the code. Although there are several reasons why a set of code requirements is positioned in the appendix, the most common reason is that the provisions are limited to a small geographic location or are of interest to only a small number of jurisdictions.

Jurisdictions have the ability to adopt any or all of the appendices based upon their own needs. However, just because an appendix has not been adopted does not lessen its value as a resource. In making decisions of interpretation of the code, as well as in evaluating alternate materials and methods, the provisions of an appendix may serve as a valuable tool in making an appropriate decision. Even in those cases where a specific appendix is not in force, the information it contains may help in administering the IBC.

Intent. Various factors are regulated that contribute to the performance of a building in **101.3** regard to the health, safety and welfare of the public. The IBC identifies several of these major factors as those addressing structural strength, egress capabilities, sanitation and other environmental issues, fire- and life-safety concerns, and energy conservation. In addition, the safety of fire fighters and emergency personnel responding to an emergency situation is an important consideration. The primary goal of the IBC is to address any and all hazards that are attributed to the presence and use of a jurisdiction's buildings and structures, and to safeguard the public from such hazards.

The intent of the code is more inclusive than most people realize. A careful reading will note that in addition to providing for life safety and safeguarding property, the code also intends that its provisions consider the general welfare of the public. This latter item, *general welfare*, is not so often thought of as being part of the purpose of a building code. However, in the case of the IBC, safeguarding the public's general welfare is a part of its intent, which is accomplished, for example, by provisions that ameliorate the conditions found in substandard or dangerous buildings. Moreover, upon the adoption of a modern building code such as the IBC, the general level of building safety and quality is raised. This in turn contributes to the public welfare by increasing the tax base and livability. Additionally, slum conditions are reduced, and the subsequent reduction of unsanitary conditions contributes to safeguarding the public welfare. For example, the maintenance requirements of Section 3401.2 apply to all buildings, and as a result, the continued enforcement of the IBC slows the development of substandard conditions. A rigorous enforcement of Section 3401.2 will actually reduce the conditions that contribute to the deterioration of the existing building stock. Thus, public welfare is enhanced by the increased benefits that inure to the general public of the jurisdiction as a result of the code provisions.

Referenced codes. A number of other codes are promulgated by the International Code **101.4** Council® (ICC®) in order to provide a full set of coordinated construction codes. Seven of those companion codes are identified in this section, as they are specifically referenced in one or more provisions of the IBC. The adoption of the IBC does not automatically include the full adoption of the referenced codes, but rather only those portions specifically referenced by the IBC.

For example, Section 1503.4 also requires that roof drainage systems comply with the *International Plumbing Code®* (IPC®). As a result, when the IBC is adopted, so are the roof drainage provisions of the IPC. The extent of the reference is roof drainage; therefore, that is the only portion of the IPC that is applicable. Broader references are also provided, such as many of the references to the *International Fire Code®* (IFC®). Section 307.1.1 requires that hazardous materials in any quantity conform to the requirements of the IFC. Although the entire IFC may not be adopted by the jurisdiction, the provisions applicable to hazardous materials are in force with the adoption of the IBC.

Section 102 *Applicability*

102.1   **General.** Where there is a conflict between two or more provisions found in the code as they relate to differences of materials, methods of construction or other requirements, the most restrictive provision will govern. Typically, the code will identify how the varying requirements should be applied. For example, the occupant load along with the appropriate factor from Section 1005.1 is used to calculate the total width required for egress stairways—often referred to as the *calculated* width. Section 1009.1 also addresses the minimum required width for a stairway based on the absolute width necessary for use of a stairway under any condition, deemed to be the *component* width. When determining the proper width required by the code, the more restrictive, or wider, stairway width would be used. See Application Example 102-1.

In addition, where a conflict occurs between a specific requirement and a general requirement, the more specific provision shall apply. Again, the IBC provisions typically clarify the appropriate requirement that is to be applied. As an example, Section 1009.4.2 limits the height of stair risers to 7 inches (178 mm) as a general requirement for stairways. However, Section 1028.11.2 allows for a maximum riser height of 8 inches (203 mm) for aisle stairs serving assembly seating areas. Because the greater riser height is only permitted for a specific stair condition, rather than for all stairways in general, it is intended to apply where those special means of egress provisions established in Section 1028 are applicable.

Occasionally it is difficult, during the comparison of two different code provisions, to determine which is the general requirement and which is the specific requirement. In some cases, both requirements are specific, but one is more specific than the other. It is important that the intent of this section be applied in reviewing the proper application of the code. Where it can be determined that one provision is more specific in its scope than the other provision, the more specific requirement shall apply, regardless of whether it is more or less restrictive in application.

Application Example 102-1

GIVEN: An occupant load of 130 assigned to each of two stairways in a sprinklered office building.

DETERMINE: The required minimum width of each stairway.

    1. Based on Section 1005.1, the minimum calculated width would be:
       0.3 inches/occupant x 130 occupants = 39 inches

    2. Based on Section 1009.1, the minimum required width would be 44 inches.

SOLUTION: Therefore, the more restrictive condition, 44 inches, would apply.

For SI: 1 inch = 25.4 mm.

CONFLICTING REQUIREMENTS

Referenced codes and standards. Differences between the code and the various standards **102.4** it references are to be expected. Unlike the companion International Codes®, there is not necessarily a conscious effort to see that the publications are completely compatible with each other. As a result, it is critical that the code indicate that its provisions are to be applied over those of a referenced standard where such differences exist. For example, the provisions of NFPA 13R addressing sprinkler systems in residential occupancies allow for the omission of sprinklers at specified exterior locations, including porches and balconies. However, the provisions of IBC Section 903.3.1.2.1 mandate sprinkler protection for such areas where specific conditions exist. In this case, the provisions of the IBC for sprinkler protection would apply regardless of the allowances contained in NFPA 13R.

Section 103 *Department of Building Safety*

This section establishes the department of building safety as the jurisdictional enforcement agency charged with administering the IBC. The term *building official*, as used in the IBC, represents the individual appointed by the jurisdiction to head the department of building safety. Although many jurisdictions utilize the title *building official* to recognize the individual in charge of the building safety department, there are many other titles that are used. These include *Chief Building Inspector*, *Superintendent of Central Inspection*, *Director of Code Enforcement* and various other designations. Regardless of the title selected for use by the individual jurisdiction, the IBC views all of these as equivalent to the term *building official*.

The building official, in turn, appoints personnel as necessary to carry out the duties and responsibilities of the department. Such staff members (deputies), including inspectors, plans examiners and other employees, are empowered by the building official to carry out those functions set forth by the jurisdiction. Where the IBC references the building official in any capacity, the code reference also includes any deputies who have been granted enforcement authority by the building official. Where an inspector or plan reviewer makes a decision of interpretation, they are assuming the role of building official in arriving at that decision. There is an expectation on behalf of the jurisdiction that such employees possess the knowledge and experience to take on this responsibility. The failure to grant appropriate authority will often result in both ineffective and inefficient results.

For those jurisdictions desiring guidelines within the text of the code for the selection of department personnel, Appendix A addresses minimum employee qualifications for various positions. Experience and certification criteria for building officials, chief inspectors, inspectors and plans examiners are set forth in this appendix chapter.

Section 104 *Duties and Powers of Building Official*

General. The IBC is designed to regulate in both a prescriptive and performance manner. **104.1** An extensive number of provisions have been intentionally established to allow for jurisdictional interpretation based on the specifics of the situation. This section establishes the building official's authority to render such interpretations of the IBC. In addition, the building official may adopt policies and procedures that will help clarify the application of the code. Although having no authority to provide variances or waivers to the code requirements, the building official is charged with interpreting and clarifying the provisions found in the IBC, provided that such decisions are in conformance with the intent and purpose of the code.

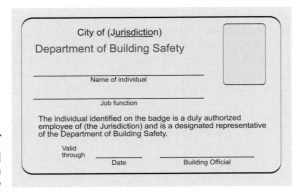

City of (Jurisdiction)

Department of Building Safety

_____
Name of individual

_____
Job function

The individual identified on the badge is a duly authorized
employee of (the Jurisdiction) and is a designated representative
of the Department of Building Safety.

Valid
through
_____   _____
Date   Building Official

Figure 104-1
**Personnel
identification
badge**

The authority to interpret the intended application of the IBC is a powerful tool available to the building official. With such authority comes a great degree of responsibility. Such interpretations must be consistent with the intent and purpose of the code. It is therefore necessary that all reasonable efforts be made to determine the code's intent in order to develop an appropriate interpretation. Various sources should be consulted to provide a broad background from which to make a decision. These could include discussions with peers, as well as information found in various educational texts and technical guides. However, it must be stressed that the ultimate responsibility for determining the appropriateness of an interpretation lies with the jurisdictional building official, and all other opinions, both verbal and written, are just that, opinions. The building official must never relinquish his or her authority to others in the administration of these very important interpretive powers. See also the discussion on alternative materials, design and methods of construction in Section 104.11.

104.4 **Inspections.** Those inspections required under the provisions found in Section 110 are to be performed by the building official or by authorized representatives of the building official. It is also acceptable that outside firms or individuals be utilized for inspections, provided such firms or individuals have been approved by the building official. This option may allow for better use of available resources. Written reports shall be provided for each inspection that is made.

104.5 **Identification.** For the benefit of all individuals involved, inspection personnel of the department of building safety are mandated to carry proper identification. The display of an identification card or badge, an example of which in shown in Figure 104-1, signifies the function and authority of the individual performing the inspection.

104.6 **Right of entry.** This section is compatible with Supreme Court decisions since the 1960s regarding acts of inspection personnel seeking entry into buildings for the purpose of making inspections. Under present case law, an inspection may not be made of a property, whether it be a private residence or a business establishment, without first having secured permission from the owner or person in charge of the premises. If entry is refused by the person having control of the property, the building official must obtain an inspection warrant from a court having jurisdiction in order to secure entry. The important feature of the law regarding right of entry is that entry must be made only by permission of the person having control of the property. Lacking this permission, entry may be gained only through the use of an inspection warrant.

If entry is again refused after an inspection warrant has been obtained, the jurisdiction now has recourse through the courts to remedy this situation. One avenue is to obtain a civil injunction in which the court directs the person having control of the property to allow inspection. Alternatively, the jurisdiction can initiate proceedings in criminal court for punishment of the person having control of the property. It cannot be repeated too strongly that criminal court proceedings should never be initiated against an owner or other person having control of the property if an inspection warrant has not been obtained. Because the consequences of not following proper procedures can be so devastating to a jurisdiction if a suit is brought against it, the jurisdiction's legal officer should always be consulted in these matters.

104.8 **Liability.** It is the intent of the IBC that the building official not become personally liable for any damage that occurs to persons or property as a result of the building official's acts so long as he or she acts in good faith and without malice or fraud. This protection is also extended to any member of the Board of Appeals, as well as any jurisdictional employee charged with enforcement of the IBC. However, there seems to be an increasing trend in the courts to find civil officers personally liable for careless acts. This section requires that the jurisdiction defend the building official or other protected party if a suit is brought against him or her. Furthermore, the code requires any judgment resulting from a suit to be assumed by the jurisdiction. However, regardless of this language in the code, the jurisdiction may elect to not defend the building official on the basis, for example, that he or she acted carelessly.

Case law regarding tort liability of building officials is constantly in a state of flux, and old doctrines may not now be applicable. Therefore, the legal officer of the jurisdiction should always be consulted when there is any question about liability.

Modifications. The provisions of this section allow the building official to make **104.10** modifications to the requirements of the code under certain specified circumstances. The building official may modify requirements if it is determined that strict application of the code is impractical and, furthermore, that the modification is in conformity with the intent and purpose of the code. Without this provision in the IBC, the building official has very little discretionary enforcement authority and, therefore, would have to enforce the specific wording in the code, no matter how unreasonable the application would be.

The code does not intend to allow the building official to issue a variance to the provisions of the code to permit, for example, the use of only a single exit where two are required. This is clearly not in conformity with the intent and purpose of the code, no matter how difficult it may be to meet the requirements. In fact, the code is very specific that any modification cannot reduce health, accessibility, structural and fire- and life-safety requirements.

Where the building official grants a modification under this section, the details of such an action shall be recorded. This document must then be entered into the files of the department of building safety. By providing a written record of the action taken and maintaining a copy of that action in the department files, the building official always has access to the decision-making process and final determination of his or her action should there be a need to review the decision.

Although it is expected that a permanent record be available for future reference when a modification is accepted, there is perhaps an even more important reason for the recording and filing of details of the approving action. The willingness to document and archive the modification action indicates the confidence of the building official in the decision that was made. A reluctance to maintain a record of the action taken typically indicates a lack of commitment to the action taken.

Alternative materials, design and methods of construction and equipment. This **104.11** section of the IBC may be one of the most important. It allows for the adoption of new technologies in materials and building construction that currently are not covered by the code. Furthermore, it gives the code even more of a performance character. The IBC thus encourages state-of-the-art concepts in design, construction and materials as long as they meet the performance intended by the code. When evaluating the alternative methods under consideration, the building official must review for equivalency in quality, strength, effectiveness, fire resistance, durability and safety. It is expected that all alternatives, once presented to the building official for review and approval, be thoroughly evaluated by the building department for compliance with this section. If such compliance can be established, the alternatives are deemed to be acceptable.

The provisions of this section, similar to those of Sections 104.1 and 104.10, reference the *intent* of the code. It is mandated that the building official, when evaluating a proposed alternate to the code, only approve its use where it can be determined that it complies with the intent of the specific code requirements. Thus, it is the responsibility of the building official to utilize those resources necessary to understand the intended result of the code provisions. Only then can the code be properly applied and enforced.

Similar to the approach taken where modifications are requested under the criteria of Section 104.10, the request for acceptance under this section should be made in writing to the building official. At a minimum, the submittal should include: (1) the specific code section and requirement, (2) an analysis of the perceived intent of the provision under review, (3) the special reasons as to why strict compliance with the code provision is not possible, (4) the proposed alternate, (5) an explanation of how the alternate meets or exceeds the intended level of compliance, and (6) a request for acceptance of the alternate material, design or method of construction.

104.11.1 **Research reports.** Whereas the provisions of Section 104.11 grant the building official broad authority in accepting alternative materials, designs and methods of construction, the process of evaluating such alternatives is often a difficult and complicated task. Valid research reports, including those termed *evaluation reports*, can address and delineate a review of the appropriate testing procedures to support the alternative as code compliant. The use of a research report may be helpful in reducing additional testing or documentation that is necessary to indicate compliance. It is important that the building official evaluate not only the information contained within the research report, but also the technical expertise of the individual or firm issuing the report. It must be noted, however, that a research report is simply a resource to the building official to assist in the decision-making process. The research report itself does not grant approval, as acceptance is still the under the sole authority of the building official.

One of the most commonly utilized research reports is the ICC Evaluation Service (ICC-ES) Evaluation Report. ICC-ES is a nonprofit, public-benefit corporation that does technical evaluations of building products, components, methods and materials. If it is found that the subject of an evaluation complies with code requirements, then ICC-ES publishes a report to that effect and makes the report available to the public. However, ICC-ES Evaluation Reports are only advisory. The authority having jurisdiction is always the final decision maker with respect to acceptance of the product, material or method in question.

104.11.2 **Tests.** The provisions of this section provide the building official with discretionary authority to require tests to substantiate proof of compliance with code requirements. The application of these provisions should be restricted to those cases where evidence of compliance is either nonexistent or involves actions considered to be impractical. Certainly, when the use of an alternative material, design or method of construction is requested under the provision of Section 104.11, test information can be quite beneficial to the building official. There may also be insufficient evidence of compliance that can be substantiated through alternate tests.

An example would be the placement of concrete that the quality-control measures (i.e., cylinder tests) did not prove to be complying with minimum strength requirements. Testing of core samples or perhaps use of nondestructive test methods might be appropriate to demonstrate compliance.

The provisions also specify that the tests be those that are specifically enumerated within the adopted construction regulations or, as an alternate, be those of other recognized national test standards. Where test standards do not exist, the building official has the authority to determine the test procedures necessary to demonstrate compliance. In addition to determining appropriate test methods or procedures, the building official is mandated to maintain records of such tests in accordance with local or state statutes.

Section 105 *Permits*

This section covers those requirements related to the activities of the building department with respect to the issuance of permits. The issuance of permits, plan review and inspection of construction for which permits have been issued constitute the bulk of the duties of the typical department of building safety. It is for this reason that the code goes into detail regarding the permit-issuance process. Additionally, the code provides detailed requirements for the inspection process in order to help ensure that the construction for which the inspections are made complies with the code in all respects.

Required. Prior to obtaining a permit, the owner of the property under consideration, or the **105.1** owner's authorized agent, must apply to the building official for any necessary permits that are required by the jurisdiction. One or more permits may be required to cover the various types of work being accomplished. In addition to building permits, which address new construction, alterations, additions, repairs, moving of structures, demolition or change in occupancy, trade permits are required to erect, install, enlarge, alter, repair, remove, convert or replace any electrical, gas, mechanical or plumbing system. It is evident that almost any work, other than cosmetic changes, must be done under the authority of a permit.

Typically, a permit is required each time a distinct activity occurs that is regulated under the code. However, certain alterations to previously approved systems can be performed under an annual permit authorized by the building official. Electrical, gas, mechanical or plumbing installations are eligible for such consideration when one or more qualified trade persons are employed by the person, firm or corporation who owns or operates the building, structure or premises where the work is to take place. In addition, the qualified individuals must regularly be present at the building or site.

Work exempt from permit. It would seem that the IBC should require permits for any type **105.2** of work that is covered by the scope of the code. However, this section provides limited applications for exempted work. This section not only exempts certain types of building construction from permits, but also addresses electrical, gas, mechanical and plumbing work that is of such a minor nature that permits are not necessary.

It is further the intent of the IBC that even though work may be exempted from a permit, such work done on a building or structure must still comply with the provisions of the code. As indicated in Section 101.2, the scope of the IBC is virtually all-inclusive. This may seem to be a superfluous requirement where a permit is not required. However, this type of provision is necessary to provide that the owner, as well as any design professional or contractor involved, be responsible for the proper and safe construction of all work being done.

A common example is a small, one-story detached accessory structure such as a storage shed. Although the code does not require a permit for an accessory building not exceeding 120 square feet (11 m^2) in floor area, all provisions in the code related to a Group U occupancy must still be followed.

Application for permit. In this section, the IBC directs that a permit must be applied for, **105.3** and describes the information required on the permit application. The permit-issuance process, as envisioned by the IBC, is intended to provide records within the code enforcement agency of all construction activities that take place within the jurisdiction and to provide orderly controls of the construction process. Thus, the application for permit is intended to describe in detail the work to be done. In this section, the building official is directed to review the application for permit. This review is not a discretionary procedure, but is mandated by the code.

The code also charges the building official with the issuance of the permit when it has been determined that the information filed with the application shows compliance with the IBC and other laws and ordinances applicable to the building at its location in the jurisdiction. The building official may not withhold the issuance of a permit if these conditions are met. As an example, the building official would be in violation in withholding the issuance of a building permit for a swimming pool because an adjacent cabana was previously constructed without a permit.

Validity of permit. The code intends that the issuance of a permit should not be construed **105.4** as permitting a violation of the code or any other law or ordinance applicable to the building. In fact, the IBC authorizes the building official to require corrections if there were errors in the approved plans or permit application at the time the permit was issued. The building official is further authorized to require corrections of the actual construction if it is in violation of the code, although in accordance with the plans. Moreover, the building official is further authorized to invalidate the permit if it is found that the permit was issued in error or in violation of any regulation or provision of the code.

Although it may be poor public relations to invalidate a permit or to require corrections of the plans after they have been approved, it is clearly the intent of the code that the approval of plans or the issuance of a permit may not be done in violation of the code or of other pertinent laws or ordinances. As the old saying goes, "Two wrongs do not make a right."

105.5   Expiration. The IBC anticipates that once a permit has been issued, construction will soon follow and proceed expeditiously until completion. However, this ideal procedure is not always the case and, therefore, the code makes provisions for those cases where work has not started, or alternately where the work, after being started, has been suspended for a period of time. In these cases, the IBC allows a period of 180 days to transpire before the permit becomes void. The code then requires that a new permit be obtained. It is assumed by the code that the department of building safety will have expended some effort in follow-up inspections of the work, etc., and, therefore, the original permit fee must be retained in order to compensate the agency for the work. The building official has the authority to grant one or more extensions of time, provided the permit holder can demonstrate a justifiable cause as to why the permit should not be invalidated. The time period for such extensions cannot exceed 180 days; however, additional extensions may be granted if approved by the building official.

There are several reasons why it is important to establish a limitation on the validity of a permit, such as purging the department files of inactive permits. Additionally, it keeps the project on track with the code edition in effect at the time of permit issuance.

Section 107 *Submittal Documents*

Plans, specifications and other construction documents, along with other applicable data, must be filed with the permit application. Such submittal documents are intended to graphically depict the construction work to be done. The IBC sets forth the necessary information that must be provided to the building official at the time of plans submittal, as well as procedures for deferred submittals.

107.1   General. In this section, the IBC directs that at the time of application for a permit, construction documents and other essential data on the project be submitted. The code requires that plans, engineering calculations and any other information necessary to describe the work to be done be filed along with the application for a permit. A statement of special inspections that may be required by Section 1705 is also to be submitted with the application for a permit. Based upon the statutes of the jurisdiction, the plans, specifications and other documents may need to be prepared by a registered design professional. Under special circumstances, the building official may also require that the submittal documents be prepared by a registered design professional even when not mandated by jurisdictional statute. The building official is permitted to waive the requirement for the filing of plans and other data, where not required by statute to be prepared by a registered design professional, provided the building official is assured that the work for which the permit is applied is of such a nature that plans or any other data are not necessary in order to indicate and obtain compliance with the code.

107.3   Examination of documents. In this section, the building official is directed to review the plans, specifications and other submittal documents filed with the permit. The building official is not at liberty to check only a portion of the plans. On this basis, the structural drawings, as well as the engineering calculations, must be checked in order for the building official to provide a full examination.

107.3.4   Design professional in responsible charge. The building official has the authority to require the designation of a design professional to act in responsible charge of the work being performed. The function of such an individual is to review and coordinate submittal documents prepared by others and, if necessary, coordinate any deferred or phased submittal items. In some cases, the design of portions of the building has been completed and may be dependent on the manufacturer of proposed prefabricated elements, such as for truss drawings. Therefore, the code specifically allows deferring the submittal of portions of plans and specifications.

Section 108　*Temporary Structures and Uses*

This section authorizes the erection of temporary buildings and structures such as those erected at construction sites. The regulation of temporary viewing stands and other miscellaneous, temporary structures would also fall under this section. The following are key provisions for temporary buildings and structures:

1. They are erected by special permit.

2. They are erected for a limited period of time.

3. They must conform to the requirements of the IBC for structural strength, fire safety, means of egress, accessibility, light, ventilation and sanitation.

Section 109　*Fees*

Permits required by the jurisdiction are not valid until the appropriate fees have been paid. The IBC provides for each individual jurisdiction to establish its own schedules for permit fees. The fees collected by the department of building safety are typically set at a level to adequately cover the cost to the department for services rendered.

Building permit valuations. The code uses the concept of valuation to establish the permit **109.3** fee. This concept is based on the proposition that the valuation of a project is related to the amount of work to be expended in the various aspects of administering the permit. Also, there should be some excess in the permit fee to cover departmental overhead. Essentially, the valuation is considered as the cost of replacing the building. The valuation also includes any electrical, plumbing and mechanical work, even though separate permits may be required for the mechanical, plumbing and electrical trades.

To provide some uniformity in the determination of valuation so that there is a consistent base for the assignment of fees, the IBC directs the building official to determine the value of the building. To assist in obtaining uniformity, the International Code Council periodically publishes "Building Valuation Data (BVD)" in its *Building Safety Journal*®. In addition, the valuation data can be accessed on ICC's website. Thus, building officials may utilize a common base in their determination of the value of buildings. However, ICC strongly recommends that all jurisdictions and other interested parties actively evaluate and assess the impact of the BVD table before utilizing it in their current code enforcement activities. As an option, any other appropriate method may also be used by the jurisdiction as the basis for determining the proper valuation of the work.

Work commencing before permit issuance. When work requiring a permit is started **109.4** without such a permit, the IBC allows the building official to assess an additional fee above and beyond the required permit fees. This fee is intended to compensate the department of building safety for any additional time and effort necessary in evaluating the work initiated without a permit. It may often be necessary for the building official to cause an investigation to be made of the work already done. The intent of the investigation is to determine to what extent the work completed complies with the code, and to describe with as much detail as possible the work that has been completed.

109.6 **Refunds.** This section authorizes the building official to establish a policy for partial or complete refunds of fees paid to the department. Although not specified in the code, there are a variety of reasons why some level of a fee refund would be appropriate. One instance would be where the permit fee is collected in error. Another reason for authorizing a refund of the fees paid would be that circumstances beyond the control of the applicant caused delays and the eventual expiration of the building permit. Typically, the building official will withhold from the refund any monies expended by the department for related administrative activities that have taken place. It should be noted that the building official, when approving a fee refund, is authorizing the disbursement of public funds. Therefore, the building official must be sure that there is good cause for the refund to the applicant.

Section 110 *Inspections*

The inspection function is arguably the most critical aspect of building department operations. An important concept views that inspections, as with the issuance of permits, that presume to give authority to violations of the code are invalid. In general, the IBC charges the permit applicant with the responsibility of ensuring that the work to be inspected remain accessible and exposed until it is approved. Any expense incurred in the process of providing for an accessible and exposed inspection, such as the removal of gypsum board or insulation, is not to be borne by the building official or jurisdiction. Therefore, it is critical that work should not proceed beyond the point where an inspection is required.

110.3 **Required inspections.** The code mandates seven typical inspections during the progress of construction of a building, as follows:

1. Footing and foundation inspection

2. Concrete slab or under-floor inspection

3. Frame inspection

4. Lath or gypsum board inspection

5. Fire-resistant penetrations inspection

6. Energy efficiency inspection

7. Final inspection

The IBC also gives the building official authority to require and make other inspections where necessary to determine compliance with the code and other laws and regulations enforced by the department. This includes those special inspections addressed in Section 1704. The need for either periodic or continuous inspection is established for construction that requires special expertise to ensure compliance. Another type of inspection is required in flood hazard areas. After the lowest floor level is established, but prior to additional vertical construction, the lowest floor elevation must be determined.

In each case for the required inspections identified in Section 110.3, the IBC is very specific as to how far the construction must have progressed prior to a request for an inspection. If it is necessary to make a reinspection because work has not progressed to the point where it is ready for inspection, the building official may charge an additional fee under the general provisions of Section 109. A reinspection fee is not specifically mandated, however, and normally would not be assessed unless the person doing the work continually calls for inspections before the work is ready for inspection. Such actions would typically cause increased costs to the jurisdiction not covered by the original building permit.

110.5 **Inspection requests.** In general, the IBC charges the individual holding the building permit with notifying the department of building safety when it is time to make an inspection. The permit holder may also authorize one or more agents for this responsibility. The code places a duty upon the permit holder or their authorized agent to have the work to be inspected accessible and exposed so it can be evaluated as to code compliance.

Approval required. The code intends that no work shall be done beyond the point where an inspection is required until the work requiring inspection has been approved. Moreover, it is intended that work requiring inspection not be covered until it has been inspected and approved. **110.6**

Section 111 *Certificate of Occupancy*

The tool that the building official utilizes to control the uses and occupancies of the various buildings and structures within the jurisdiction is the certificate of occupancy. The IBC makes it unlawful to use or occupy a building or structure unless a certificate of occupancy has been issued for that use. Furthermore, the code imposes the duty of issuing a certificate of occupancy upon the building official when he or she is satisfied that the building, or portion thereof, complies with the code for the intended use and occupancy.

Prior to use or occupancy of the building, the building official shall perform a final inspection as addressed in Section 110.3.10. If no violations of the code and other laws enforced by the department of building safety are found, the building official is required to issue a certificate of occupancy. Figure 111-1 illustrates the information that must be provided on the certificate of occupancy.

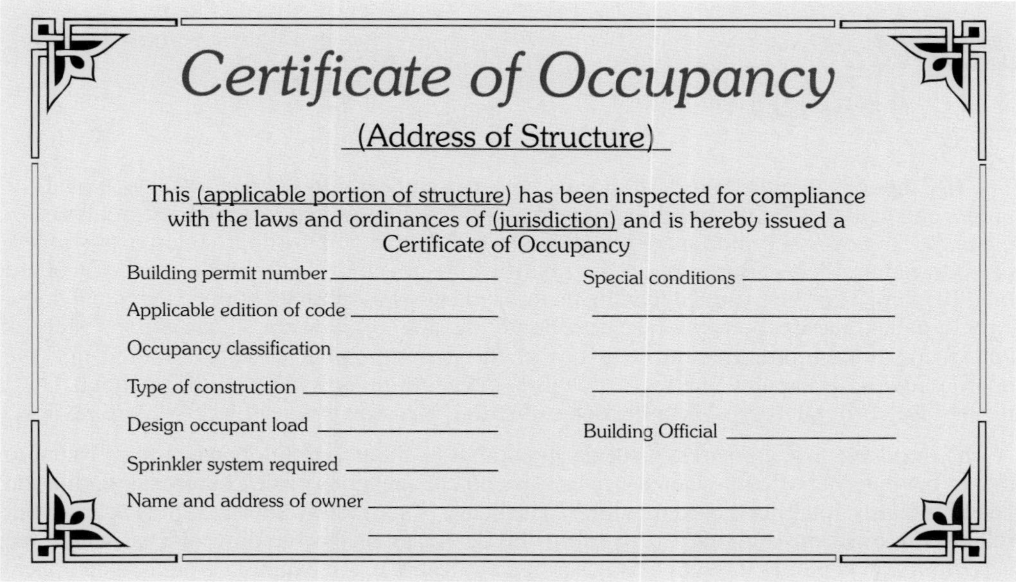

Certificate of Occupancy

(Address of Structure)

This (applicable portion of structure) has been inspected for compliance with the laws and ordinances of (jurisdiction) and is hereby issued a Certificate of Occupancy

Building permit number ____________ Special conditions ____________

Applicable edition of code ____________

Occupancy classification ____________

Type of construction ____________

Design occupant load ____________ Building Official ____________

Sprinkler system required ____________

Name and address of owner ____________

Figure 111-1 Certificate of occupancy

Where a portion of a building is intended to be occupied prior to occupancy of the entire structure, the building official may issue a temporary certificate of occupancy. This situation would occur where partial occupancy is requested prior to completion of all work authorized by the building permit. Prior to issuance of a temporary certificate of occupancy, it is critical that the building official ascertain that those portions to be occupied provide the minimum levels of safety required by the code. In addition, the building official shall establish a definitive length of time for the temporary certificate of occupancy to be valid.

In essence, the certificate of occupancy certifies that the described building, or portion thereof, complies with the requirements of the code for the intended use and occupancy, except as provided for existing buildings undergoing a change of occupancy as regulated by Chapter 34. However, any certificate of occupancy may be suspended or revoked by the building official under one of three conditions: (1) where the certificate is issued in error, (2) where incorrect information is supplied to the building official or (3) where it is determined that the building or a portion of the building is in violation of the code or any other ordinance or regulation of the jurisdiction. The most common reason for suspending or revoking a certificate of occupancy is that the building is being used for a purpose other than that intended when approval for occupancy was granted. When a permit is issued, plans reviewed, inspections made, and a certificate of occupancy given, there is an expectation that the building will be used for specific activities. As such, the hazards associated with such activities can be addressed. However, where an unanticipated use occurs, there is the potential that the necessary safeguards are not in place to address the related hazards. An unauthorized use can be intentional or unintentional; however, it is no less a concern until the use is discontinued or the necessary remedies are in place.

Section 112   *Service Utilities*

The building official is authorized to control the connection release for any service utility where the connection occurs to a building or system regulated by the IBC. Such authority also includes the temporary connection of the service utility. Perhaps even more important, the building official is also granted authority for the disconnection of service utilities where it has been determined that an immediate hazard to life or property exists. As in all administrative functions, it is important that due process be followed.

Section 113   *Board of Appeals*

The IBC intends that the board of appeals have the authority to hear and decide appeals of orders and decisions of the building official relative to the application and interpretations of the code. However, the code specifically denies the authority of the board relative to waivers of code requirements. Based upon the qualification level set forth by the code for board membership, it can be assumed that the board is intended as fundamentally a technical body. Any broader authority may place the board outside of its area of expertise, such as addressing internal administrative issues of the department. It should be noted that the board's role is to hear and decide on appeals of decisions made by the building official. Until the building official makes his or her determination, the issue is not subject to board review.

The importance of a board of appeals should not be taken lightly. Its role is equivalent to that of the building official when it comes to technical questions placed before them, thus the need for highly-qualified board members. The board is not merely advisory in its actions, but rather is granted the authority to overturn the decisions of the building official where the determination is within the scope as described by the code. An example would be an appeal requesting the use of an alternate method of construction previously denied by the building official. The alternate method would be permitted if found by the board to be equivalent or better than that set forth in the code. Where the board of appeals is desired to take on a more expansive role, such as code adoption functions or contractor licensing oversight, those duties should be specifically granted by the jurisdiction. More detailed information on the qualifications and duties of a board of appeals is found in IBC Appendix B.

Section 114   *Violations*

The provisions of this section establish that violations of the code are considered unlawful and such violations shall be abated. Where necessary, a notice of violation may be served by the building official to the individual responsible for the work that is in violation of the code. Further action may be taken where there is a lack of compliance with the notice. The jurisdiction shall establish penalties based upon the various specified violations.

Section 115 *Stop Work Order*

The stop work order is a tool authorized by the IBC to enable the building official to demand that work on a building or structure be temporarily suspended. Intended to be utilized only under rare circumstances, this order may be issued where the work being performed is dangerous, unsafe or significantly contrary to the provisions of the code. A stop work order is often a building official's final method in obtaining compliance on issues of extreme importance. All other reasonable avenues should be considered prior to the issuance of such an order.

The stop work order shall be a written document indicating the reason or reasons for the work to be suspended. It shall also identify those conditions where compliance is necessary before the cited work is allowed to resume. All work addressed by the order shall immediately cease upon issuance. It is important that the stop work order be presented to either the owner of the subject property, the agent of the owner or the individual doing the work. Only after all issues have been satisfied may work continue. Because of the potential consequences involved with *shutting down* a construction project, it is critical that the appropriate procedures be followed. Although a stop work order can be issued at any time during the construction process, the most common application of the provision is where work has commenced before, or without, the issuance of a building permit. The order is given to notify those affected parties that all work be stopped until a valid permit is obtained. An example of a stop work order is shown in Figure 115-1.

STOP WORK

DEPARTMENT OF BUILDING SAFETY

NOTICE

This building has been inspected and
☐ Footing/foundation
☐ Concrete slab/underfloor
☐ Framing
☐ Lath/gypsum board
☐ Energy efficiency
☐ Rough electrical
☐ Rough gas
☐ Rough plumbing
☐ Rough mechanical
☐ Final inspection

IS NOT ACCEPTABLE

Please call _____
before any further work is done.

_____
Date   Inspector

Do not remove this notice.

- - - - - - - - - - - - - - - - - -
DETACH and bring this portion of card with you.

LOCATION: _____

Date _____

Building Official

Phone

Figure 115-1
Stop work order

Section 116 *Unsafe Structures and Equipment*

The provisions of this section are intended to define what constitutes an unsafe building and unsafe use of a building. Unsafe buildings and structures are considered public nuisances and require repair or abatement. The abatement procedures are indicated in this section, including the creation of a report on the nature of the unsafe condition. Written notice shall be provided and the method of service of such notice is specified in Section 116.4. Where restoration or repair of the building is desired, the provisions of Chapter 34 for existing buildings are appropriate.

KEY POINTS

- The IBC provides minimum standards to safeguard the health, safety and welfare of the public.

- Rendering interpretations of specific code provisions is the responsibility of the building official, to be based on the purpose and intent of the code.

- Modifications to specific IBC provisions may be acceptable where strict application of the code is impractical.

- Alternative designs, materials and methods of construction to those detailed in the code are to be evaluated by the building official based upon an equivalency to the prescribed regulations.

- Tests may be mandated by the building official in order to verify compliance with the code.

- The issuance of permits, plan review and inspection of construction for which permits have been issued constitutes the bulk of the duties of the typical department of building safety.

- The code intends that the issuance of a permit should not be construed as allowing a violation of the code nor any other local ordinance applicable to the building.

- Submittal documents must be appropriately submitted to the building official for review and approved again prior to receiving a building permit.

- The building official has the authority to require the designation of a registered design professional to function as an individual to review and coordinate submittal documents prepared by others.

- A certificate of occupancy, granted by the building official, indicates that the structure is lawfully permitted to be occupied.

- The board of appeals provides a mechanism for individuals to challenge the interpretation of the building official on technical issues in the IBC.

DEFINITIONS

Section 201 *General*

A number of definitions are applicable specifically to the *International Building Code®* (IBC®) and may not have an appropriate definition for code purposes in the dictionary. Therefore, definitions are found throughout the IBC to assist the user in the proper application of the requirements. The IBC provides definitions in Chapter 2 of terms that are generally used in a number of varied locations throughout the code. In addition, there are also definitions within most of the chapters of the IBC that are intended to apply primarily within that chapter. In order to determine whether or not a definition for a specific item is contained within the IBC, Chapter 2 must be examined. If the specific definition under review is found in a specific chapter other than Chapter 2, Section 202 will identify the appropriate reference. Although infrequent, the definitions of some terms are contained within the text of the requirement. For example, the definition of *day care* is implied in the description of Group E occupancies. Other frequently used and significant terms are undefined (i.e., 1-hour fire-resistance-rated construction), and their meaning can be discerned only from their context. There are numerous definitions in Chapter 2, but only selected definitions are included in this commentary.

An important feature of this section is the requirement that ordinarily accepted meanings be utilized for definitions that are not provided in the code. Such meanings are based upon the context within which the term or terms appear. The code defines terms that have specific intents and meanings insofar as the code is concerned, and leaves it up to the user to apply all undefined terms in the manner in which they are ordinarily used.

Section 202 *Definitions*

AMBULATORY HEALTH CARE FACILITY. Facilities where individuals are provided with medical care on less than a 24-hour basis are classified as Group B occupancies. However, two separate definitions highlight the fact that there are two unique types of persons that occupy such facilities. The important difference involves the self-preservation capabilities of the individuals. Where the occupants are capable of self-preservation (the ability to respond to emergency situations without physical assistance from others), the building is considered an "outpatient clinic" as defined in Section 304.1.1. If such self-preservation cannot be accomplished due to the application of sedation or similar procedures, then the facility is by definition an "ambulatory health care facility" as described in Chapter 2. The need for separate definitions is due to special provisions in Section 422 that specifically regulate those types of facilities where individuals are temporarily incapable of self-preservation. Ambulatory health care facilities are more highly regulated than outpatient clinics in regard to smoke compartmentation, automatic sprinkler system protection, and fire alarm system requirements.

APPROVED. Throughout the code, the term *approved* is used to describe a specific material or type of construction, such as approved automatic flush bolts mentioned in Section 1008.1.9.3, Item 3, or an approved barrier in vertical exit enclosures addressed in Section 1022.7. Where *approved* is used, it merely means that such design, material or method of construction is acceptable to the building official (or other authority having jurisdiction), based on the intent of the code. It would seem appropriate that the building official base his or her decision of approval on the result of investigations or tests, if applicable, or by reason of accepted principles.

APPROVED SOURCE. One provision mandating the use of an approved source is found in Section 104.11.1, which specifically identifies the use of valid research reports as an acceptable method the building official can utilize to evaluate alternative methods and materials of construction. It is expected that the authors of such reports be technically competent and appropriately experienced in the subject under consideration. The building official is designated as the individual solely charged with ascertaining that the person, firm or corporation providing the technical evaluation meets the necessary qualifications.

ATTIC. Several provisions apply to the attic area of a building, such as those relating to ventilation of the attic space. In order to fully clarify that portion of a building defined as an attic, Section 202 identifies an attic as that space between the ceiling beams at the top story and the roof rafters. An attic designation is appropriate only if the area is not considered occupiable. Where this area has a floor, it would be defined as a story. A common misuse of terminology is the designation of a space as a *habitable* or *occupiable* attic. Such a designation is inappropriate insofar as once such a space is utilized for some degree of occupancy, it is no longer deemed an attic.

BUILDING OFFICIAL. Regardless of title, the individual who is designated by the jurisdiction as the person who administers and enforces the IBC is considered by the code to be the building official. In addition, all other individuals who have been given similar enforcement authority, such as plans examiners and inspectors, are also considered building officials to a limited degree under the IBC. A further discussion of the duties and responsibilities of the building official is found in the commentary on Section 104.

COURT. Open and unobstructed to the sky above, an exterior area is considered a court where it is enclosed on at least three sides by exterior walls of the building or other enclosing elements, such as a screen wall. Regulations for courts, including those used for egress purposes, are found throughout the code. Examples of courts are shown in Figure 202-1. Although the IBC does not mandate a minimum depth for consideration as a court, it is expected that certain design and structural features of the building that create minor exterior wall offsets would not require designation as a court. The determination of the presence of a court under such conditions is subject to the building official's discretion.

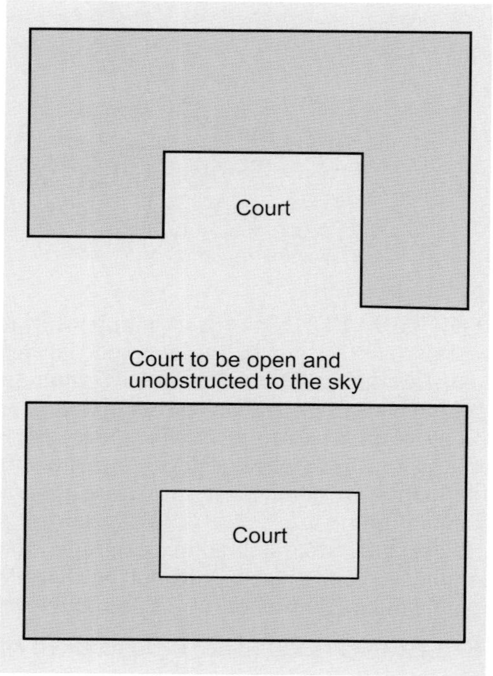

Court

Court to be open and unobstructed to the sky

Court

Figure 202-1
Definition of courts

DWELLING UNIT AND DWELLING. A dwelling unit is considered a single unit that provides living facilities for one or more persons. Dwelling units include permanent provisions for living, sleeping, eating, cooking and sanitation, thus providing a complete independent living arrangement. A dwelling unit, while typically addressed in the IBC as a portion of a Group R-2 occupancy, may also be classified as Group R-1 or R-3. A dwelling is a building that contains either one or two dwelling units. Dwellings are typically regulated under the provisions of the *International Residential Code*® (IRC®), as noted in the exception to Section 101.2.

HABITABLE SPACE. An area within a building, typically a residential occupancy, used for living, sleeping, eating or cooking purposes would be considered habitable space. Those areas not considered to meet this definition include bathrooms, closets, hallways, laundry rooms, storage rooms and utility spaces. Obviously, habitable spaces as defined in this section are those areas usually occupied, and as such are more highly regulated than their accessory use areas. Although typical, it is not necessary that a room or area be finished in order to be considered habitable space. It is not uncommon for a dwelling unit to have a large basement that is not completely finished-out. Nevertheless, the basement may be used as living space, particularly for children who use it as a playroom. Such a basement would be considered habitable space, as the definition is simply based upon the use of the room or area.

HIGH-RISE BUILDING. A high-rise building is defined as one having floor levels used for human occupancy located more than 75 feet (22 860 mm) above the lowest level of fire department vehicle access as illustrated in Figure 202-2. Most moderately large and larger cities have apparatus that can fight fires up to about 75 feet (22 860 mm): thus, the fire can be fought from the exterior. Any fires above this height will require that they be fought internally. Also, in some circles, 75 feet (22 860 mm) is considered to be about the maximum height for a building that could be completely evacuated within a reasonable period of time. Thus, the fire department's capability plus the time for evacuation of the occupants constitute the criteria used by the IBC for defining a high-rise building.

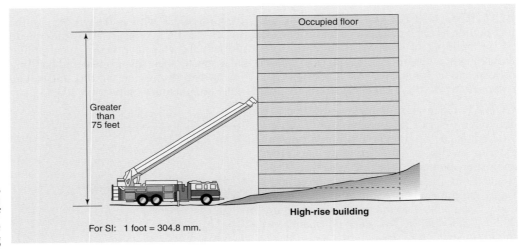

Figure 202-2
Definition of high-rise building

OCCUPIABLE SPACE. A number of provisions in the code apply only to those spaces, rooms or areas typically occupied during the course of a building's use. This definition clarifies that an occupiable space is intended for human occupancy, and as such is provided with means of egress, as well as light and ventilation facilities. Occasionally the code refers to the term *normally occupied* space. For example, Section 1022.3 limits openings in exit enclosures to those necessary for exit access from *normally occupied spaces*. Although not defined in the code, these spaces are generally occupied for extended periods of time during the building's use. There is an expectation that a fire or other hazardous condition would be quickly identified and addressed, rather than go unnoticed for an extended time. Examples of those spaces that would not be considered *normally occupied* include storage rooms, mechanical equipment rooms and toilet rooms.

SLEEPING UNIT. The single required characteristic of a sleeping unit is that it is used as the primary location for sleeping purposes. The room or space that has sleeping facilities may also provide for eating and living activities. It could have a bathroom or a kitchen but not both, as this would qualify it as a dwelling unit. Guestrooms of Group R-1 hotels and motels would typically be considered sleeping units. Sleeping units are also commonly found in congregate living facilities, such as dormitories, sorority houses and fraternity houses, and regulated as Group R-2 occupancies.

Group R occupancies are not the only types of uses where sleeping units are located. Several of the varied uses classified as Group I occupancies also contain resident or patient sleeping units. The proper designation of these spaces as sleeping units is important in the application of Section 420 mandating the separation of sleeping units in Group I-1 occupancies, as well as addressing the appropriate accessibility provisions of Chapter 11.

STORY. Although seemingly quite obvious, the definition of a story is that portion of a building from a floor surface to the floor surface or roof above. In the case of the topmost story, the height of a story is measured from the floor surface to the top of the ceiling joists, or to the top of the roof rafters where a ceiling is not present. The critical part of the definition of a story involves the definition of *story above grade plane* as described in the following discussion.

It is not uncommon for a roof level to be utilized for purposes other than weather protection or mechanical equipment. A roof patio, garden or sports area is sometimes provided in order to utilize as much of the building as possible. Although an occupied roof does not meet the definition of "story," there are certain provisions in the IBC that would be applicable due to the fact occupants can be present. For example, an occupied roof must be provided with a complying means of egress designed for the anticipated occupant load of the roof level. Required fire alarm system protection should also be extended to such occupied roofs. However, the roof level would not be considered as an additional story above grade plane for the purpose of allowable building height, nor would it be considered part of the building area for allowable area purposes. A careful analysis should be made when determining which provisions are applicable to an occupied roof.

STORY ABOVE GRADE PLANE. Throughout the code, the number of qualifying stories in a building is a contributing factor to the proper application of the provisions. As an example, a building's allowable types of construction are based partly on the limits in story height placed on various occupancy groups. In this case, the code is limiting construction type based on the number of stories above grade plane. The code defines a story above grade plane as any story having its finished floor surface entirely above grade plane. However, floor levels partially below the grade at the building's exterior may also fall under this terminology. The critical part of the definition involves whether or not a floor level located partially below grade is to be considered a story above grade plane. There are two criteria that are important to the determination if a given floor level is to be considered a story above grade plane:

1. If the finished floor level above the level under consideration is more than 6 feet (1829 mm) above the grade plane as defined in Section 502.1, the level under consideration is a story above grade plane, or

2. If the finished floor level above the level under consideration is more than 12 feet (3658 mm) above the finished ground level at any point, the floor level under consideration shall be considered a story above grade plane.

Where either one of these two conditions exists, the level under consideration is to be considered a story above grade plane.

Conversely, if the finished floor level above the level under consideration is 6 feet (1829 mm) or less above the grade plane, and does not exceed 12 feet (3658 mm) at any point, the floor level under consideration is not considered a story above grade plane. By definition in Section 502.1, it is regulated as a basement. Figures 202-3 and 202-4 illustrate the definitions of "Story," "Basement" and "Story above grade plane."

Although the criteria for establishing the first story above grade plane in Item 2 indicates that such a condition occurs where the 12-foot (3658 mm) limitation is exceeded, the application of this provision is not that simple. It is not the intent of the code to classify a story that is completely below grade except for a small entrance ramp or loading dock as a *story above grade plane*, provided there is no adverse effect on fire department access and staging. An analysis of the impact of such limited elevation differences is necessary to more appropriately apply the code's intended result.

WALKWAY, PEDESTRIAN. Described as a walkway used exclusively as a pedestrian trafficway, a pedestrian walkway provides a connection between buildings. A pedestrian walkway may be located at grade, as well as above ground level (bridge) or below grade (tunnel). The provisions addressing pedestrian walkways are optional in nature and utilized primarily to allow for the consideration of the connected buildings as separate structures. Regulations for pedestrian walkways and tunnels are found in Section 3104.

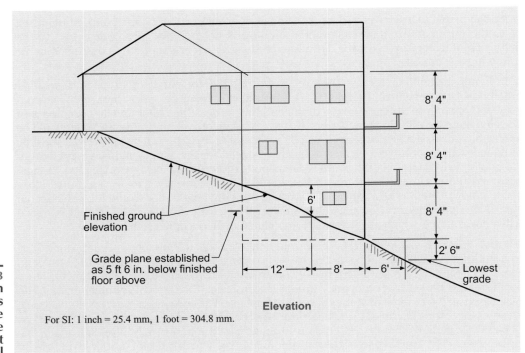

Finished ground
elevation

Grade plane established
as 5 ft 6 in. below finished
floor above

8' 4"

8' 4"

6'

8' 4"

2' 6"

Lowest
grade

12'

8'

6'

Elevation

For SI: 1 inch = 25.4 mm, 1 foot = 304.8 mm.

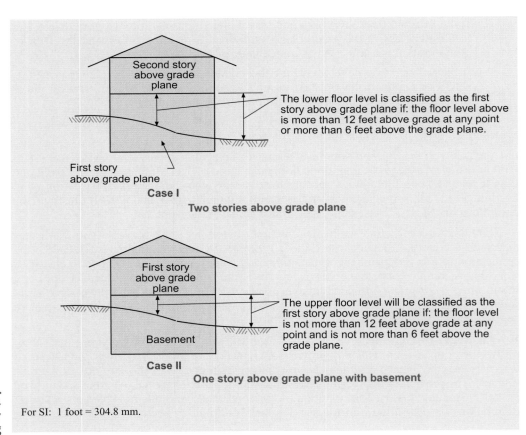

Second story
above grade
plane

First story
above grade plane

The lower floor level is classified as the first
story above grade plane if: the floor level above
is more than 12 feet above grade at any point
or more than 6 feet above the grade plane.

Case I

Two stories above grade plane

First story
above grade
plane

Basement

The upper floor level will be classified as the
first story above grade plane if: the floor level
is not more than 12 feet above grade at any
point and is not more than 6 feet above the
grade plane.

Case II

One story above grade plane with basement

For SI: 1 foot = 304.8 mm.

YARD. Used throughout the code to describe an open space at the exterior of a building, a yard must be unobstructed from the ground to the sky and located on the same lot on which the building is situated. A court, which is bounded on three or more sides by the exterior walls of the building, is not considered a yard. Both a yard and a court are expected to provide adequate openness and natural ventilation so that the accumulation of smoke and toxic gases will not occur.

It is not intended that exterior areas devoted to parking, landscaping or signage be prohibited to qualify as a yard, provided access to and from the building is available and maintained for both the occupants and fire department personnel. It is also important to recognize that the code provisions sometimes require a *yard* and at other times an *open space,* as well as references to *fire separation distance.* Although the differences may appear to be subtle, each term is applied somewhat differently.

Figure 202-5
Pedestrian walkway

KEY POINTS

- The IBC defines terms that have specific intents and meanings insofar as the code is concerned.

- Those definitions found in Chapter 2 are terms that are generally used in a number of varied locations throughout the code.

- Ordinarily accepted meanings are to be utilized for definitions that are not provided in the code, based upon the context within which the term or terms appear.

- *Story above grade plane* is the most critical term in the identification of floor levels within a building.

USE AND OCCUPANCY CLASSIFICATION

Chapter 3 of the code considers the risk to occupants of buildings as well as the probability of property loss. For the most part, the level of risk to occupants within a building is dependent on the number of occupants, the density of such occupants, their age, their capability of self-preservation and their familiarity with, and control over, the conditions to which they are subjected. With regard to protection against property loss, the code considers not only the internal hazards but also exposures to and from adjacent buildings and the combined hazard of buildings in a neighborhood. Regulations are based on previous experience with adjustments for current technology. In developing regulations, consideration is given to the rate of fire spread, its intensity and duration. The code recognizes the performance that may be expected for materials common in today's marketplace and recognizes materials and methods of construction that are economically feasible. Finally, the code considers the facilities that are necessary to operate and maintain a building, which cannot impinge on the health and general welfare of a community's citizens. The bulk of the requirements in the balance of the code are greatly dependent on the proper application of this chapter.

Section 302 *Classification*

302.1 **General.** Every structure, or portion of a structure, must be classified with respect to its use by placing it into one of the 26 specific occupancy groups identified in the code. These groups are utilized throughout the *International Building Code*® (IBC®) to address everything from building size to fire-protection features. The occupancy groups are organized into 10 categories of a more general nature, representing the following types of uses: assembly, business, educational, factory/industrial, high-hazard, institutional, mercantile, residential, storage and utility/miscellaneous.

The provisions of Section 302.1 direct the building official to:

    1. Classify all buildings into one or more of the 26 groups identified in the IBC.

The occupancy classification is typically established by the design professional during the code analysis phase. Most of the time, the designer's determination is consistent with that of the building department. However, where there is disagreement as to the proper classification of the various uses within the building, it is the building official's responsibility to make the final decision. This authority is granted in Section 104.1 dealing with the interpretive powers of the building official. Although the IBC lists in some detail the uses allowed within a specific occupancy classification, the building official will at times also be called upon to judge whether or not a selected classification is appropriate under specific conditions. Assigning occupancy classification often not only depends on the use, but also on the extent and intensity of that use. A use may be so incidental to the overall occupancy that its effect on fire and life safety is negligible. As an example, the administrative office area in a high school performs a business-type function, but such a use is so incidental to the general operation and activities of the school that assigning it to a separate occupancy group would quite probably be unproductive. Therefore, the building official's judgment will often be relied upon to classify occupancies that could potentially fall into more than one group.

    2. Address any room or space within the building that will be occupied at different times for different purposes.

Although an uncommon occurrence, a building space may at times be used for an activity that is considerably different from its typical use. It is important that the hazards associated with that different use also be addressed. The code is basically asking that such a space be assigned multiple occupancy classifications, with the requirements of each assigned occupancy group to be applied. For practical purposes, the classification of the space should probably be based upon the more restrictive of the occupancies involved. This would account for most of the requirements that would be in place. Any additional requirements that would be applicable because of the other occupancy classification would then be layered on top of the other provisions. Another method would be to establish the occupancy classification of the major use, that which has the greatest occurrence, with a layering of the other occupancy requirements on top. Whatever the procedure, it is important that all anticipated uses, and the hazards these uses pose to the building's occupants, are taken into account.

A common example of a space utilized for various functions is a high school gymnasium. The space takes on various occupancy classifications based upon the varied activities that occur, including Group E (physical education classroom), Group A-3 (community activities), Group A-4 (spectator gymnasium) and Group M (weekend craft shows). The classification of the space, which will result in the necessary safeguards being put in place, requires a comprehensive review of the anticipated activities and the hazards involved. A seasonal change in occupancy is another occurrence that must be considered. The creation of a *haunted house* for Halloween activities in a space typically utilized for other purposes is not uncommon. Regardless of the occupancy classification assigned, it is important that all of the anticipated uses be identified in order to apply the necessary code requirements.

   3. Regulate buildings having two or more distinct occupancy classifications under the provisions of Section 508.

Many buildings cannot simply be classified under a single designation. A hotel, considered a Group R-1 occupancy, typically has assembly spaces classified as Group A. In addition, Group M and B occupancies may be present. Each distinct occupancy will be regulated based upon the specific hazards that the individual uses create. The relationship between one occupancy and another is also very important. Where multiple occupancy conditions exist, the provisions of Section 508 are applicable.

There are two approaches to assigning occupancy classifications where buildings have multiple uses. One approach is to evaluate the building as individual areas, assigning classifications specific to the use that is under evaluation. Once this process is complete, a re-evaluation should occur to determine which classifications can be revised to reflect that of the major use. Another option is to initially classify the building as a single occupancy. Then each anticipated use that cannot be adequately addressed under the major classification will be assigned its own classification. Whatever approach is used, the goal is to make sure that the code provisions that are intended to address the anticipated uses of the space, and their potential hazards, are put in place.

There is an expectation that small support and circulation elements be included within the occupancy classification of the area in which they are located. This would include toilet rooms, storage closets, mechanical equipment rooms and corridors. These spaces do not take on a unique classification unless they pose a unique hazard that can only be addressed with a different classification. A small closet of 20 square feet (1.86 m²) would certainly not be considered a Group S occupancy where it is a portion of a Group B office space. On the other hand, a 3,600-square-foot (334.5 m²) storage room could hardly be considered as merely an extension of the Group B. At some point, the use of space and its relationship to other spaces in the building provide for a need to assign a separate classification.

   4. Classify a use into the group that the occupancy most nearly resembles, based on life and fire hazard, when the use is not described specifically in the code.

The code intends to divide the many uses possible in buildings and structures into 10 separate groupings where each group by itself represents a broadly similar hazard. The perils contemplated by the occupancy groupings are of the fire- and life-safety types and are broadly divided into two general categories: those related to people and those related to property. The people-related hazards are divided further by activity, by number of occupants, their ages, their capability of self-preservation and the individual's control over the conditions to which he or she is subjected. The property-related hazards are divided further by the quantity of combustible, flammable or explosive materials and by whether such materials are in use or in storage.

The uses to which a building may be put are obviously manifold, and as a result the building official will, on more than one occasion, either find or be presented with a use that will not conveniently fit into one of the occupancy classifications outlined in the code. As indicated previously in this commentary, under these circumstances the IBC directs the building official to place the use in that classification delineated in the code that it most nearly resembles based on its life and fire risk. This requirement gives the building official broad authority to use judgment in the determination of the hazard of the affected group and, as a result of this evaluation, determine the occupancy classification that the hazards of the use most nearly resemble.

Occasionally, there may be a question as to which classification is to be assigned to a specific use. The owner of a building and the building official may have a difference of opinion as to the proper occupancy classification, or the building official may face a use that appears to fit into one of the code-described groups, but after further analysis it is determined that the hazards representative of the code-defined group are not present in the use proposed. In such situations, the building official should use his or her authority to place the use in the occupancy classification that it most nearly resembles based on its life- and fire-hazard characteristics. It must be remembered that the purpose of occupancy classification is solely to have the ability to regulate for the hazards associated with the building's expected use.

Section 303 *Assembly Group A*

The important item to be considered in this section is the description of an assembly occupancy as "the use of a building or structure, or portion thereof, for the gathering of persons for purposes such as civic, social or religious functions, recreation, food or drink consumption, or awaiting transportation." The description of an assembly occupancy is further defined by the numerous examples of Group A uses listed in Section 303.

Where the area used for assembly purposes is accessory to another occupancy in the building and contains an occupant load of fewer than 50 persons or less than 750 square feet (70 m^2) in floor area, the room or space can merely be considered an extension of the other occupancy. As an example, a break room with an occupant load of 30 in a large manufacturing facility could simply be considered a portion of the Group F occupancy. As an option, a Group B classification can be assigned to the small assembly space. If the assembly use having less than 50 persons is not accessory to another use, such as a small free-standing chapel or a café located in a strip shopping center, it too is to be classified as Group B. As a result, a Group A occupancy classification is only assigned to an assembly use containing 50 or more occupants.

Exception 4 permits the classification of those assembly areas accessory to a Group E occupancy as part of the Group E. However, the application of the Group E classification for the entire educational facility is based on the assembly areas being subsidiary to the school function. This would seem to indicate that the users of the accessory assembly spaces are limited to students, teachers, relatives of students, administrators and others directly involved in educational activities. A typical example of an accessory assembly space is a library or media center that is used almost exclusively by students of the school. On the other hand, gymnasiums and auditoriums located in high school buildings are often used for community functions and other outside activities such as sports tournaments, craft shows and community theater productions that have no relationship to normal educational uses. In such cases, a classification of Group A is more appropriate based upon these unrelated uses. Even in those situations where a Group E classification is appropriate for accessory assembly uses, it is important that the assembly accessibility provisions of Chapter 11 be applied. In addition, the assembly means of egress provisions set forth in Section 1028 should also be utilized in designing the means of egress system for the assembly areas, regardless of occupancy classification.

The fifth exception is intended to allow small and moderately-sized educational rooms and auditoriums accessory to places of religious worship to be classified as a portion of the major occupancy rather than individually. This would result in their classification as part of the overall Group A-3 occupancy. The exception is simply a design allowance that can be utilized to eliminate or reduce any potential mixed occupancy conditions. If the designer wishes to classify such spaces individually, such as using a Group E occupancy classification for any religious educational rooms, such a classification is also permissible.

The concerns unique to assembly uses are based primarily on two factors: the large occupant loads and the concentration of those occupant loads into very small areas. Both conditions must exist to warrant a Group A classification. For example, a Group M big-box store also has the potential for a sizable occupant load; however, the anticipated density of occupants within the store is not as concentrated as those densities regulated under the Group A provisions. Assembly uses are further divided based upon other factors unique to the activities that take place.

A factor involving human behavior in theaters classified as Group A-1 assembly rooms is the fact that in many cases the occupants are not familiar with their surroundings and the lighting level is usually low. Thus, when an emergency arises, the occupants may perceive the danger to be greater than presented, and panic may occur because of the fear of not being able to reach an exit for escape. In addition, the concentration of occupants in such uses is quite dense. The presence of a stage and its distinctive hazards that occur in some Group A-1 occupancies cause unique concerns, addressed by the special provisions of Section 410.

Group A-2 occupancies include uses primarily intended for the consumption of food or drink, and include dining rooms, cafeterias, restaurants, cafes, nightclubs, taverns and bars. The fire record in occupancies of this type is not very good, based in part on the delay in responding to a fire or other emergency incident. Because of the common presence of loose tables and chairs, aisles are often difficult to maintain, resulting in obstructions to egress travel. Overcrowding conditions, low-lighting levels and the consumption of alcoholic beverages also increase the risks associated with many of these types of occupancies. The gaming floor and associated areas of casinos are also typically classified as Group A-2 based in part upon the congestion and distractions often encountered.

Occupancies classified as Group A-3 have varying degrees of occupant density, numerous types and numbers of furnishings and equipment, and fire loading that can vary from low to high. The hazards for uses in this category are similar to most of those of the Group A-1 and A-2 occupancies. Where a use does not conveniently fit into one of the other four Group A classifications, a Group A-3 designation is typically appropriate. The classification of an assembly occupancy as a Group A-3 is also common where multiple assembly uses are likely to occur. For example, a meeting room at a hotel is used for various functions, including seminar presentations, dining activities, trade shows and wedding receptions. Although these functions may have different Group A designations when viewed individually, as a group they pose a hazard level that can be appropriately addressed with the Group A-3 classification. Therefore, most multipurpose rooms are simply classified as Group A-3 occupancies.

Occupancies classified as Groups A-4 and A-5 are similar in nature, with the controlling difference being that Group A-5 occupancies are outdoor structures. Therefore, the fire hazard for Group A-5 occupancies is less than for Group A-4 uses, and significantly lower than the other assembly occupancies. It is also expected that there be little to no smoke accumulation under fire conditions in the assembly areas of a Group A-5 occupancy. However, there still exist the hazards of crowding a large number of occupants within a relatively small space. The hazard of panic is assumed to be a large portion of the overall concern for Group A-5 structures. Generally, associated spaces such as concession stands, locker rooms, storage areas, press boxes and toilet rooms are included as a portion of the

Group A-5 classification. However, where uses within the building create conditions more hazardous than anticipated by the Group A-5 designation, such uses must be classified according to their individual characteristics. For example, an enclosed restaurant within the stadium that seats 400 people should be appropriately classified as a Group A-2 occupancy.

A review of the uses specifically listed as Group A occupancies indicates that although some are general in nature, such as libraries and arenas, others are specific to a room or area within a building (courtroom, lecture hall and waiting areas in transportation terminals). This concept of identifying uses is not limited to Group A occupancies, but rather is consistent throughout all of the occupancy group descriptions. It is critical that the building official thoroughly evaluate the uses that are anticipated and assign occupancy classifications based on the hazards that have been identified.

Section 304 *Business Group B*

The most common use classified as a Group B occupancy is an office building, or a portion of a building containing office tenants or office suites. The portions of such business occupancies where records and accounts are stored are also considered part of the Group B use. Examples of uses involving office, professional or service-type transactions are listed in the code.

Airport traffic control towers are considered Group B occupancies and are regulated under the special-use provisions of Section 412.3. Carwash structures and motor vehicle showrooms are also considered business occupancies. If the space contains a limited number of vehicles that are present in a very controlled condition, it is anticipated that the fire risk is limited, and classification as a Group B occupancy is appropriate.

Medical offices, including both outpatient clinics and ambulatory health care facilities, are classified as Group B occupancies. By definition, an outpatient clinic is not expected to serve patients who will be temporarily rendered incapable of self-preservation due to their treatment. On the other hand, an ambulatory health care facility is expected to have one or more individuals present who are temporarily rendered incapable of self-preservation due to the application of nerve blocks, sedation or anesthesia. Although both types of facilities would be classified as Group B occupancies, the unique concerns applicable to ambulatory health care facilities are addressed in the special provisions of Section 422.

Educational occupancies above the 12th grade, including college classrooms and training rooms for adult education, are considered Group B uses. This designation is not necessarily incumbent on the number of students (occupants) in the room. However, because a lecture hall in a college classroom building would fall into two different occupancy classifications, Groups A-3 and B, it is important that the concerns of both classifications be considered. As a part of the overall building analysis, a Group B classification would typically be appropriate. It can be viewed as simply an extension of the Group B uses that occur throughout the building. This might include using the Group B criteria for construction type, mixed-occupancy provisions and other more general categories. On the other hand, a Group A classification would seem to be necessary in evaluating the means of egress and fire protection requirements. The code is silent to this type of analysis; however, the intent should be to regulate the building in a manner consistent with the hazards that are anticipated.

Where not exceeding the maximum quantities of hazardous materials allowed by the code in Section 307, testing and research laboratories may be considered Group B occupancies. Where the allowable quantities are exceeded, such laboratories would be classified as Group H.

Section 305 *Educational Group E*

Group E occupancies are educational uses for students through the 12th grade, where the use houses six or more students at any one time. Under certain conditions, facilities used for day-care services are also considered Group E occupancies.

To be classified as a Group E occupancy, a day-care operation providing educational, supervision or personal care services must operate on less than a 24-hour basis. In other words, full-time care facilities cannot be considered Group E occupancies. The number of children housed in a day-care operation classified as Group E is not limited; however, where the number of children is five or fewer the use is typically classified as a Group R-3. Additional requirements for a Group E day-care facility are that rooms used for infant/toddler care ($2^1/_2$ years of age or younger) shall be located on the level of exit discharge and each of such rooms must have an exit door directly to the exterior of the building. See Figure 305-1. The provisions addressing children over the age of $2^1/_2$ years are found in Section 305.2, whereas infant/toddler care in Group E occupancies is addressed in the exception to Section 308.5.2.

There is a unique feature involved in educational occupancies—the use of school buildings for assembly purposes outside the scope of the educational use. For example, many school auditoriums are used for community theater and other productions to which the public at large is invited. Also, the school gymnasium in many cases is utilized for neighborhood recreation activities or sporting events where all ages of occupants are present. It is not uncommon for school auditoriums and gymnasiums to be rented out to groups for special functions. When these additional uses are anticipated, there is adequate reason that they be classified as Group A occupancies. Therefore, on account of these multipurpose uses in many school buildings, it is necessary that the code requirements applicable to all expected uses be enforced in order to satisfy the safety requirements for each use. See the additional discussion in Section 303.

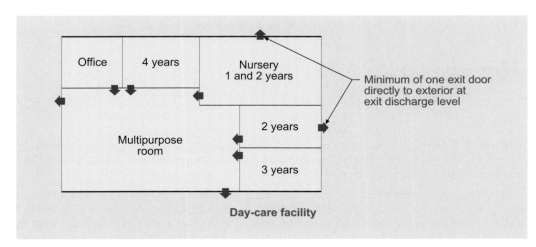

Day-care facility

Figure 305-1
Group E classification

Section 306 *Factory*
Group F

Although the potential hazard and fire severity of the multiple uses in the Group F occupancy classification is quite varied, these uses share common elements. The occupants are adults who are awake and generally have enough familiarity with the premises to be able to exit the building with reasonable efficiency. Public occupancy is usually quite limited, and most occupants are aware of the potential hazards the use creates. Group F occupancies are generally regarded as factory and industrial uses. The degree of hazard between the uses is very broad, and therefore the occupancy is divided into two categories.

Many of the Group F-1 uses contain some degree of hazardous material as a necessary part of the manufacturing process. However, where the amount of hazardous material does not exceed the maximum allowable quantities set forth in Table 307.1(1) or 307.1(2), the lower classification of Group F is appropriate. Because of the similarity between the names of the uses in Group F-1 occupancies and those in Group H occupancies, care must be exercised when determining the appropriate classification, and operators of Group F-1 uses should be apprised of the limitations on the quantities of hazardous materials that are allowed.

Some of the activities specifically listed as Group F occupancies also occur in a limited sense as accessory functions, and as such are not to be classified as Group F. For example, *food processing* is identified as a Group F-1 occupancy, but this is not to say that a kitchen serving a restaurant, school or worship building should be classified as such. Kitchens are typically considered to be classified as a portion of the major occupancy that they serve. The food processing operations designated as Group F-1 occupancies primarily include large factories that produce canned or packaged items in bulk.

The hazard from uses in Group F-2 occupancies is very low; in fact, the activities are deemed as among the lowest hazard groups in the code. It is assumed that the fabrication or manufacturing of noncombustible materials will pose little if any fire risk to the building or its occupants. Foundries would be considered Group F-2 occupancies, as would facilities used for steel fabrication or assembly. Manufacturing operations producing ceramic, glass or gypsum products are also included in this classification.

Section 307 *High-hazard*
Group H

High-hazard Group H occupancies are characterized by an unusually high degree of explosion, fire or health hazard as compared to typical commercial and industrial uses. The identification of hazardous occupancies is provided in this section.

There is one common feature about all Group H occupancies—they are designated as Group H based on excessive quantities of hazardous materials contained therein. Where the quantities of hazardous material stored or used in a building exceed those set forth in Section 307, a Group H classification is warranted. On the other hand, where such quantities are not exceeded, a Group H classification is not appropriate.

Because of the technical nature of the operations and materials found in Group H occupancies, a number of specific terms are defined by the IBC. The definitions are intended to assist the code user in applying the provisions of this chapter, as well as other portions of the code relating to high-hazard uses.

Group H-1 occupancies are those buildings containing high-explosion hazard materials. Materials that have the potential for detonation must be housed in buildings regulated in a very special manner, and designed and constructed unlike any other occupancies described in the code. Examples of detonable materials include explosives and Class 4 oxidizers.

Group H-2 generally includes those occupancies that contain materials with hazards of accelerated burning or moderate explosion potential, including materials with deflagration hazards. Common occupancies included in this category are those operations where flammable or combustible liquids are being used, mixed or dispensed. The potential for a hazardous incident is increased because of the materials' exposure to the surrounding area. Occupancies containing combustible dusts are also considered Group H-2, as dusts in suspension, or capable of being put into suspension, in the atmosphere are a deflagration hazard.

Buildings containing materials that present high-fire or heat-release hazards are classified as Group H-3 occupancies. Where flammable or combustible liquids are present in such occupancies, they must be stored in normally closed containers or utilized in low-pressure systems. Because of the enclosed nature of these liquids, the hazard level is not nearly as severe as it is for Group H-2 occupancies. Other hazardous materials such as organic peroxides and oxidizers, based upon their hazard classification, may also be used or stored in Group H-3 occupancies.

Group H-4 occupancies are those containing health-hazard materials such as corrosives and toxics. Section 307.2 defines "Health hazards" as those "chemicals for which there is statistical significant evidence that acute or chronic health effects are capable of occurring in exposed persons." Quite often, a material considered a health hazard also possesses the characteristics of a physical hazard. It is important that all hazards of materials be addressed.

Occupancies classified as Group H-5 are those uses containing semiconductor fabrication facilities, including the ancillary research and development areas. The Group H-5 category was created in order to address the explosive and highly toxic materials used in semiconductor fabrication by providing specific requirements for the particular operations conducted, while at the same time providing a level that allows reasonable transaction of the fabrication process.

A more complete commentary on hazardous materials is provided under the discussion of Section 414.

High-hazard Group H. The concept of maximum allowable quantities of hazardous materials as the basis for occupancy classification is further extended through the use of control areas as regulated by Section 414.2.

307.1

Maximum allowable quantities. Occupancy classifications of buildings containing hazardous materials are based on the *maximum allowable quantities* concept. Tables 307.1(1) and 307.1(2), together with their appropriate footnotes, identify the maximum amounts of hazardous materials that may be stored or used in a control area before the area must be designated as a Group H occupancy. The maximum quantities of hazardous materials permitted in non-Group H occupancies vary for different states of materials (solid, liquid or gas) and for different situations (storage or use). The allowable quantities are also varied based on protection that is provided, such as fire-extinguishing systems and storage cabinets.

Control areas. Areas in a building that are designated to contain less than or equal to the maximum allowable quantities of hazardous materials and that are properly separated from other areas containing hazardous materials are called *control areas*. Any combination of hazardous materials, up to the maximum allowable quantities, is permitted in a control area. A control area may be an entire building or only a portion of the building. It can be part of a story, an entire story or even include multiple stories.

The control-area method is based upon the concept of fire-resistive compartmentation. It regulates quantities of hazardous materials per compartment (control area), rather than per building. The limit for the entire building, using control areas, is then established by limiting the total number of control areas allowed per building and the quantities of hazardous materials that are located in each control area. The control-area concept was introduced in an effort to regulate buildings of different sizes in a consistent manner. It is based on a premise that the storage and use of limited quantities of hazardous materials (not exceeding maximum allowable quantities) in areas that are separated from each other by fire-resistance-rated separations do not substantially increase the risk to the occupants or change the character of the building to that of a hazardous occupancy, subject to a limitation on the number of control areas. The fire-resistive separations are relied on to minimize the risk of having multiple control areas involved simultaneously during an emergency.

The occupancy classification of a control area is the same as the occupancy classification of the portion of the building in which the control area is located. There is no special occupancy designation for a control area. For example, a control area in a mercantile occupancy is merely part of the Group M occupancy. Further discussion of control areas is addressed in the commentary of Section 414.2.

Increased quantities. Given this basic understanding of maximum allowable quantities and control areas, the various options in the code for increasing the quantities of hazardous materials within a building are as follows:

1. Buildings are generally allowed to have up to the basic maximum allowable quantities of hazardous materials without restriction with respect to separations or protection. In this case, the entire building is designated as a control area. The boundaries of the control area are the boundaries of the building (i.e., exterior walls, roof and foundation). See Figure 307-1.

2. Using the footnotes to Tables 307.1(1) and 307.1(2), the maximum allowable quantities can often be increased by adding sprinklers throughout the building or by using approved storage cabinets, safety cans or other code-approved enclosures to protect the hazardous materials. It is important that the increases identified in the footnotes only be utilized where applicable.

3. Two other options are available to further increase the quantities of hazardous materials in any building:

   3.1. Provide additional control areas, or

   3.2. Construct the building as required for a Group H occupancy.

4. Assuming additional control areas are used, each additional control area must be separated from all other control areas by minimum 1-hour fire barriers, or 2-hour fire barriers if required by Section 414.2.4. Vertical isolation of control areas must be accomplished by floors having a minimum 2-hour fire-resistance rating. Under limited conditions, the floor construction of the control area separation may be reduced to 1 hour. Its application is limited to fully sprinklered two- or three-story buildings of Type IIA, IIIA or VA construction. In all cases, construction supporting such floors shall have an equivalent fire-resistance rating. A designated percentage of the maximum allowable quantities of hazardous materials is allowed in each control area per Table 414.2.2. See Figure 307-2.

   The permitted number of control areas decreases vertically through the building, as does the quantities of hazardous materials per control area. As hazardous materials are located higher and higher above ground level, they become more difficult to address under emergency conditions. As with many other conditions regulated by the code, a key factor is the ability of the fire department to access the incident area. The higher the hazardous materials are located in the building, the more restrictive the provisions become, owing to the limitations on fire department access and operations.

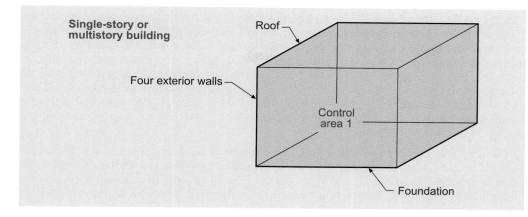

Figure 307-1
Control area boundaries for one control area

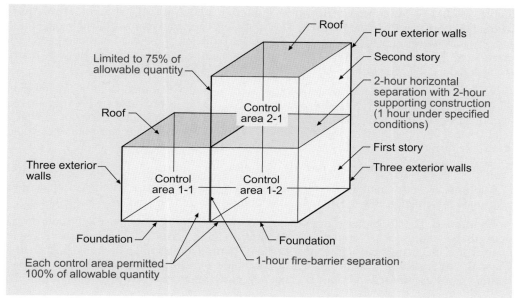

Figure 307-2
Multistory areas

Storage and use. One other fundamental concept involved in applying the maximum allowable quantities is *situation of material*. The maximum allowable quantities in the code are based on three potential situations: storage, use-closed and use-open.

Though not defined by the code, the term *storage* is generally considered to include materials that are idle and not immediately available for entering a process. The term *not immediately available* can be thought of as requiring direct human intervention to allow a material to enter a process or, alternatively, as using approved supervised valving systems that separate stored material from a process. In the case of liquids and gases, storage is generally considered to be limited to materials in closed vessels (not open to the atmosphere). For example, materials kept in closed containers such as drums or cans are in storage because deliberate action (opening the drum or can) would be required to use the material. However, when a container or tank is connected to a process, the question arises of whether the material in the container or tank is in storage or in use.

In general, the quantity of material that would be considered to be in use is the quantity that could normally be expected to be involved in a process, or that could reasonably be expected to be released or involved in an incident as a result of a process-related emergency. Consider, for example, a process having hazardous materials that are piped from an underground storage tank outside of a building to a dispensing outlet within a building. Because the tank is connected to a process within the building, it could be argued that the contents of the tank are available for use in the building (see definition of "Use" in Section

415.2) and that the amount should be counted toward the maximum allowable quantities. However, if an approved, reliable arrangement of valving is provided between the supply and the point where the material is dispensed, it would be reasonable to conclude that the quantity on the supply side of such valving that is outside of the building would be unlikely to impact incidents occurring within the building and, therefore, need not be counted toward the maximum allowable quantities. This reliable arrangement of valving can be considered an interruption of the connection between the confined material (storage) and the point where material is placed into action or made available for service, as discussed in the definition of "Use." See Figure 307-3.

The difference between use-closed and use-open is basically whether the hazardous material in question is exposed to the atmosphere during a process, with the exception that gases are defined as always being in closed systems when used insofar as they would be immediately dispersed (unless immediately consumed) if exposed to the atmosphere without some means of containment.

Table 307.1(1)—Maximum Allowable Quantity per Control Area of Hazardous Materials Posing a Physical Hazard. This table sets forth maximum allowable quantities for physical hazard materials. All three situations (storage, use-closed and use-open) are considered. For the specific case of gases, maximum allowable amounts are all listed under storage and use-closed because the definition of "Use" includes all gases.

With two exceptions, any combination of materials or situations listed in this table is allowed in each control area. These two exceptions are (1) as provided by Footnote h, the aggregate of IA, IB and IC flammable liquids, and (2) as provided by Footnote b, which requires that aggregate quantities of materials in both use and storage must not exceed the allowable quantity for storage.

Specific footnotes to the table provide the following information:

Footnote a. This footnote references Section 414.2 for the use of control areas. For additional information, see the discussion of control areas in Sections 307.1 and 414.2.

Footnote b. This footnote requires quantities of materials that are in use to be counted as both storage and use when comparing quantities to those permitted. For example, a single control area in a manufacturing facility would be permitted up to 30 gallons (116.25 L) of a Class II combustible liquid in use and a maximum of 120 gallons (454.2 L) in storage without being considered a Group H occupancy. However, a total of 150 gallons (567.75 L)

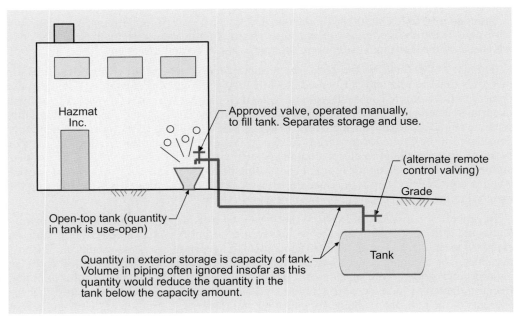

Figure 307-3
Example of storage versus use

would be prohibited. The total of both use and storage is limited to 120 gallons (454.2 L), with no more than 30 gallons (116.25 L) permitted in use.

Footnote c. This footnote exempts small size containers of certain consumer products that are considered to present minimal hazards based on the types of materials and the container sizes.

Footnote d. This footnote allows certain materials to have exempt amounts doubled when stored or used in sprinklered buildings. Compounding with the increases provided by Footnote e is allowed when both footnotes are applicable. Materials and situations referencing both Footnotes d and e can receive four times the listed maximum allowable quantity when both footnotes are applied. See Application Example 307-1.

Footnote e. This footnote allows exempt amounts for certain materials in storage to be doubled when approved storage cabinets, gas cabinets, safety cans, etc., as applicable, are employed. Also, see Footnote d.

Footnote f. This footnote allows certain materials to be stored or used in unlimited quantities in sprinklered buildings. When the building is fully sprinklered, it has the effect of classifying the building as other than a Group H occupancy. Application of this footnote is limited to the storage or use of Class IIIB combustible liquids or Class I oxidizers. An unlimited quantity of each of these materials is permitted in a fully sprinklered building without requiring a designation of Group H.

Footnote g. This footnote limits storage and use of certain materials to sprinklered buildings. Where a non-Group H classification is desired and the quantity does not exceed the maximum allowable, the building is required to be sprinklered throughout.

Footnote h. Where flammable liquids are concerned, the maximum allowable quantities are regulated both individually and cumulatively. To be considered an occupancy other than Group H, the control area must not contain more than the maximum allowable quantities for each type of flammable liquid, Class IA, IB and IC, as well as the combination of such limits established by Table 307.1(1).

Footnote i. The threshold of 660 gallons (2498 L) is commonly utilized in the *International Fire Code* where regulating fuel oil storage and piping systems. This footnote allows for additional quantities above those set forth in the table.

Footnote k. Substantial increases in the maximum allowable quantities are permitted for Class 3 oxidizers used for maintenance and sanitation purposes. A common application of the increased quantities occurs in health-care facilities.

Footnote l. If the net weight of the pyrotechnic composition of fireworks is known, that weight is utilized in applying the limitations of the table. Otherwise, 25 percent of the gross weight of the fireworks is to be used, including the packaging materials.

Application Example 307-1

GIVEN: A fully-sprinklered Group F-1 storage building housing Class II combustible liquids. The Class II liquids are all stored in approved safety cans. The entire building is a single control area.

DETERMINE: The maximum allowable quantity of the Class II liquids in storage in order to maintain the Group F-1 classification.

SOLUTION:

Basic MAQs per Table 307.1(1)	120 gallons
Sprinkler increase per footnote d (100%)	+ 120 gallons
	240 gallons
Safety can increase per footnote e (100%)	+ 240 gallons
Total of maximum permitted for Group F-1 classification	480 gallons

Footnote o. Cotton is almost exclusively pressed and stored as densely-packed baled cotton meeting the weight and dimension requirements of ISO 8115. In this form, the fibers are not easily ignitable, and the regulation by this table as a hazardous material is deemed unnecessary.

Footnote p. This note further clarifies the application of Exception 3 to Section 307.1. Where liquid or gaseous fuel is utilized in the operation of machinery or equipment, including vehicles, the quantities are not to be included in the determination of maximum allowable quantities.

Table 307.1(2)—Maximum Quantity per Control Area of Hazardous Materials Posing a Health Hazard. This table is similar in nature to Table 307.1(1), except that the maximum allowable quantities listed in Table 307.1(2) are for health hazard materials. For discussions of specific footnotes, see the above discussion of Table 307.1(1).

307.8 **Multiple hazards.** As previously noted, most hazardous materials possess the characteristics of more than one hazard. This section requires that all hazards of materials must be addressed. For example, if a material possesses the hazard characteristics of a Class 2 oxidizer and a corrosive, the material would be regulated under the provisions for both Group H-3 and H-4 occupancies. The more restrictive provisions of each occupancy must be satisfied.

Section 308 *Institutional Group I*

Group I occupancies are institutional uses and in the IBC are considered to be basically of two broad types. First are those facilities where individuals are under supervision and care because of physical limitations of health or age. The second category includes those facilities in which the personal liberties of the occupants are restricted.

In both types, the occupants are either restricted in their movements or require supervision in an emergency, such as a fire, to escape the hazard by proceeding along an exit route to safety. There is actually a third category in which the occupants enjoy mobility and are reasonably free of constraints but do require a measure of professional care and are asleep for a portion of the day.

As Group I occupancies are people-related occupancies, the primary hazard is from the occupants' lack of free mobility needed to extricate themselves from a hazardous situation. On the other hand, the hazard from combustible contents is typically very low and, as a result, the occupancy requirements for Group I occupancies are essentially based on the limited free mobility of the occupants. Also, the occupants of most Group I occupancies are usually institutionalized for 24 hours or longer and therefore are asleep at some point during their stay. Thus, the protection requirements of the code are more comprehensive than in almost any other people-related occupancy. It should be noted that institutional occupancies classified as Group I-1 or I-4 may be classified as Group R-3 where care is provided for five or fewer persons. As an option for such small institutional uses, the structure need only comply with the *International Residential Code*® (IRC®). The code typically recognizes that such small occupant loads in educational or institutional environments can be adequately addressed for fire and life safety through the provisions for dwelling units.

Although the Group I-3 classification is only applicable where six or more individuals are restrained or secured, there is no indication as to the proper classification where five or fewer persons are involved. Where the intended detention or restraint occurs as an accessory use within some other occupancy, classification would be based on that of the major occupancy. For example, up to five individuals could be restricted in areas such as interrogations rooms for alleged shoplifters in a covered mall building, jewelry viewing rooms for customers of a retail store and time-out rooms in a school, without classifying that portion of the building a Group I-3 occupancy. Further discussion on allowances for locking devices for such spaces can be found in the commentary on Section 1008.1.9.3, Exception 1.

Group I-1. The occupants housed in buildings classified as Group I-1 occupancies are ambulatory but live in a supervised environment where personal-care services are provided. There is no need for the occupants to receive any physical staff assistance should a fire or other emergency exist. Types of uses included in this category include half-way houses, drug-rehabilitation facilities, assisted-living facilities, convalescent-care facilities and group homes. Several of the listed uses may rather be considered Group I-2 or I-3 occupancies if the residents are incapable of self-preservation because of injury, illness or incarceration. For example, an alcohol treatment center may provide lockdown for a number of persons under care. Where this number exceeds five, a Group I-3 classification would be more appropriate than a Group I-1. The same concern occurs where a convalescent facility provides care for individuals who are not able to respond to emergency conditions without physical assistance. Again, once the threshold of five persons incapable of self-preservation is exceeded, the proper classification would not be Group I-1. **308.2**

The specific classification is based on the number of residents. In this case, a Group I-1 is the proper classification for more than 16 occupants. The threshold of 17 or more is based on the number of supervised individuals who reside in the facility and does not include associated staff members. Where the number of residents is between six and 16, a Group R-4 classification is appropriate. Those supervised residential facilities that provide personal care services for five or fewer occupants can be considered either a Group R-3 occupancy or be designed and constructed under the provisions of the IRC.

Group I-2. The primary feature that distinguishes the Group I-2 occupancy from the others is that it is a health-care facility in which the patients are, in general, nonambulatory. This classification includes hospitals, detoxification facilities, mental hospitals and nursing homes. The nursing homes included in this category are deemed to provide intermediate care or skilled nursing care. Nurseries for the full-time care of infants (under the age of $2^1/_2$ years) are included in this classification, as the code assumes that the very young require the same protection as is provided for those individuals whose capability of self-preservation is severely restricted. Where health care is provided for a limited time period, such as at an out-patient health care clinic, a classification of Group I-2 is not appropriate. In such cases, a Group B classification is warranted, even in those cases where some of the patients are incapable of self-preservation. **308.3**

Group I-3. The uses of Group I-3 occupancies encompass mental hospitals, jails, prisons, reformatories, detention centers and other buildings where the personal liberties of the inmates are similarly restricted. For guidance on the classification of detention facilities having occupant loads of five or less, see the general discussion of Section 308. The classification of Group I-3 buildings shall also include one of five occupancy conditions. Several provisions specific to Group I-3 occupancies vary based upon which condition is anticipated, such as the manner of subdividing resident housing areas. The conditions are described as follows: **308.4**

1. The highest level of freedom assigned to a Group I-3 occupancy is considered Condition 1. Free movement is permitted throughout the sleeping areas and the common areas, including access to the exterior for egress purposes. A facility classified as Condition 1 is permitted to be constructed as a Group R occupancy, mostly likely a Group R-2. There may also be cases where it is more appropriate to classify the use as some occupancy other than Group R. For example, an industrial building included within a Condition 1 facility would most probably be classified as a Group F-1 occupancy.

2. Condition 2 buildings permit free movement between smoke compartments; however, access to the exterior for egress purposes is restricted because of locked exits.

3. Access between smoke compartments is not permitted in Condition 3 occupancies, except for the remote-controlled release of locked doors for necessary egress travel.

Movement within each individual smoke compartment is permitted, including access to individual sleeping rooms and group activity spaces.

4. Condition 4 buildings restrict the movement of occupants to their own space, with no freedom to travel to other sleeping areas or common areas. Movement to other sleeping rooms, activity spaces and other compartments is controlled through a remote release system.

5. The lack of freedom provided in Condition 5 facilities is consistent with that of Condition 4. However, staff-controlled manual release is necessary to permit movement throughout other portions of the building.

308.5 **Group I-4, day-care facilities.** Where custodial care is provided for persons for periods less than 24 hours, an occupancy classification of Group I-4 is appropriate. The code further restricts this category by limiting the care to individuals other than parents, guardians or relatives. This occupancy classification is appropriate for day-care facilities with children no older than $2^1/_2$ years of age, as well as older persons who are deemed to be incapable of self-preservation. The need for supervision and personal-care services is the primary factor that contributes to this type of use being classified as an institutional occupancy.

Day-care facilities for children above the age of 30 months are considered educational occupancies (Group E). Where certain conditions are met, such a facility caring for infants/toddlers (30 months or less in age) may also be classified as Group E. Where the care activities are for adults who can physically respond to an emergency situation without physical assistance, an institutional classification is not appropriate. Although the code mandates the assignment of an occupancy classification of Group R-3 regardless of the number of occupants being accommodated in the facility, it should be noted that Section 310.1 limits a Group R-3 classification for adult care facilities to those conditions where no more than five persons are accommodated. Where the occupant load of such adult care facilities exceeds five, a more representative classification (such as Group B or A-3) is possibly warranted. In all cases, the exception is limited only to those adult care facilities where the occupants are capable of responding to an emergency situation without physical assistance from others.

Section 309 *Mercantile Group M*

The mercantile uses listed in this section are mostly self-explanatory. For the most part, occupants of this type of use are ambulatory adults, with younger children supervised by parents or other adults. Although the occupancies may contain a variety of combustible goods, the possibility of ignition is limited. High-hazard materials may be present in small quantities, but not enough of the hazard material is present to be considered a Group H occupancy. For the limitations on hazardous materials in a Group M occupancy, see the discussion on Section 414.

As a sales operation, a service station whose primary function is the fueling of motor vehicles is considered a Group M use. This classification would also apply to the canopy constructed over the pump islands. By assigning the structures associated with the fueling of motor vehicles a classification the same as that for other sales operations, such as convenience stores, there is no question as to the proper application of the code. Through the design and construction of a motor-vehicle service station in conformance with Section 406.5 and the *International Fire Code*®(IFC®), there is no distinct uncontrolled hazard that would cause separate and unique occupancies to be assigned. Conversely, where service or repair activities are involved, a Group S-1 classification is warranted. This would include operations limited to the exchange of parts, such as tire and muffler shops, as well as service-oriented activities (oil change and lubrication work).

Section 310 *Residential Group R*

Group R occupancies are residential occupancies and are characterized by:

1. Use by people for living and sleeping purposes.
2. Relatively low potential fire severity.
3. The worst fire record of all structure fires.

The basic premise of the provisions in this section is that the occupants of residential buildings will be spending about one-third of the day asleep and that the potential for a fire getting out of control before the occupants are awake is quite probable. Furthermore, once awakened, the occupants will be somewhat confused and disoriented, particularly in hotels.

Residential Group R. The residential subdivisions are based on occupant density as well as **310.1** the permanency of the occupants. Therefore, hotels, motels and similar uses in which the residents are essentially transients are separated from apartment houses. The reason for this is the occupants' lack of familiarity with their surroundings. This in turn leads to confusion and disorientation when a fire occurs while the occupants are asleep. Because of this key difference, hotels and motels are considered Group R-1 occupancies whereas apartment houses are designated as Group R-2. The Group R-1 classification also includes transient boarding houses and bed-and-breakfast establishments where the number of transient occupants is often quite low. Such facilities are permitted to be classified as Group R-3 where the occupant load does not exceed 10. A similar allowance is permitted for congregate living facilities of a nontransient nature, such as convents and monasteries, permitting classification as a Group R-3 where the occupant load is 16 or less persons. A higher threshold was chosen for this application due to the nontransient nature of the occupants and the resulting familiarity with their surroundings.

Included in the Group R-2 occupancy classification are dormitories, fraternity and sorority houses, convents and monasteries. These types of uses are considered congregate living facilities. By definition, they contain one or more sleeping units where the residents share bathroom and/or kitchen facilities. An occupancy classification of Group R-2 is only appropriate where the occupant load of the facility is 17 or more persons. A lesser number will result in a Group R-3 classification. The Group R-2 classification is also appropriate for live/work units as regulated by Section 419. Live/work units include those dwelling units or sleeping units where a significant portion of the space includes a nonresidential use operated by the tenant.

Group R-3 occupancies are generally limited to mixed-use buildings containing one or two dwelling units, as well as those small facilities used for adult or child care. It is expected that the occupant load of a Group R-3 occupancy will be quite low. Typically, dwellings would not be classified as Group R-3 occupancies, as they will be regulated by the IRC. Only where the dwelling falls outside the scope of the IRC will the Group R-3 classification for such structures be appropriate. For example, a four-story dwelling would be regulated as an R-3 occupancy, as would a single dwelling unit located above a small retail store.

Assisted living facilities may be classified as Group R-4 occupancies, provided the number of residents under residential care does not exceed 16. Where the number of residents is five or less, the use is considered a Group R-3 occupancy.

Where the use of a Group R-3 or R-4 occupancy consists of adult or child care activities, the IRC may be utilized, provided the building falls under the scoping provisions of the IRC. It should be noted that Group R-4 occupancies constructed under the provisions of the IRC are required to be protected by either an NFPA 13 or 13R automatic sprinkler system. As a general rule, facilities utilized for residential care or assisted living purposes are permitted to be designed and constructed solely under the provisions of the IRC. However, sprinkler protection is also mandated to supplement the IRC requirements to ensure that all such buildings housing Group R-4 uses be sprinklered.

In general, this designation includes storage occupancies that are not highly hazardous and uses related to the storage, servicing or repair of motor vehicles. Such storage uses are classified into two divisions based on the hazard level involved. Group S-1 describes those buildings used for moderate-hazard storage purposes, whereas low-hazard uses make up the Group S-2 classification.

Before discussing the two different types of storage uses, consideration should be given to the classification of borderline uses. An exclusionary rule is used to assist in determining those moderate-hazard storage uses that are to be classified as Group S-1. For example, a Group S-1 occupancy is used for storage of materials not classified as a Group S-2 or H occupancy. The building official will often be called upon to decide which classification is most appropriate when a use can fall within the two Group S occupancy classifications. As guidance in making this decision, it is usually more appropriate to choose the most restrictive occupancy, which is the Group S-1. This is particularly true when the area and height of the building in question is far below the allowable. Classifying the use into the more restrictive category would allow the building to be protected to a higher level and address the worst-case situations that might occur. By classifying the use into the least restrictive category, it would typically reduce the required controls, causing a potential problem where the building operator chooses to store combustible materials within the building.

311.2 Moderate-hazard storage, Group S-1. Group S-1 occupancies are typically used for the storage of combustible commodities. A complete list of all products allowed in this use would be very lengthy; however, many of the more common storage items are identified by the code. In addition, repair garages for motor vehicles are considered Group S-1 occupancies. In general, buildings classified in this manner would be used for the storage of commodities that are manufactured within buildings classified as Group F-1 occupancies. Commodities that constitute a high physical or health hazard, and exceed the maximum allowable quantities set forth in Section 307, would be stored in the appropriate Group H occupancy.

311.3 Low-hazard storage Group S-2. Group S-2 occupancies include the storage of noncombustible commodities, as well as open or enclosed parking garages. Buildings in which noncombustible goods are packaged in film or paper wrappings or cardboard cartons, or stored on wooden pallets are still considered Group S-2 occupancies. This also includes any products that have minor amounts of plastics, such as knobs, handles or similar trim items. It is important, however, that the commodities being stored are essentially noncombustible, insofar as the provisions that regulate Group S-2 occupancies are based on an anticipated minimal fire load.

This section covers those utility occupancies that are not normally occupied by people, such as sheds and other accessory buildings, carports, small garages, fences, tanks and towers, and agricultural buildings. The fire load in these structures and uses varies considerably but is usually not excessive. Because they are normally not occupied, the concern for fire load is not very great, and as a group these uses constitute a low hazard. It is also important to note that a Group U occupancy is not expected to have any public use.

Group U occupancies can generally be divided into two areas. The first includes those buildings that are accessory to other major-use structures. Although these accessory-use buildings will at times be occupied, the time period for occupancy is typically limited to short intervals. The second type of Group U occupancy is those miscellaneous structures that cannot be properly classified into any other listed occupancy. The structures are not intended to be occupied, but must be classified in order to regulate any hazards they may pose to property or adjoining structures and persons.

If the jurisdiction has adopted Appendix C, then it will govern the design and construction of agricultural buildings that come under its purview; however, many urban jurisdictions do not adopt this appendix chapter. In this case, should an occasional agricultural building be constructed, it would be regulated by Section 312.

KEY POINTS

- Proper occupancy classification is a critical decision in determining code compliance.

- Uses are classified by the code into categories of like hazards, based on the risk to occupants of the building as well as the probability of property loss.

- Group A occupancies include rooms and buildings with an occupant load of 50 or more, utilized for the gathering together of persons for civic, social or religious functions; recreation, food or drink consumption; or similar activities.

- The hazards unique to assembly uses are based primarily on the large occupant loads and the concentration of occupants into very small areas.

- Business uses, such as offices, are classified as Group B occupancies and are considered moderate-hazard occupancies.

- Group E occupancies are limited to schools for students through 12th grade and most day-care operations.

- Manufacturing occupancies, classified as Group F, are defined based upon whether or not the materials being produced are combustible or noncombustible.

- Group H occupancies are heavily regulated because of the quantities of hazardous materials present in use or storage.

- Where amounts of hazardous materials are limited in control areas to below the maximum allowable quantities, the occupancy need not be considered a Group H.

- Both physical hazards and health hazards are addressed under the requirements for Group H occupancies.

- Institutional occupancies, classified as Group I, are facilities where individuals are under supervision and care because of physical limitations of health or age, or that house individuals whose personal liberties are restricted.

- Group M occupancies include both sales rooms and motor fuel-dispensing facilities.

- Residential Group R occupancies are partially regulated based upon occupant load or number of units, as well as the occupants' familiarity with their surroundings.

- Group S occupancies for storage are viewed in a manner consistent with Group F manufacturing uses.

- Group U occupancies are utility in nature and are seldom, if ever, occupied.

Build Safer Communities

New ANSI-Approved Standards for High Winds

REFERENCED IN THE 2009 IBC AND IRC!

ICC 500-2008: ICC/NSSA STANDARD FOR THE DESIGN AND CONSTRUCTION OF STORM SHELTERS from ICC and the National Storm Shelter Association provides design requirements for the main wind-resisting structural system and components and cladding of storm shelters, and provides basic occupant life safety and health requirements. (50 pages)   #7026S08

REFERENCED IN THE 2009 IBC AND IRC!

ICC 600-2008: STANDARD FOR RESIDENTIAL CONSTRUCTION IN HIGH-WIND REGIONS provides an adoptable, contemporary set of prescriptive requirements to improve structural integrity and performance of homes in wind climates of 100 to 150 mph. (175 pages)   #7027S08

ORDER NOW! 1-800-786-4452 | www.iccsafe.org/store

Quantity discounts available.

SPECIAL DETAILED REQUIREMENTS BASED ON USE AND OCCUPANCY

This chapter provides specific detailed regulations for those types of buildings and uses that have very unique characteristics. The uses in Chapter 4, though encompassing only a very small fraction of the uses commonly encountered, require special consideration. Some of the provisions address conditions that could occur in various occupancy classifications such as covered mall buildings, high-rise buildings and underground buildings. Concerns associated with motor vehicle-related uses, hazardous occupancies and institutional uses are specifically addressed. Special elements within a building, such as stages, platforms, motion picture projection rooms and atriums, are also regulated by Chapter 4.

In all cases it should be remembered that the provisions found in Chapter 4 deal in a more detailed manner with uses and occupancies also addressed elsewhere in the code. Some of the provisions in this chapter may be more restrictive than the general requirements of the code, whereas others may be less restrictive. The general rules found in other areas of the *International Building Code®* (IBC®) will govern unless modifications from this chapter are utilized.

Section 402   *Covered Mall and Open Mall Buildings*

Provisions for covered mall buildings included in the IBC set forth specific code requirements for a specific building type. Provisions in this section only apply to covered mall buildings having a height of not more than three levels at any one point and not more than three stories above grade plane. Furthermore, the provisions are only those that are considered to be unique to covered mall buildings. For those features that are not unique, the general provisions of the code apply. Covered mall buildings that comply in all respects with other provisions of the code are not required to comply with these provisions. It should be noted that foyers and lobbies of office buildings, hotels and apartment buildings should not be considered covered mall buildings.

Figure 402-1
Open mall

individual tenant spaces are to be based upon the occupant load as typically determined. Figure 402-2 depicts the method for determination of the occupant load in a tenant space having a retail area, a storage room, an office and a bathroom. Although the total tenant space contains 1,500 square feet (139.3 m²) of floor area, each individual use has a designated occupant load based upon the appropriate factor from Table 1004.1.1. In this example, the occupant load would be calculated at 44.

Not only must the occupant load be determined for each individual tenant space, but the occupant load for the entire covered mall building must also be determined. It is highly unlikely that all tenant spaces will be fully occupied at the same time in a covered mall building. Therefore, a different method is used to determine the number of occupants from which to base the means of egress from the mall itself. This occupant load is to be determined based on the gross leasable area of the covered mall building, excluding any anchor buildings and those tenant spaces having an independent means of egress, with the occupant-load factor determined by the following formula:

Occupant load factor = (0.00007) (gross leasable area) + 25

As a result, the net effect is that the total occupant load computed for the covered mall building will be something less than the summation of the occupant loads previously determined for each individual tenant space.

The occupant load factor used for egress purposes shall not exceed 50, nor is it ever required to be less than 30. Where there is a food court provided within the covered mall building, the occupant load of the food court is to be added to the occupant load of the covered mall building as previously calculated in order to determine the total occupant load.

In utilizing several examples, assume a building contains 600,000 square feet (55 740 m²) of gross leasable area. The occupant load factor, when calculated, would be 67. However, a factor of 50 would be used in determining an occupant load of 12,000. Should a food court be present that seats 600 occupants, the occupant load of 12,000 would be increased accordingly. Where a covered mall building contains 100,000 square feet (9290 m²) of gross leasable area, the occupant-load factor would be 32. A factor of 32 would then be used to calculate the occupant load of the covered mall building, which would be 3,125 occupants. As the provision is applied to a smaller covered mall building, an occupant load factor of 30 will be used when the gross leasable area of the covered mall building is less than 71,500 (6642.3 m²) square feet. As a final note, because anchor buildings are not considered a part of a covered mall building, their occupant load shall not be included in computing the total number of occupants for the mall.

Number of means of egress. Figure 402-2 also depicts the requirements of Section 402.4.2 for the determination of the number of exits from the tenant space. Based on an occupant load of 44, the provisions of Chapter 10 would require only one means of egress from this tenant space. Therefore, the number of means of egress complies with the code. However, if the distance, x, exceeds 75 feet (22 860 mm), two means of egress would be required even though the occupant load is less than 50.   **402.4.2**

Arrangements of means of egress. The provisions of this section are unique to the covered mall building and are depicted in Figure 402-3. The limitations described in this section encompass a large number of occupants, and this section prevents those occupants from having to traverse long portions of the mall to reach a means of egress. The provisions also prevent the overcrowding of the mall, such as if a large number of patrons from these uses were to be discharged into the mall at the same time and some distance from a means of egress.   **402.4.3**

In securing the intent of Section 402, the provisions of Section 402.4.3.1 also repeat the requirement that means of egress for anchor buildings shall be provided independently from the mall exit system. Furthermore, the mall shall not egress through the anchor buildings. Moreover, the termination of a mall at an anchor building where no other means of egress has been provided except through the anchor buildings shall be considered to be a dead end, which is limited in accordance with Section 402.4.5.

402.4.4 Distance to exits. Figure 402-4 depicts the multifaceted provisions of this section as follows:

1. The first case is illustrated for travel within the tenant space and includes the provisions applicable to tenant spaces A and B. For tenant space A, the diagram depicts the application of the code for a tenant space with a closed front with only a swinging exit door to the mall. The entrance to the mall would be the point at which occupants from the tenant space pass through the exit door from the tenant space to the mall.

   Tenant space B represents the condition for an open storefront using a security grille instead of a standard exit door. The entrance to the mall in this case is the point at which occupants of the tenant space pass by an imaginary plane that is common to both the tenant space and the pedestrian mall. The location of the assumed required clear exit width along the open front of tenant space B may be placed at any point along the front, and its location would depend only on that which would render the least-restrictive application of the provisions.

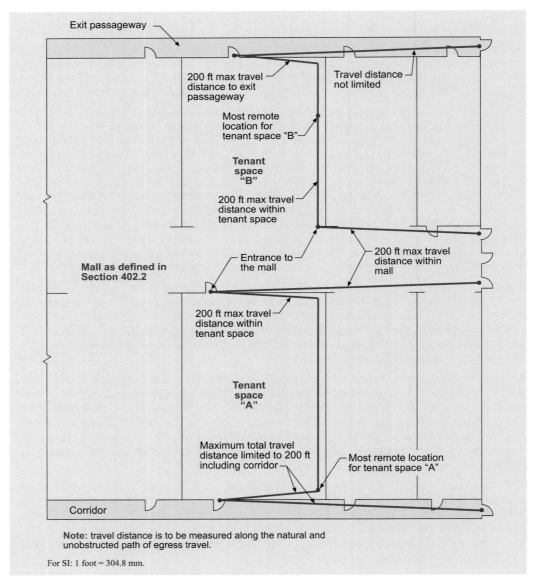

Note: travel distance is to be measured along the natural and unobstructed path of egress travel.

For SI: 1 foot = 304.8 mm.

Figure 402-4
Travel distance

For either tenant space A or B, the code permits the travel distance within the tenant space to the entrance to the mall to be a maximum of 200 feet (60 960 mm).

2. After the occupants exit from a tenant space into the mall, the code permits another 200 feet (60 960 mm) of exit travel distance to one of the exit elements described in Section 1002. This travel limitation also applies to all other locations in the mall where occupants may be located when an exiting condition occurs.

It can be seen from this discussion, plus the perusal of Figure 402-4, that the travel distances permitted for a covered mall building are generally more liberal than those permitted by Section 1016.1. This liberalized and increased travel distance is based on the rationale that travel within a mall will be within an area where special fire protection features are provided.

3. Another limitation of these provisions regarding travel distance within covered mall buildings is also illustrated by Figure 402-4. In this instance, if the path of travel is through a secondary exit from tenant space B to an exit (in this case an exit passageway), travel distance is not limited once the exit is reached.

4. In the case of exiting via the corridor as depicted for tenant space A, the total travel distance is limited to 200 feet (60 960 mm). This limitation is based on the consideration that a corridor does not offer as much protection as either an exit (such as an exit passageway) or the mall.

Access to exits. This section uses the same approach as does Chapter 10 of requiring that **402.4.5** the means of egress be arranged so that the occupants may go in either direction to a separate exit. However, in this section the dead end is measured in a manner similar to that of Exception 3 of Section 1018.4. Figure 402-5 shows the manner in which dead-end conditions are measured and limited.

Regardless of the occupant load served, the minimum width of an exit passageway or corridor from a mall is to be 66 inches (1676 mm). The exit passageway shown in Figure 402-6 must be at least 66 inches (1676 mm) wide. The main entrances shown in the same figure are not subject to this requirement insofar as the exit width limitations of Section 1005.1 will most often require a greater exit width.

Another exiting provision that is unique to the covered mall building regards exit passageway enclosures. Section 402.4.6 allows mechanical and electrical equipment rooms, building service areas, and service elevators to open directly into exit passageways, provided the minimum 1-hour fire-resistance-rated separation is maintained.

Mall width. With its added life-safety systems, the mall may be considered a corridor **402.5** without meeting the requirements of Section 1018.1 when the mall complies with the conditions of this section as depicted in Figure 402-7. In this case, the code requires that the minimum mall width be 20 feet (6096 mm), and this typical cross section shows that the minimum required width may be divided so that a clear width of 10 feet (3048 mm) is provided separately on each side of any kiosks, vending machines, benches, displays, etc., contained in the mall. In addition, food court seating in the mall would have to be located so as not to encroach upon any required mall width. Understandably, the mall width shall also accommodate the occupant load immediately tributary thereto.

Types of construction. Where covered mall buildings, including their anchor buildings, **402.6** are constructed of other than Type V construction, they may be of unlimited area. This allowance is applicable only where the entire building, including the anchor buildings and attached parking structures, are surrounded by permanent open space at least 60 feet (18 288 mm) in width and the anchor buildings are no more than three stories in height above grade plane. See Figure 402-8. For those anchor buildings exceeding three stories, the general height and area limitations of Chapter 5 are applicable, including appropriate increases for frontage and the presence of an automatic sprinkler system. In all cases, the minimum type of construction required for any open or enclosed parking garage is that mandated by Section 406. The allowance provided in Section 402.6.1 for reduced open space surrounding covered mall buildings, including their associated parking garages and anchor stores, is consistent with that for other unlimited area buildings as permitted by Section 507.5 since a covered mall building contains similar characteristics of those buildings addressed in Section 507. They are limited in height, contain similar occupancies, are fully sprinklered, and require significant open space surrounding the building. A limit is placed on the occupancies contained within the covered mall building to maintain consistency with the occupancy groups addressed in Section 507.5. The reduction in required open space is not permitted where the covered mall building or anchor stores include Group E, H, I or R occupancies.

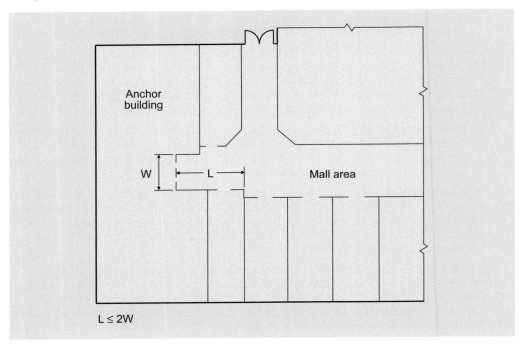

Figure 402-5
Dead-end
mall criteria

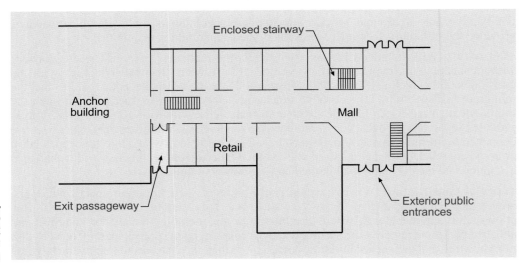

Figure 402-6
Exit
passageway
width

402.7 **Fire-resistance-rated separation.** Under the concept of covered mall buildings, there is no requirement for a fire separation between tenant spaces and the mall. Similarly, the food court needs no separation from adjacent tenant spaces or the mall. The hazards presented by an attached parking garage, however, must be addressed through the separation provided by a minimum 2-hour fire-resistance-rated fire barrier.

Tenant spaces must be separated from each other by fire partitions complying with Section 709. In addition to protecting one tenant from the activities of a neighbor, the tenant separation requirements for malls are also intended to assist in the goal of restricting fire to the area of origin. There is no requirement, however, for the separation of tenant spaces from the mall itself.

As a general rule, the anchor building is viewed as a separate building from the covered mall building. Therefore, a fire wall must be used to provide the necessary fire-resistive separation. However, only a fire barrier is required where the anchor building is no more than three stories in height and its use is consistent with one of those identified in the definition of "Covered mall building" in Section 402.2. Although some type of 2-hour fire-resistance-rated separation is mandated between an anchor building and the mall, openings in such a separation typically need no fire-protection rating. Anchor buildings, other than Group R-1 sleeping units, constructed of Type I or II noncombustible construction may have unprotected openings into the mall.

Interior finish. The interior finish requirements for tenant spaces and anchor buildings are **402.8** regulated for interior finishes based upon their specific occupancy classification in accordance with Section 803.9. The common areas, including the mall and exits, are to have wall and ceiling finishes that have a minimum Class B flame spread rating.

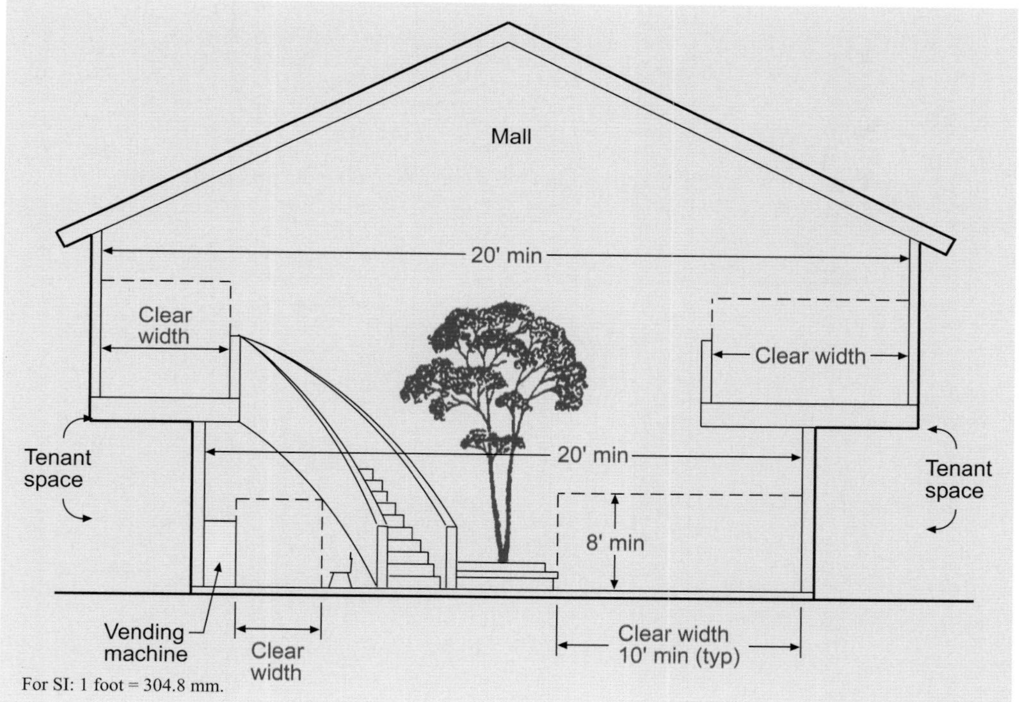

For SI: 1 foot = 304.8 mm.

**Figure 402-7
Mall width
requirements**

Automatic sprinkler system. The automatic fire-sprinkler system is the primary means of **402.9** fire protection for the covered mall building. The system is required throughout all portions of the covered mall building other than open parking garages. Additionally, the code requires a standpipe system in accordance with Section 905. Because of the reliance placed on the sprinkler system, this section requires the following additional safeguards:

1. The code requires that the sprinkler system be complete and operative throughout all of the covered mall building before occupancy of any of the tenant spaces. In

those areas that are unoccupied, an alternate protection method may be approved by the building official.

2. The mall and the tenant spaces shall be protected by separate sprinkler systems, except that the code will permit tenant spaces to be supplied by the same system as the mall, provided they can be independently controlled.

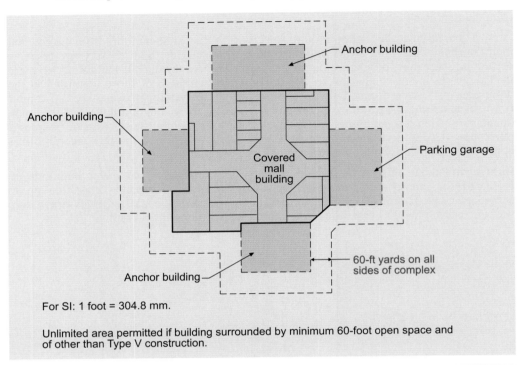

For SI: 1 foot = 304.8 mm.

Figure 402-8
Covered mall building

Unlimited area permitted if building surrounded by minimum 60-foot open space and of other than Type V construction.

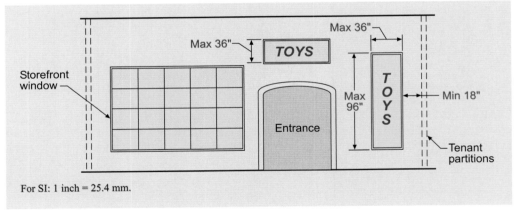

Figure 402-9
Limitations for plastic signs and panels in malls

For SI: 1 inch = 25.4 mm.

402.10 **Smoke control.** A smoke-control system need not be provided for a covered mall building unless an atrium is provided that connects at least three floor levels of the building. In covered mall buildings of one or two stories, no smoke control system is required. In addition, a smoke control system is not required for an open mall building due to the lack of a roof over the mall area. The minimum 20-foot (6096 mm) width mandated for an open mall, extending from the floor to the roof, provides an equivalent level of protection as a smoke control system in the maintenance of a tenable environment in the mall area.

402.11 **Kiosks.** Kiosks and similar structures, both temporary and permanent in nature, are regulated for construction materials and fire protection owing to their presence in an established egress path. Such structures shall be noncombustible or constructed of fire-retardant-treated wood, complying foam plastics or complying aluminum composite materials. Active fire protection is provided by required fire suppression and detection devices. Kiosks are also limited in size, and their relationship to other kiosks is regulated. Multiple kiosks can be grouped together, provided their total area does not exceed 300 square feet (28 m^2). At that point, a separation of at least 20 feet (6096 mm) is required from another kiosk or grouping of kiosks.

Security grilles and doors. Quite often, mall tenants wish to have the dividing plane **402.13** between the mall and the tenant space completely open during business hours. Horizontal sliding or vertical security grilles or doors are usually placed across this opening. This section permits their use, provided they do not detract from safe exiting from the tenant space into the mall. To secure that intent, the code requires four limitations outlined in this section.

Plastic panels and plastic signs. In this section, the IBC limits plastic panels and plastic **402.16** signs because they are within an exitway (the mall) and they are combustible (even though of approved plastic). It is important to note that the percentage of wall covered is based on the area common to each single tenant space. Thus, for a tenant space whose common wall with the mall is 60 feet (18 288 mm) wide and 11 feet (3353 mm) high, the total area is 660 square feet (61.3 m^2). As the code permits 20 percent of that area to be of plastic panels or signs, the sum of all of the plastic signs and panels on the common wall is limited to 132 square feet (12.3 m^2). Figure 402-9 illustrates the code limitations for plastic signs and panels.

This section also requires that the use of foam plastic in signs be based on testing in accordance with UL 1975, *Fire Tests for Foamed Plastics Used for Decorative Purposes*.

Section 403 *High-rise Buildings*

This section encompasses special life-safety requirements for high-rise buildings. The comparatively good fire record notwithstanding, particularly in office buildings, fires in high-rise buildings have prompted government at all levels to develop special regulations concerning life safety in high-rise buildings. The potential for disaster that is due to the large number of occupants in high-rise buildings has resulted in the provisions included in this section.

The high-rise building is characterized by several features:

1. It is impractical, if not impossible, to completely evacuate the building within a reasonable period of time.

2. Prompt rescue will be difficult, and the probability of fighting a fire in upper stories from the exterior will be low.

3. High-rise buildings are occupied by large numbers of people, and in certain occupancies the occupants may be asleep during an emergency.

4. A potential exists for stack effect. The stack effect can result in the distribution of smoke and other products of combustion throughout the height of a high-rise building during a fire.

   The provisions in this section are designed to account for the features described above.

Applicability. Although a high-rise building can be defined in accordance with the special **403.1** features just described, the IBC elects to define a high-rise building in Section 202 as one having floors used for human occupancy located more than 75 feet (22 860 mm) above the lowest level of fire department vehicle access.

This section identifies those types of buildings and structures to which the provisions for high-rise buildings do not apply. Included in this group of structures are aircraft-traffic control towers, open parking garages, Group A-5 occupancies, special industrial occupancies and buildings with an occupancy in Group H-1, H-2 or H-3.

403.2 Construction. Primarily because the sprinklered high-rise building is provided with an increased level of fire protection supervision and control, the IBC permits certain modifications of the code requirements, which are sometimes referred to as trade-offs. The trade-offs for construction type are considered to be justified on the basis that the sprinkler system, although a mechanical system, is highly reliable because of the provisions that require supervisory initiating devices and water-flow initiating devices for every floor. In addition, a secondary on-site supply of water is mandated for those high-rise buildings subject to a moderate to high level of seismic risk.

This section permits some degree of reduction in the required fire-resistance ratings of building elements required to be protected on account of type of construction. In the evaluation of the maximum allowable height and area permitted for the building, the original construction type would remain applicable.

In addition to the reductions permitted for building elements identified in Table 601 based on the building's type of construction, the fire-resistance rating of shaft enclosures may be reduced to 1 hour where sprinklers are installed within the shafts at the top and at alternate floor levels.

The reduction in fire-resistance ratings is not applicable in all cases. In all high-rise buildings of Type IA construction, the 3-hour rating for structural columns supporting floors must be maintained. The critical role of columns in the structural integrity of a high-rise structure during fire conditions mandates that their fire resistance not be lessened. For buildings that exceed 420 feet (128 m) in height, no reduction to the required fire resistance of any building elements is permitted. In addition, the required vertical shaft protection cannot be reduced from the general 2-hour requirement. The increased risk of catastrophic damage associated with these very tall buildings requires an increased level of fire resistance.

403.3 Automatic sprinkler system. The automatic fire-sprinkler system required by Section 403.3 must be completely reliable, as must the other life-safety systems. As part of that reliability effort, a high-rise building must be provided with a secondary water supply where required by Section 903.3.5.2. In Seismic Design Category C, D, E or F, an on-site secondary water supply shall be provided, with a supply of water equal to the hydraulically calculated sprinkler design demand, including the hose stream requirement, for a duration of at least 30 minutes. As fires can (and do) break out as a result of earthquake damage to the various mechanical systems within a building, it is imperative that the reliability of the sprinkler system be such that any resulting fires can be automatically extinguished.

403.4 Emergency systems. Among the more important life-safety features required by the code are the alarm and communications systems required by this section. Where it is expected that people will be unable to evacuate the building, it is imperative that they be informed as to the nature of any emergency that may break out, as well as the proper action to take to exit to a safe place of refuge. Furthermore, a system is necessary in most cases to provide for communication between the fire officer in charge at the scene and the fire fighters throughout the building. Section 911 provides details of the required fire command center utilized by the fire department to coordinate fire suppression and rescue operations.

In order to provide for efficient and reliable communications among fire fighters, police officers, medical personnel and other emergency responders, an emergency responder radio communications system must be installed in all high-rise buildings. The details for such systems are established in Section 510 of the *International Fire Code*® (IFC®) and, more specifically, IFC Appendix J, *Emergency Responder Radio Coverage*. Although Appendix J is not mandatory, it can be adopted by jurisdictions by reference if the jurisdiction does not have its own criteria for the in-building performance of its analog or digital radio system. Section 510.2 specifies that acceptable radio coverage is satisfied when 95 percent of all areas on each floor of the building meet the signal strength requirements in Sections 510.2.1 and 510.2.2 for signals transmitted into and out of a building.

In addition, one of the fire department's duties during a fire event is to expel the smoke after the fire has occurred. Three methods for smoke exhaust are now available: through natural means, through the use of mechanical air-handling equipment, or through an equivalency approach. Where the method of natural ventilation is used for the removal of smoke, openable windows or panels are required to be distributed around the perimeter of each floor level. The fire departments can open the appropriate windows as necessary and provide pressurization through the use of fans. The use of fixed tempered glass panels is also acceptable if they are not coated in a manner that will modify the natural breaking characteristics of the glass. Where mechanical air-handling equipment is used for smoke removal purposes, the building's HVAC system is equipped with appropriate dampers at each floor that are arranged in a manner that will stop the recirculation of air through the use of 100 percent fresh air intake and outside exhaust. The panel for controlling this system is to be located in the building's fire command center. In addition, the building official has the authority to approve any other means of smoke removal provided it accomplishes the intended goal of the prescriptive mechanical or natural ventilation approaches described by the code. See Figure 403-1. It must be noted that this smoke exhaust system is for fire department use only and not intended to be a part of the occupant-related life-safety systems placed in high-rise buildings.

Furthering the intent of the IBC that life-safety systems in high-rise buildings be highly reliable, the code requires that the power supply to the life-safety systems be regulated by the appropriate provisions of NFPA 70, more specifically, Articles 700 and 701. The basis of the reliability is that the building's power be automatically transferable to a standby or emergency power system in the event of the failure of the normal power supply. Those standby power loads required by the code include power and lighting for the fire command center, ventilation and automatic fire detection equipment for smokeproof enclosures, and elevators. Some of the requirements and details for the standby power system are depicted in Figure 403-2.

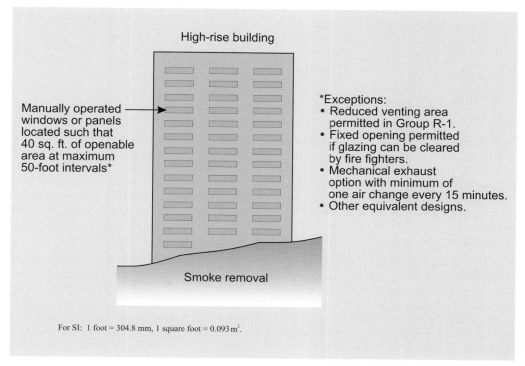

High-rise building

Manually operated windows or panels located such that 40 sq. ft. of openable area at maximum 50-foot intervals*

*Exceptions:
• Reduced venting area permitted in Group R-1.
• Fixed opening permitted if glazing can be cleared by fire fighters.
• Mechanical exhaust option with minimum of one air change every 15 minutes.
• Other equivalent designs.

Smoke removal

For SI: 1 foot = 304.8 mm, 1 square foot = 0.093 m².

Figure 403-1
Smoke removal

The code further requires that lighting for exit signs, means of egress illumination and elevator car lighting be automatically transferable to an emergency power system capable of operation within 10 seconds of the failure of the normal power supply. Additionally, all emergency voice/alarm communications systems, automatic fire detection systems, fire alarm systems and electrically powered fire pumps are to be provided with emergency power. Figure 403-3 illustrates one means of transfer to emergency power.

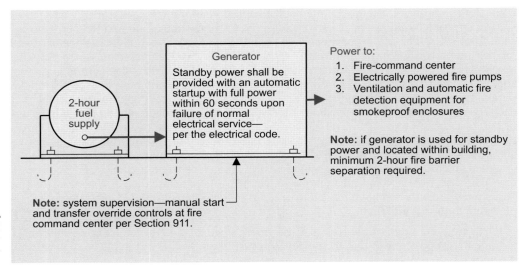

Figure 403-2
Standby
power

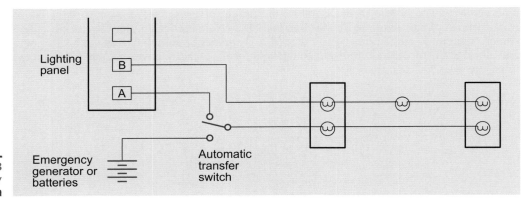

Figure 403-3
Emergency
lighting system

403.5.1   **Remoteness of exit stairway enclosures.** The general requirements for separation of exit or exit access doorways as established in Sections 1015.2.1 and 1015.2.2 are supplemented in this section by adding a minimum required separation distance between the exit enclosures. In addition to maintaining a minimum separation between the doors to the exit enclosures of one-third the length of the overall diagonal dimension of the area served, the exit enclosures must be located at least 30 feet apart or not less than one-fourth the diagonal dimension, whichever is less. See Figure 403-4. If three or more exit enclosures are mandated, at least two of the exit enclosures must be separated as indicated.

403.5.2   **Additional exit stairway.** During a fire that requires a full evacuation of a building of extensive height, the fire-fighting operations will reduce the capacity of the egress system. The extended period of time needed to fully evacuate a very tall building means that people will still be evacuating while full fire-fighting operations are taking place. Sound high-rise fire-fighting doctrine provides that the fire department take control of one stair, the one most appropriate to the circumstances of the given fire condition. This can result in a significant reduction in egress capacity of the stairway system. For example, in order to conduct suppression activities in a building with two required stairs of the same width, one-half of the exit capacity is unavailable while the building is still being evacuated. An additional stair is required so that egress capacity will be maintained through the time that full evacuation is complete.

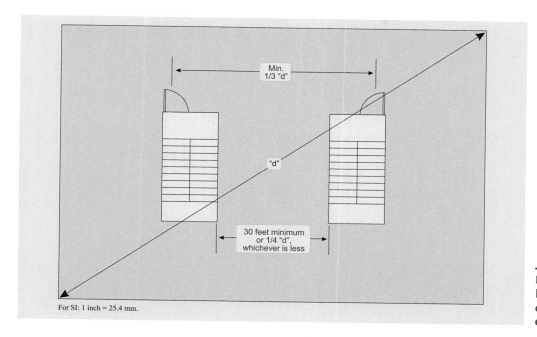

Figure 403-4
**Remoteness
of exit
enclosures**

It is important to note that this additional stair is not required to be a dedicated fire department stair. The fire department should be able to choose the stair which is most appropriate for the actual fire event. As a result, it will be necessary for emergency responders to manage evacuation flow to the available stairs.

The application of this requirement is limited to only those buildings over 420 feet (128 m) in height. In addition, it does not apply to Group R-2 occupancies due to the limited occupant load of such uses. In determining if the egress width is provided by the stairway system, it must be assumed that the widest of the stairways is the one that is unavailable for means of egress travel. The remaining stairways must be sized to accommodate the total required egress width. The additional stairway's sole purpose is to provide additional egress capacity. Therefore, other means of egress design issues, such as travel distance and exit separation, are not regulated. See Application Example 403-1.

The additional exit stairway is not required where occupant evacuation elevators are provided in accordance with Section 3008. The availability of elevators for evacuation purposes provides for a reasonable alternative to an additional stairway.

Stairway door operation. In those cases where it is impractical to totally evacuate the occupants from the building through the stairway system, it must be possible to move the occupants to different floors of the building that are safe by way of the stairway system. For this to happen, the doors to the stair enclosures must either be unlocked or be designed for automatic unlocking from the fire command station. **403.5.3**

The IBC further requires that a telephone or other two-way communication system (such as a two-way system with speaker and microphone) be located at every fifth floor in each required stair enclosure for those cases where the stair enclosure doors are to be locked. Moreover, the code requires that this communication system be connected to an approved station that is constantly attended. Thus, anyone trapped in the stairway during a nonfire emergency may call for help without traversing more than two levels. In the case of office buildings and apartment houses, the attended station may be considered the office of the building as long as the office has continuous attendance by responsible individuals who are familiar with the life-safety systems. For hotel buildings, the most likely choice for the attendance station will probably be the hotel telephone operators, and, again, they must be trained to assist the persons trapped within the stair enclosure.

Application Example 403-1

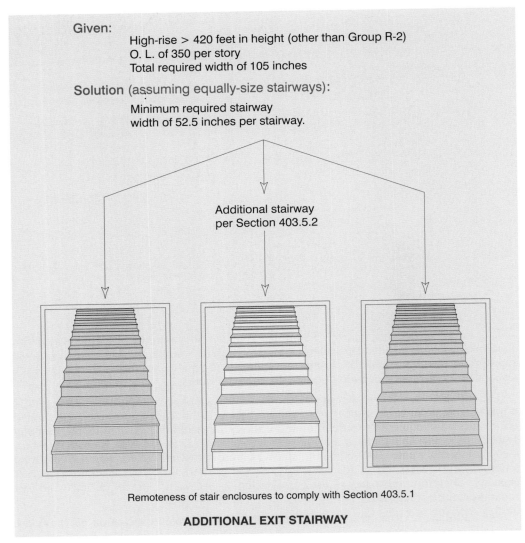

Given:

High-rise > 420 feet in height (other than Group R-2)
O. L. of 350 per story
Total required width of 105 inches

Solution (assuming equally-size stairways):

Minimum required stairway width of 52.5 inches per stairway.

Additional stairway per Section 403.5.2

Remoteness of stair enclosures to comply with Section 403.5.1

ADDITIONAL EXIT STAIRWAY

403.5.4 Smokeproof exit enclosures. Those exit enclosures in a high-rise building that serve floor levels located more than 75 feet (22 860 mm) above the lowest level of fire department vehicle access must be designed as smokeproof enclosures. Exit stairways that do not serve floors above the height indicated are not regulated by this section. Section 1022.9 regulates the access, extension and termination relating to the utilization of smokeproof enclosures and pressurized stairways as a part of the means of egress system. Section 909.20 provides the construction and ventilation criteria for smokeproof enclosures, as well as establishing stair pressurization as an acceptable alternative.

403.5.5 Luminous egress path markings. In high-rise buildings, increased visibility for travel on stairways and through exit passageways is important due to the extreme conditions that may be encountered under emergency conditions. The use of photoluminescent or self-illuminating materials to delineate the exit path is required in high-rise buildings housing Group A, B, E, I, M and R-1 occupancies. Specific requirements related to these egress path markings are set forth in Section 1027 and include the regulation of striping on steps, landings, and handrails; perimeter demarcation lines on the floor and walls, including their transition; acceptable materials; and illumination periods. An example of such markings is shown in Figure 403-5.

**Figure 403-5
Luminous
egress path
markings**

Elevators. In order to facilitate the rapid deployment of fire fighters, a fire service access elevator is now required in high-rise buildings that have an occupied floor more than 120 feet (36 576 mm) above the lowest level of fire department vehicle access. Usable by fire fighters and other emergency responders, the specific requirements for the elevator are set forth in Section 3007. There are a number of key features that allow fire fighters to use the elevator for safely accessing an area of a building that may be involved in a fire. A complying lobby is required adjacent to the elevator hoistway opening, creating a protected area from which to stage operations. Access to standpipe hose valves is required, as are two-way communication features. A more comprehensive discussion of the requirements for fire service access elevators is found in the analysis of Section 3007.

403.6

The use of elevators as an evacuation element for occupants of a high-rise building is possible provided the elevators are in compliance with the requirements established in Section 3008. The controls and safeguards provided in Section 3008 create a suitable environment to allow complying elevators to be used for occupant self-evacuation purposes. The presence of such elevators does not reduce the general means of egress requirements established in Chapter 10, however, the additional exit stairway mandated for high-rise buildings by Section 403.5.2 is no longer required. It is important to note that the installation of occupant evacuation elevators in high-rise buildings is not mandated by the code, however such elevators are permitted for use for occupant self-evacuation and may be utilized as an alternative to the additional stairway requirement.

Section 404 *Atriums*

This section was developed to fill a need for code provisions applicable to the trends in the architectural design of buildings where the designer makes use of an atrium. Prior to the early 1980s, building codes did not provide for atriums, and, moreover, atriums were prohibited because of the requirements for protection of vertical openings. They were, however, permitted on an individual basis, usually under the provisions in the administrative sections of the code permitting alternate designs and alternate methods of construction. The general concept of alternate protection is to provide for both the equivalence of an open court and at the same time provide protection somewhat equivalent to shaft protection to prevent products of combustion from being spread throughout the building via the atrium.

An atrium is considered "an opening connecting two or more stories other than enclosed stairways, elevators, hoistways, escalators, plumbing, electrical, air-conditioning or other equipment, which is closed at the top and not defined as a mall." This section permits large unprotected vertical openings through floors without the need for a shaft enclosure. The use of atriums is permitted in all buildings other than those classified as Group H occupancies.

Note that most cases where two floors are open to each other do not create atrium conditions. That is because Exception 7 of Section 708.2 permits two floors to be open to each other where seven conditions are met. The atrium provisions are typically only utilized for open two-story spaces where they cannot fully comply with the conditions of Exception 7 and are too large to qualify as a mezzanine as permitted by Exception 9. Addressed in Exception 5, the atrium provisions are provided only as another exception to the requirements of Section 708.2. In general, any opening through a floor/ceiling assembly is required to be protected by a complying shaft enclosure. However, a number of exceptions permit alternative methods of compliance where applicable, such as floor dampers and through-penetration firestop systems. Compliance with the atrium provisions of Section 404 is simply another permitted option to the use of a shaft enclosure to protect a floor/ceiling opening.

404.3 **Automatic sprinkler protection.** One of the basic requirements for atriums is that the building be provided with an automatic sprinkler system throughout. See Figure 404-1. Two exceptions modify this general requirement. Those areas of the building adjacent to or above the atrium are not required to be sprinklered if appropriately separated from the atrium. This separation must consist of minimum 2-hour fire barriers, horizontal assemblies or both. In addition, sprinkler protection is not required at an atrium ceiling located more than 55 feet (16 764 mm) above the atrium floor.

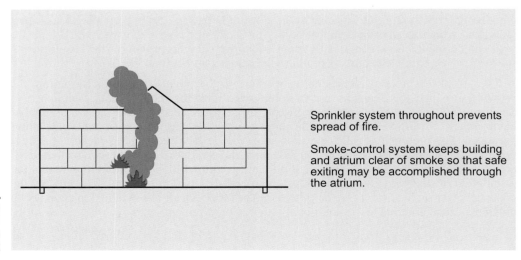

Sprinkler system throughout prevents spread of fire.

Smoke-control system keeps building and atrium clear of smoke so that safe exiting may be accomplished through the atrium.

Figure
404-1
**Atrium
concept**

Smoke control. Another major component of the life-safety system for a building **404.5** containing an atrium is the required smoke-control system. The design of the smoke-control system is to be in accordance with Section 909. Although the exhaust method is typically used as the means of accomplishing smoke control, the code would not prohibit the use of the airflow or pressurization methods where shown to be suitable. One of these methods is often used where the ceiling height makes it difficult to maintain the smoke layer at least 6 feet (1829 mm) above the floor of the means of egress.

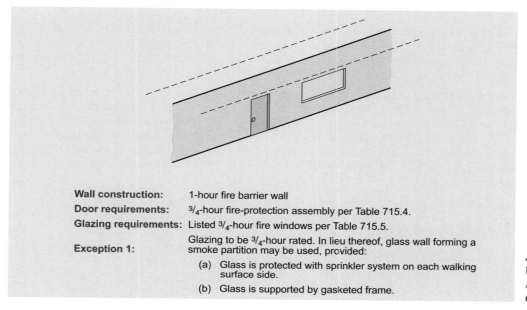

Wall construction: 1-hour fire barrier wall
Door requirements: ³/₄-hour fire-protection assembly per Table 715.4.
Glazing requirements: Listed ³/₄-hour fire windows per Table 715.5.

Exception 1:   Glazing to be ³/₄-hour rated. In lieu thereof, glass wall forming a smoke partition may be used, provided:

    (a) Glass is protected with sprinkler system on each walking surface side.

    (b) Glass is supported by gasketed frame.

Figure 404-2
Atrium enclosure

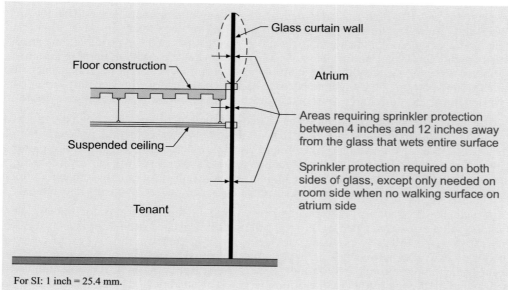

Glass curtain wall

Floor construction

Atrium

Areas requiring sprinkler protection between 4 inches and 12 inches away from the glass that wets entire surface

Suspended ceiling

Sprinkler protection required on both sides of glass, except only needed on room side when no walking surface on atrium side

Tenant

For SI: 1 inch = 25.4 mm.

Figure 404-3
Glass protection

An exception eliminates the requirement for smoke control in those atriums that connect only two stories. However, as previously addressed, most situations where two floors are open to each other are not regulated under the provisions of Section 404. Typically, Exception 7 of Section 708.2 is utilized to permit an opening between two floor levels without requiring compliance with any of the atrium provisions.

404.6 **Enclosure of atriums.** With some exceptions, an enclosure separation is required between the atrium and the remainder of the building. See Figure 404-2. The basic requirement is for a 1-hour fire-resistance-rated fire barrier with openings protected in accordance with Tables 715.4 and 715.5. This degree of enclosure, in addition to the other special conditions of Section 404, is intended to provide protection somewhat equivalent to the otherwise mandated shaft protection. Two alternative methods of atrium separation are described in Exceptions 1 and 2. The special sprinkler-wetted glass enclosure as depicted in Figure 404-3 provides a prescriptive method of achieving equivalency. In addition, the separation may consist of a $^3/_4$-hour-rated glass-block wall assembly.

The separation between adjacent spaces and the atrium may be omitted on a maximum of any three floor levels, provided the remaining floor levels are separated as provided in this section. In computing the atrium volume for the design of the smoke-control system, the volume of such open spaces shall be included.

404.9 **Travel distance.** The code does not specifically address open or unenclosed stairways where provided within the atrium space. Under those limited circumstances where a two-story atrium occurs, Exception 1 of Section 1022.1 and Exceptions 3 and 4 of Section 1016.1 allow for an unenclosed stairway to be used as a required means of egress. However, where the atrium connects three or more stories, only Exception 4 would permit an unenclosed stairway, but only where the stairway is not considered a required means of egress.

Section 405 *Underground Buildings*

Structures that have floor levels well below ground level, and thus the level of access and egress from the exterior, present special hazards to both the building occupants and fire personnel. Much like high-rise buildings, underground buildings can create difficult egress conditions as well as pose many problems for the fire department in their rescue and suppression activities. Fundamental to the protection features of this type of building are the requirements for Type I noncombustible construction and the installation of an automatic sprinkler system. A standpipe system is also required. For clarification, only the underground portion of the structure needs to be of Type I construction, and only those floor levels at the highest discharge level and below need to be sprinklered. See Figure 405-1.

The basic criteria for consideration as an underground building is that a floor level used for human occupancy be located more than 30 feet (9144 mm) below the finished floor of the lowest level of exit discharge. Exempted are sprinklered dwellings; parking garages having fire suppression systems; fixed guideway transit systems such as subways; stadiums, arenas and similar assembly uses; those buildings where only a very limited amount of floor area would qualify by the definition; and mechanical spaces that are typically unoccupied.

A valuable concept in fire protection is utilized in the provisions for underground buildings that extend even deeper into the ground. Where an occupied floor level is located more than 60 feet (18 288 mm) below the finished floor of the lowest level of exit discharge, at least two compartments of approximately equal size must be created. The compartmentation must extend throughout the underground portion of the structure, up to and including the highest level of exit discharge. The separation between the two areas is intended to allow for horizontal egress travel to a refuge area if necessary, while also permitting the use of the compartment as a staging area for fire-suppression activities. A smoke barrier is required as the separation element, with door openings also protected in a manner to restrict smoke leakage. Other openings and penetrations are strictly limited. Air supply and exhaust systems, where provided, must be independent of the other compartments. Where the underground portion of the building is served by elevators, each compartment must have access to at least one elevator. An elevator lobby, enclosed by a smoke barrier, may be used to allow a single elevator to serve more than one compartment. Doors into the elevator lobby shall be gasketed, have a drop sill and be automatic closing by smoke detection. See Figure 405-2.

Smoke control is also an important part of the overall fire-protection package. By limiting the spread of smoke to only the originating area of the fire, the remainder of the underground building should be provided with acceptable egress paths. Each compartment shall be provided with its own smoke-control system and manual fire-alarm system. A complying communications system is also an integral part of the fire- and life-safety concept for underground buildings.

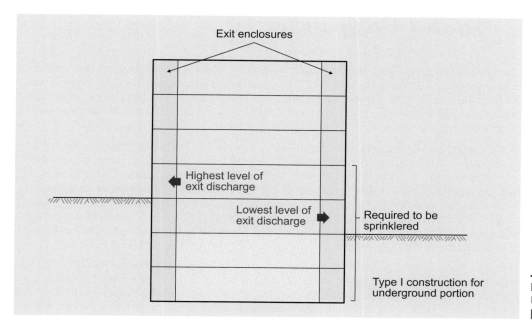

Figure 405-1
Underground buildings

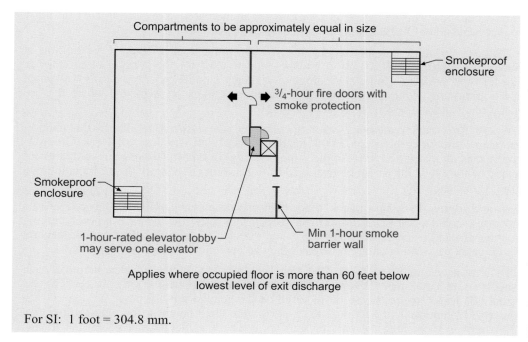

Figure 405-2
Compartmentation of underground buildings

For SI: 1 foot = 304.8 mm.

Stairways serving the floor levels of an underground building are to be smokeproof enclosures, with at least two means of egress from each floor level. If multiple compartments are formed, each compartment must have at least one exit, with a second egress path available into an adjoining compartment. It is mandatory that multiple exits be provided, and within enclosures designed to resist the penetration of smoke.

Both standby and emergency power shall be provided to specific loads identified by this section. See the discussion of Section 403.4.

Section 406 *Motor-vehicle-related Occupancies*

Although uncommon, fire hazards related to motor vehicles are a concern, particularly where associated with other occupancies. The code regulates occupancies containing motor vehicles, whether they be parked, under repair or being fueled. The hazards are primarily related to the fuel in the vehicles, as the overall fire loading related to vehicle occupancies is typically quite low.

406.1 **Private garages and carports.** In order to secure the low fire hazard intended by the Group U occupancy classification, the code initially limits buildings in this category to 1,000 square feet (93 m^2) in floor area and one story in height. However, the code goes on to develop exceptions to the 1,000-square-foot (93 m^2) limitation for those private garages that are not used for repair or fuel-dispensing operations. In such cases, a limit of 3,000 square feet (279 m^2) is established for the Group U occupancy. Because it is doubtful that a garage containing a repair or fuel-dispensing use would receive a Group U classification, the maximum permissible floor area would almost always be 3,000 square feet (279 m^2).

For a mixed-occupancy building such as a Group R-2 apartment house with a Group U private parking garage the exterior wall and opening protection for the Group U occupancy are required to be the same as required for the major occupancy in the building (Group R-2 in the example). Thus, the increase in area is justified on the basis of restricting the use of the building and increasing the exterior wall and opening protection. Where this provision of the code is utilized, it also limits the allowable floor area of the total building to that of the major occupancy. This would generally allow for an increased allowable area, as the *unity formula* provision of Section 508.4.2 would not be applicable.

The code also allows an increase in area to 3,000 square feet (279 m^2) if the building is limited to a Group U as its only occupancy and is used for the parking of private- or pleasure-type vehicles. Under such single-occupancy conditions, the exterior wall must have a minimum 1-hour fire-resistance rating and openings are regulated where the fire separation distance is less than 5 feet (1524 mm).

Where fire walls complying with Section 706 are utilized to divide the parking occupancy into floor areas of 3,000 square feet (279 m^2) or less, multiple Group U occupancies are permitted to be in the same building. In this situation, each compartment created by the fire walls would be restricted in size in order to classify the entire building a Group U occupancy.

406.2 **Parking garages.** Parking garages, other than private garages, will fall into one of two categories, either open or enclosed. The special characteristics of each type of parking structure are addressed in Sections 406.3 and 406.4. This section addresses the general requirements for parking garages, whether they be open or enclosed.

To allow access to the garage to other than high-profile vehicles, the clear height of each level is to be at least 7 feet (2134 mm). Note that the minimum height of the means of egress system is 7 feet, 6 inches (2286 mm), based on the general provisions of Section 1003.2. However, Exception 7 to Section 1003.2 allows the clear height to be reduced to 7 feet (2134 mm) in those areas of parking garages used for vehicular and pedestrian traffic.

Guards must also be provided in accordance with the general provisions of Section 1013. All parking areas more than 12 inches (305 mm) above adjacent levels shall be provided with vehicle barriers at the ends of parking spaces and drive lanes. The height of the vehicle barriers cannot be less than 2 feet, 9 inches (835 mm). Where the parking garage is connected directly to a room containing a fuel-fired appliance, a vestibule is mandated to separate the two spaces. This provides at least two doorways to isolate the equipment from the vehicle area. The vestibule is not necessary where the appliance ignition sources are placed at least 18 inches (457 mm) above the floor. The *International Mechanical Code*® (IMC®) and/or *International Fuel Gas Code*® (IFGC®) should also be consulted for those requirements regulating the installation of mechanical equipment within parking garages.

Open parking garages. Studies and tests of fires in open parking garages have shown that, **406.3** in addition to a low fire loading, the potential for a large fire is exceedingly remote. Based on this data, the IBC establishes special provisions for open parking garages in this section, which in general are less restrictive than those for enclosed parking structures addressed in Section 406.4. The key is that the open parking garage is well ventilated naturally, and as a result, the products of combustion dissipate rapidly and do not contribute to the spread of fire.

To secure the proper amount of openness, the code, as illustrated in Figure 406-1, specifies the following:

1. The building must have openings on at least two sides.

2. The openings must be uniformly distributed along each side.

3. The area of openings in the exterior walls on any given tier must be at least equal to 20 percent of the wall area of the total perimeter of each tier.

4. Unless the required openings are uniformly distributed over two *opposing* sides of the building, the aggregate length of openings considered to provide natural ventilation shall constitute a minimum of 40 percent of the wall length of the perimeter of that tier.

5. The area of openings in the interior walls must be at least 20 percent of the area of the interior walls with openings uniformly distributed.

As a general rule, the maximum allowable height and area of open parking garages is calculated in the same manner as for other buildings. Classified as a Group S-2 occupancy, the provisions of Chapter 5 would apply. However, where the open parking garage contains no uses other than the parking of private motor vehicles, the specific size limitations of Table 406.3.5 and Section 406.3.6 take effect. Because the potential fire severity of an open parking structure used solely for vehicle parking is extremely low, the code permits area and height limitations in excess of those for other Group S-2 occupancies. For example, a stand-alone open parking garage of Type IIB construction would be permitted a floor area of 50,000 square feet (4645 m²) per tier based on Table 406.3.5, with a height limit of eight tiers for a ramp-access garage. For such an open parking garage exceeding three stories in height, the total area of the multistory building is not limited to three times that for a one-story building, as is required by Section 506.4.1, Item 3, but rather can be computed as the permitted area per tier times the number of tiers. Therefore, in the example just given, the total area permitted would be 400,000 square feet (37 160 m²) for a stand-alone Type IIB open parking garage. The maximum height in tiers has been limited somewhat arbitrarily by the code, based on the length of time it would take for fire-department personnel to reach the top of the structure for fire-suppression purposes.

The area and height increases above the tabular limits listed in Table 406.3.5 for single-use open parking garages are those outlined in Section 406.3.6, and are basically keyed to the provision of more natural ventilation area than the minimum required by the code. For unlimited-area buildings permitted by this section, see Figure 406-2.

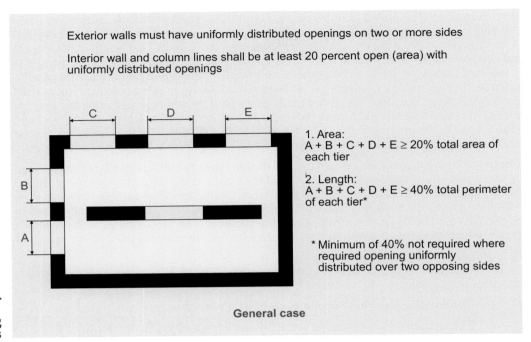

Exterior walls must have uniformly distributed openings on two or more sides

Interior wall and column lines shall be at least 20 percent open (area) with uniformly distributed openings

1. Area:
A + B + C + D + E ≥ 20% total area of each tier

2. Length:
A + B + C + D + E ≥ 40% total perimeter of each tier*

* Minimum of 40% not required where required opening uniformly distributed over two opposing sides

General case

Figure 406-1
Open parking garages

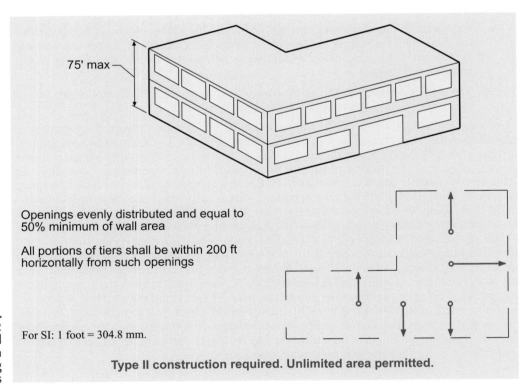

75' max

Openings evenly distributed and equal to 50% minimum of wall area

All portions of tiers shall be within 200 ft horizontally from such openings

Figure 406-2
Unlimited area open parking garages

For SI: 1 foot = 304.8 mm.

Type II construction required. Unlimited area permitted.

In the classification of a Group S-2 parking structure as an open parking garage, the code identifies the following prohibitions:

1. There shall be no automobile repair work performed in the building.

2. There shall be no parking of buses, trucks or similar vehicles.

3. There shall be no partial or complete closing of the required exterior wall openings by tarpaulins or by any other means.

4. There shall be no dispensing of fuel.

The intent of these limitations is to further ensure low fire loading, low possibility of fire spread and natural cross ventilation.

Enclosed parking garages. Any vehicle parking garage that does not meet the criteria of an **406.4** open parking garage or a Group U private garage is to be regulated under the general provisions for a Group S-2 occupancy. Table 503 will limit the height and floor area of an enclosed parking garage, with an allowance for use of the roof for parking purposes. Ventilation of an enclosed parking garage must be provided in accordance with the IMC.

Motor fuel-dispensing facilities. Because most of the hazards involved with a **406.5** fuel-dispensing operation are due to the storage and dispensing of flammable liquids, the majority of regulations are addressed by the IFC. The primary provisions of this section apply to canopies that are placed over the gasoline pumps for the purpose of customer convenience. Because of the potential exposure of gasoline and vehicle fires during fuel-dispensing operations, the canopies and supports over pumps are required by this section to be of noncombustible construction or, alternatively, constructed of fire-retardant-treated wood, complying heavy-timber members or be of 1-hour fire-resistance-rated combustible construction. Occasional combustible materials may be used in or on a canopy under limited conditions. The allowance for approved plastic panels installed in canopies over motor-vehicle pumps is intended to isolate the combustible plastic materials from other buildings so that if the materials become ignited, they will not present an exposure problem to other buildings.

To avoid damage to vehicles and canopies, the height of canopies must not be less than 13 feet, 6 inches (4115 mm). The 13-foot, 6-inch (4115 mm) dimension should provide adequate clearance for recreational vehicles.

Repair garages. In the IFC, a repair garage is defined as any building or portion thereof that **406.6** is used to service or repair motor vehicles. The potential exists for a moderate fire hazard that is due to the presence of various combustible and flammable liquids such as solvents, cleaning products and gasoline. During repair operations, it is also not uncommon for ignition sources to be present. It is this combination that creates the highest level of hazards that are addressed by Section 406. Classified as Group S-1 occupancies, special concerns for repair garages are primarily regulated through the IFC.

The presence of a repair garage in a building with different types of uses is addressed no differently than other mixed occupancy conditions. The provisions of Section 508.1 are applicable, allowing the option of using the *accessory occupancies, nonseparated occupancies* or *separated occupancies* method for addressing the multiple occupancy groups in the building.

Garages used for the repair of vehicles powered by nonodorized gases, such as hydrogen and nonodorized liquid natural gas, are to be provided with an approved flammable gas-detection system. The design of the gas detection will provide for activation of the safeguards at the point the level of flammable gas exceeds 25 percent of the lower explosive limit. At this point, the system must initiate audible and visual alarm signals in the garage, deactivate the garage's heating systems and activate any mechanical ventilation interlocked with gas detection. Similar functions should take place where there is a failure of the detection system.

Section 407 *Group I-2*

In institutional occupancies, particularly those classified as Group I-2, it is important to balance the fire-safety concerns with the functional concerns of the health-care operations. This section modifies the general code provisions in an effort to achieve such a balance.

407.2 **Corridors.** Corridors are intended to provide a direct egress path adequately separated from hazards in adjoining spaces. However, in hospitals, nursing homes and other Group I-2 occupancies, a number of necessary modifications are provided to facilitate the primary functioning of these types of health-care facilities. These modifications recognize the special needs of these occupancies to provide the most efficient and effective health-care services. See Figure 407-1.

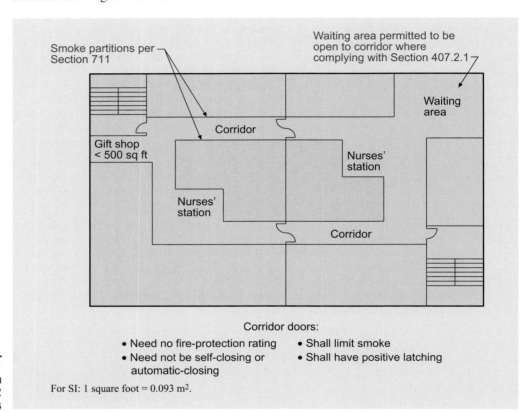

Corridor doors:
- Need no fire-protection rating
- Need not be self-closing or automatic-closing
- Shall limit smoke
- Shall have positive latching

For SI: 1 square foot = 0.093 m².

Figure 407-1
Corridors in Group I-2 occupancies

In order to provide appropriate waiting spaces for visitors, Section 407.2.1 allows such waiting spaces to be unseparated from the corridors. One reason for this is to permit the waiting areas and similar spaces to be so located as to permit direct visual supervision by health-care facility staff. In exchange for the elimination of the corridor separation, certain conditions are imposed on the location of such waiting spaces. Although the scoping language only includes waiting areas and similar spaces, the primary criteria limiting those spaces that can be open to a corridor are identified in Item 1. Health care facilities will often create alcoves adjacent to the corridor for the temporary storage of medical supplies, linen carts, food carts, etc., that are necessary to the daily functions of the facility. Without the alcoves, the corridors would be obstructed by these uses. Therefore, the code makes an allowance for such spaces. Allowances are also made for areas associated with the treatment of mental-health patients. Provided the areas are under continuous supervision by facility staff, they may be open to the corridor where six conditions are met.

Similarly, Section 407.2.2 makes provisions for the location of nurses' stations and similar spaces necessary for doctors' and nurses' charting and communications in positions that need not be separated from the corridors. Essentially, these special-use areas are permitted to be located in the corridor. When this arrangement occurs, however, it is necessary that the construction surrounding the nurses' station be as required for corridors.

Corridor walls. Walls enclosing corridors and other spaces permitted by Section 407.2 to **407.3** be open into corridors are intended to provide a relatively smoke-free environment during the relocation of patients during a fire emergency. Therefore, such walls must be constructed in accordance with the provisions of Section 711 as smoke partitions. The walls may extend either tight to the floor or roof deck above, or extend tight to the ceiling, provided the ceiling is also constructed to limit smoke transfer.

Corridor doors protecting those spaces adjacent to the corridor are not required to have a fire-protection rating, nor are they required to be self-closing assemblies. They must, however, be able to limit the transfer of smoke through the opening but need not be tested for air leakage under UL 1784. One of the most controversial issues relative to the arrangement of health-care facilities such as hospitals and nursing homes is the matter of the installation of door closers on doors to patient sleeping rooms. The health-care industry has long believed it is more important to the proper delivery of health-care services that the doors to patient rooms not be self-closing and therefore constantly closed. In recognition of this special need, self-closing or automatic-closing devices are not required on corridor doors. Positive latching is required, however, and roller latches are not considered acceptable latching hardware. Where positive latching is not desired, typically where sliding doors are installed at patient or treatment rooms, the common corridor arrangement cannot be utilized. In such instances, the spaces could be designed as health-care suites under the provisions of Section 1014.2.2. Corridor-type configurations within such suites are not subject to the requirements of Section 407.3.

Locking devices may be arranged so that they are readily operable from the patient-room side and are readily operable by the facility staff from the opposite side. This special arrangement permits keys or other limited access methods to be utilized for the patient rooms. However, egress from the patient rooms shall be unrestricted unless such rooms are in mental-health facilities.

Smoke barriers. Evacuation of a building such as a hospital or nursing home is a virtual **407.4** impossibility in the event of a fire, particularly in multistory structures. Horizontal evacuation, on the other hand, is possible with a properly trained staff. As a result, the code makes provisions for horizontal compartmentation as illustrated in Figure 407-2, so that if necessary, patients can be moved from one compartment to another. This intent is secured by this section wherein, under most conditions, each story of a Group I-2 occupancy is required to be divided into at least two approximately equal compartments by a smoke barrier constructed in accordance with Section 710. Limited by floor area and travel distance, each compartment shall be sized to permit the housing of patients from adjoining smoke compartments. It is expected that in multistory buildings, the floor construction also provides for smoke compartmentation vertically. As such, the concept of smoke resistance must be considered relative to vertical openings and penetrations, including vertical exit enclosures and shaft enclosures. A more detailed analysis of the requirements for vertical smoke compartmentation in multistory Group I-2 occupancies is found in the discussion of Section 712.9.

Automatic sprinkler system. Owing primarily to the very limited mobility of patients in **407.5** this type of occupancy, the code requires every compartment containing patient sleeping units to be equipped with an automatic fire-sprinkler system. Approved quick-response or residential sprinklers shall be provided throughout the smoke compartments, not just in the patient sleeping units.

Secured yards. It is not uncommon that a secured exterior area or yard be provided for **407.8** Group I-2 occupancies, particularly where the facility specializes in the treatment of mental disabilities such as Alzheimer's disease. Where such fencing and locked gates prohibit the continuation of the exit discharge to the public way, the use of safe dispersal areas is acceptable. To adequately provide for temporary refuge, the safe dispersal area must be sized to accommodate the occupant load of the egress system it serves. In all cases, the entire dispersal area must be located at least 50 feet (15 240 mm) from the building.

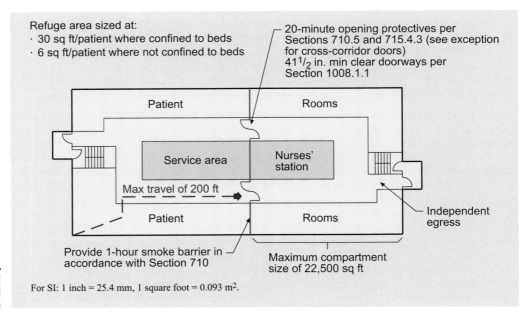

Refuge area sized at:
· 30 sq ft/patient where confined to beds
· 6 sq ft/patient where not confined to beds

20-minute opening protectives per Sections 710.5 and 715.4.3 (see exception for cross-corridor doors) 41$^{1}/_{2}$ in. min clear doorways per Section 1008.1.1

Patient   Rooms

Service area   Nurses' station

Max travel of 200 ft

Patient   Rooms

Independent egress

Provide 1-hour smoke barrier in accordance with Section 710

Maximum compartment size of 22,500 sq ft

**Figure 407-2
Hospital
compartmentation**

For SI: 1 inch = 25.4 mm, 1 square foot = 0.093 m^2.

Section 408   *Group I-3*

The concerns for both security and fire safety must be balanced when it comes to Group I-3 detention facilities. Special consideration must be given to the secured areas without sacrificing an unreasonable degree of fire- and life-safety for the occupants. This section addresses the unique conditions that occur in these types of buildings.

Section 408.3 modifies the general requirements for the means of egress found in Chapter 10. A major difference is the allowance for glazing in the doors and walls of required exit stairways, provided a number of conditions are met. As would be expected, the most dramatic variation from the general requirements has to do with the locking hardware. The requirements vary based upon the nature of the detention occupancy. Reference must be made to the occupancy conditions of Section 308.4 to determine the appropriate egress criteria.

Similar to the provisions of Section 407 for Group I-2 occupancies, smoke compartments must be created where the occupant load per story is 50 or more. Additionally, regardless of occupant load, floor levels utilized as sleeping areas must be divided into a minimum of two compartments. More than two smoke compartments may be necessary on any floor level where the dictated travel distances cannot be provided or where the occupant load of the compartment is excessive. No more than 200 occupants can be assigned to a single compartment. The refuge area must be sized to accommodate the total number of residents that may be contained within the compartment. Independent egress is needed from each compartment so that it is not necessary to travel back into the compartment where travel originated.

An important feature of the Group I-3 provisions is the allowance for multiple levels of residential housing to be open without an enclosure. Through the safeguards provided, it is possible to provide increased security by opening up the multiple housing areas to a single common area where visual supervision is more easily accomplished. It is important that independent egress to an exit be provided from each level. The limit of 23 feet (7010 mm) between the lowest and highest floor levels, as well as the required egress directly out of each story, provide additional qualifications that must be met in order to eliminate the required vertical enclosure protection.

As an additional allowance for security purposes, the fire-protection rating is no longer required for security glazing installed in 1-hour fire barriers, fire partitions and smoke barriers that may be present. Rather, equivalent protection is provided through compliance with four specific conditions addressing the glazing and its frame. The use of security glazing is necessary in such facilities to track and contain inmate movement for the protection of other inmates and administrative personnel. Three of the most common types of fire separations are addressed: fire barriers, fire partitions and smoke barriers. The new methodology is not applicable to fire walls, nor is it permissible where the fire separation wall has a required fire-resistance rating of more than 1 hour. The conditions imposed on the security glazing limit the area of each individual glazed panel, mandate sprinkler protection that will wet the entire glazing surface on both sides, regulate the gasketed frame for deflection, and prohibit the installation of obstructions between the sprinklers and the glazing.

Section 409 *Motion-picture Projection Rooms*

Prior to the 1970s, building codes addressed the subject of motion picture projection rooms based on the hazard of the cellulose nitrate film being used at that time. Actually, production of cellulose nitrate films ceased around 1950, although its use continued thereafter. In fact, even today, some cellulose nitrate film is used at film festivals and special occasions requiring the projection of historically significant films that are still imprinted on cellulose nitrate film. Where this type of film continues to be utilized or stored, it will be regulated under the provisions of NFPA 40. Although the provisions in the codes since 1970 are based on the use of safety film, some of the protection requirements for cellulose nitrate film have been retained in the present requirements, such as ventilation requirements for the projection room.

The intent of the current provisions regulating motion picture projection rooms is to provide safety to the occupants of a theater from the hazards consequent on the light source where electric arc, xenon or other light-source projection equipment is used. Although not used to any extent today, electric-arc projection lamps emit hazardous radiation. Xenon lamps, which have been highly prevalent as projection lamps, emit ozone. As a result, the provisions of Section 409 are based on the lamps used for projection of the film rather than the type of film to be used, as long as the film is not nitrate based.

The provisions intend to isolate the projection room so that it does not present a danger to the theater audience. As the room is designed for the projection of safety film, there is no intent to provide a special fire-resistive enclosure, and fire protection of openings between the projection room and the auditorium is not required. However, due to the projection lamps, it is the intent of the code to provide an emission-tight separation so that any opening should be sealed with glass or other approved material such that emissions from the projection lamps will not contaminate the auditorium.

Section 410 *Stages and Platforms*

The provisions in Section 410 are continuously reviewed in an attempt to bring the code requirements in line with the present methods and technologies regarding the use of stages and platforms. Although the basic provisions for life safety remain essentially unchanged, occasional modifications have been made that are due to the need to accommodate state-of-the-art performances.

410.2 **Definitions.** Although the definitions in this section are complete and reasonably understandable, there are terms unique to the performing arts that are not generally understood, such as fly gallery, gridiron and pinrail. The distinctions between the definitions of a stage and a platform are also very important because of the specific requirements for each. The primary difference between a stage and a platform is the presence of overhead hanging curtains, drops, scenery and other stage effects. The amount of combustible materials associated with a stage are typically greater than for a platform. Thus, the fire-severity potential is much higher.

410.3 **Stages.** An assembly occupancy considered among the most hazardous is a Group A-1 containing a large occupant load and a performance stage. The hazard created by the stage is the presence of combustibles in the form of hanging curtains, drops, leg drops, scenery, etc., which in the past have been the source of ignition for disastrous fires in theaters. Modern stages also have an increased hazard from special effects such as pyrotechnics, utilized in so-called *spectaculars*.

Where the stage height exceeds 50 feet (15 240 mm), the fire hazard is even greater because the fly area that is usually above the stage is a large blind space containing combustible materials that have a fuel load considerably greater than that normally associated with an assembly occupancy. Many of the construction requirements for stages are depicted in Figure 410-1.

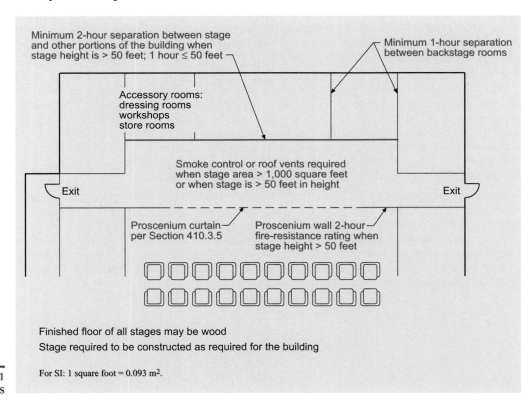

Finished floor of all stages may be wood

Stage required to be constructed as required for the building

For SI: 1 square foot = 0.093 m².

Figure 410-1
Stages

Stage construction. In addition to the features shown in Figure 410-1, any stage may have **410.3.1** a finished floor of wood, provided construction of the stage floor or deck is in compliance with this section. As the area above and at the sides of stages can be filled with combustible materials that can be moved both vertically and horizontally, such as curtains, drops, leg drops, scenery and other stage effects, the code requires that such stages be constructed of the same materials as required for floors for the type of construction of the building and separated from the balance of the building.

Proscenium wall. Where the stage height exceeds 50 feet (15 240 mm), measured from the **410.3.4** lowest point on the stage floor to the highest point of the roof or floor deck above, a proscenium wall must be provided. The proscenium wall is intended by the IBC to provide a complete fire separation between the stage and the auditorium. Extending from the foundation continuously to the roof, the wall is to have a minimum fire-resistance rating of 2 hours.

Proscenium curtain. Because the opening in the proscenium wall described in Section **410.3.5** 410.3.4 is too large to protect with any usual type of fire assembly, the code requires that it be protected with a fire-resistive fire curtain or water curtain. Where a fire curtain is installed, it must comply with the provisions for fire safety curtains set forth in NFPA 80 *Fire Doors and Other Opening Protectives.* A fire curtain or water curtain is not required where a complying smoke control system or natural ventilation is provided. The purpose of the proscenium curtain protection is to provide occupants with additional time to exit the assembly seating area if there is a fire in the stage area. With the benefits afforded by an engineered smoke control system or natural ventilation, the occupants should be equally or better protected from the hazards of fire than with a proscenium curtain or water curtain. By providing a performance-based alternative to a proscenium curtain, more design options are available where the use of fire safety curtains is considered impractical or causes obstructions of the production. It is important to note that the elimination of the proscenium curtain is not permitted if the provisions of Section 1028.6.2 are being utilized for a decrease in the required egress widths of the assembly seating area.

The requirement for a complying fire curtain is triggered solely by the proscenium wall provisions of Section 410.3.4. Where a proscenium wall is fire-resistance rated for a different purpose, such as a bearing wall in a Type IB building, the fire curtain is not required.

Stage ventilation. The Iroquois Theater Fire in 1903 was directly responsible for the **410.3.7** requirement for automatic vents in the roofs of theater stages. Because of the presence of large amounts of combustible materials, excessive quantities of smoke will accumulate in and above the stage area unless it is automatically vented or removed by a smoke-control system. The removal of smoke is necessary for fire fighting as well as the prevention of panic by drawing off the smoke so that it will not infiltrate the theater auditorium.

The maximum floor area of stages that is permitted without the installation of venting is 1,000 square feet (93 m²). The stage area to be considered includes the performance area and adjacent backstage and support areas not separated from the performance area by fire-resistance-rated construction. In addition, stages must be equipped with smoke-removal equipment or roof vents where they are greater than 50 feet (15 240 mm) in height. If either of these two conditions exist, stage ventilation is required. The detailed requirements for smoke vents in the IBC are intended to provide reliability and a reasonable assurance that after many years of operation the vents will operate when needed.

Platform construction. Materials used in the construction of permanent platforms must be **410.4** consistent with those materials permitted based on the building's type of construction. Therefore, in noncombustible buildings, the platforms must be of noncombustible construction. However, in buildings of Type I, II and IV construction, the use of fire-retardant-treated wood is permitted where all of the following conditions are met:

1. The platform is limited in height to 30 inches (762 mm) above the floor.
2. The floor area of the platform does not exceed one-third the floor area of the room in which it is located.

3.   The platform does not exceed 3,000 square feet (279 m^2) in floor area.

In those situations where the concealed area below the platform is to be used for storage or any purpose other than equipment, wiring or plumbing, the floor construction of the platform is to be fire-resistance rated for a minimum of 1 hour. Otherwise, no protection of the platform floor is necessary.

As it is often impractical to construct temporary platforms of fire-resistive materials, the code permits temporary platforms to be constructed of any materials, but restricts the use below the platform to that of electrical wiring or plumbing to operate platform equipment. Therefore, no storage of any kind is permitted beneath temporary platforms, because of the potential for a fire to start and spread undetected.

410.5   **Dressing and appurtenant rooms.** Not only must a stage exceeding 50 feet (15 240 mm) in height be separated from the adjoining seating area by a minimum 2-hour fire-resistance-rated proscenium wall, but such a separation is also required between the stage and all other portions of the building, including all related backstage areas. Dressing rooms, property rooms, workshops, storage rooms and all other areas must be separated from the stage with minimum 2-hour fire-resistance-rated fire barriers and/or horizontal assemblies, and all openings must be appropriately protected. A minimum 1-hour fire-resistance-rated separation is required where the stage height does not exceed 50 feet (15 240 mm).

In addition to their required fire separation from the stage, dressing rooms and all other related backstage areas must be separated from each other. One-hour fire-resistance-rated fire barriers and/or horizontal assemblies, along with opening protectives, satisfy the minimum requirements. The hazards caused by the significant fire loading that occurs in conjunction with stages are greatly reduced through the use of compartments.

410.6   **Automatic sprinkler system.** One of the special areas mentioned in Table 903.2.11.6 that requires a suppression system is stages. The general requirement mandates the sprinklering of not only the stage area but also all support and backstage areas serving the stage. An automatic sprinkler system is an effective tool in limiting the exposure of a fire to the area of origin. Sprinklers are not required for stages with both small floor areas and low roof heights. Under such conditions, the amount of combustibles in the stage area is typically very limited.

Section 411   *Special Amusement Buildings*

Amusement buildings are usually classified as Group A occupancies but should be classified as Group B where the occupant load is less than 50. The major factors contributing to the loss of life in fires within amusement buildings has been the failure to detect and extinguish the fire in its incipient stage, the ignition of synthetic foam materials and subsequent fire and smoke spread, and the difficulty of escape. Provisions for the detection of fires, the illumination of the exit path and the sprinklering of the structures are required to protect the occupants in such structures. However, amusement buildings or portions thereof without walls or a roof are not required to comply with this section, provided they are designed to prevent smoke from accumulating in the assembly areas. Approved smoke-detection and alarm systems are also required in amusement buildings. A provision of Section 411.7 is that on the activation of the system as described, an approved directional exit marking system shall activate in those areas where the configuration of the space is such as to disguise the path and make the egress route not readily apparent.

Section 412 *Aircraft-related Occupancies*

Because of the unique nature of occupancies related to aircraft manufacture, repair, storage and even flight control, provisions have been developed to address the special conditions that may exist. Although the various uses fall into different occupancy classifications, they all have one thing in common—they are related to aircraft. Additional requirements related directly to aviation facilities are found in Chapter 11 of the IFC.

Airport-traffic control towers. These provisions are intended to reconcile the differences **412.3** between the life-safety needs of air-traffic control towers and the life-safety requirements in the body of the code. The life and property loss in these towers has been very small even though they have not complied completely with all of the code requirements in the past. In developing these provisions, consideration was given to the inherent qualities of the use, which makes general requirements of the IBC inappropriate. For example, air-traffic control personnel are required to undergo medical examinations to ensure they are of sound body and mind. Recognition was also given to the life-safety record of these uses and specific limitations, which are imposed on the allowable size, type of construction, etc. The provisions also require automatic fire-detection systems. Because of the critical nature of the facility, a standby power system is required for towers over 65 feet (19 812 mm) in height.

Aircraft hangar. Aircraft hangars are intended to be classified as Group S-1 occupancies. **412.4** All aircraft hangars are to be located at least 30 feet (9144 mm) from any public way or lot line, providing adequate spatial separation for neighboring areas. Otherwise, their exterior walls must have a minimum 2-hour fire-resistance rating. Because of the concerns about below-grade spaces under any facility where flammable and combustible liquids are commonly present, the code requires the hangar floor over a basement to be liquid and air tight with absolutely no openings. Floor surfaces must also be sloped to allow for drainage of any liquid spills.

Residential aircraft hangars. Where a private aircraft hangar is accessory to a dwelling, it **412.5** is classified as a residential aircraft hangar, provided it meets the criteria of this section. Where the hangar is less than 20 feet (6096 mm) in height and less than 2,000 square feet (186 m^2) in floor area, it is considered to be no greater a hazard than any private garage housing several motor vehicles.

  The fire separation between a dwelling and an attached hangar is to be at least a 1-hour fire-resistance-rated fire barrier. Self-closing doors between the dwelling and the hangar are the only permitted openings, and each door must have a minimum 4-inch-high (102 mm) noncombustible sill. Two means of egress from the hangar are required, only one of which may pass through the dwelling. Smoke alarms shall be installed within the hangar, and the mechanical and DWV (drainage, waste and vent) systems installed for the hangar shall be independent of the dwelling's systems. Every reasonable effort is being made to isolate the residential aircraft hangar from the dwelling to reduce the likelihood that a fire in the hangar will be a life-safety threat to occupants of the dwelling.

Aircraft paint hangars. The hazards involved with the application of flammable paint or **412.6** other liquids cause aircraft painting operations to be highly regulated. Where the quantities of flammable liquids exceed the exempt quantities listed in Table 307.1(1), such hangars are classified as Group H-2 occupancies. They must be built of noncombustible construction, provided with fire suppression per NFPA 409 and ventilated in the manner prescribed by the IMC. Where the amount of flammable liquids within the hangar does not exceed the maximum allowable quantities set forth in Table 307.1(1), the classification is most appropriately an S-1 occupancy, and the provisions of this section do not apply.

412.7 **Heliports and helistops.** Helistops are differentiated from heliports by the presence of refueling facilities, maintenance operations, and repair and storage of the helicopters; thus, helistops pose similar hazards to those posed by aircraft repair hangars. The minimum size of a helicopter landing area is addressed, as are requirements for construction features and egress. Where heliports and helistops are constructed in compliance with the provisions of this section, they may be erected on buildings regulated by this code.

Section 413 *Combustible Storage*

Any occupancy group containing high-piled stock or rack storage is subject to the provisions of the IFC as well as the IBC. Chapter 23 of the IFC regulates combustible storage based on a variety of conditions, including the type of commodities stored, as well as the method of storage and the size of the storage area.

This section also specifically addresses any concealed spaces within buildings, including attics and under-floor spaces, that are used for the storage of combustible material. Where combustible storage occurs in areas typically considered unoccupiable, the storage areas are to be separated from the remainder of the building by 1-hour fire-resistance-rated construction on the storage side. The protective membrane need only be applied on the storage side insofar as the location of the hazard has been identified as the storage area only. Openings are to be protected with self-closing door assemblies that are either of noncombustible construction or are a minimum $1^3/_4$-inch (45 mm) solid wood. This separation is not necessary in Group R-3 and U occupancies. In addition, those combustible storage areas protected with sprinkler systems need not be separated. The provisions are not intended to apply to those storage rooms that are constructed and regulated as usable spaces within the building.

Section 414 *Hazardous Materials*

Figure 414-1 outlines the process for determining the code requirements that are a function of the quantities of hazardous materials stored or used. The outline is useful for both design and review. To begin, one must determine the hazardous processes and materials involved in a given occupancy and gain a thorough understanding of the operations taking place. Once the hazardous processes and materials have been identified, it is necessary to classify the materials based on the categories used by the code.

Section 414.1.3 provides the means for the building official to acquire outside technical assistance to assist in the review of a project. Such assistance is often critical in assuring appropriate decisions are made.

Classifying materials is a subjective science, requiring judgment decisions by an expert familiar with the characteristics of a particular material to categorize it within the categories used by the IBC and IFC. Accordingly, material classifications must be determined by qualified individuals, such as industrial hygienists, chemists or fire-protection engineers. Though some jurisdictions employ individuals qualified to make these determinations, most jurisdictions rely on outside experts acceptable to the jurisdiction to submit a report detailing classifications compatible with the system used by the code.

Often, a permit applicant will attempt to submit a cadre of Material Safety Data Sheets (MSDS) as a means of identifying material classifications. Though these may contain the information necessary to determine the proper classification, they do not normally contain a complete designation of classifications that is compatible with the system used by the IBC. Therefore, MSDSs are not normally acceptable as a sole means of providing material classifications to a jurisdiction.

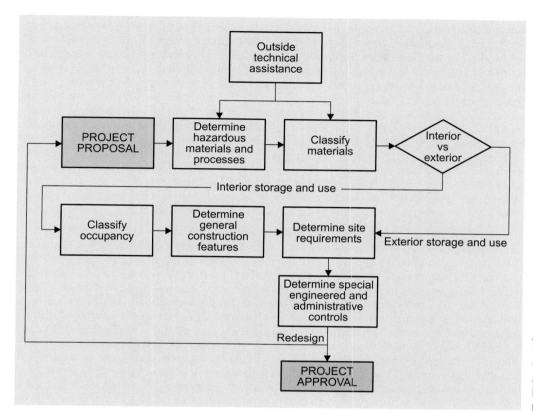

**Figure 414-1
Code
approach to
hazardous
materials**

The building official should understand that it is not the responsibility of the jurisdiction to provide classifications for hazardous materials. Rather, it is the responsibility of the permit applicant to provide material classification information. In this way, potential liability of the jurisdiction for improper classification of materials is avoided.

In the classification system used by the *International Codes*®, hazardous materials are generally divided into two major categories, physical and health hazards, and 12 subcategories as follows:

Physical Hazards

   Explosives and fireworks

   Combustible dusts and fibers

   Flammable and combustible liquids

   Flammable solids and gases

   Organic peroxides

   Oxidizers

   Pyrophoric materials

   Unstable (reactive) materials

   Water-reactive materials

   Cryogenic liquids

Health Hazards

Highly toxic and toxic materials

Corrosives

414.1.3 **Information required.** A report is required to allow the building department to evaluate the presence of hazardous materials within the proposed building based on the criteria established by the IBC. Since Tables 307.1(1) and 307.1(2) are critical in the evaluation of buildings containing hazardous materials, information is needed in order to properly utilize the tables. Such information must include the maximum expected quantities of each material in use and/or storage conditions, those fire-protection features that are to be in place, and any use of control areas for isolation of the materials. The submission of a technical report is necessary to allow the jurisdiction to perform a code compliance evaluation. The requirement for a technical report gives jurisdictions the benefit of expert opinions provided by knowledgeable persons in the particular hazard field of concern. Technical reports are required to be prepared by an individual, firm or corporation acceptable to the jurisdiction, and must be provided without charge to the jurisdiction. Where the quantities of hazardous materials are such that a Group H occupancy is warranted, floor plans must be submitted to the building official identifying the locations of hazardous contents and processes.

414.2 **Control areas.** As addressed previously in the discussion of Section 307, areas in a building that are designated to contain the maximum allowable quantities of hazardous materials in use, storage, dispensing or handling are considered control areas. At a minimum, 1-hour fire barriers shall be used to separate control areas from each other. Where required by Table 414.2.2 for the fourth story above grade plane and all stories above, a minimum 2-hour fire-resistance rating is required for such fire barriers. Openings in fire barriers are to be protected in accordance with Section 715. As a general rule, all floor construction that forms the boundaries of control areas is to have a minimum 2-hour fire-resistance rating. Building elements structurally supporting the 2-hour floor construction shall have an equivalent fire-resistance rating. There is an allowance for those two-story and three-story sprinklered buildings that are primarily of 1-hour fire-resistive construction (Types IIA, IIIA and VA), which permits 1-hour floor construction of the control area and the supporting construction. It is apparent that a considerable level of fire separation must be achieved in order to increase the quantity of hazardous materials in non-Group H buildings. An example of this provision is illustrated in Figure 414-2.

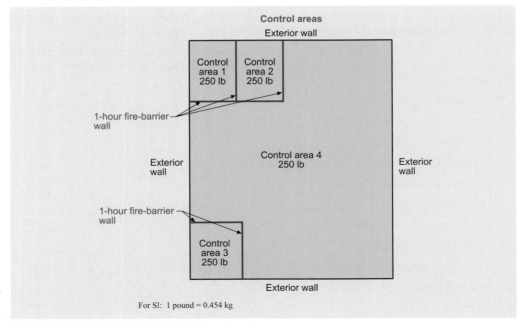

Figure 414-2
Floor plan

For SI: 1 pound = 0.454 kg

Explosion control. Table 414.5.1 indicates, based on material, the explosion control **414.5.1** methods that must be provided where hazardous materials exceed the allowable quantities specified in Table 307.1(1). Explosion control is also required in any structure, room or space occupied for purposes involving explosion hazards. Once some type of explosion control is required, Section 911 of the IFC must be referenced to identify the details for controlling explosion hazards.

Spill control, drainage and containment. The intent of this section is the prevention of **414.5.5** the accidental spread of hazardous material releases to locations outside of containment areas. Applicable to rooms, buildings or areas used for the storage of both solid and liquid hazardous materials, the specifics for spill control, drainage and containment are contained in the IFC.

Weather protection. In order to be considered outside storage or use in the application of **414.6.1** the IFC, hazardous material storage or use areas must be primarily open to the exterior. If it is necessary to shelter such areas for weather protection purposes, the enclosure and its location are limited by the following requirements:

1. No more than one side of the perimeter of the area may be obstructed by enclosing walls and structural supports unless the total obstructed perimeter is limited to 25 percent of the structure's total perimeter.

2. The minimum clearance between the structure and neighboring buildings, lot lines or public ways shall be equivalent to that required for outside storage or use areas without weather protection.

3. Unless increased by the provisions of Section 506, the maximum area of the overhead structure shall be 1,500 square feet (140 m^2).

4. The structure must be constructed of approved noncombustible materials.

Section 415 *Groups H-1, H-2, H-3, H-4 and H-5*

The provisions of this section apply to those buildings and structures where hazardous materials are stored or used in amounts exceeding the maximum allowable quantities identified in Section 307. Applied in concert with the IFC, the requirements address the concerns presented by the high level of hazard as compared to other uses. For a further discussion, see the commentary on Section 414.

Fire separation distance. This section, along with Table 415.3.1, provides regulations that **415.3** limit the locations on property for Group H occupancies and establish minimum percentages of perimeter walls of Group H occupancies required to be located on the building exterior. Based on the specific Group H occupancy involved, the building must be set back a minimum distance from lot lines as shown in Figure 415-1. As illustrated in Figure 415-2, the distance is measured from the walls enclosing the high-hazard occupancy to the lot lines, including those on a public way. An exception to this method of measurement occurs where two buildings are on the same site and an assumed imaginary line is placed between them under the provisions of Section 705.3. In such a situation, the assumed line is to be ignored in the application of this section. This provides for a reasonable spatial separation as the required separation distances for explosive conditions are much greater than typically required when regulating exterior wall and opening protection. The specific provisions in this section also require that Group H-2 and H-3 occupancies included in mixed-use buildings have 25 percent of the perimeter wall of the Group H occupancy on the exterior of the building. The access capability for fire personnel is greatly enhanced where the hazardous conditions are located in such a manner that allows for exterior fire-fighting operations. Exceptions are provided for smaller, liquid use, dispensing and mixing rooms; liquid storage rooms; and spray booths. See Figure 415-3. It should be noted that where a detached building, required by Table 415.3.2, is located on the lot in accordance with this section, wall construction and opening protection is not regulated based on location on the lot. A minimum fire separation distance of 50 feet (15 240 mm) is required for such buildings. Therefore, the exterior wall and opening requirements of Table 602 have no application.

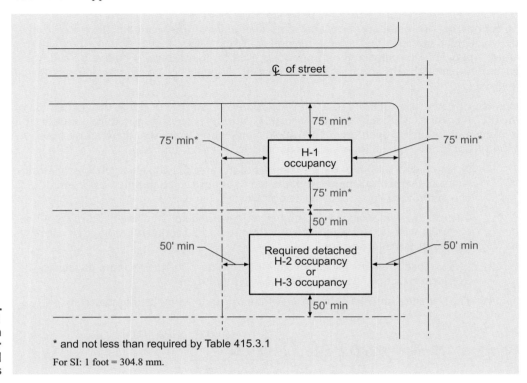

Figure 415-1
Location on property for the detached buildings

* and not less than required by Table 415.3.1

For SI: 1 foot = 304.8 mm.

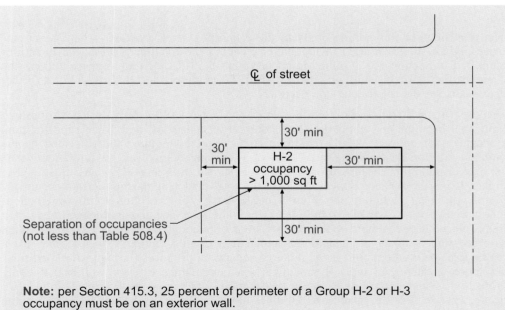

Figure 415-2
Location on property for mixed occupancies that include a Group H-2 occupancy

Note: per Section 415.3, 25 percent of perimeter of a Group H-2 or H-3 occupancy must be on an exterior wall.

For SI: 1 foot = 304.8 mm, 1 square foot = 0.093 m².

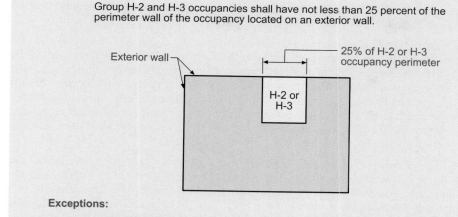

Group H-2 and H-3 occupancies shall have not less than 25 percent of the perimeter wall of the occupancy located on an exterior wall.

Exceptions:

- Liquid use, dispensing and mixing rooms that comply with the IFC and NFPA 30 and are not more than 500 square feet in area
- Liquid storage rooms that comply with the IFC and NFPA 30 and are not more than 1,000 square feet in area
- Spray booths complying with the IFC

For SI: 1 square foot = 0.093 m².

Figure 415-3
Perimeter wall
on exterior

Special provisions for Group H-1 occupancies. Because of the extreme hazard presented **415.4** by Group H-1 occupancies, this section requires that such occupancies be used for no other purpose, and it prohibits basements, crawl spaces and underfloor spaces where flammable or explosive material might gather. Roofs are required to be of lightweight construction so that, in case of an explosion, they will rapidly vent with minimum destruction to the building. In addition, thermal insulation is sometimes required to prevent heat-sensitive materials from reaching decomposition temperatures.

This section also requires that Group H-1 occupancies that contain materials possessing health hazards in amounts exceeding the maximum allowable quantities for health-hazard materials in Table 307.1(2) also meet the requirements for Group H-4 occupancies. This provision is parallel to Section 307.8, which requires multiple hazards classified in more than one Group H occupancy to conform to the code for each of the occupancies classified.

Special provisions for Group H-2 and H-3 occupancies. Group H-2 and H-3 **415.5** occupancies containing large quantities of the more dangerous types of physical hazard materials are considered to present unusual fire or explosion hazards that warrant a separate and distinct occupancy in a detached building used for no other purpose, similar to the requirements for a Group H-1 occupancy. The threshold quantities for requiring detached Group H-2 and H-3 occupancies are set forth in Table 415.3.2.

This section also requires water-reactive materials to be protected from water penetration or liquid leakage. Fire-protection piping is allowed in such areas in recognition of both the integrity of fire-protection system installations and the need to protect water-reactive materials from exposure fires.

Group H-2. Both this section and the IFC are to be used in the regulation of buildings **415.6** containing the following hazardous materials operations:

1. Combustible dusts, grain processing and storage.

2. Flammable and combustible liquids.

3. Liquefied petroleum gas distribution facilities.

4. Dry cleaning plants.

The hazards presented in these operations, through either the storage, use, handling, processing or transporting of hazardous materials, are unique enough to require special provisions, both in this section and in the IFC.

415.7 **Groups H-3 and H-4.** This section identifies several specific issues in Group H-3 and H-4 occupancies. Group H gas rooms shall be isolated from other areas of the building by minimum 1-hour fire barrier and/or horizontal assemblies. Highly toxic solids and liquids must also be separated from other hazardous material storage by fire barriers and/or horizontal assemblies having a minimum 1-hour fire-resistance rating, unless the highly toxic materials are stored in approved hazardous material storage cabinets. A related provision requires liquid-tight noncombustible floor construction in areas used for the storage of corrosive liquids and highly toxic or toxic materials.

415.8 **Group H-5.** The Group H-5 occupancy category was created to standardize regulations for semiconductor manufacturing facilities. This section provides the specific regulations for these occupancies. The H-5 category requires engineering and fire-safety controls that reduce the overall hazard of the occupancy to a level thought to be equivalent to a moderate hazard Group B occupancy. Accordingly, the areas permitted for Group H-5 occupancies are the same as for Group B occupancies.

The code requires that special ventilation systems be installed in fabrication areas that will prevent explosive fuel to air mixtures from developing. The ventilation system must be connected to an emergency power system. Furthermore, buildings containing Group H-5 occupancies are required to be protected throughout by an automatic fire-sprinkler system and fire and emergency alarm systems. Fire and emergency alarm systems are intended to be separate and distinct systems, with the emergency-alarm system providing a signal for emergencies other than fire. This section also provides requirements for piping and tubing that transport hazardous materials that allow piping to be located in exit corridors and above other occupancies subject to numerous, stringent protection criteria. The provisions for Group H-5 occupancies are correlated with companion provisions in Chapter 18 of the IFC.

Table 415.3.1—Minimum Separation Distances for Buildings Containing Explosive Materials. The IFC is typically referenced for the minimum separation distances for buildings containing explosive materials that are classified as Group H-1, H-2 or H-3 occupancies. Where the IFC does not specify such minimum separation distances, the use of Table 415.3.1 is mandated. This table also sets forth the minimum distances to lot lines from buildings containing detonable materials that are required by Table 415.3.2 to be detached. Separations from inhabited buildings and other buildings containing explosives that are subject to a sympathetic explosive reaction are also specified. The distances in the table may be reduced when adequate barricades are installed in accordance with Footnote d. Furthermore, the distances may be adjusted based on the TNT equivalency of explosive materials involved. TNT equivalency for some materials is published; however, for many materials, TNT equivalency must be evaluated based on actual testing, calculations performed by a chemist or both. Alternately, the full values in the table can be used if TNT equivalency cannot be reliably estimated. To use TNT equivalency, the weight of the material in question is multiplied by the ratio of relative blast energy with respect to TNT. The resultant weight is used when applying the table to determine the required separation distances.

For those occupancies required by Table 415.3.2 to have detached storage based on quantities of materials that do not possess explosive characteristics, a minimum distance of 50 feet (15 240 mm) should be used as indicated in Section 415.3.1, Item 3.

The distances in this table are derived from the American Table of Distances for Storage of Explosive Materials, which is included in C.F.R. Titles 27 and 30.

One significant difference between application of the American Table of Distances as published in the C.F.R. versus IBC Table 415.3.1 is that measurements to inhabited buildings under the American Table of Distances allow the use of vacant space on adjacent lots. Table 415.3.1, on the other hand, requires separation distances to be measured only on the lot containing the building housing explosive materials. This will usually have the effect of the IBC requiring greater distances between buildings on adjacent properties. Also, see discussion in Section 415 for Group H-1 occupancies in this chapter.

Table 415.3.2—Required Detached Storage. This table establishes the threshold quantities of hazardous materials requiring detached buildings. Once the quantities listed in the table are exceeded, a detached building is required. The limitations placed on detached buildings required by Table 415.3.2 and classified as Group H-2 or H-3 are essentially the same as those applied to H-1 occupancies. The detached building must contain no other occupancy classification, be limited to one story in height, and have no basement, crawl space or similar under-floor area. Though the table only specifically addresses storage amounts, conditions involving use should be treated in a similar manner to that prescribed by Footnote b of Table 307.1(1), which requires *use* quantities to be aggregated with *storage* quantities when comparing to exempt amounts. See Application Example 415-1.

Application Example 415-1

GIVEN: A manufacturing operation requires up to 300 pounds of a Class 3 water reactive (in use).

DETERMINE: The maximum amount of the Class 3 material that can be stored in the building without detached storage being required.

Group H-2

Group F-1 Manufacturing

Group H-2 Detached storage

Required where aggregate quantity in use and storage exceeds 2,000 pounds

Limited to 2,000 pounds of Class 3 water reactives

2,000 lb	total permitted in mixed-occupancy building
- 300 lb	in use
1,700 lb	maximum quantity allowed in storage

For SI: 1 pound = 0.454 kg.

Otherwise, a single occupancy Group H-2 detached building is required.

Section 416 *Application of Flammable Finishes*

This section applies to those buildings used for the spraying of flammable finishes such as paints, varnishes and lacquers. In addition, Chapter 15 of the IFC contains extensive requirements for these types of operations. The IFC addresses a variety of spraying arrangements, each of which is specifically defined and regulated. These include spray rooms, spray booths, spraying space and limited spraying space. The IBC provides limited construction provisions only for those arrangements determined to be spray rooms, as well as ventilation and surfacing requirements for all spraying spaces. An automatic fire-extinguishing system is mandated for all areas where the application of flammable finishes occur, including all spray, dip and immersing spaces and storage rooms.

The occupancy classification of buildings, rooms and spaces utilized for flammable finish application is not specifically addressed. Certainly, where the quantity of hazardous materials exceeds the maximum amounts established by Table 307.1(1) or 307.1(2), a Group H classification is warranted. However, where the maximum amounts are not exceeded, the analysis would be no different than for other types of uses. In a manufacturing building, a Group F-1 occupancy classification would be appropriate. Spraying operations within a vehicle repair garage would most likely be considered part of the Group S-1 classification. A spray room, designed and constructed to house the spraying of flammable finishes, must be adequately separated from the remainder of the building. The enclosure must consist of fire barriers, horizontal assemblies or both, each having a minimum fire-resistance rating of 1 hour. Spraying rooms must be frequently cleaned; thus, all of the interior surfaces must be smooth and easily maintained. The smooth surfaces also allow for the free passage of air in order to maintain efficient ventilation. The room construction must also be tight in order to eliminate the passage of residues from the room, which should be easily accomplished because of the fire separation required. Spraying spaces not separately enclosed shall be provided with noncombustible spray curtains to restrict the spread of vapors.

Section 417 *Drying Rooms*

Where the manufacturing process requires the use of a drying room or dry kiln, the room or kiln containing the drying operations must be of noncombustible construction. It must also be constructed in conformance with the specific and general provisions of the code as they relate to the special type of operations, processes and materials that are involved. Clearance between combustible contents that are placed in the dryer and any overhead heating pipes must be at least 2 inches (51 mm). In addition, methods are addressed to insulate high-temperature dryers from adjacent combustible materials.

Section 418 *Organic Coatings*

Defined in Section 2002.1 of the IFC, organic coatings are those compounds that are applied for the purpose of obtaining a finish that is protective, durable and decorative. Used to protect structures, equipment and similar items, organic coatings provide a surface finish that resists the effects of harsh weather. The concern for occupancies where organic coatings are manufactured or stored is based primarily upon the presence of flammable vapors. As such, this type of use is highly controlled, both by the IBC and the IFC.

The manufacturing of organic coatings creates a high probability that flammable vapors will be present. Therefore, buildings where such materials are manufactured shall be without basements or pits because of the heavier-than-air nature of the vapors. In addition, no other occupancies are permitted in buildings used for the manufacture of organic coatings. The processing of flammable or heat-sensitive material must be done in a noncombustible or detached structure. Tank storage of flammable and combustible liquids inside a building must also be located above grade. In order to isolate the various hazard areas, the storage tank area must be separated from the remainder of the processing areas by minimum 2-hour fire barriers and/or horizontal assemblies. Because of the extreme hazards involved with nitrocellulose storage, it must also be separated by 2-hour fire-resistance-rated fire barriers and/or horizontal assemblies, or preferably located on a detached pad or in a separate structure.

Section 419 *Live/work Units*

An increasingly popular concept of building use combines a residential unit with a small business activity. Residential live-work units typically include a dwelling unit along with some public service business, such as an artist's studio, coffee shop or chiropractor's office. There may be a small number of employees working within the residence and the public is able to enter the work area of the unit to acquire service. Live/work units are a throwback to 1900 era community planning where residents could walk to all of the needed services within their neighborhood. These types of units began to re-emerge in the 1990s through a development style known as "Traditional Neighborhood Design." More recently, adaptive reuse of many older urban structures in city centers incorporated the same live/work tools to provide a variety of residential unit types. Provisions specifically addressing live/work units recognize the uniqueness of this type of use.

By definition, a "live/work unit" is primarily residential in nature but has a sizable portion of the space devoted to nonresidential activities. Often service-related in nature, the nonresidential portion is limited in several respects. The unit itself, including both the residential and nonresidential portions, is limited to 3,000 square feet (279 m^2) in total floor area. In addition, the nonresidential activities cannot take up more than 50 percent of the unit's total floor area. The portion dedicated to nonresidential use must be located on the first floor of the unit, or where applicable, on the unit's main floor level. In addition to the unit's residents, a limit of 5 workers or employees is permitted at any one time. The intent is that the provisions for live/work units are not applicable to home offices. More specifically, the provisions exempt residential units with office uses from the requirements for live/work units where the office area is less than 10 percent of the area of the dwelling unit or sleeping unit. An overview of the limitations is shown in Figure 419-1.

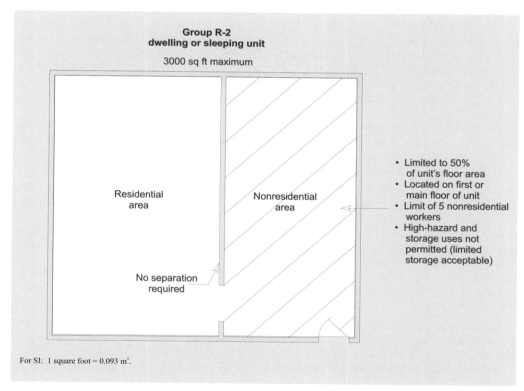

For SI: 1 square foot = 0.093 m^2.

**Figure 419-1
Live/work
unit**

The occupancy classification of a live/work unit is Group R-2 based upon the primary use of the unit. Although differing uses are typically classified based upon the characteristics of the varying uses involved and considered as mixed-occupancy conditions, in this case a single classification is considered acceptable. The potential hazards created due to the nonresidential uses are addressed through the special requirements of Section 419 that are to be applied in addition to those required due to the Group R-2 classification. Since live/work units are regulated as single-occupancy conditions, the provisions of Section 508 for mixed-occupancy buildings do not apply. In addition to the other limitations on use of a live/work unit, storage uses and those activities involving hazardous materials are prohibited. The increased fire load found in many storage uses is not considered in the live/work provisions, nor is the potential physical or health hazard that is due to the use or storage of hazardous materials. A very small amount of storage is permitted if it is deemed to be accessory to the nonresidential use.

Even though a live/work unit is classified as a Group R-2 occupancy, there are several issues where the residential and nonresidential portions are regulated independently. Structural loading conditions, accessibility features and ventilation rates are all to be based on the individual function of each space within the unit. In all other cases, the provisions applicable to a Group R-2 occupancy are to be applied to the entire live/work unit.

A similar provision is contained in the 2009 *International Residential Code*® (IRC®) to address nonresidential uses that occur within single-family dwellings and in dwelling units of two-family dwellings and townhouses. In such cases, the live/work unit would be regulated by the IRC except for those special provisions set forth in IBC Section 419. In addition, a live/work unit would not be classified as a Group R-2 occupancy where located in a building designed and constructed under the scope of the IRC.

Section 420 *Groups I-1, R-1, R-2 and R-3*

In residential-type uses, it is important that any fire conditions created in one of the dwelling units or sleeping units does not spread quickly to any of the other units. As residential fires are the most common of fire incidents, it is critical that neighboring units be isolated from the unit of fire origin. The need for an adequate level of fire-resistance is enhanced because of the lack of immediate awareness of fire conditions when the building's occupants are sleeping. The provisions are applicable to hotels and other Group R-1 occupancies, apartment buildings, dormitories, fraternity and sorority houses, and other types of Group R-2 occupancies, and between dwelling units of a Group R-3 two-family dwelling. Sleeping units and dwelling units of a supervised residential care facility classified as Group I-1 must also be provided with such fire-resistive separations.

Where dwelling units or sleeping units are adjacent to each other horizontally, the minimum required separation is a fire partition. The wall serving as a fire partition is regulated by Section 709 and typically must have a minimum 1-hour fire-resistance rating. A reduction to a $^1/_2$-hour fire partition is permitted under the special conditions set forth in Exception 2 of Section 709.3. Where dwelling or sleeping units are located on multiple floors of a building, they must be separated from each other with minimum 1-hour fire-resistance-rated horizontal assemblies as described in Section 712. An allowance for a $^1/_2$-hour reduction, similar to that permitted for fire partitions, is also available under specified conditions.

In addition to the required separation between adjoining units, dwelling units and sleeping units must also be separated by complying fire partitions and/or horizontal assemblies from other adjacent occupancies. See Figure 420-1. Applicable only in mixed-occupancy buildings, this requirement takes precedence over the allowances in Sections 508.2 and 508.3 for accessory occupancies and nonseparated occupancies, respectively. Even in those cases where the mixed-occupancy provisions of Section 508.2 or 508.3 are applied, the separation requirements of Section 420 must be followed. In those cases where the separated occupancy provisions of Section 508.4 are utilized, the more restrictive fire-resistive rating must be applied.

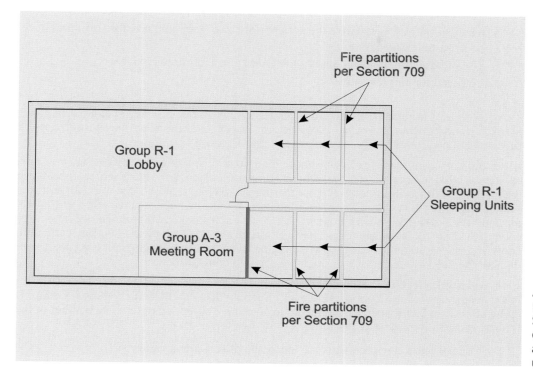

Fire partitions
per Section 709

Group R-1
Lobby

Group R-1
Sleeping Units

Group A-3
Meeting Room

Fire partitions
per Section 709

Figure 420-1
Separation of dwelling units and sleeping units

It should be noted that the separation of dwelling units and sleeping units from other types of spaces in the building only applies if those spaces are of a different occupancy than that of the residential units. For example, a separation is not required between a Group R-1 sleeping unit in a hotel and the adjacent hotel lobby if the lobby is classified as a portion of the Group R-1 occupancy.

Section 422 *Ambulatory Health Care Facilities*

Ambulatory health care facilities, commonly referred to as ambulatory surgery centers or day surgery centers, are defined in Chapter 2 as a building or portion of a building "used to provide medical, surgical, psychiatric, nursing or similar care on a less than 24-hour basis to individuals who are rendered incapable of self-preservation." Defined as Group B occupancies, such facilities are generally regarded as moderate in hazard level due to their office-like conditions. However, additional hazards are typically present due to presence of individuals who are temporarily rendered incapable of self-preservation due to the application of nerve blocks, sedation or anesthesia. While the occupants may walk in and walk out the same day with a quick recovery time after surgery, there is a period of time where a potentially large number of people could require physical assistance in case of an emergency that would require evacuation or relocation.

Although classified as a Group B occupancy in the same manner as an outpatient clinic or other health care office, an ambulatory health care facility poses distinctly different hazards to life and fire safety, such as:

- Patients incapable of self-preservation require rescue by other occupants or emergency responders
- Medical staff must stabilize the patient prior to evacuation, possibly resulting in delayed staff evacuation
- Use of oxidizing medical gases such as oxygen and nitrous oxide
- Potential for surgical fires

As a result of the increased hazard level, additional safeguards have been put in place. Smoke compartments must be provided in larger facilities and the installation of fire-protection systems is typically mandated. See Figure 422-1.

Ambulatory care facilities having more than 10,000 square feet (929 m^2) of floor area must be subdivided into at least two smoke compartments by smoke barriers in accordance with Section 710. Additional smoke compartments may be required due to travel distance limitations. Any point within a smoke compartment must be no more than 200 feet (60 960 mm) in travel distance from a smoke barrier door. Each smoke compartment must be large enough to allow for 30 square feet (2.8 m^2) of refuge area for each nonambulatory patient. In addition, at least one means of egress must be available from each smoke compartment without the need to return back through the original compartment.

As a general rule, Group B occupancies do not require a sprinkler system based solely on their occupancy classification. However, Section 903.2.2 mandates that a Group B ambulatory care facility be provided with an automatic sprinkler system when either of the following conditions exist at any time:

- Four or more care recipients are incapable of self-preservation, or
- One or more care recipients that are incapable of self-preservation are located at other than the level of exit discharge.

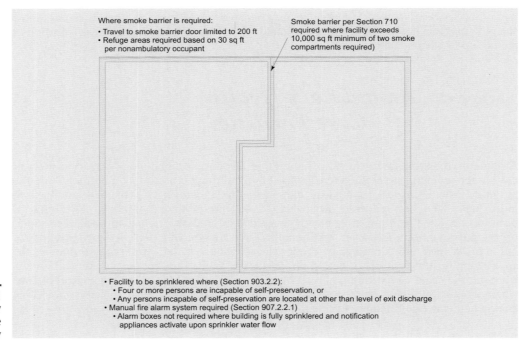

Figure 422-1
Ambulatory health-care facility

In addition, the fire alarm requirements are more stringent than those of other Group B occupancies. Section 907.2.2 requires the installation of a manual fire alarm system in all Group B fire areas containing an ambulatory health care facility. The manual fire alarm boxes are not required if the building is fully sprinklered and the occupant notification appliances activate upon sprinkler water flow.

Section 423 *Storm Shelters*

ICC-500, *ICC/NSSA Standard on the Design and Construction of Storm Shelters*, establishes minimum requirements for structures and spaces designated as hurricane, tornado or combination shelters. The standard addresses the design of such shelters from the perspective of the structural requirements for high wind conditions, and addresses minimum requirements for the interior environment during a storm event. Although the IBC does not mandate that storm shelters be provided, it does regulate their design if they are constructed.

KEY POINTS

- Special uses such as covered mall buildings, atriums, high-rise buildings, underground buildings and parking garages are so unique in the type of hazards presented that specialized regulations are provided in the IBC.

- A covered mall building or open mall building consists of various tenants and occupants, as well as the common pedestrian area that provides access to the tenant spaces.

- For those features that are not unique to a covered mall building, the general provisions of the code apply.

- The means of egress provisions for a covered mall building are typically more liberal than those for other buildings.

- High-rise buildings are characterized by the difficulty of evacuation or rescue of the building occupants, the difficulty of fire-fighting operations from the exterior, high occupant loads and potential for stack effect.

- There are a number of provisions for high-rise buildings that are less restrictive than the general requirements, including the reduction in fire resistance for certain building elements.

- The special allowance for a reduction in construction type is not applicable to any high-rise building exceeding 420 feet (128 m) in height.

- Smoke detection, alarm systems and communications systems are important characteristics of a high-rise building.

- Occupant egress and evacuation, as well as fire department access, are addressed in high-rise buildings through provisions for stairway enclosure remoteness, an additional stairway, luminous egress path markings, a fire service access elevator and occupant evacuation elevators.

- The use of the atrium provisions is typically limited to those multistory applications where compliance with the shaft enclosure requirements or exceptions of Section 708.2 is not possible.

- Buildings containing atriums, high-rise buildings, and covered mall buildings must be provided with automatic sprinkler systems throughout.

- Another component of the life-safety system for a building containing an atrium is a required smoke-control system.

- An underground building is regulated in a manner similar to that for a high-rise building, as the means of egress and fire department access concerns are similarly extensive.

- Fundamental to the protection features for an underground building are the requirements for Type I construction for the underground portion and the installation of an automatic sprinkler system.

- Private garages and carports are regulated to a limited degree based upon the hazards associated with the parking of motor vehicles.

- Special provisions for open parking garages are typically less restrictive than those for enclosed parking structures because of the natural ventilation that is available.

- In Group I-2 and I-3 occupancies, the functional concerns of the health-care operations must be balanced with the fire-safety concerns.

- Stages exceeding 50 feet (15 240 mm) in height present additional risks that are due to the expected presence of high combustible loading such as curtains, scenery and other stage effects.

- Under-floor areas and attic spaces used for the storage of combustible materials must be isolated from other portions of the building with fire-resistance-rated construction.

- Hazardous materials that are used or stored in any quantity are subject to regulation by Sections 307, 414, 415 and the IFC.

- An increase in the maximum allowable quantities of hazardous materials in a building not classified as a Group H is permitted through the proper use of control areas.

- High-hazard occupancies are highly regulated because of the hazardous processes and materials involved in such occupancies.

- Special provisions are applicable to the construction, installation and use of buildings for the spraying of flammable finishes in painting, varnishing and staining operations.

- Special allowances and conditions are applicable to live/work units where a dwelling unit or sleeping unit includes a significant amount of nonresidential use operated by the tenant.

- Dwelling units and sleeping units must be separated from each other through the use of fire partitions and/or horizontal assemblies.

- Health care offices where care is provided to individuals who are rendered incapable of self-preservation are considered to be ambulatory health care facilities.

- Storm shelters, where provided for safe refuge from hurricanes, tornados and other high-wind events, must be constructed in accordance with ICC-500.

CHAPTER

5

GENERAL BUILDING HEIGHTS AND AREAS

Chapter 5 provides general provisions that are applicable to all buildings. These include requirements for allowable floor area, including permitted increases for open spaces and for the use of automatic sprinkler systems; unlimited area buildings; and allowable height of buildings with acceptable increases. Buildings containing multiple uses and occupancies are regulated through the provisions for incidental use areas, accessory occupancies, nonseparated occupancies and separated occupancies. Miscellaneous topics addressed in Chapter 5 include premises identification and mezzanines.

In addition to the general provisions set forth in Chapter 5, there are several special conditions under which the specific requirements of Chapter 5 can be modified or exempted, including the horizontal building separation allowance and unique provisions for buildings containing a parking garage.

Section 501 *General*

501.2 **Address numbers.** In this section, the *International Building Code*® (IBC®) intends that buildings be provided with plainly visible and legible address numbers posted on the building or in such a place on the property that the building may be identified by emergency services such as fire, medical and police. The primary concern is that responding emergency forces may locate the building without going through a lengthy search procedure. In furthering the concept, the code intends that the approved street numbers be placed in a location readily visible from the street or roadway fronting the property if a sign on the building would not be visible from the street. Regardless of the sign's location, the minimum height of letters or numbers used in the address is to be at least 4 inches (102 mm) and have a color in contrast to the color of the background itself.

Section 502 *Definitions*

AREA, BUILDING. The term *building area* describes that portion of the building's floor area to be utilized in the determination of whether or not a structure complies with the provisions of Chapter 5 for allowable building size. It is not to be confused with the term *floor area*, which is the basis for occupant-load determination in Chapter 10 for means of egress evaluation, nor the term *fire area* as addressed in Chapter 7.

The definition of building area is the area included within the surrounding exterior walls of the building, and the definition further states that the floor area of a building or portion thereof not provided with surrounding exterior walls shall be the usable area under the horizontal projection of the roof or floor above. The intent of this latter provision is to address where a building may not have exterior walls or may have one or more sides open without an enclosing exterior wall. Examples would include a canopy covering pump islands at a service station, or the drive-through area of a fast food restaurant. Where a column line establishes the outer perimeter of the usable space under the roof, it is also typically the extent of building area. Beyond the column line, the overhead cover is simply viewed as a projection. See Figure 502-1. If all of the area beneath the roof above can be considered usable space, then the building area is measured to the leading edge of the roof above. See Figure 502-2.

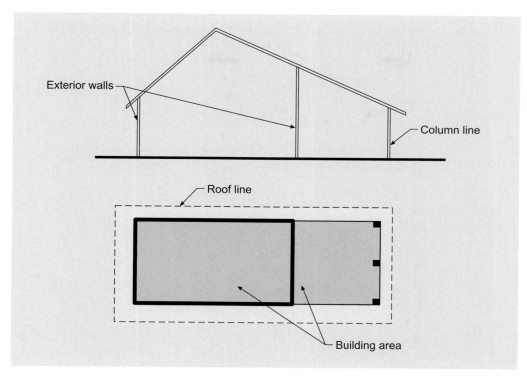

Figure 502-1
Building area

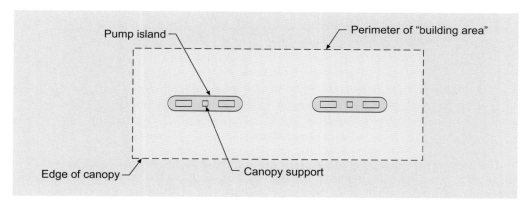

Figure 502-2
Building area

BASEMENT. A basement is considered to be any floor level that does not meet the definition of "Story above grade plane" as established in Chapter 2. There are limited provisions in the code that are specifically applicable to basements. One such significant requirement is established in Section 903.2.10.1.3 mandating the sprinklering of basements where adequate exterior openings are not provided. In short, the code regulates a below-ground floor level based upon its qualification as a story above grade plane.

GRADE PLANE. The code indicates that the grade plane is a reference plane representing the average of the finished ground level adjoining the building at its exterior walls. Under conditions where the finished ground level slopes significantly away from the exterior walls, that reference plane is established by the lowest points of elevation of the finished surface of the ground within an area between the building and lot line, or where the lot line is more than 6 feet (1829 mm) from the building, between the building and a line 6 feet (1829 mm) from the building. Where the slope away from the building is minimal (typically provided only to drain water away from the exterior wall) the elevation at the exterior wall provides an adequate reference point.

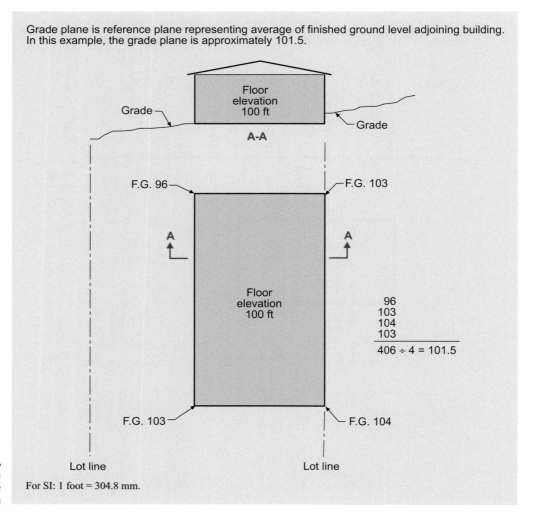

Grade plane is reference plane representing average of finished ground level adjoining building. In this example, the grade plane is approximately 101.5.

Grade

Floor elevation 100 ft

Grade

A-A

F.G. 96

F.G. 103

A

A

Floor elevation 100 ft

96
103
104
103
406 ÷ 4 = 101.5

F.G. 103

F.G. 104

Lot line

Lot line

**Figure 502-3
Grade plane
calculation**

For SI: 1 foot = 304.8 mm.

The method for calculating grade plane can vary based upon the site conditions. Where the slope is generally consistent as it passes across the building site, it may only require the averaging of a few points along the exterior wall of a rectangular-shaped building, as illustrated in Figure 502-3. Where the slope is inconsistent or retaining walls are utilized, or where the building footprint is complex, the determination of grade plane can be more complicated. In such cases, a more exacting method for calculating the grade plane must be utilized. In addition, where fire walls are present, the elevation points should be taken at the intersections of the fire wall and the exterior walls.

This definition is important in determining the number of stories above grade plane within a building as well as its height in feet. In some cases, the finished surface of the ground may be artificially raised with imported fill to create a higher grade plane around a building so as to decrease the number of stories or height in feet. The code does not prohibit this practice, and as long as a building meets the code definition and restriction for height or number of stories, the intent of the code is met. See Figure 502-4.

It is important to note that for the vast majority of buildings, it is not necessary to precisely calculate the grade plane. In such buildings, a general approximation of grade plane is sufficient to appropriately apply the code. A detailed calculation is only necessary in those limited situations where it is not obvious how the building is to be viewed in relationship to the surrounding ground level.

HEIGHT, BUILDING. Once the elevation of the grade plane has been calculated, it is possible to determine the building's height. This height is measured vertically from the grade plane to the average height of the highest roof surface. Examples of this measurement are shown in Figure 502-5.

Where the building is stepped or terraced, it is logical that the height is the maximum height of any segment of the building. It may be appropriate under certain circumstances that the number of stories in a building be determined in the same manner. Because of the varying requirements of the code that are related to the number of stories, such as means of egress, type of construction, fire resistance of shaft enclosures, etc., each case should be judged individually based on the characteristics of the site and construction. In addition to those factors better related to the number of stories, other items to consider are fire department access, location of exterior exit doors, routes of exit travel and types of separation between segments.

Figure 502-6 illustrates one example in which the height of the building and number of stories are determined for a stepped or terraced building. In the case of a stepped or terraced building, the language *total perimeter* is used to define the situation separating the first story from a basement and is intended to include the entire perimeter of the segment of the building. Therefore, in the cross section of Figure 502-7, the total perimeter of the down-slope segment would be bounded by the retaining wall, the down-slope exterior wall, and the east and west exterior walls. In the case illustrated, the building has three stories and no basement for the down-slope segment. The measurement for the maximum height of the building would be based upon the maximum height of the down-slope segment.

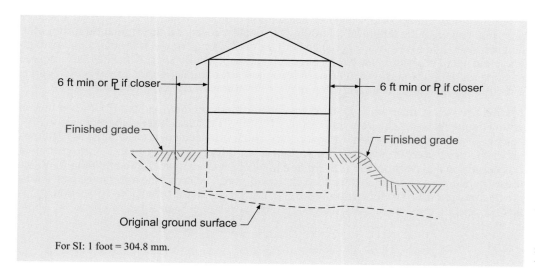

6 ft min or ℞ if closer

6 ft min or ℞ if closer

Finished grade

Finished grade

Original ground surface

For SI: 1 foot = 304.8 mm.

Figure 502-4
Use of built-up soil to raise finished grade

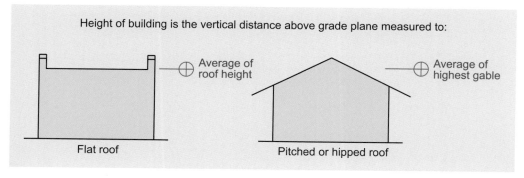

Height of building is the vertical distance above grade plane measured to:

⊕ Average of roof height

⊕ Average of highest gable

Flat roof

Pitched or hipped roof

Figure 502-5
Height of building

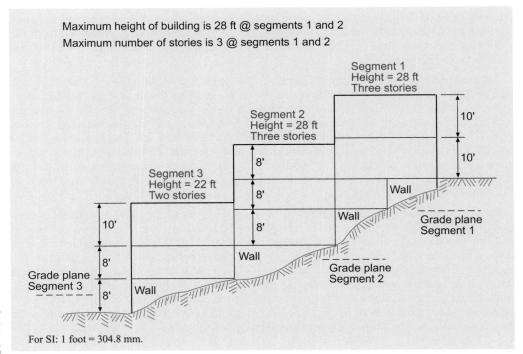

Maximum height of building is 28 ft @ segments 1 and 2
Maximum number of stories is 3 @ segments 1 and 2

Segment 1
Height = 28 ft
Three stories

Segment 2
Height = 28 ft
Three stories

Segment 3
Height = 22 ft
Two stories

10'

10'

8'

8'

8'

10'

8'

8'

Wall

Wall

Wall

Wall

Grade plane
Segment 1

Grade plane
Segment 2

Grade plane
Segment 3

Figure 502-6
**Terraced
building**

For SI: 1 foot = 304.8 mm.

Similar to an unnecessarily detailed calculation of grade plane, there is seldom a need to precisely calculate the height of a building. Typically, a general determination of building height is adequate to ensure compliance with the code. For example, it is not necessary to go into great detail evaluating the average roof elevation of a built-up roof that has a low degree of slope for drainage purposes. The need for a more exacting determination of roof height is directly related to any uncertainty that may occur in reviewing for code compliance.

MEZZANINE. A mezzanine is merely a code term for an intermediate floor level placed within a room between the floor and ceiling of a story. Typically limited in floor area to $33^1/_3$ percent of the area of the room or space into which it is located, a mezzanine is regulated under the provisions of Section 505. A floor level fully complying with the provisions of this definition and Section 505 is permitted to be considered a mezzanine in order to utilize those special provisions applicable to such a condition. The use of the mezzanine provisions is a design option, as an elevated floor level complying with the provisions of Section 505 is always permitted to be considered an additional story rather than a mezzanine.

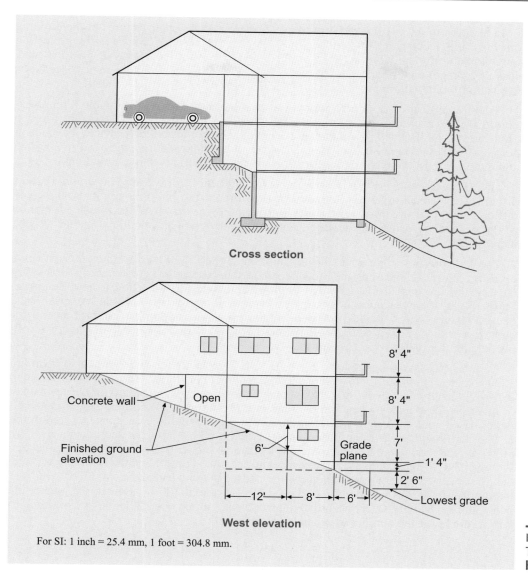

Cross section

Concrete wall

Open

8' 4"

8' 4"

7'

Finished ground
elevation

6'

Grade
plane

1' 4"

2' 6"

12' 8' 6'

Lowest grade

West elevation

For SI: 1 inch = 25.4 mm, 1 foot = 304.8 mm.

Figure 502-7
**Three-story
building**

Section 503 *General Building Height and Area Limitations*

The IBC regulates the size of buildings in order to limit to a reasonable level the magnitude of a fire that potentially may develop. The size of a building is controlled by its floor area and height, and both are limited by the IBC. Whereas floor-area limitations are concerned primarily with property damage, life safety is enhanced as well by the fact that in the larger building there are typically more people at risk during a fire. Height restrictions are imposed to address egress concerns and fire department access limitations.

The essential ingredients in the determination of allowable areas are:

1. The amount of combustibles attributable to the use that determines the potential fire severity.

2. The amount of combustibles in the construction of the building, which contributes to the potential fire severity.

In addition to the two factors just itemized, there may be other features of the building that have an effect on area limitations. These include the presence of built-in fire protection (an automatic fire-sprinkler system), which tends to prevent the spread of fire, and open space (frontage) adjoining a sizable portion of the building's perimeter, which decreases exposure from adjoining properties and provides better fire department access.

A desirable goal of floor-area limitations in a building code is to provide a relatively uniform level of hazard for all occupancies and types of construction. A glance through Table 503 of the IBC will reveal that, in general, the higher hazard occupancies have lower permissible areas for equivalent types of construction and, in addition, the less fire-resistant and more combustible types of construction have more restrictive area limitations.

The IBC also limits the maximum height and number of stories based on similar reasons discussed for area limitations. In addition, the higher the building becomes, the more difficult access for fire fighting becomes. Furthermore, the time required for the evacuation of the occupants increases; therefore, the fire resistance of the building should also be increased.

The code presumes that when the height of the highest floor used for human occupancy exceeds 75 feet (22 860 mm), the life-safety hazard becomes even greater because most fire departments are unable to adequately fight a fire above this elevation from the outside. Furthermore, the evacuation of occupants from the building is often not feasible. Thus, Section 403 prescribes special provisions for these high-rise buildings. Similar concerns for buildings with occupied floors well below the level of exit discharge are addressed in Section 405 for underground buildings.

Coming back to this section, the code specifies in Table 503 both the maximum allowable height in feet (mm) and the maximum number of stories. The maximum height in feet is regulated solely by the building's construction type, with no regard for the occupancy or multiple occupancies located in the building. However, the maximum height in stories varies based upon the occupancy group involved. Where multiple occupancies are located in the same building, and the provisions of Section 508.4 for separated occupancies are utilized, each individual occupancy can be located no higher than set forth in the table. See Figure 503-1. Where the nonseparated occupancies provisions of Section 508.3 are applied, the most restrictive height limitations of the nonseparated occupancies involved will limit the number of stories in the entire building. See Figure 503-2. In general, the greater the potential fire- and life-safety hazard, the lower the permitted overall height in feet (mm), as well as the fewer the number of permitted stories.

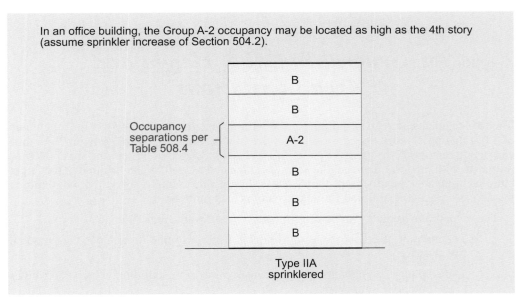

In an office building, the Group A-2 occupancy may be located as high as the 4th story (assume sprinkler increase of Section 504.2).

Occupancy separations per Table 508.4

B
B
A-2
B
B
B

Type IIA sprinklered

Figure 503-1
Height limitations–separated occupancies

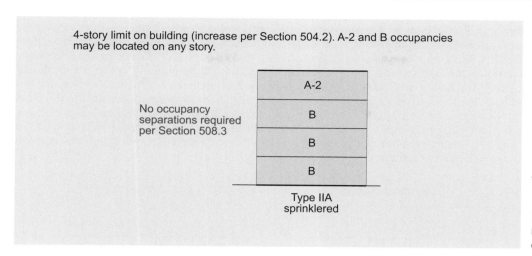

4-story limit on building (increase per Section 504.2). A-2 and B occupancies may be located on any story.

No occupancy separations required per Section 508.3

A-2

B

B

B

Type IIA sprinklered

Figure 503-2
Height limitations– nonseparated occupancies

General. Height and area limitations for buildings are set forth in Table 503. As indicated on the table, the height limits are expressed as both stories and feet above the grade plane. For allowable area purposes, the numbers in the table refer to the building area (in square feet), as defined in Section 502.1, per story. Therefore, based on occupancy group and type of construction, Table 503 identifies the limitations on building size permitted by the IBC. Examples of the use of this table, without any permitted increases, are shown in Figure 503-3. **503.1**

Group B
VB construction
9,000 sq ft per story
2 stories
40 ft

Group S-2
IIIB construction
26,000 sq ft per story
3 stories
55 ft

Group I-2
IIA construction
15,000 sq ft per story
2 stories
65 ft

Group A-2
VA construction
11,500 sq ft per story
2 stories
50 ft

For SI: 1 foot = 304.8 mm, 1 square foot = 0.093 m².

Figure 503-3
Allowable area and height

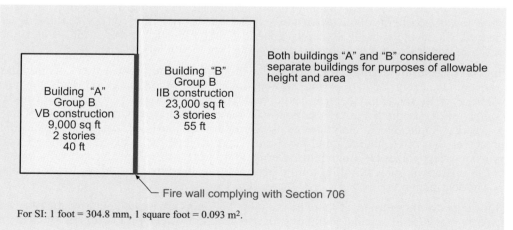

Building "A"
Group B
VB construction
9,000 sq ft
2 stories
40 ft

Building "B"
Group B
IIB construction
23,000 sq ft
3 stories
55 ft

Both buildings "A" and "B" considered separate buildings for purposes of allowable height and area

Fire wall complying with Section 706

For SI: 1 foot = 304.8 mm, 1 square foot = 0.093 m².

Figure 503-4
Fire walls

The IBC establishes in Table 503 what is commonly referred to as tabular values for height and area. By use of the term *tabular*, the code recognizes that the building in question has no features that might be considered to improve the overall fire hazard (such as a fire-sprinkler system or adjacent open areas); alternatively, where such features exist, they provide no benefit for the situation under consideration. An example is found in Section 508.2.1 for accessory occupancies areas, where the floor area of the accessory occupancy cannot exceed the tabular values in Table 503. Although the building being analyzed may qualify for height and area increases established elsewhere in Chapter 5, they are not considered in the evaluation of accessory occupancy area compliance. For the more typical application of the provision, this section states that the height of the building and the area of any floor within the building shall not be greater than the tabular (basic) area specified in Table 503, unless the building is entitled to height or area increases, which are described in other sections of Chapter 5.

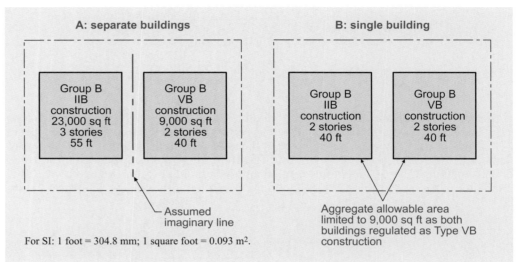

Figure 503-5
Buildings on the same lot

In this section, the IBC indicates that fire walls, in addition to exterior walls, create separate buildings when evaluating for allowable height and area. Defined and regulated under the provisions of Section 706, the function of a fire wall is to separate one area of a building from another with a fire-resistance-rated vertical separation element. Where a fully-complying fire wall is provided, it provides two compartments, one on each side of the wall, which may each be considered under the IBC to be separate buildings. Multiple fire walls may be utilized to create a number of separate buildings within a single structure. One of the resulting benefits of the use of a fire wall is that the limitations on height and area are then addressed individually for each separate building created by fire walls within the structure, rather than for the structure as a whole. See Figure 503-4.

503.1.1 Special industrial occupancies. This special provision exempts certain types of buildings from both the height limitations and the area limitations found in Table 503. Thus, the type of construction is not limited, regardless of building height or area. It is also not necessary to comply with the provisions of Section 507 for unlimited area buildings to utilize this provision. Applicable to structures housing low-hazard and moderate-hazard industrial processes that often require quite large areas and heights, the relaxation of the general provisions recognizes the limited fire severity, as well as the need for expansive buildings to house operations such as rolling mills, structural metal-fabrication shops, foundries and power distribution.

503.1.2 Buildings on the same lot. Where two or more buildings are located on the same lot, they may be regulated as separate buildings, in a manner consistent with buildings situated on separate parcels of land. See Figure 503-5A.

As an option, multiple buildings on a single site may be considered one building, provided the limitations of height and area based on Table 503 are met. The height of each building and the aggregate area of all buildings is to be considered in the determination.

Under this method, the provisions of the code applicable to the aggregate building shall also apply to each building individually. See Figure 503-5B. Further regulations for buildings on the same lot are discussed in the commentary for Section 705.3.

Section 504 *Building Height*

Because automatic fire-sprinkler systems have exhibited an excellent record of in-place fire suppression over the years, the IBC allows height increases as well as area increases, where an automatic fire-sprinkler system is installed throughout the building. The code permits an increase of one story in the number of stories, and 20 feet (6096 mm) in building height, where the building is provided with an automatic fire-sprinkler system throughout. These increases are directly applied to Table 503. See Figure 504-1. It should be emphasized that this increase applies both to an increase in the number of stories and also an increase of the height limit in feet (mm).

There are basically four variations to the general requirements for height and story increases:

1. Such increases are not permitted for Group I-2 occupancies of Type IIB, III, IV or V construction, or for Groups H-1, H-2, H-3 and H-5 occupancies. These occupancies present unusual hazards that limit their heights even where a sprinkler system is present. The increases may also not be taken where the provisions of Table 601, Note d, for 1-hour fire-resistance rating substitution are utilized.

2. One-story aircraft manufacturing buildings and hangars may be of unlimited height when sprinklered and surrounded by adequate open space. Such uses require very large structures and through the safeguards provided, should be adequately protected.

3. For Group R buildings provided with an NFPA 13R sprinkler system, the increases in height and number of stories apply only up to a maximum of 60 feet (18 288 mm) or four stories, respectively. The limitation of four stories and 60 feet for buildings sprinklered with a 13R system cannot be exceeded under any circumstances. See Figure 504-2. In those residential buildings where an NFPA 13, rather than an NFPA 13R system, is installed, the limitations of 60 feet (18 288 mm) and four stories do not apply.

4. Roof structures such as towers and steeples may be of unlimited height when constructed of noncombustible materials, whereas combustible roof structures are limited in height to 20 feet (6096 mm) above that permitted by Table 503. See Figure 504-3. In all cases, such roof structures are to be constructed of materials based on the building's type of construction. These requirements are not based on the presence of the sprinkler system. Additional requirements for roof structures can be found in Section 1509.

As an important note, except for those buildings provided with an NFPA 13R system, the increases in building height and number of stories permitted by this section for a sprinklered building may be taken in addition to those floor area increases permitted by Sections 506.2 and 506.3.

Section 505 *Mezzanines*

A mezzanine is defined in Section 502 as an intermediate floor level within a room or space. As long as the area of the mezzanine is limited in size, an intermediate floor without enclosure causes no significant safety hazard. The occupants of the mezzanine by means of sight, smell or hearing will be able to determine if there is some emergency or fire that takes place either on the mezzanine or in the room in which the mezzanine is located. However, once portions of or all of the mezzanine is enclosed, or the mezzanine exceeds one-third the area of the room in which it is located, life-safety problems such as occupants not being aware of an emergency or finding a safe exit route from the mezzanine become important. Therefore, the code places the restrictions encompassed in this section on mezzanines to ameliorate the life-safety hazards that can be created.

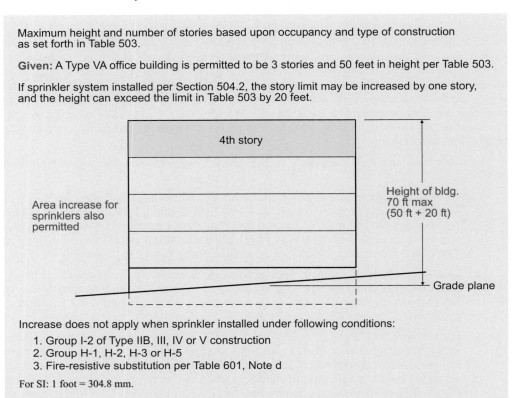

Maximum height and number of stories based upon occupancy and type of construction as set forth in Table 503.

Given: A Type VA office building is permitted to be 3 stories and 50 feet in height per Table 503.

If sprinkler system installed per Section 504.2, the story limit may be increased by one story, and the height can exceed the limit in Table 503 by 20 feet.

Increase does not apply when sprinkler installed under following conditions:

1. Group I-2 of Type IIB, III, IV or V construction
2. Group H-1, H-2, H-3 or H-5
3. Fire-resistive substitution per Table 601, Note d

For SI: 1 foot = 304.8 mm.

Figure 504-1
Allowable height increase

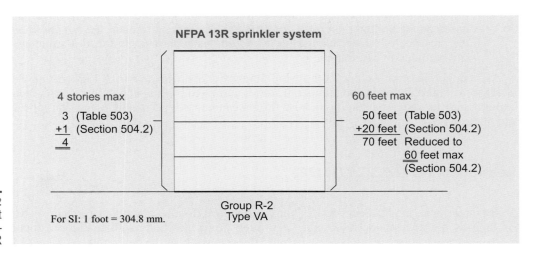

NFPA 13R sprinkler system

4 stories max

3 (Table 503)
+1 (Section 504.2)
4

60 feet max

50 feet (Table 503)
+20 feet (Section 504.2)
70 feet Reduced to
60 feet max
(Section 504.2)

Group R-2
Type VA

For SI: 1 foot = 304.8 mm.

Figure 504-2
Height increase– Group R

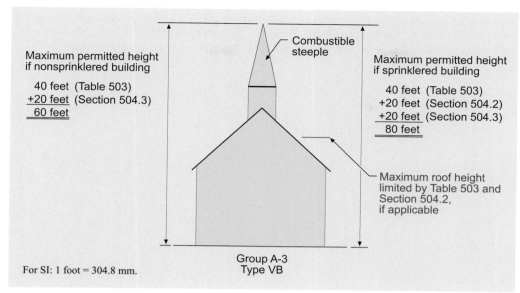

Combustible steeple

Maximum permitted height if nonsprinklered building

40 feet (Table 503)
+20 feet (Section 504.3)
60 feet

Maximum permitted height if sprinklered building

40 feet (Table 503)
+20 feet (Section 504.2)
+20 feet (Section 504.3)
80 feet

Maximum roof height limited by Table 503 and Section 504.2, if applicable

For SI: 1 foot = 304.8 mm.

Group A-3
Type VB

Figure 504-3
Roof
structures

General. By virtue of the conditions placed on mezzanines in Section 505, a complying **505.1** mezzanine is not considered to create additional building area or an additional story for the purpose of limiting building size. The floor area of a complying mezzanine need not be added to the area of the floor below for the purpose of limiting building area by Section 503. This allowance essentially provides for free floor area in the comparison of the total actual area to the total allowable area. As previously mentioned, complying mezzanines also do not contribute to the actual number of stories in relationship to the allowable number of stories permitted by Sections 503 and 504. The limitations imposed on mezzanines are deemed sufficient to permit such benefits.

In contrast to the above allowances, the floor area of mezzanines must be included as a part of the aggregate floor area in determining the fire area as defined in Section 902.1. Because the size of a fire area is based on a perceived level of fire loading present within the building, the contribution of a mezzanine's fire load to the fire loading in the room in which the mezzanine is located cannot be overlooked. Figure 505-1 depicts the proper use of these provisions. The clear height above and below the floor of the mezzanine is also regulated at a minimum height of 7 feet (2134 mm).

Area limitations. There is no limit on the number of mezzanines that may be placed within **505.2** a room; however, the total floor area of all mezzanines must typically not exceed one-third the floor area of the room in which they are located. See Figure 505-2. As illustrated in Figure 505-3, any enclosed areas of the room in which the mezzanine is located are not to be utilized in the calculations for determining compliance with the one-third rule.

Where two specific conditions exist, the aggregate floor area of mezzanines may be increased up to two-thirds of the floor area of the room below. First, the building must contain special industrial processes as identified in Section 503.1.1, and second, the building shall be of Type I or Type II construction. Intermediate floor levels are very common in buildings of this kind because of the nature of their operations. By limiting the increased mezzanine size to noncombustible buildings housing primarily noncombustible processes, fire safety is not compromised.

A second exception also permits an increase in allowable mezzanine size, up to a maximum of one-half of the area or room in which the mezzanine is located. See Figure 505-4. The increased size takes into consideration the enhancements of noncombustible construction, automatic sprinkler system protection and occupant notification. By limiting construction to Type I or II, there is no contribution to the fire hazard that is due to the materials of construction. The automatic sprinkler protection will increase the potential for limiting fire spread. In addition, the occupant notification system increases occupant awareness of a fire condition and allows for evacuation during the early fire stages.

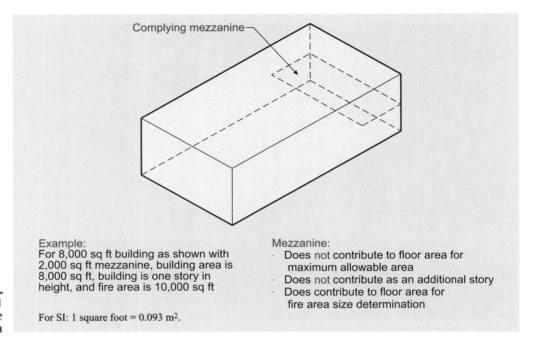

Complying mezzanine

Example:
For 8,000 sq ft building as shown with 2,000 sq ft mezzanine, building area is 8,000 sq ft, building is one story in height, and fire area is 10,000 sq ft

For SI: 1 square foot = 0.093 m².

Mezzanine:
· Does not contribute to floor area for maximum allowable area
· Does not contribute as an additional story
· Does contribute to floor area for fire area size determination

Figure 505-1
Mezzanine
height and area

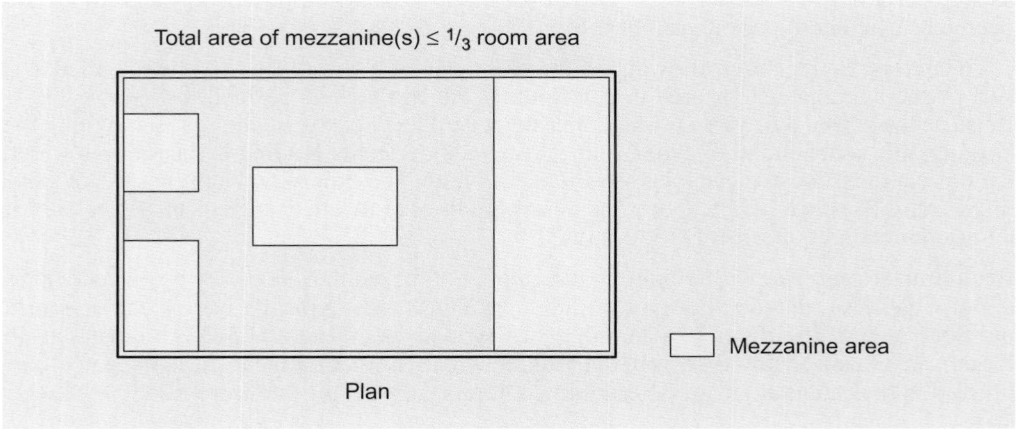

Total area of mezzanine(s) ≤ ¹/₃ room area

Mezzanine area

Plan

Figure 505-2
Mezzanine
area

505.3 **Egress.** The value in considering an elevated floor level as a mezzanine is threefold: A mezzanine does not contribute to building area, is not counted as an additional story, and the means of egress system is not as highly regulated. A single means of egress is permitted from a mezzanine if the common path of egress travel does not exceed the limitations of Section 1014.3. This is similar to the approach taken for egress from any room or space; however, the occupant load limitations of Section 1015.1 need not be applied. A more logical approach would be the application of the exception as the general rule—that a single means of egress is permitted in those cases where compliance with Section 1015.1 is possible. When measuring the maximum travel distance considered a common path, the distance traveled on an unenclosed stairway shall be included and measured in the plane of the tread nosing. Because a mezzanine is not considered an additional story, stairways serving only the mezzanine are typically permitted to be unenclosed.

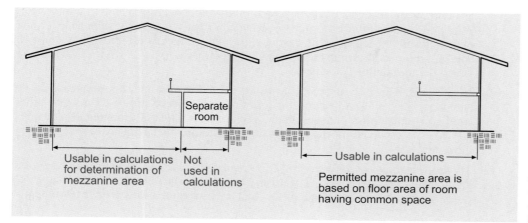

Usable in calculations for determination of mezzanine area

Not used in calculations

Separate room

Usable in calculations

Permitted mezzanine area is based on floor area of room having common space

Figure 505-3
Mezzanine area

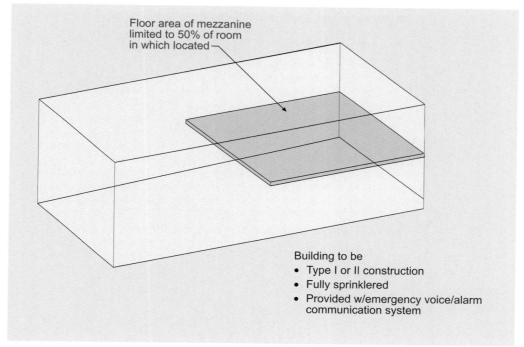

Floor area of mezzanine limited to 50% of room in which located

Building to be
- Type I or II construction
- Fully sprinklered
- Provided w/emergency voice/alarm communication system

Figure 505-4
Maximum floor area of mezzanines

Openness. As a general rule, a mezzanine must be open to the room in which it is located. Any side that adjoins the room will be considered open if it is unobstructed, other than by walls or railings not more than 42 inches (1067 mm) in height, or columns and posts. There are, however, five exceptions to the requirement for openness that may result in most mezzanines being permitted to be partially or completely enclosed. If in compliance with any one of the five exceptions, the mezzanine need not be enclosed.

505.4

1. Illustrated in Figure 505-5, this criterion is that the enclosed area contains a maximum occupant load of 10 persons. The limitation on occupant load is based upon the aggregate area of the enclosed space. This exception is consistent with other provisions of the code that relax the requirements where the occupant load is

expected to be relatively small. This exception may result in the enclosure of an entire mezzanine or just a portion of it.

2. As shown in Figure 505-6, a mezzanine may also be fully or partially enclosed if it has two means of egress, one of which gives direct access to an exit component. While one means of egress must be provided directly to an exit, such as a vertical exit enclosure or exterior exit stairway, the second means of egress may enter an adjacent room as shown in the drawing or may be a stairway down into the room in which the mezzanine is located.

3. Also depicted in Figure 505-5, up to 10 percent of the mezzanine area may be enclosed. This is usually done for toilet rooms, closets, utility rooms and other similar uses that must, of necessity, be enclosed. As long as the aggregate area does not exceed 10 percent of the area of the mezzanine, the enclosure is permitted by the code.

4. Those mezzanines used for control equipment in industrial buildings may be glazed on all sides. This exception is necessary because of the delicate nature of much of today's control equipment and the fact that it may require a dust-free environment.

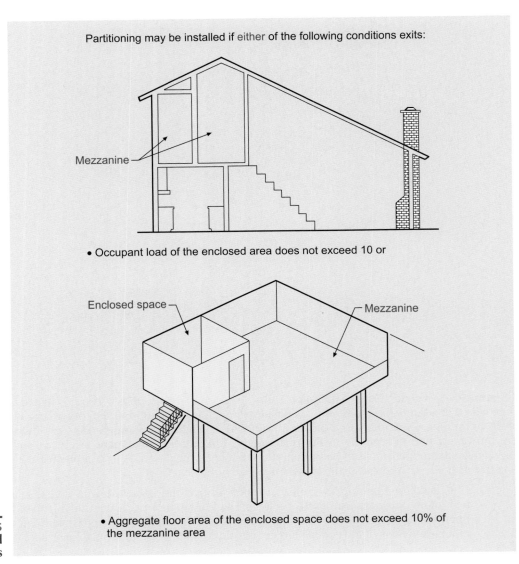

Figure 505-5
Enclosed mezzanines

5. This exception allows enclosure of the entire mezzanine, provided specified egress and sprinkler provisions are met. Similar to the conditions of Exception 2, this exception requires a minimum of two means of egress from the mezzanine. However, it differs in that none of the egress travel is required to reach an exit component on the mezzanine level. Applicable only to buildings that are sprinklered throughout, the exception allows for all required egress from an enclosed mezzanine to occur on unenclosed stairways leading to the floor level below. The exception is not intended to apply to Group H and I occupancies or buildings more than two stories in height above grade plane. See Figure 505-7.

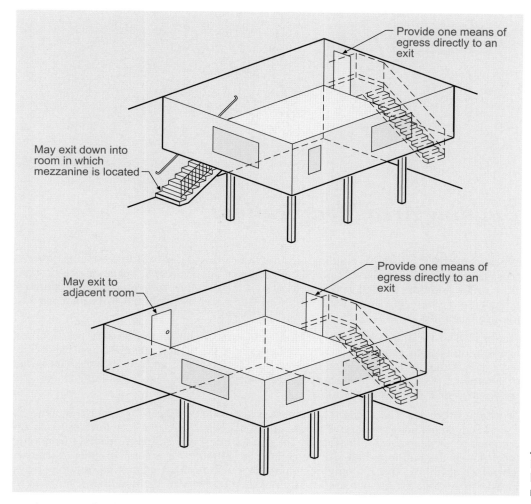

Figure 505-6
Enclosed mezzanines

Equipment platform. In buildings containing platforms that house equipment, such platforms need not be considered stories or mezzanines, provided they conform with the provisions of this section regulating platform size, extent of automatic sprinkler protection and guards. In addition, the equipment platforms cannot serve as any portion of the exiting system from the building. Complying platforms are not deemed to be additional stories, do not contribute to the building floor area, and furthermore, need not be included in determining the size of the fire area.

505.5

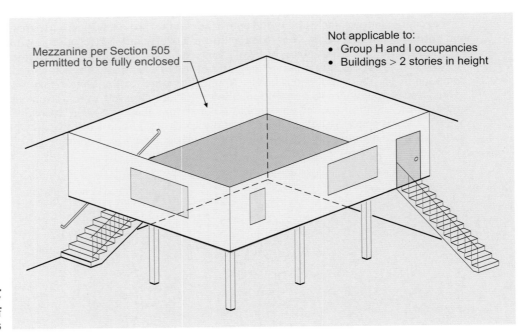

Mezzanine per Section 505
permitted to be fully enclosed

Not applicable to:
- Group H and I occupancies
- Buildings > 2 stories in height

Figure 505-7
Enclosure of
mezzanines

Section 506 *Building Area Modifications*

Whereas the basic allowable building area per floor is regulated by Table 503, increases to those areas are permitted based upon the presence of adequate open space on one or more sides of the building, as well as the protection of the structure with an automatic sprinkler system. In addition, the overall allowable building area is permitted to be increased in multistory buildings.

506.1 **General.** The formula for the calculation of the maximum allowable area per floor (in square feet) is additive, determined from the sums of the tabular area based on Table 503, any increase that is due to building frontage per Section 506.2, and any increase that is due to automatic fire-sprinkler protection as established in Section 506.3. A simple example is shown in Application Example 506-1.

506.2 **Frontage increase.** The initial requirement of the code, insofar as a frontage increase is concerned, is that it adjoin or have access to a public way. Thus, the structure could extend completely between side lot lines and to the rear lot line, be provided with access from only the front of the building, and still potentially be eligible for a small frontage increase. Therefore, it follows that if a building is provided with frontage consisting of public ways and/or open space for an increased portion of the perimeter of the building, some benefit should accrue based on better access for the fire department. Also, if the yards or public ways are wide enough, there will be a benefit that is due to the decreased exposure from adjoining properties.

Because of the beneficial aspects of open space adjacent to a building, the IBC permits increases in the tabular areas established from Table 503 based on the amount of open perimeter and width of the open space and public ways surrounding the building. For any open space to be effective for use by the fire department, it is mandated that it be accessed from a public way or a fire lane so that the fire department will have access to that portion of the perimeter of the building that is adjacent to open space. See Figure 506-1.

Open space and public ways—what can and can't be used. In addition to allowances for public ways, the IBC uses the term *open space* where relating to frontage increases in the

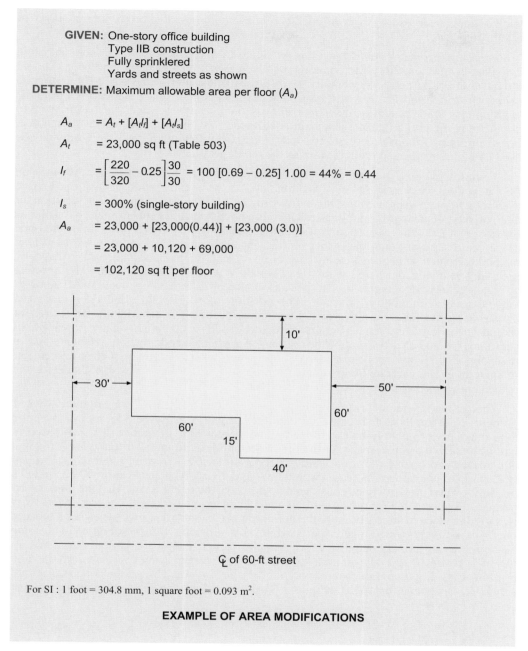

Application Example 506-1

GIVEN: One-story office building
Type IIB construction
Fully sprinklered
Yards and streets as shown
DETERMINE: Maximum allowable area per floor (A_a)

A_a = A_t + [$A_t I_f$] + [$A_t I_s$]

A_t = 23,000 sq ft (Table 503)

I_f = $\left[\dfrac{220}{320} - 0.25\right]\dfrac{30}{30}$ = 100 [0.69 – 0.25] 1.00 = 44% = 0.44

I_s = 300% (single-story building)

A_a = 23,000 + [23,000(0.44)] + [23,000 (3.0)]

= 23,000 + 10,120 + 69,000

= 102,120 sq ft per floor

10'

30'   50'

60'

60'

15'

40'

Ҫ of 60-ft street

For SI : 1 foot = 304.8 mm, 1 square foot = 0.093 m².

EXAMPLE OF AREA MODIFICATIONS

determination of allowable floor areas. Although the term *open space* is not specifically defined in the IBC, the definition of a *yard* is an open space unobstructed from the ground to the sky that is located on the lot on which the building is situated. It is logical that this definition is consistent with the intended description of open space. This definition seems to preclude the storage of pallets, lumber, manufactured goods, home improvement materials or any other objects that similarly obstruct the open space. However, it would seem reasonable to permit automobile parking, low-profile landscaping, fire hydrants, light standards and similar features to occupy the open space. These types of obstructions can be found within the public way, so their allowance within the open space provides for consistency. Because a yard must be unobstructed from the ground to the sky, open space widths should be measured from the edge of roof overhangs or other projections as shown in Figure 506-2.

Regarding the use of public ways for providing frontage increases, the width of public way that should be used for determining area increases seems to cause confusion. Should the full width of the public way or only the distance to the centerline be used? The confusion evolves from Section 702.1, which states that fire separation distance is measured from the building face to the centerline of a street, alley or public way. However, the requirement to use the centerline is limited to fire separation distance and is not applicable to Section 506.2. For determining frontage increases for open space, the full width of the public way may be used by buildings located on both sides of the public way.

The following type of question is also sometimes asked: "Why can't I use the big open field next door for area increases?" Section 506.2.2 specifically mandates that open space used for a frontage increase must be on the same lot as the building under consideration, or alternatively, dedicated for public use. There is a good reason for this limitation, insofar as the owner of one parcel lacks control over a parcel owned by another and, thus, the open space can disappear when the owner of "the big open field" decides to build on it. One method by which some jurisdictions have allowed such large open spaces to be used is by accepting joint use of shared yards. It is typically necessary that a recorded restrictive covenant be executed to ensure that the shared space will remain open and unoccupied as long as it is required by the code. The creation of a no-build zone does not seem unreasonable insofar as the aim is to maintain open spaces between buildings. Any covenant should be reviewed by legal counsel to be sure it will accomplish what is intended. In addition, it should clearly describe the reason and applicable code section so that any future revisions or deletions may be considered if the owners wish to terminate such an agreement. In such an event, each building should be brought into current code compliance, or the agreement would be required to remain in effect.

Whereas use of a public way as open space is permitted by the IBC, other publicly-owned property is generally not, because the building official usually has no control over the long-range use of publicly-owned property, and there is little assurance that such property will be available as open space for the life of the building. Remember that what is today's publicly-owned open parking lot could become tomorrow's new city hall, and the open space used to justify area increases would no longer exist. Whereas Section 506.2.2 allows publicly-owned property to be considered open space, the intent is such that the property be permanently dedicated for public use and maintained as unobstructed. The term *public way* was used in place of streets because the definition in Chapter 10 allows the use of a broader range of publicly-owned open space while still allowing the building official some discretion as to the acceptability of a particular parcel. *Public way* usually conjures up visions of streets and alleys, but how about other open spaces such as power line right-of-ways, flood-control channels or railroad rights-of-way? Many such open spaces are generally acceptable, provided there is a good probability that they will remain as open

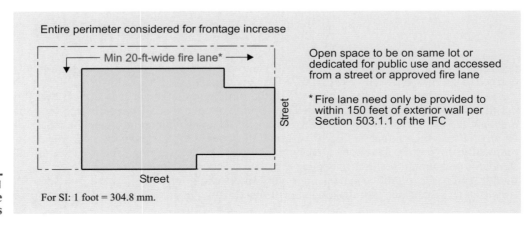

Figure 506-1
Open space access

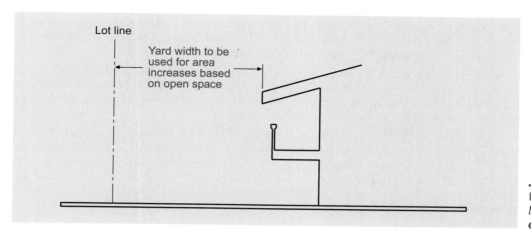

Figure 506-2
**Measurement
of open space**

space during the life of the building for which they will serve. Power lines and flood-control channels are usually good bets for longevity, but railroad routes are often abandoned and, therefore, may not be as good a bet. There is also an expectation that the public way is maintained in an unobstructed condition to allow for fire department access, which potentially would disallow the use of waterways and similar features. If the public way does not provide for fire department access, its use for a frontage increase is prohibited. It should be noted that the definition for public way requires any such public parcel of land, other than a street or alley, to lead to a street. Figure 506-3 provides a visual summary of open space and public ways that could be used for open-space area increases.

How much increase? In the case where public ways or open space adjoin more than 25 percent of the building's perimeter, the code permits an increase in the building area per story as shown in Table 503. The amount of the increase is based upon the percentage of open perimeter having a width of at least 20 feet (6096 mm). By utilizing the formula shown below, the area increase that is due to frontage (I_f) can be determined by:

$$I_f = [F/P - 0.25] \, W/30$$

WHERE:

I_f = Area increase due to frontage (percent)

F = Building perimeter that fronts on a public way or open space having 20-foot-minimum (6096 mm) open width

P = Perimeter of entire building

W = Width of public way or open space in accordance with Section 506.2.1

Based upon this method of calculation, the maximum area increase permitted will typically be 75 percent, as shown in Application Example 506-2. This is based on the general requirement of Section 506.2.1 that requires a value of 30 feet (9144 mm) to be used for the value W in those cases where W exceeds 30 feet (9144 mm). As this figure illustrates, the entire perimeter of the building must adjoin a public way or open space having a width of at least 30 feet (6 m). Where less than the entire perimeter has adequate open area, the area increase for frontage will be reduced as illustrated in Application Example 506-3.

Where the open space at the building's perimeter is between 20 feet (6096 mm) and 30 feet (9144 mm) in width, the code permits the use of the weighted average of such width in relation to the entire perimeter. This approach allows for the width W in Equation 5-2 to be more representative of the availability of open space around the building, rather than basing the frontage increase on simply the smallest open space of 20 feet (6096 mm) or more.

Whereas in most cases, the increase available based on the weighted average method is minimal, it does provide for some degree of allowable area adjustment. An example of calculating *W* by weighted average is shown in Application Example 506-4.

Whereas 75 percent is generally the largest allowable frontage increase, a greater area increase is permitted for those buildings that comply with all of the requirements for unlimited area buildings as described in Section 507 other than compliance with the 60-foot (18 288 mm) open space or public way requirement.

A maximum frontage increase of just less than 150 percent can be achieved based on the entire perimeter being open with a minimum width of slightly less than 60 feet (18 288 mm). Once 60 feet (18 288 mm) of accessible open space and public ways is obtained for 100 percent of the building's perimeter, the provisions of Section 507 are applicable and the frontage increase formula is not to be used. An example of the increased frontage increase is shown in Application Example 506-5, which also includes the calculated increase based on weighted average.

How must access be provided to an open space? The IBC provides no details as to the degree of fire department accessibility required in order to consider open space for an allowable area increase; it only mandates that access be provided from a street or approved fire line. It is clearly not the intent of the provisions to mandate a street or fire line completely around a building in order to acquire the maximum frontage increase. However, fire personnel access from such streets or fire lanes is necessary. Although it is not a requirement to provide access around a building for fire department apparatus, other than that required by IFC Section 503.1.1, the frontage increase is based on the ability of fire personnel to physically approach the building's exterior under reasonable conditions. For example, where the space adjacent to the building is heavily-forested or steeply-sloped, the frontage increase addressed in Section 506.2 is not permitted. The presence of a lake or similar water feature next to a building would also prohibit an area increase. The evaluation of each individual building and its site conditions is necessary to properly apply the code for fire department access.

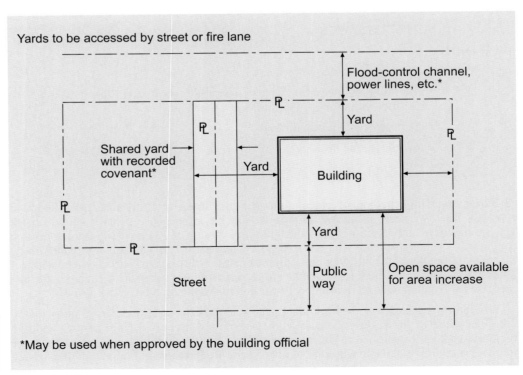

Figure 506-3
Yards and public ways available for area increases

GIVEN: Building width, yards and street as shown
DETERMINE: Frontage increase for area modification

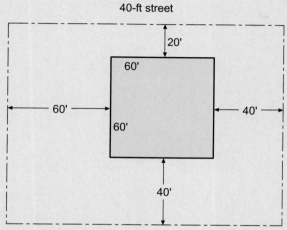

40-ft street

20'

60'

60'

60'

40'

40'

For SI: 1 foot = 304.8 mm.

$$I_f = \left[\frac{F}{P} - 0.25\right]\frac{W}{30}$$

WHERE:

I_f = Area increase due to frontage (percent)

F = Building perimeter that fronts on a public way or open space having 20 feet (6090 mm) open minimum width

P = Perimeter of entire building

W = Minimum width of public way and/or open space

I_f $= \left[\frac{240}{240} - 0.25\right]\frac{40}{30}$

       $= [1 - 0.25]\frac{30^*}{30}$

       $= [0.75]\ 1.0$

       = 75% increase

I_f = 0.75

* W cannot exceed 30 feet

MAXIMUM FRONTAGE INCREASE

Application Example 506-3

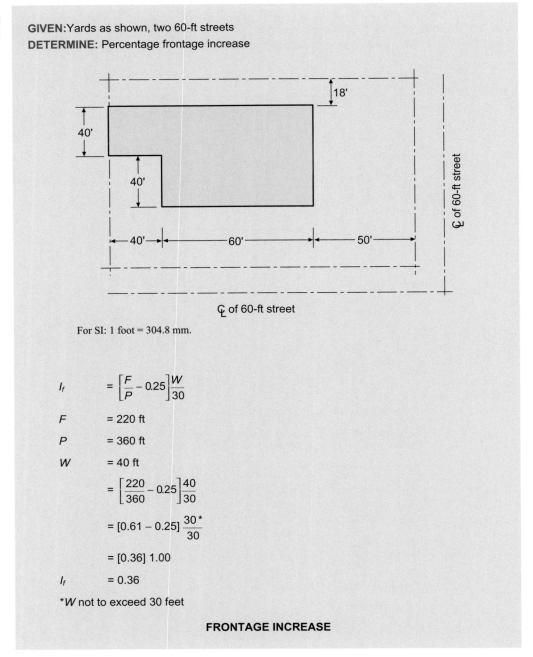

GIVEN: Yards as shown, two 60-ft streets

DETERMINE: Percentage frontage increase

For SI: 1 foot = 304.8 mm.

$$I_f = \left[\frac{F}{P} - 0.25\right]\frac{W}{30}$$

F = 220 ft

P = 360 ft

W = 40 ft

$$= \left[\frac{220}{360} - 0.25\right]\frac{40}{30}$$

$$= [0.61 - 0.25]\frac{30^*}{30}$$

$$= [0.36]\,1.00$$

I_f = 0.36

*W not to exceed 30 feet

FRONTAGE INCREASE

506.3 Automatic sprinkler system increase. Because of the excellent record of automatic sprinkler systems for the early detection and suppression of fires, the IBC allows quite large floor area increases where an automatic fire sprinkler system is installed throughout the building. In this case, the maximum allowable area of a one-story building based on Table 503 may be increased by an additional 300 percent, and for a building of two or more stories in height, the tablular allowable area may be increased by an additional 200 percent. This restriction of permitting a smaller increase in area for multistory buildings protected by an automatic fire sprinkler system is based on the assumption by the code that the fire department suppression activities are still going to be required even where an automatic fire sprinkler system is installed. Therefore, a multistory building presents more problems to the fire department than a one-story building, and a smaller increase in area is permitted. These increases for sprinkler protection, like any for increased frontage, are to be added to the tabular area found in Table 503. Examples of the automatic sprinkler system increase are illustrated in Application Example 506-6.

The IBC permits area increases for buildings protected throughout with an NFPA 13 sprinkler system, but not those buildings protected with an NFPA 13R system. In addition, the increase is not applicable to Group H-1, H-2 or H-3. The code considers that the conditions requiring the installation of sprinkler systems in these three high-hazard occupancies are such that the sprinkler system should not also be used to increase the allowable area. In a mixed occupancy building, this restriction is only applicable to the Group H-2 and H-3 portions. Other occupancies within the building are permitted to utilize the automatic sprinkler system increase. Similar to the limitations on allowable height increases for sprinklered buildings, an area increase is not permitted where the sprinkler system is used for the substitution of 1-hour construction as permitted by Note d of Table 601.

Application Example 506-4

GIVEN: A building fronted by 60-foot street and three yards as shown

DETERMINE: The quantity W to be used in the calculation of I_f (area increased due to frontage)

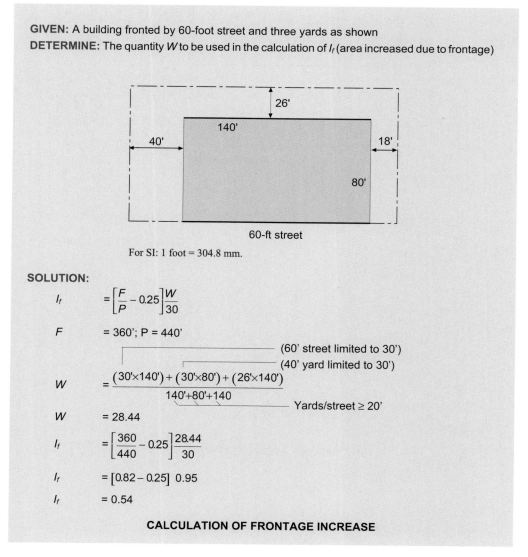

60-ft street

For SI: 1 foot = 304.8 mm.

SOLUTION:

$$I_f = \left[\frac{F}{P} - 0.25\right]\frac{W}{30}$$

F = 360'; P = 440'

(60' street limited to 30')
(40' yard limited to 30')

$$W = \frac{(30' \times 140') + (30' \times 80') + (26' \times 140')}{140' + 80' + 140}$$

Yards/street ≥ 20'

W = 28.44

$$I_f = \left[\frac{360}{440} - 0.25\right]\frac{28.44}{30}$$

$$I_f = [0.82 - 0.25]\ 0.95$$

$$I_f = 0.54$$

CALCULATION OF FRONTAGE INCREASE

What other conditions affect the determination of allowable floor areas? As expected, there are myriad situations that can arise involving the determination of the allowable area for a building. It should be noted that, as illustrated in Application Example 506-7, the introduction of a fire wall in a large-area building will result in the loss of a portion of the open space at the building perimeter. Application Example 506-8 illustrates the permitted increase for an open space shared by two buildings on the same lot.

Application Example 506-5

GIVEN: Fully-sprinklered, one-story retail sales building yards as shown, 40-ft street, fully sprinklered

DETERMINE: Percentage increase for area purpose (I_f)

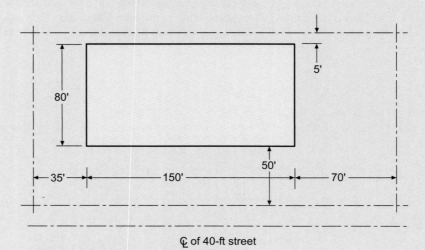

℄ of 40-ft street

For SI: 1 foot = 304.8 mm.

$$= 100 \left[\frac{F}{P} - 0.25 \right] \frac{W}{30}$$

F = 310 ft

P = 460 ft

W = 35 ft

$$I_f = \left[\frac{310}{460} - 0.25 \right] \frac{35}{30}$$

$$= [0.67 - 0.25]\, 1.17^*$$

$$= [0.42]\, 1.17$$

I_f = 0.49

Using the weighted average concept:

$$W = \frac{150(60') + 80(35') + 80(60')}{310}$$

W = 53.22

I_f = [0.42] 1.77*

I_f = 0.74

*Cannot exceed 2.0 per Section 506.2.1

ALLOWABLE AREA DETERMINATION

- Area increase for sprinklered single-story building to be 300% of area in Table 503 (I_s = 3.0)
- Area increase for sprinklered multistory building to be 200% of area in Table 503 (I_s = 2.0)
- Area increase permitted with height increase

EXAMPLE

GIVEN: Group B occupancy single-story
            Type VB construction
            No open yards available
Find: Total allowable area
            Basic allowable area = 9,000 sq ft (Table 503)
            Sprinkler increase (I_s) = 27,000 sq ft (3.0 × 9,000)
            Total allowable area = 36,000 sq ft
GIVEN: Same situation, however two stories in height
Find: Total allowable area
            Basic allowable area = 9,000 sq ft (Table 503)
            Sprinkler increase (I_s) = 18,000 sq ft (2.0 × 9,000)
            Total allowable area = 27,000 sq ft/floor

AREA INCREASE FOR SPRINKLERS

For SI: 1 square foot = 0.093 m^2.

Single occupancy buildings with more than one story. Where a single-occupancy building includes more than one story above grade plane, each story must be analyzed for allowable area purposes. In addition, calculations may need to be performed in order to evaluate if the building is within the total allowable building area. The aggregate area of all stories in the building, other than the basement, must not exceed the total allowable building area.

506.4

The code intends that a basement that does not exceed the allowable floor area permitted for a one-story building need not be included in determining the total allowable area of the building. This provision is a holdover from many years back when basements were most commonly used for service of the building. Today, it is not uncommon to find basements occupied for the same uses as the upper floors; consequently, the office building with three stories above grade plane and a basement as illustrated in Figure 506-4 is permitted by the IBC to have an area of four times that allowed for a one-story building of like occupancy and construction type. There apparently has been no adverse experience for cases of this type (most likely because of the fire-sprinkler protection required in many basements), and therefore the code provision appears to be satisfactory. The code does not address how to handle a basement that exceeds the area permitted for a one-story building. However, as the

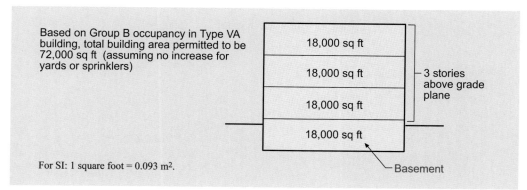

Based on Group B occupancy in Type VA building, total building area permitted to be 72,000 sq ft (assuming no increase for yards or sprinklers)

18,000 sq ft
18,000 sq ft — 3 stories above grade plane
18,000 sq ft
18,000 sq ft
Basement

For SI: 1 square foot = 0.093 m2.

**Figure 506-4
Allowable
area for
basement**

code does not permit any story to exceed that permitted for a one-story building, it would seem logical that a basement should also be limited to the same area. If anything, a fire in a basement is more difficult to fight than one in an above-ground story. Thus, it does not seem reasonable or appropriate to permit a single basement with an area exceeding that allowed for a one-story building. This section also does not detail the method of regulating allowable areas where multiple basements are present. However, the provisions of the exception to Section 506.4 indicate that only a single level of basement is exempt from inclusion in the building area calculations. Additional requirements for basements extending well below grade level may be found in Section 405 for underground buildings.

Application Example 506-7

AREA INCREASES WITH FIRE WALL

GIVEN: BUILDING A: Group S-2 Occupancy
                                      Type IIB Construction
                                      One Story

                 BUILDING B: Group B Occupancy
                                        Type IIB Construction
                                      One Story

DETERMINE: Maximum allowable floor area for Building A and Building B

SOLUTION: When the separation wall is a fire wall, each building is to be evaluated individually, and Building A would not be eligible for an area increase for open space. This is because Section 503.1 considers that each portion of a building separated by a fire wall is a separate building. As separate buildings, buildings A and B do not reflect similar conditions. Building B is bounded on the right side by a 50-foot-wide yard and at the front by a 90-foot-wide open space consisting of a 30-foot yard and 60-foot street (public way). However, Building A is bounded by open space only at the front, which provides only a 25 percent frontage. The right side is occupied by another building (Building B).

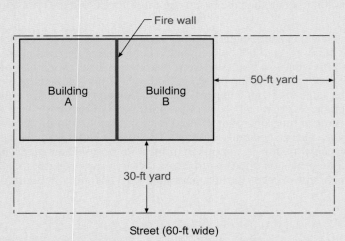

For SI: 1 foot = 304.8 mm.

Building A
    No increases for open space
    Allowable Area = 26,000 square feet
Building B
    25 percent increase (assuming exterior walls of equal length)
    Allowable Area = 23,000 square feet + 5,750 square feet = 28,750 square feet

Area determination. In addition to the floor area limits placed on each story of a building, the entire building is also limited in size. For multistory buildings other than those consisting of only two stories, the IBC permits the total combined floor area to be three times the total allowable area permitted per floor. Two-story structures are limited in size to a total combined floor area of twice that permitted per floor. Examples of these limitations are shown in Figure 506-5.   **506.4.1**

The general limitation on allowable building area is not applicable to unlimited area buildings in compliance with Section 507. In addition, four-story residential occupancies provided with an NFPA 13R automatic sprinkler system are permitted additional allowable area beyond that permitted under the general provisions. As the installation of a 13R sprinkler system in a residential occupancy does not provide for an increase in allowable area for sprinklered buildings, and the use of an NFPA 13R system is limited to buildings no more than four stories in height, it is considered appropriate to permit the maximum allowable area per story for each of the stories in the residential building. This would include four-story structures as depicted in Figure 506-6.

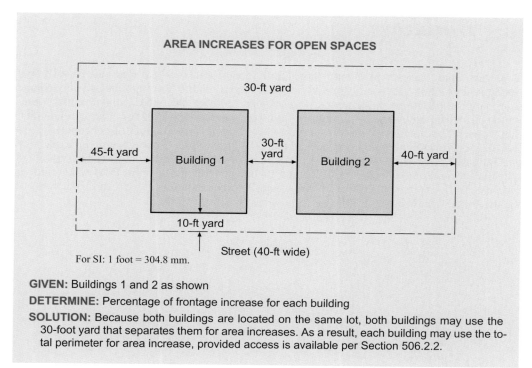

AREA INCREASES FOR OPEN SPACES

For SI: 1 foot = 304.8 mm.

Application Example 506-8

GIVEN: Buildings 1 and 2 as shown

DETERMINE: Percentage of frontage increase for each building

SOLUTION: Because both buildings are located on the same lot, both buildings may use the 30-foot yard that separates them for area increases. As a result, each building may use the total perimeter for area increase, provided access is available per Section 506.2.2.

Mixed occupancies. For allowable area purposes, the introduction of multiple occupancies into the building requires an additional level of scrutiny. The methods for determining allowable area compliance in mixed-occupancy buildings differ based upon the manner in which the various occupancies are addressed. Where the accessory occupancy method is utilized, the allowable area of the main occupancy will govern. Where occupancies are regulated under the nonseparated occupancy method, the allowable area is limited to that of the most restrictive occupancy involved. In the case of separated occupancies, the unity formula must be used to determine if the allowable area is exceeded. Further discussion can be found in the narratives of Section 508 for each of the three mixed-occupancy methodologies.   **506.5**

Where a single-story mixed-occupancy building is being evaluated for allowable area compliance, the process is addressed in Section 508.1. In similar fashion, a mixed-occupancy building with more than one story above grade plane must be analyzed on a story-by-story basis and each individual story must be shown to be in compliance with the applicable provisions of Section 508.1. In all cases, regardless of the building's number of stories, each individual story must comply for allowable area purposes. For those buildings that are two and three stories in height, if each such story complies with the allowable area limitations, then the entire building complies. However, the approach to buildings four or more stories in height above grade plane differs from that used for two-story and three-story conditions. Again, each story in the building must comply individually. In addition, the aggregate sum of the actual areas of each story divided by the allowable areas of such stories cannot exceed three. The procedure is illustrated in Application Example 506-9.

In all cases for mixed-occupancy conditions, reference is made to the provisions of Section 508.1 which identify the three options for mixed-occupancy determination (accessory occupancies, nonseparated occupancies and separated occupancies). Since each story is to be evaluated independently in a multistory condition, it is possible for the designer to utilize different mixed-occupancy options on various stories within the building.

Section 507 *Unlimited Area Buildings*

There are many cases where very large undivided floor areas are required for efficient operation in such facilities as warehouses and industrial plants. Through the use of adequate safeguards, the IBC recognizes this necessity and allows unlimited areas for these uses under various circumstances. Large open floor space is also desirable for other applications; therefore, such allowances are also permitted for business and mercantile occupancies, as well as specific assembly and educational uses. The use of this section is typically intended to eliminate fire-resistive construction of the building that would be mandated based on the area limitations of Section 503. Contrary to the general philosophy that as a building increases in floor area the allowable types of construction become more restrictive, many of the unlimited area uses permit the use of any construction type.

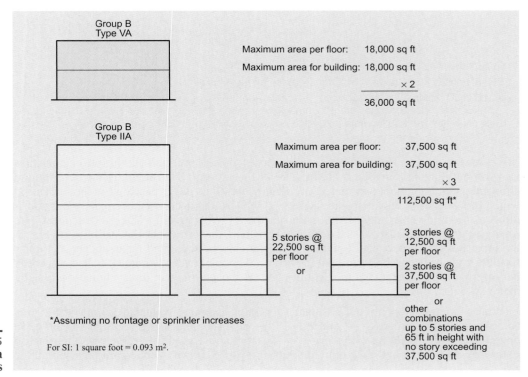

**Figure 506-5
Building area
limitations**

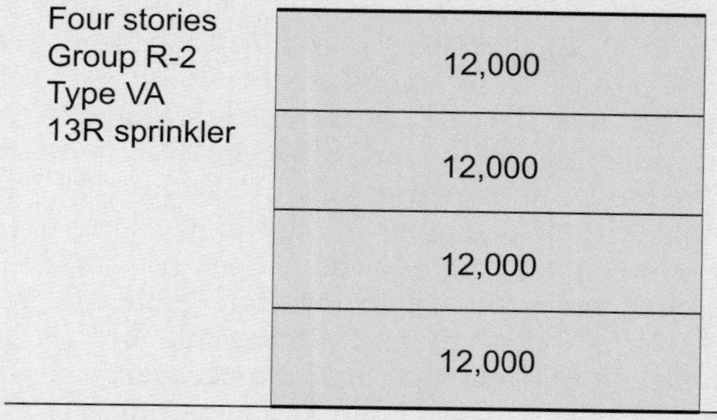

Height:
3 stories (Table 503)
+1 story increase (Section 504.2)
4 stories permitted*

Four stories
Group R-2
Type VA
13R sprinkler

| 12,000 |
| 12,000 |
| 12,000 |
| 12,000 |

Area: (Assume no frontage increase)
12,000 sq ft (Table 503)
× 4 stories (Section 506.4, Ex. 2)
48,000 sq ft permitted for total building (no story to exceed 12,000 sq ft

For SI: 1 foot = 304.8 mm, 1 square foot = 0.093 m².

*Maximum height of 60 feet (Section 504.2)

Figure 506-6
Group R area determination

Application Example 506-9

GIVEN: A fully-sprinklered four-story Type IIA Hotel, containing a Group A-2 restaurant, Group A-3 meeting rooms, and Group M retail stores. The floor areas of each occupancy are as shown. Inadequate frontage provides for no area increase.

DETERMINE: If the building complies with the allowable area provisions of Chapter 5 if the occupancies are separated under the provisions of Section 508.4 (separated occupancies).

A-2 8,000 sq ft	R-1 38,000 sq ft	
	R-1 46,000 sq ft	
	R-1 46,000 sq ft	
A-3 24,000 sq ft	R-1 8,000 sq ft	M 14,000 sq ft

SOLUTION FOR TOTAL BUILDING AREA:
Allowable area per occupancy
A-2:   15,500 + 15,500 (200%) = 46,500
A-3:   15,500 + 15,500 (200%) = 46,500
M: 21,500 + 21,500 (200%) = 64,500
R-1:   24,000 + 24,000 (200%) = 72,000
Sum of ratios calculation per story

1st story $\frac{24,000}{46,500} + \frac{8,000}{72,000} + \frac{14,000}{64,500} = 0.52 + 0.11 + 0.22 = 0.85 \leq 1.0$

2nd story $\frac{46,000}{72,000} = 0.64 \leq 1.0$

3rd story $\frac{46,000}{72,000} = 0.64 \leq 1.0$

4th story $\frac{8000}{46,500} + \frac{38,000}{72,000} = 0.17 + 0.53 = 0.70 \leq 1.0$

Each story complies
Sum of ratios calculation for building
0.85 + 0.64 + 0.64 + 0.70 = 2.83 ≤ 3.0
Entire building also complies

ALLOWABLE AREA OF MULTISTORY MIXED-OCCUPANCY BUILDING

For SI: 1 square foot = 0.093 m².

Historically, structures constructed under the provisions for unlimited area buildings have performed quite well in regard to fire and life safety. A number of occupancy groups, particularly those relating to institutional, residential and high-hazard occupancies, are excluded from the benefits derived from the provisions for unlimited area buildings. Such occupancies pose unacceptable risks that are due to their unique characteristics. As a general rule, only those occupancy classifications specifically identified in this section are permitted to be housed in buildings allowed to be unlimited in area by Section 507. For example, Group I occupancies are not specifically permitted by any of the provisions addressing unlimited area buildings. Therefore, it would appear no amount of Group I is permitted in such structures. There is, however, one method that allows for a limited degree of such prohibited occupancies. Section 508.2.3 indicates that the allowable area for an accessory occupancy is to be based upon the allowable area of the main occupancy. If the main occupancy is permitted by Section 507 to be in an unlimited area building, the accessory occupancy should also enjoy the same benefit. Figure 507-1 illustrates this condition.

It is also not uncommon for two or more occupancies regulated under the provisions of Section 507 to be located within the same building. For example, assume a one-story building contains a Group M furniture store and its associated S-1 warehouse. Because both Group M and S-1 occupancies are permitted in an unlimited area building complying with Section 507.3, both occupancies are permitted to be located in the same unlimited area building. See Figure 507-2. In addition, no fire-resistive separation is required between the two occupancies based upon the allowances for separated occupancies in Table 508.4.

507.2 **Nonsprinklered, one-story.** This section addresses a Group F-2 or a S-2 occupancy in a one-story building of any type of construction. Both Group F-2 and S-2 occupancies by definition are low-hazard manufacturing or storage uses, which the code considers to be low fire risks. Fire risk is further reduced by requiring that the building be surrounded by yards or streets with a minimum width of 60 feet (18 288 mm). The relatively low fire loading expected in such occupancies is why the code does not require the installation of an automatic fire sprinkler system for this application. The use of this provision is applicable to buildings of all construction types; however, it is anticipated that the structure mirror the contents in the absence of combustible elements.

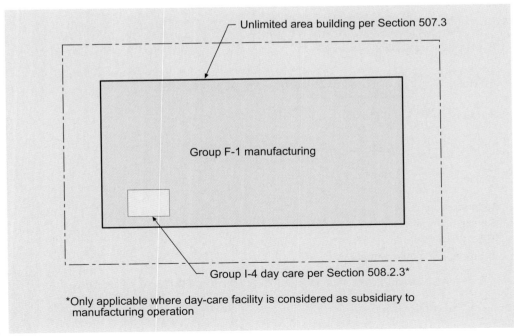

Figure 507-1
Accessory occupancy in an unlimited area building

Unlimited area building per Section 507.3

Group F-1 manufacturing

Group I-4 day care per Section 508.2.3*

*Only applicable where day-care facility is considered as subsidiary to manufacturing operation

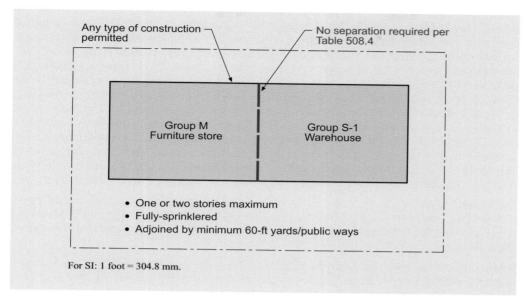

Any type of construction permitted

No separation required per Table 508.4

Group M
Furniture store

Group S-1
Warehouse

- One or two stories maximum
- Fully-sprinklered
- Adjoined by minimum 60-ft yards/public ways

For SI: 1 foot = 304.8 mm.

Figure 507-2
**Mixed-occupancy
unlimited area
buildings**

Sprinklered, one-story. Specific moderate-hazard occupancies, limited to Groups B, F, M and S, are permitted in single-story buildings of unlimited area where the building is completely surrounded by streets or yards not less than 60 feet (18 288 mm) in width and the entire structure is protected by an automatic fire sprinkler system. Limited areas of Groups H-2, H-3 and H-4 may be located in such buildings classified as Group F or S, when located and constructed under the limitations found in Section 507.8. Additionally, the limitation of one story does not apply where the building is of Type I or II construction, is utilized for rack storage, and is not intended for public access. This unlimited area storage facility, required to conform with Chapter 23 of the IFC, is permitted to be of any height. **507.3**

In most applications, the use of the unlimited area provisions simply means that the type of construction is not regulated, regardless of the size of the building's floor area. The code assumes that the amount of combustibles and, consequently, the potential fire severity is relatively moderate. In addition, the protection provided by the automatic fire sprinkler system plus the fire-department access furnished by the 60-foot (18 288 mm) yards or streets surrounding the building reduce the potential fire severity to such a level that unlimited area is reasonable.

One-story Group A-4 occupancies are also permitted to be of unlimited area where a sprinkler system is provided throughout and a minimum 60-foot (18 288 mm) open space surrounds the building. However, because of the increased risk posed by the anticipated high number and concentration of occupants in such a structure, the construction type of the building is limited to Type I, II, III or IV. It is reasonable to assume that Type V construction is inappropriate for this type of assembly use where the benefits of unlimited floor area are provided. The automatic sprinkler system required in an unlimited area building housing a Group A-4 occupancy may be omitted in those specific areas occupied by indoor participant sports, including tennis, skating, swimming and equestrian activities. Such an omission mandates that exit doors from the participant sports areas lead directly to the outside, and the installation of a fire-alarm system with manual fire-alarm boxes is required.

Group A-1 and A-2 occupancies are also permitted occupancies when in compliance with the general limitations of Section 507.3 plus four additional criteria as established in Section 507.3.1. The building must be classified as Type I, II, III or IV construction. In addition, the Group A assembly occupancies must be separated with fire barriers from other occupancies within the building, in accordance with the separated occupancy provisions of Section 508.4.4. For example, in an unlimited area retail sales building, a Group A-2 restaurant would be required to be separated from the Group M sales area by a minimum 2-hour fire-resistance-rated fire barrier. No reduction in the minimum required fire-resistance rating of the fire barrier is permitted for the presence of an automatic sprinkler system. Each individual Group A tenant would also be limited in area by the provisions of Section 503.1, which would include any applicable increases to Table 503 permitted by Sections 506.2 and 506.3. No additional height increases would apply to the Group A occupancies because of the specific limitation of one story for the entire building. A fourth requirement mandates that all required means of egress from the assembly spaces exit directly to the exterior of the building. See Figure 507-3. Application of this exception does not require the assembly use to be accessory to the major use of the building, nor is the assembly floor area limited to 10 percent of the floor area. As an additional note, it would seem appropriate that the provision also extend to Group A-3 occupancies, as such uses are typically considered equal or lesser in hazard level to the Group A-1 and A-2 classifications.

507.4 Two-story. In Groups B, F, M and S, the unlimited-area provisions also apply to structures that are two stories in height. Minimum 60-foot (18 288 mm) open space or public ways must surround the building, and an automatic fire sprinkler system is required throughout the structure.

507.5 Reduced open space. There may be situations where the full 60 feet (18 288 mm) of open space or public ways surrounding an unlimited area building cannot be obtained or is undesirable. The IBC permits a reduction in the required open space under very specific conditions as illustrated in Figure 507-4. In no case may the permanent open space be reduced to less than 40 feet (12 192 mm) in width. By limiting the amount of reduced open space, requiring a high degree of exterior wall fire resistance, and mandating opening protectives in all openings in the exterior wall facing the reduced open space, the code provides protection equivalent to full 60-foot (18 288 mm) yards or public ways. A final point: The permitted reduction in open space is only applicable to specific portions of Section 507. The reduced open space is prohibited when applying the provisions of Sections 507.7, 507.8, 507.9 and 507.10.

507.6 Group A-3 buildings of Type II construction. Although most assembly occupancies are viewed as relatively high-hazard because of the concerns associated with the number and concentration of the occupants, certain types of uses classified as Group A-3 occupancies

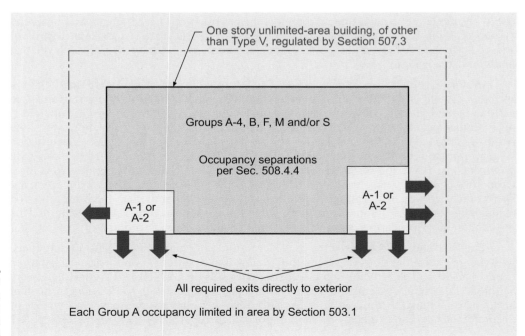

Figure 507-3
Group A occupancies in unlimited-area buildings

are considered moderate-hazard. Therefore, it is possible to utilize the unlimited area provisions for specific Group A-3 uses where the specified criteria are met.

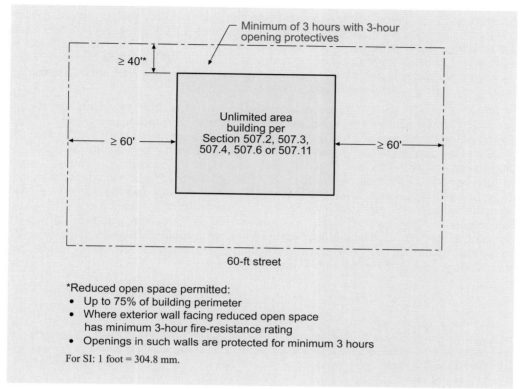

Minimum of 3 hours with 3-hour opening protectives

≥ 40'*

≥ 60'

Unlimited area building per Section 507.2, 507.3, 507.4, 507.6 or 507.11

≥ 60'

60-ft street

*Reduced open space permitted:
- Up to 75% of building perimeter
- Where exterior wall facing reduced open space has minimum 3-hour fire-resistance rating
- Openings in such walls are protected for minimum 3 hours

For SI: 1 foot = 304.8 mm.

Figure 507-4
Reduced open space

Such buildings allowed to be of unlimited area are limited to one story in height, must be of Type II noncombustible construction and contain only those specific types of assembly uses listed in the code. By limiting the types of buildings to places of religious worship, gymnasiums (without spectator seating), lecture halls and similar uses, it is anticipated that the fire loading is relatively low. Buildings such as libraries, museums and similar uses pose a higher risk that is due to the large amount of combustibles expected to be present. The potential for combustible loading is further reduced by the prohibition of a stage as a part of the use, although a platform is acceptable. Installation of an automatic sprinkler system is mandated, as is the presence of a minimum 60-foot (18 288 mm) open space around the building.

Group A-3 buildings of Type III and IV construction. Buildings of Type III or IV construction are also granted unlimited area status when housing specified Group A-4 occupancies as also identified in Section 507.6. The requirements for sprinkler protection and adequate open space are also applicable. In addition, the assembly use must be located relatively close to the exterior ground level to expedite the exiting process. As a part of this requirement, any elevation change from the building to the grade level must be accomplished by ramps rather than stairs. This further provides for an efficient means of egress from the assembly building. The additional limitation regarding floor level location is mandated due to the combustible nature of the building's construction. **507.7**

Group H occupancies. Because many large industrial operations, both manufacturing and warehousing, have a need to utilize a limited quantity of high-hazard materials in some manner, it is necessary that Group H-2, H-3 and H-4 occupancies be permitted to a small degree in Group F and S unlimited area buildings. Because of the allowances given to **507.8**

buildings of unlimited area, it is critical that the high-hazard occupancies be strictly limited in floor area and adequately separated from the remainder of the building.

There are two factors that limit the allowable floor area of the permitted Group H occupancies; the building's type of construction and the location of the high-hazard uses in relationship to the building. Where high-hazard occupancies are located on the perimeter of the building, fire-department access is enhanced, and exposure from interior areas is reduced. Accordingly, the permitted floor area of the Group H occupancies located on the building's perimeter is considerably greater than that allowed for such high-hazard uses completely surrounded by the unlimited-area building.

Where Group H-2, H-3 and H-4 occupancies are located at the perimeter of the unlimited-area Group F or S building, their size is restricted by the area limitations of Table 503 as modified by Section 506.2. However, in no case may such floor area exceed 10 percent of the area of the entire building. In a condition where the high-hazard occupancy is totally enclosed by the unlimited area building, the size of the occupancy is limited to only 25 percent of the area limitations specified in Table 503. Both of these conditions are shown in Application Example 507-1. The example also illustrates that multiple Group H occupancies that are not located at the perimeter of the building are limited in size based on the aggregate floor area of such occupancies. Similarly, where multiple Group H occupancies do occur on the building's perimeter, the maximum permitted size is also based on the total of all such occupancies. In all situations, the appropriate fire-barrier assemblies mandated by Table 508.4 must be provided.

Application Example 507-1

GIVEN: A 130,000 square-foot Group F-1 of Type IIB construction having unlimited area under the provisions of Section 507.3. One H-3 storage room is located on the building's perimeter. Multiple H-3 storage rooms are located such that they are not located along an exterior wall.

DETERMINE: The maximum allowable floor areas for the H-3 storage rooms.

Maximum aggregate area of interior rooms "A" and "B" 14,000 sq ft (Table 503) x 25% = **3,500** sq ft maximum

Maximum allowable floor area of perimeter room "C"

14,000 sq ft (Table 503)	10% × 130,000 sq ft
+ 3,500 sq ft (Section 506.2)	= 13,000 sq ft maximum
17,500 sq ft maximum	

or

Therefore, limited to **13,000** sq ft

UNLIMITED AREA GROUP F OR S BUILDING WITH GROUP H OCCUPANCIES

For SI: 1 square foot = 0.093 m^2.

Aircraft paint hangar. The provisions of Section 412.6 address aircraft painting operations **507.9** where the amount of flammable liquids in use exceeds those maximum allowable quantities listed in Table 307.1(1). Classified as Group H-2, aircraft paint hangars must be fully suppressed and of noncombustible construction. Such one-story hangars may be unlimited in floor area where complying with Section 412.6, provided they are surrounded by public ways or yards having a width of at least one and one-half times the height of the building.

Group E buildings. Because of the various fire- and life-safety concerns associated with **507.10** educational occupancies, buildings housing uses classified as Group E are typically not eligible for consideration as unlimited-area buildings. Only when the following six criteria are met does the IBC permit the area of a Group E educational building to be unlimited:

1. The building is limited to one story in height.

2. The building is of Type II, IIIA or IV construction.

3. Two or more means of egress are provided from each classroom.

4. At least one means of egress from each classroom is a direct exit to the exterior of the building.

5. An automatic sprinkler system is provided throughout the building.

6. The building is surrounded by open space at least 60 feet (18 288 mm) in width.

Motion-picture theaters. Because of their limited combustible loading, motion-picture **507.11** theaters are granted unlimited floor areas in a manner relatively consistent with other moderate-hazard uses. This specific allowance is not extended to the other uses classified as Group A-1, such as performance theaters, because of their higher fire-severity potential.

In order to address the concerns related to the high-density, high-volume occupant loads often encountered in motion-picture theaters, unlimited area is only permitted where the building is of Type II noncombustible construction. This restriction further limits the fire load contained within the building construction. In concert with Section 507.3, a fire sprinkler system must be installed throughout, and minimum 60-foot (18 288 mm) open areas must completely surround the building.

The application of this provision differs from the allowance granted in Section 507.3.1. That provision permits any Group A-1 occupancy, including motion picture theaters, to be located in an unlimited area building complying with Section 507.3, provided the limitations of the exception are met. However, it does not allow Group A-1 occupancies themselves to be unlimited in area. On the other hand, this section permits a Group A-1 theater complex to be unlimited in area, provided it is fully-sprinklered, of Type II construction and is surrounded by adequate open space.

Covered mall buildings and anchor stores. The provisions of Section 402.6 for unlimited **507.12** area covered mall buildings are referenced for convenience purposes. Note that although the reduction in open space permitted by Section 507.5 is not applicable to covered mall buildings, a similar reduction is permitted by Section 402.6.1.

Section 508 *Mixed Use and Occupancy*

Multiple uses commonly occur within a single building. Each use creates its own distinct hazards, many of which are addressed by the code. However, many of the hazards are similar in nature, which allows the varied uses to be grouped into categories that recognize the common concerns. These categories are identified in Chapter 3 as occupancy groups. Where two or more occupancy groups share a single building, it is necessary to evaluate their relationship to each other as a mixed-occupancy condition. This section provides various methods to address such relationships in regard to occupancy classification, allowable height and area, and fire-resistance-rated separation.

508.1 General. A mixed-occupancy condition exists where two or more distinct occupancy groups are determined to exist within the same building. In fact, it is quite common for a building to contain more than one occupancy group. For example, hotel buildings of various sizes not only house the residential sleeping areas, but may contain administrative offices, retail and service-oriented spaces, parking garages and, in many cases, restaurants, conference rooms and other assembly areas. Each of these uses typically constitutes a distinct and separate occupancy as far as Chapter 3 of the IBC is concerned. Because this situation is not uncommon, the code specifies requirements for buildings of mixed occupancies. Under such circumstances, the designer has available several methodologies (accessory occupancies, nonseparated occupancies and separated occupancies) to address the mixed-occupancy concerns. The methods that have been established represent a hierarchy of design prerogatives that may be utilized at the discretion of the design professional. Although compliance is required with only one of the three mixed-occupancy methods, it is acceptable to utilize two or even all three methods in the same building as shown in Figure 508-1. The common format utilized in presenting the requirements for each of the methods allows for a comparison of the provisions. This should assist in determining the most appropriate method, or methods, for the building under consideration. A simple comparison of the three mixed-occupancies methods is shown in Figure 508-2.

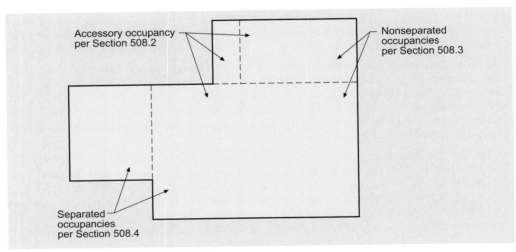

Figure 508-1
Aggregate accessory occupancies

	Accessory Occupancies Section 508.2	Nonseparated Occupancies Section 508.3	Separated Occupancies Section 508.4
Occupancy Classification	Individually classified	Individually classified	Individually classified
Allowable Area	Based on allowable area of main occupancy	Based on most restrictive of occupancies within building	Determined such that sum of the ratios cannot exceed 1.0
Allowable Height	Based on tabular values of Table 503	Based on most restrictive of occupancies within building	Based on general provisions of Section 503.1
Separation	No separation required	No separation required	Separation as required by Table 508.4
Special Conditions	1. Subsidiary to main occupancy 2. Aggregate area ≤ 10% of story 3. Aggregate area ≤ value in Table 503 4. Not applicable to Groups H-2, H-3, H-4 and H-5	1. Most restrictive provisions of Ch. 9 apply to entire building 2. Not applicable to Groups H-2, H-3, H-4 and H-5	

Figure 508-2
Summary of mixed occupancy methods

It is important to recognize that there is no relationship between the mixed occupancy provisions of Section 508.3 and the fire area concept utilized in Section 903.2 for automatic sprinkler systems. Compliance with any of the three mixed-occupancy methods does not relieve the responsibility to comply with Section 901.7 and Table 707.3.9 regarding the proper separation of fire areas. An example is shown in Figure 508-3.

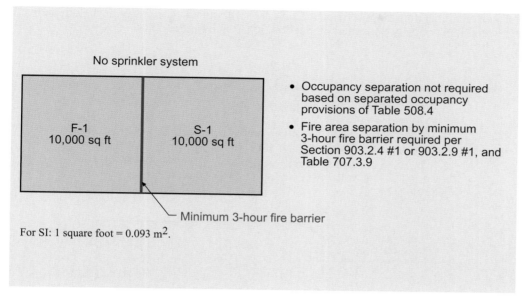

No sprinkler system

F-1
10,000 sq ft

S-1
10,000 sq ft

- Occupancy separation not required based on separated occupancy provisions of Table 508.4
- Fire area separation by minimum 3-hour fire barrier required per Section 903.2.4 #1 or 903.2.9 #1, and Table 707.3.9

Minimum 3-hour fire barrier

For SI: 1 square foot = 0.093 m².

Figure 508-3
Occupancy separation vs fire area separation

Accessory occupancies. Those minor uses in a building that are not considered consistent **508.2** with the major occupancy designation can potentially be considered accessory occupancies. They often are necessary or complimentary to the function of the building's major use, but have few characteristics of the major occupancy in regard to fire hazards and other concerns. Therefore, accessory occupancies must each be assigned to an occupancy group established in Chapter 3 based upon their own unique characteristics. While maintaining the philosophy of a mixed occupancy building, this section permits such relatively small accessory uses to be considered merely a portion of the major occupancy for fire separation purposes. A good example would be a lunchroom seating 120 persons and located in a large manufacturing facility. Whereas the individuals using the lunchroom are generally the same individuals who work elsewhere in the factory, the hazards encountered while they are occupying the lunchroom are quite different from those created in the Group F-1 manufacturing environment. Therefore, the lunchroom must be appropriately classified as a Group A-2 occupancy, creating a mixed-occupancy condition. However, through compliance with the accessory occupancy provisions of this section, the need for a fire-resistance-rated separation between the lunchroom and manufacturing area is eliminated. In fact, no physical separation of any kind would be mandated. It is important to note, however, that in spite of the absence of a fire separation, the two areas would maintain their unique occupancy classifications. They would continue to be classified as a Group A-2 and F-1, respectively, and the building would be considered a mixed occupancy.

As previously indicated, consideration as an accessory occupancy is only possible where the use is subsidiary to the main occupancy of the building. There are several additional criteria that must also be met in order to utilize the accessory occupancy method. The occupancy under consideration cannot exceed 10 percent of the floor area of any floor of the building, nor more than that permitted by Table 503 without area increases for frontage and sprinkler protection. See Figure 508-4. It is specified that the 10 percent limitation is based on the aggregate floor areas of all accessory occupancies, not individually. Multiple minor uses that cumulatively make up more than 10 percent of the total floor area could pose a hazard that the code does not anticipate. There are unique situations—such as where minor uses are adequately separated spatially or are of such different types of uses that their aggregate area is not relevant—where the regulation of minor uses as individual areas could potentially be considered. See Figure 508-5. The application of these limits is subject to the interpretation of the building official, based on conditions unique to each building under consideration. The general limitations applied to accessory occupancies are shown in Application Example 508-1. As an additional limitation, the height of any accessory occupancy cannot be located higher than that specified in Table 503 without adjustment for the height increase for sprinklers typically permitted by Section 504.2. See Figure 508-6.

Application Example 508-1

General criteria for consideration as accessory occupancy:

√   1.   **Accessory to major use.** Conference room's primary function limited to use by employees of factory.

√   2.   **Does not exceed 10% of floor area.** Conference room does not exceed 7,225 square feet, 10% of total floor area.

√   3.   **Does not exceed tabular allowable area.** Conference room does not exceed 9,500 square feet allowed by Table 503 for Group A-3 occupancy of Type IIB construction.

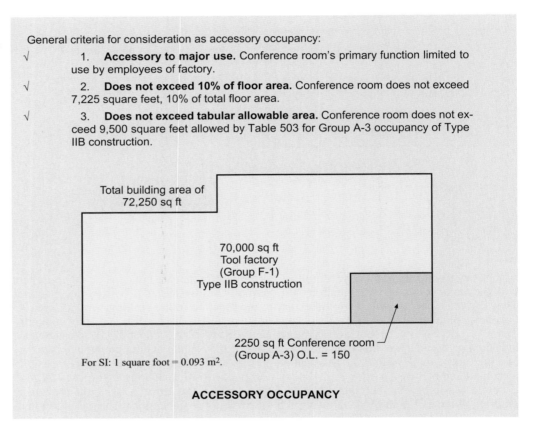

Total building area of 72,250 sq ft

70,000 sq ft
Tool factory
(Group F-1)
Type IIB construction

2250 sq ft Conference room
(Group A-3) O.L. = 150

For SI: 1 square foot = 0.093 m².

ACCESSORY OCCUPANCY

Although the primary allowance provided by the accessory occupancy provisions is the lack of a required fire separation between the accessory occupancy and the remainder of the building, there is also a potential benefit that is due to the manner in which allowable area is regulated. The code calls for the allowable area of the accessory occupancy to be based upon the main occupancy of the building. This approach typically allows for a greater allowable area than would be permitted under the conditions for both nonseparated occupancies and separated occupancies. Another important benefit provides for occupancies not normally permitted in unlimited area buildings, as regulated by Section 507, to be located in such buildings. For example, a Group A-2 lunchroom considered an accessory occupancy may be located in a two-story Group B unlimited area building complying with Section 507.4 with no separation required between the two occupancies. This allowance, along with the limitations for accessory occupancies, is shown in Application Example 508-2.

Three exceptions indicate those conditions related to accessory occupancies under which some degree of fire-resistance-rated or smoke-resistant separation is mandated. The first exception indicates that the fire separation cannot be eliminated where the accessory occupancy is classified as Group H-2, H-3, H-4 or H-5. Where such occupancies occur in a mixed occupancy building, it is necessary to apply the separated occupancy provisions of Section 508.4. Exception 2 clarifies that any separation required for those spaces designated as incidental accessory occupancies must be provided in accordance with Section 508.2.5. The third exception mandates that the separation elements addressed in Section 420 for buildings containing dwelling units and sleeping units be provided.

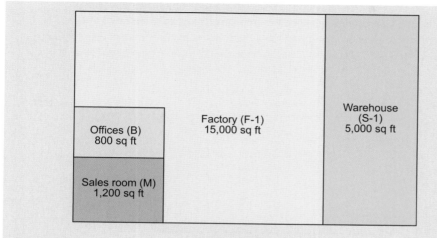

Notes:
- Building area of 22,000 sq ft
- Areas of offices and sales room, both individually and combined, are less than 10 percent of floor area (3.6% + 5.5% = 9.1%); therefore, permitted to be considered accessory occupancies
- Area of warehouse exceeds 10 percent of building floor area (23%); therefore, not permitted to be considered accessory occupancy

For SI: 1 square foot = 0.093 m².

Figure 508-4
Aggregate accessory occupancies

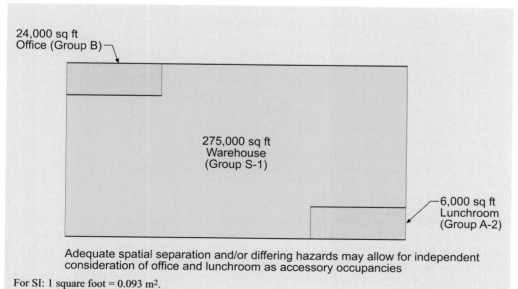

Adequate spatial separation and/or differing hazards may allow for independent consideration of office and lunchroom as accessory occupancies

For SI: 1 square foot = 0.093 m².

Figure 508-5
Individual accessory occupancies

Application Example 508-2

GIVEN: A 75,000 square-foot one-story office building housing a 3750 square foot employee lunchroom with an occupant load of 250 persons. The building is fully sprinklered, is of Type VB construction and qualifies for unlimited area under the provisions of Section 507.3.

DETERMINE: Application of the accessory occupancy provisions of Section 508.3.1 for the lunchroom.

1. Is the accessory occupancy subsidiary to the building's major occupancy?

   Yes, the lunchroom is intended to serve the employees of the office space.

2. Is the accessary occupancy no more than 10% of the floor area of the story?

   Yes, 10% of 75,000 sq ft = 7500 sq ft maximum; the lunchroom is 3750 sq ft.

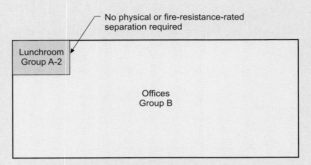

For SI: 1 square foot = 0.093 m².

3. Is the accessory occupancy no larger than the tabular values in Table 503?

   Yes, the tabular value for a Group A-2 of VB construction is 6000 sq ft; the lunchroom is 3750 sq ft.

4. What is the occupancy classification of the lunchroom?

   Group A-2, based on the individual classification of the use.

5. How are the other requirements of the IBC applied?

   The provisions for each occupancy are applied only to that specific occupancy.

6. What is the allowable height and area of the building?

   The building's allowable height and area are based on the major occupancy involved; in this case it is Group B. Based upon the criteria of Section 507.4, the building is permitted to be unlimited in area and limited to two stories in height.

7. What is the allowable height of the lunchroom?

   One story, based on Table 503 for Group A-2 in a Type VB building. The lunchroom must be located on the first story.

8. What is the minimum required separation between the lunchroom and the office?

   There is no fire-restrictive or physical separation required, owing to compliance with the provisions of Section 508.2 for accessory occupancies.

508.2.5 Separation of incidental accessory occupancies. There are times where the hazards associated with a particular use do not rise to the level of requiring a different occupancy classification; however, such hazards must still be addressed because of their impact on the remainder of the building. These uses are identified as incidental accessory occupancies, which are generally regulated independently of the mixed-occupancy provisions. Incidental accessory occupancies are uniquely addressed through the use of fire-resistance-rated separations or fire-extinguishing system protection.

What is an incidental accessory occupancy? There are occasionally one or more rooms or areas in a building that pose risks not typically addressed by the provisions for the general occupancy group under which the building is classified. However, such rooms or areas may functionally be an extension of the primary use. These types of spaces are considered in the IBC to be incidental accessory occupancies and regulated according to their hazard level. These areas are not ever intended to be considered different occupancies, creating a mixed-use condition, but rather are classified in accordance with the main occupancy of the portion of the building where the incidental accessory occupancy is located. There is no specific definition for an incidental accessory occupancy, as it is simply described as any of those rooms or areas listed in Table 508.2.5. If it is not listed, it is not considered an incidental accessory occupancy for code purposes.

The designation and regulation of incidental accessory occupancy does not apply to those areas within and serving a dwelling unit. Otherwise, the special hazards that may be found in buildings of various uses and occupancies are addressed through the construction of a fire barrier and/or horizontal assembly separating the incidental accessory occupancy from the remainder of the building, the installation of an automatic fire-extinguishing system in the incidental accessory occupancy, or, in special cases, both the fire separation and fire-extinguishing system.

Incidental accessory occupancies are listed in Table 508.2.5. Most of the rooms or areas identified in the table are regulated where located in any of the occupancy groups established by the code, other than dwelling units as previously noted. A few of the incidental accessory occupancies are to be regulated only where located within a specific occupancy or a limited number of occupancies.

It is common for many of the listed incidental accessory occupancies to be unoccupied for extended periods of time, creating the potential for a fire to grow unnoticed. Oftentimes, combustible or hazardous materials are present in such areas. Because of the potentially high fuel load and lack of constant supervision, spaces such as furnace rooms, machinery rooms, laundry rooms, and waste collection rooms are selectively considered incidental use areas. Other uses, such as paint shops, laboratories and vocational shops, may cause concern to the point where they too must be protected or separated from other areas of the building.

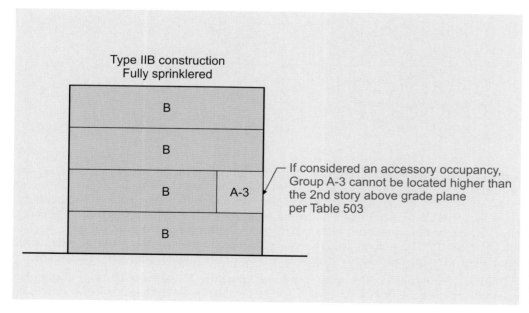

Type IIB construction
Fully sprinklered

B

B

B   A-3

If considered an accessory occupancy, Group A-3 cannot be located higher than the 2nd story above grade plane per Table 503

B

Figure 508-6
Maximum height of accessory occupancy

How are incidental accessory occupancies regulated? The fire-resistance-rated separations required by Table 508.2.5 are to be fire barriers and/or horizontal assemblies, typically having a minimum fire-resistant rating of 1 hour. For incinerator rooms, paint shops, and rooms containing fire pumps, the minimum required rating is greater. Where an automatic fire-extinguishing system provides the necessary protection, it need only be installed within the incidental accessory occupancy under consideration. Examples of the requirements are illustrated in Figure 508-7.

As shown in Figure 508-8, there is a variety of combinations regarding the separation and protection methods identified in Table 508.2.5. It should be noted that where an automatic fire-extinguishing system is utilized without a fire barrier as the protective element, the incidental accessory occupancy must still be separated from the remainder of the building. This separation need only consist of construction capable of resisting the passage of smoke. Although not required to have a fire-resistant rating, partitions must either extend to the underside of the floor or roof deck above, or to the underside of a fire-resistance-rated floor/ceiling or roof/ceiling assembly. Doors are to be self-closing or automatic-closing upon detection of smoke, with no air transfer openings or excessive undercuts. See Figure 508-9. The room must be tightly enclosed, providing for the containment of smoke while assisting in the heat increase necessary to activate the fire-extinguishing system.

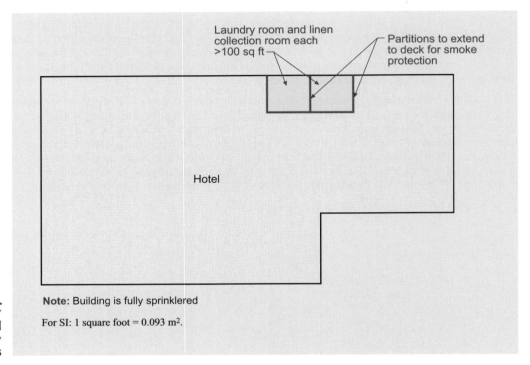

Figure 508-7
Incidental accessory occupancies

Note: Building is fully sprinklered

For SI: 1 square foot = 0.093 m².

Table 508.2.5 Incidental Accessory Occupancies. Incidental accessory occupancies are limited to those rooms or spaces listed in this table. The listed rooms have been selected for inclusion because of the increased hazard they present to the other areas of the building. However, it is recognized that the degree of hazard is such that a separate occupancy classification is not warranted. In fact, such a classification may be overly restrictive. The intent of the fire separation and fire protection requirements is to provide safeguards because of the increased hazard level presented by the incidental accessory occupancy.

Furnace rooms and boiler rooms. The hazard potential for fuel-fired heating equipment is addressed once the thresholds established in the table have been exceeded. It should be noted that the limitations are based on individual pieces of equipment, rather than the aggregate amounts from all equipment within the space. For example, the requirements of Table 508.2.5 are not applicable where there are two furnaces within the furnace room, each with an input rating of 300,000 Btu per hour (87 900 watts). Because no furnace exceeds the 400,000 Btu/hr (117 200 watts) threshold, there is no fire barrier separation or fire-extinguishing system protection required.

Room or Area[1]	1-hour Separation or Fire-extinguishing System	1-hour Separation	2-hour Separation	1-hour Separation and Fire-extinguishing System	2-hour Separation and Fire-extinguishing System
Furnace rooms	X				
Boiler rooms	X				
Refrigerant machinery rooms	X				
Laboratories	X				
Vocational shops	X				
Laundry rooms	X				
Waste and linen collection rooms	X				
Group I-3 cells		X			
Group I-2 waste and linen collection rooms		X			
Hydrogen cut-off rooms[2]		X	X		
Battery system storage areas[2]		X	X		
Fire pump rooms[3,4]			X	X	
Paint shop[3]			X	X	
Incinerator rooms					X

[1] See Table 508.2 for specifics of each use or area
[2] Varies based on type of occupancy
[3] Option of two methods
[4] Varies based on building height

Figure 508-8
Incidental use area requirements

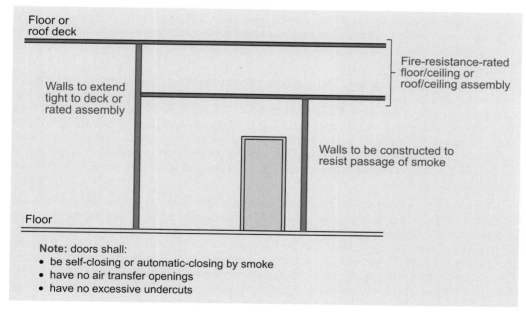

Floor or roof deck

Walls to extend tight to deck or rated assembly

Fire-resistance-rated floor/ceiling or roof/ceiling assembly

Walls to be constructed to resist passage of smoke

Floor

Note: doors shall:
• be self-closing or automatic-closing by smoke
• have no air transfer openings
• have no excessive undercuts

Figure 508-9
Incidental accessory occupancy smoke separations

The regulated furnace or boiler is anticipated to be located within a room isolated from the remainder of the building. More specifically, the furnace room or boiler room must either be separated from other portions of the building with a complying fire barrier and/or horizontal assembly, or isolated by construction capable of resisting the passage of smoke and provided with a fire-extinguishing system. In either case, it is not permissible to eliminate the equipment enclosure. Because the intent of the requirement is to address the hazards associated with specified boilers and furnaces, the prescribed degree of separation and/or protection must always be provided.

Hydrogen cut-off rooms. Hydrogen cut-off rooms are defined in Section 421.2 as rooms or spaces that are intended exclusively to house a gaseous hydrogen system. Special requirements applicable to such rooms are set forth in Section 421. Where the quantities of materials would cause a hydrogen cut-off room to be classified as a Group H occupancy, the separation requirements of Section 508.4 for separated occupancies will apply rather than those of Table 508.2.5.

The reference in the table exempting those rooms classified as Group H is not strictly limited to hydrogen cut-off rooms. In fact, where the quantities of hazardous materials in any of the rooms or areas designated by Table 508.2.5 exceed those permitted by Section 307.1 causing classification as a Group H occupancy, the use of the table is not appropriate. In a mixed-occupancy building, all spaces with a Group H classification must be separated from the remainder of the building in accordance with Section 508.4 for separated occupancies.

Paint shops. The provisions of Section 416 control the construction, installation and use of rooms for spraying paints, varnishes or other flammable materials used for painting, varnishing, staining or similar purposes. The paint shops regulated by Table 508.2.5 are those rooms where the same types of spraying operations occur. As a general rule, a minimum 1-hour fire-resistive enclosure is mandated by Section 416.2 to isolate the spraying operations from the remainder of the building. However, in all but Group F occupancies, Table 508.2.5 mandates a higher degree of protection.

Laboratories and vocational shops. In educational buildings, particularly secondary schools, it is common to find multiple laboratories and vocational shops that are an extension of the educational function. Because of the presence of some quantities of hazardous materials in such laboratories, as well as in those labs associated with Group I-2 occupancies, the code mandates some degree of separation and/or protection. Where the quantities of hazardous materials warrant a Group H classification, the use of this table is inappropriate. Vocational shops in schools also pose a hazard to the remainder of the building that is due to the hazardous processes and combustible materials involved. Where no such hazards exist, such as in a computer lab or design lab, the table is not intended to apply.

508.3 Nonseparated occupancies. This section presents another of the available methods addressing the relationship between different occupancies in a mixed-occupancy building. Under the specific conditions of this methodology, fire-resistance-rated separations are not mandated between adjacent occupancies. The fundamental concept behind this provision assumes that if the building is designed in part to address the most restrictive and most hazardous conditions that are expected to occur based upon the occupancies contained in the building, a fire-resistance-rated separation is not needed. In fact, no physical separation of any type is required.

Utilizing the nonseparated-occupancy method, the building must be individually classified for each unique occupancy that exists. The height and area limitations for those occupancies will be used to determine the required type of construction for each occupancy, with the most restrictive type of construction required for the entire building. In addition, the most restrictive fire-protection system requirements (automatic sprinkler systems and fire alarm systems) that apply to an occupancy in the building shall apply to the entire structure. For the application of other code provisions, each individual occupancy will be regulated by only the specific requirements related to that occupancy. A special condition mandates that if a high-rise building is regulated under the nonseparated-occupancy provisions, even those portions of the building that are not considered high-rise must comply with the high-rise requirements. See Figure 508-10. Another important limitation is that the use of the nonseparated occupancy method is not permitted for Group H-2, H-3, H-4 and H-5 occupancies.

The approach to understanding the rationale for the nonseparated-occupancy method may be better understood by viewing the structure as multiple single-use facilities. In evaluating the building described in Application Example 508-3, assume that the building is entirely a Group B occupancy. Based upon that assumption, determine its maximum allowable height and area, as well as any required fire protection features. Now evaluate the building as if it were entirely a Group E occupancy, again addressing the maximum allowable height and area, along with the requirements for fire protection systems. By applying the most restrictive height, area and fire protection provisions of both Group B and E occupancies to the entire building, the conditions for nonseparated occupancies can be determined.

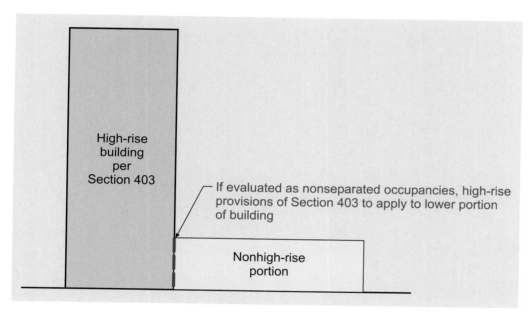

Figure 508-10
Nonseparated occupancies in a high-rise building

Separated occupancies. Using this method of addressing mixed-occupancy buildings, the code directs that each portion of the building housing a separate occupancy be individually classified and comply with the requirements for that specific occupancy. Furthermore, the code intends that each pair of occupancies be evaluated through Table 508.4 as to the relationship of the hazards involved, often mandating a fire-resistance-rated separation between them. Measured from the grade plane, the allowable height is limited based on the type of construction of the building as shown in Table 503. See Figure 508-11.

508.4

For allowable floor area considerations in a building having multiple occupancies, the code uses a formula that is very similar to the interaction formula used in structural engineering where two different types of stress are imposed on a member at the same time.

In the case of a mixed-occupancy building, the code uses this type of formula for the calculation of the allowable building area for each floor. For example, if there are three different occupancies in a building, the formula is as follows:

$a1/A1 + a2/A2 + a3/A3 \leq 1.0$

WHERE: $a1$, $a2$ and $a3$ represent the actual areas for the three separate occupancies, and $A1$, $A2$ and $A3$ represent the allowable areas for the three separate occupancies

See Application Examples 508-4 and 508-5 for examples of this computation.

This formula essentially prorates the areas of the various occupancies so that the sum of percentages must not exceed 100 percent.

It is also appropriate to utilize a variation of this formula for the determination of the allowable area for the total building as evidenced by the provisions of Section 506.5.2. See Application Example 506-9. For all practical purposes, the need to evaluate the building as a whole is only necessary in buildings of four or more stories above grade plane. For two- and three-story buildings, if each floor is compliant for allowable area purposes, the entire building will always comply.

Application Example 508-3

GIVEN: A mixed-occupancy building of Type VB construction, housing both Group B and Group E occupancies. Assume no allowable height or area increases are available.

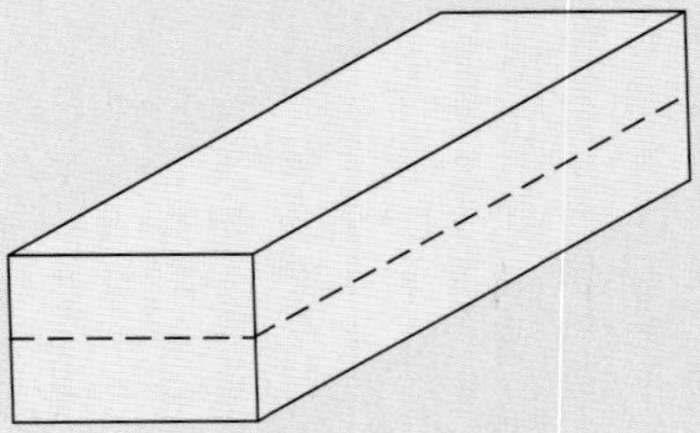

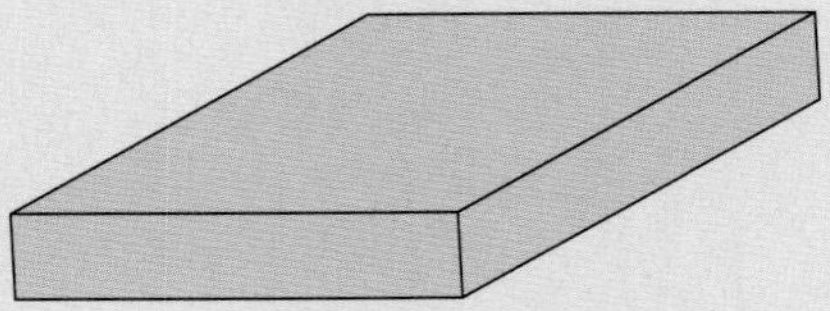

Group B

- 9,000 sq ft/story max
- 2-story max
- No fire protection systems required

Group E

- 9,500 sq ft/story max
- 1-story max
- Manual fire alarm system required

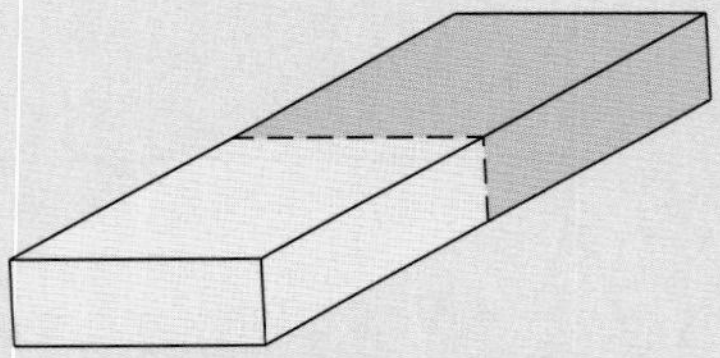

Group B/E as nonseparated occupancy

- 9,000 sq ft/story max
- 1-story max
- Manual fire alarm system required throughout building

For SI: 1 square foot = 0.093 m².

DETERMINE: The limitations that apply if the occupancies are to be considered nonseparated. As the more restrictive requirements for each occupancy must apply to the entire building, the following limitations are imposed in order to eliminate any form of occupancy separation:

SOLUTION:

Maximum allowable area:	9,000 sq ft based on Group B
Maximum allowable height:	1 story based on Group E
Fire protection features:	Manual fire alarm system required throughout building, based on Group E

NONSEPARATED OCCUPANCIES

Building:
- Fully-sprinklered per NFPA 13
- Type VA construction
- Contains Group A-3 and B occupancies

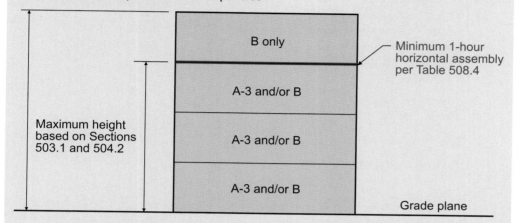

Minimum 1-hour horizontal assembly per Table 508.4

B only

A-3 and/or B

A-3 and/or B

A-3 and/or B

Maximum height based on Sections 503.1 and 504.2

Grade plane

Figure 508-11
Maximum height of separated occupancies

Application Example 508-4

GIVEN: A one-story building housing day care classified as Group E, offices and a conference center. The building is of Type V-A construction. No yards or sprinklers are available for area increase purposes. Floor areas are as follows:

Office (B)	4,500 square feet
Assembly (A-3)	3,000 square feet
E	6,000 square feet

DETERMINE: If the building area is within the allowable area.

SOLUTION: In accordance with Section 508.3.3.2:

$$\frac{\text{Actual area of office}}{\text{Allowable area of office}} + \frac{\text{Actual area of assembly}}{\text{Allowable area of assembly}} + \frac{\text{Actual area of E}}{\text{Allowable area of E}} \leq 1$$

$$\frac{4,500}{18,000} + \frac{3,000}{11,500} + \frac{6,000}{18,500} \overset{?}{\leq} 1$$

$$.25 + .26 + .32 \overset{?}{\leq} 1$$

$$.83 < 1$$

For SI: 1 square foot = 0.093 m². Building is within the allowable area.

DETERMINING ALLOWABLE AREAS FOR A MIXED-OCCUPANCY BUILDING

Application Example 508-5

GIVEN: A one-story manufacturing building (Group F-1) containing a Group H-3 storage room. Frontage is available on only two sides allowing for a 25% allowable area increase. The building is of Type IIIB construction and fully sprinklered. The manufacturing area is 44,000 square feet, and the Group H-3 storage room is 2,200 square feet in floor area.

DETERMINE: If the building area is within the allowable area.

SOLUTION: $\dfrac{\text{Actual F}-1}{\text{Allowable F}-1} + \dfrac{\text{Actual H}-3}{\text{Allowable H}-3} \leq 1.0$

F-1 =	12,000	Tabular area		H-3 =	13,000	Table 503
	+3,000	Frontage increase			+3,250	Section 506.2
	+36,000	Sprinkler increase			+ 0	Section 506.3 Exc. 2
	51,000	Total allowable area			16,250	

$$= \frac{44,000}{51,000} + \frac{2,200}{16,250}$$

$$= 0.863 + 0.135 = 0.998 \leq 1.0$$

∴ Building is within allowable area

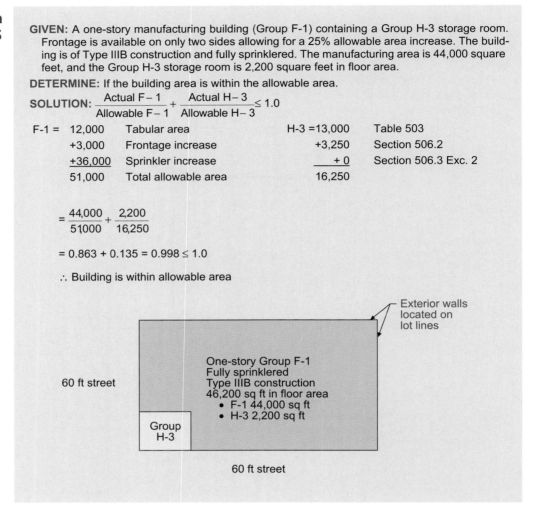

Exterior walls located on lot lines

60 ft street

One-story Group F-1
Fully sprinklered
Type IIIB construction
46,200 sq ft in floor area
• F-1 44,000 sq ft
• H-3 2,200 sq ft

Group H-3

60 ft street

Table 508.4 Required Separation of Occupancies. The code has established several alternative methods for addressing mixed-occupancy buildings regarding fire separations between the various occupancies involved. Both Sections 508.2.4 and 508.3.3 for accessory occupancies and nonseparated occupancies, respectively, do not require any fire-resistance-rated separation between occupancies. However, Table 508.4, establishing fire separations under the separated occupancy method of Section 508.4, indicates varying degrees of separation. Fire-resistive separations of 1, 2, 3 and 4 hours are selectively required based on the occupancies involved. In some cases, however, no fire separation is mandated. The intent of the table is to provide for relative separation requirements based primarily on dissimilar risk. The fire-resistance ratings, including the lack of such required ratings in many circumstances, appropriately recognize the degree of dissimilarity between the various occupancies.

In a general sense, the following logic was utilized in formulating the table. High-hazard (Group H) occupancies are required to be separated from each other and from all other occupancies. Ordinary or moderate-hazard commercial/industrial (Groups B, F-1, M and S-1) occupancies require no separation from each other; however, they are required to be separated from all other occupancies. People-intensive (Groups A and E) occupancies also require no occupancy separation between each other but must be separated from all other occupancies except for fully-sprinklered Group F-2 and S-2 occupancies. Group R occupancies require no separation from other Group R occupancies, but such separations are mandated between all other occupancy classifications. Similar criteria apply to the Group I occupancies with a modification for Group I-2. The philosophy of the provisions set forth in the table dictates increased fire-resistance ratings on the basis of greater inherent dissimilar risk. Where Table 508.4 mandates some degree of occupancy separation, an occupancy shall be physically separated from the other occupancy through the use of fire barriers, horizontal assemblies, or a combination of both vertical and horizontal fire-resistance-rated assemblies.

In some cases, the rationale behind the use of fire-resistance-rated separations between incompatible occupancies concerns itself with the amount of combustibles encompassed in the adjoining occupancies and is termed in fire-protection circles as *fire loading*. Thus, if the amount of combustibles or fire loading in one occupancy is quite high while there is a limited fire load anticipated in the other occupancy, some degree of fire separation between the two distinct occupancies is necessary. However, the relationship of fire loads is not the only factor in determining an appropriate occupancy separation. In some cases, the separation is specified to be of 1 hour or more in duration mainly because of what the code implies to be incompatibility between the activities that occur within the two occupancies. For example, the code requires a separation of 1 hour between a Group I-2 occupancy and a Group S-2 occupancy in a fully-sprinklered building. The limited amount of combustibles in neither occupancy does not justify a fire-resistive separation; however, because of the presumed incompatibility between the two occupancies, the 1-hour separation is considered to be justified.

As a general rule, a reduction of the fire-resistance ratings in Table 508.4 by 1 hour is permitted in buildings equipped throughout with an automatic sprinkler system. The potential of a sizable fire spreading throughout a building is minimal under fully-sprinklered conditions.

The table includes several exceptions to the fire-resistance ratings separating different occupancies where the anticipated fire severity for a specific occupancy is less than that anticipated by Table 508.4 for the general occupancy classification. The fire-resistance ratings set forth by the table are further modified under the following conditions:

1. The required separation for storage occupancies may be reduced by 1 hour where the storage is limited to the parking of private or pleasure vehicles, but may never be less than 1 hour.

2. No physical or fire-resistance-rated separation is mandated between a commercial kitchen and the dining area it serves.

It is necessary to again emphasize that there is no relationship between the separated occupancy provisions of Section 508.4 and the fire area concept utilized in Section 903.2 for automatic sprinkler systems. Compliance with Table 508.4 does not relieve the responsibility to comply with Section 901.7 and Table 707.3.9 regarding the proper separation of fire areas. An example is shown in Figure 508-3.

Section 509 *Special Provisions*

The provisions of this section allow for modifications or exceptions to the general provisions for building heights and areas as regulated by Chapter 5 of the IBC. These special provisions are viewed as specific in nature, and, based upon Section 102.1, take precedence over any general provisions that may apply. Because this section permits, rather than requires, the use of the special conditions established in Section 509, the provisions are optional. Much like the application of the mezzanine provisions of Section 505, only where the designer elects to utilize the special allowances does this section apply.

It is evident that several of the provisions overlap in their scope. For example, Sections 509.2, 509.4 and 509.7 all address a potential condition where an open parking garage is located below a Group R occupancy. It is the choice of the designer which of the three methods to use where such a condition exists, based upon the benefits and consequences of each method. Or, as stated above, none of the methods need to be applied. The requirements could simply be based on the general provisions of Chapter 5 for building height and area.

509.2 **Horizontal building separation allowance.** This section is one of several that contain provisions that might be considered the only exceptions to the principle that a fire wall is strictly a vertical element without horizontal offsets. In Item 1, the code makes provisions for the use of a minimum 3-hour fire-resistance-rated horizontal assembly as an equivalent construction feature, in certain aspects, to a fire wall. This methodology is often referred to as "podium" or "pedestal" buildings.

The provisions create, in effect, an exception that allows a Group S-2 parking garage, limited-size Group A, B, M and/or R occupancies, along with operational areas, in the basement(s) and/or first story above grade plane of a building to be considered a separate building for specific purposes. The floors above such parking must only house Group A, B, M, R and/or S occupancies. As these occupancies are quite common in Type V construction, the typical application is to grant the maximum number of stories for Type V construction without the penalty of sacrificing one story for the garage. Distinct buildings are also created for area limitations and fire wall continuity.

It is fairly common in terrain that has a rolling or hillside character to erect apartment houses and small office/retail buildings with a garage in the basement or first story. Because of the slope of the ground surface, the lowest level is usually partially within the ground; therefore, the walls are normally designed as reinforced concrete or reinforced masonry-retaining walls. The construction of the lowest level is thereby easily able to conform to the code construction requirements for a Type IA building.

If this lowest level is classified as the first story above grade plane, it would typically be included in the number of total stories permitted by the code. However, as depicted in Figure 509-1, the first story would not be included where the level below the 3-hour horizontal separation is of Type IA construction.

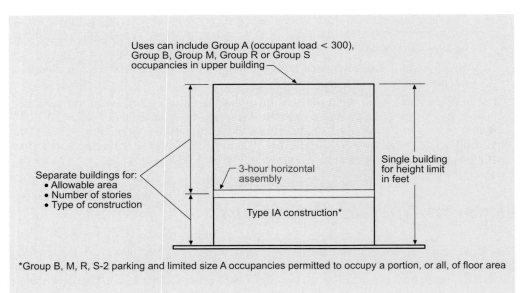

Uses can include Group A (occupant load < 300), Group B, Group M, Group R or Group S occupancies in upper building

Separate buildings for:
• Allowable area
• Number of stories
• Type of construction

3-hour horizontal assembly

Single building for height limit in feet

Type IA construction*

Figure 509-1
Horizontal building separation

*Group B, M, R, S-2 parking and limited size A occupancies permitted to occupy a portion, or all, of floor area

The code lists seven conditions that must be met in order to take advantage of these provisions. The first condition regulates the fire-resistance rating of the horizontal assembly. The second condition identifies the maximum height at which the horizontal separation can be located. Thirdly, the minimum type of construction for portions below the horizontal assembly is established. The fourth condition addresses the methods for protection of openings that will occur in the fire-resistance-rated horizontal assembly. Fifth, only Group A occupancies having an occupant load of less than 300, and Groups B, M, R and S are permitted to be located above the horizontal assembly. The sixth condition limits the use of the building below the horizontal assembly to a Group S-2 parking garage; multiple Group A occupancies, each with an occupant load under 300; Group B, M, and R occupancies; and/or entry lobbies, mechanical rooms, storage rooms and similar areas that are necessary for operation of the building. An often-forgotten provision is the seventh condition, requiring that the overall height in feet (mm) of both buildings not exceed the height limits set forth in Section 503 for the least type of construction that occurs in the building.

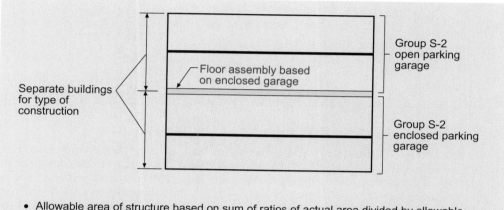

- Allowable area of structure based on sum of ratios of actual area divided by allowable area, which cannot exceed 1
- Enclosed parking garage of Type I or II construction and equivalent to open parking garage construction type
- Height and number of floors above basement limited per Table 406.3.5
- Floor separating open and enclosed garages based on enclosed garage requirement
- Enclosed parking garage limited to parking and small accessory uses

**Figure 509-2
Open/enclosed
parking**

509.3 **Group S-2 enclosed parking garage with Group S-2 open parking garage above.** The provisions of this section are similar in nature to those of Section 509.2, insofar as the two different parking uses in a single structure, one located above the other, may be considered two separate and distinct buildings for the purpose of determining the type of construction. Five specific conditions must be met in order for an open parking garage, located above an enclosed parking garage, to be regulated on its own for construction type. Details of this special situation are shown in Figure 509-2.

509.4 **Parking beneath Group R.** Consistent with the concepts expressed by Sections 509.2 and 509.3, the provisions of this section address residential uses located above a first story used as a parking garage. Where parking is limited to the first story, the number of stories used in the determination of the minimum type of construction may be measured from the floor above the garage. The construction type of the parking garage and the floor assembly between the residential area and the garage are further regulated. See Figure 509-3.

509.5 **Group R-1 and R-2 buildings of Type IIIA construction.** Typically applicable to hotels and apartment houses, this section permits an increase in height, both in stories and feet (mm), where fire walls are utilized to create compartments having a maximum floor area of 3,000 square feet (279 m²). The fire walls must have a minimum fire-resistance rating of 2 hours. Where there is a basement under the first floor, the first-floor construction must be at least 3 hours.

Group R-1 and R-2 occupancies are limited by Table 503 to four stories and 65 feet (19 825 mm) in Type IIIA construction. This provision permits an increase to six stories and 75 feet (22 860 mm), based primarily on the benefits derived from dividing the structure into small compartments separated by fire-resistance-rated construction.

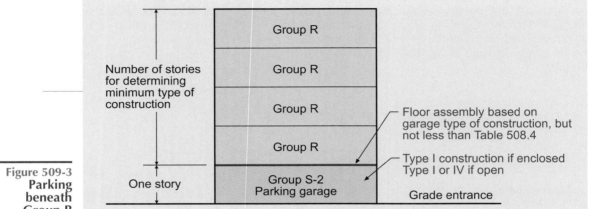

Figure 509-3
**Parking
beneath
Group R**

509.6 **Group R-1 and R-2 buildings of Type IIA construction.** Because the Group R-1 and R-2 buildings addressed in this section are of noncombustible, rather than combustible, construction, it is not necessary to divide the structure into small compartments to receive a height increase as was the case in Section 508.5. Rather, the substantial increase from four stories to nine stories, and from 65 feet (19 825 mm) to 100 feet (30 480 mm), is based on three other conditions that must occur:

1. An open area of at least 50 feet (15 240 mm) must be maintained from the Group R-1 and R-2 building to any other buildings on the same lot and from all lot lines.

2. A minimum 2-hour fire-resistance-rated fire wall must segregate the means of egress.

3. The floor construction of the first floor requires a minimum fire-resistance rating of $1^1/_2$ hours.

509.7 **Open parking garage beneath Groups A, I, B, M and R.** The excellent fire-safety record for open parking garages is the basis for this modification in the general provisions for allowable floor area and allowable height. Where located below assembly, institutional, business, mercantile or residential occupancies, an open parking garage is regulated for height and area by Section 406.3. Those permitted occupancies located above the parking garage are independently regulated by Section 503 for height and area. The only exception requires that the height of the portion of the building above the open parking garage, both in feet (mm) and stories, be measured from the grade plane.

The details of construction type are applicable to each of the occupancies involved; however, the structural-frame members shall be of fire-resistance-rated construction according to the most restrictive fire-resistive assemblies of the occupancy groups involved. Egress from the areas above the parking garage shall be isolated from the garage, with the level of protection at least 2 hours.

Because the provisions of Section 509.2 can also address open parking garages below similar occupancies, the application of this section may be limited. There are several minor differences between the two provisions; however, the general concept remains consistent.

509.8 **Group B or M with Group S-2 open parking garage.** A desirable feature in high-density areas is to have offices and/or retail stores on the first floor of open parking structures. This provision allows for the type of construction for the ground floor, as well as any basement used for Group B or M uses, to be evaluated separately from that of the open parking garage above, provided the uses are properly separated and the egress from the garage is independent from that of the first floor (and basement if applicable). This provision reverses the conditions addressed by other provisions of Section 509 where the parking garage is located below other occupancies. The resulting benefit, shown in Figure 509-4, provides for a potential reduction in the type of construction by permitting the evaluation of allowable floor areas independently for the open parking garage and the Group B and/or M occupancies.

Multiple buildings above Group S-2 parking garages. Where the varying provisions of **509.9**
Section 509 are utilized to create separate buildings above and below a complying horizontal separation, it is acceptable for two or more buildings to be located above the separation while only one building (a parking garage), is located below. For example, a condominium building is permitted to be regulated as a separate building from an adjacent office building even though both are located above a single parking facility designed under the special provisions of Sections 509.2, 509.3 or 509.7 as applicable. An example is shown in Figure 509-5.

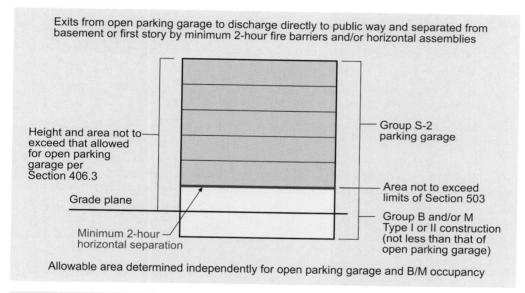

Exits from open parking garage to discharge directly to public way and separated from basement or first story by minimum 2-hour fire barriers and/or horizontal assemblies

Height and area not to exceed that allowed for open parking garage per Section 406.3

Grade plane

Minimum 2-hour horizontal separation

Group S-2 parking garage

Area not to exceed limits of Section 503

Group B and/or M Type I or II construction (not less than that of open parking garage)

Allowable area determined independently for open parking garage and B/M occupancy

Figure 509-4
Group B and M uses below an open parking garage

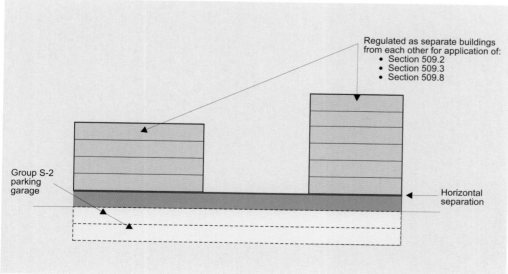

Regulated as separate buildings from each other for application of:
• Section 509.2
• Section 509.3
• Section 509.8

Group S-2 parking garage

Horizontal separation

Figure 509-5
Separate buildings above horizontal separation

KEY POINTS

- Table 503 provides the basic allowable area per floor for all occupancies and types of construction.

- The maximum height of a building is regulated by Table 503 for both the total height in feet and the maximum height in stories.

- Buildings of three or more stories in height are permitted to be three times the allowable area permitted for a single floor.

- Stepped or terraced buildings should be reviewed carefully to determine the height, as often each segment can be viewed independently of the others.

- Mezzanines are defined by the code in a very specific manner.

- The area of the basement typically does not need to be included in the total allowable area of the building.

- The installation of automatic sprinkler systems throughout a building typically provides for a sizable increase in the allowable area and building height.

- Sufficient open yards and public ways may be used to increase the allowable area.

- Through the use of adequate safeguards, the IBC allows certain types of buildings and occupancies to have unlimited floor areas.

- Incidental accessory occupancies, classified as a part of the building where the area is located, shall be separated, protected or both.

- In mixed occupancy buildings, three different methods (accessory occupancies, nonseparated occupancies and separated occupancies) are available for addressing occupancy classification, allowable height and area, and separation.

- The special provisions of Section 509 provide for alternative approaches to the specific height and area requirements of Chapter 5.

TYPES OF CONSTRUCTION

As its title implies, this chapter develops requirements for the classification of buildings by type of construction. In addition to identifying fire-resistance rating requirements for the major building elements, the *International Building Code* (IBC®) regulates exterior walls for fire resistance based upon their fire separation distance. The use of combustible materials in otherwise noncombustible buildings is also addressed.

Section 602 *Construction Classification*

Since early in the last century, the fire protection required for the various types of construction has been based on hourly fire-endurance ratings as established by the American Society for Testing and Materials. Prior to this time, fire-resistance requirements were developed by specifying the type and thickness of materials used.

Many of the concepts in previous building codes that have carried over to today were developed from the reports issued by the committee known as the Department of Commerce Building Code Committee, which was appointed by Herbert Hoover, then Secretary of Commerce. The committee was also dubbed the *Little Hoover Commission* and was appointed to investigate building codes. This was an outgrowth of the findings of the Senate Committee on Reconstruction and Production, which was appointed in 1920 to study the various factors entering into the recovery of our economy from the depression of the early 1920s. Although the committee studied a wide-ranging set of those institutions and groups affected by the economy, it was especially interested in construction. During its tenure, the committee held numerous hearings and expressed the following sentiment at their conclusion: "The building codes of this country have not been developed on scientific data, but rather on compromise; they are not uniform in practice and in many instances involve an additional cost of construction without assuring more useful or more durable buildings." Thus, the stage was set for improvement in building regulations, and the timing was especially favorable for the model codes to take advantage of the reports of the Department of Commerce Building Code Committee.

The IBC classifies construction into five basic categories, listed in a somewhat descending order from the most fire resistant to the least fire resistant. These five types are based on two main groupings, noncombustible (required) construction (Types I and II) and combustible (permitted) construction (Types III, IV and V). It will be noted that the various types of construction within the five categories are further subdivided based on fire protection and represented as follows:

1. Noncombustible, protected—Type IA, IB and IIA construction.

2. Noncombustible, unprotected—Type IIB construction.

3. Combustible and/or noncombustible, protected—Types IIIA, IV and VA.

4. Combustible and/or noncombustible, unprotected—Types IIIB and VB.

Although Types III, IV and V are commonly considered combustible construction, the use of noncombustible materials, either in part or throughout the building, is certainly acceptable. The reference to combustible construction more simply indicates that such construction is acceptable in Types III, IV and V but not mandated. A perusal of Table 503 will show the reader that the IBC considers Type II, III and IV buildings to be of comparable protection. For example, Types IIA, IV and, to some degree, IIIA are permitted the same approximate areas and heights for most occupancy classifications. The same is also true for Types IIB and IIIB.

Differing from the concept of mixed-occupancy buildings, the code does not permit a building to be considered to have more than one type of construction. In simple terms, classification of a building for construction type is based on the *weakest link* concept. If a building does not fully conform to the provisions of Chapter 6 for type of construction classification, it must be classified into a lower type into which it does conform. Unless specifically permitted elsewhere by the code, the presence of any combustible elements regulated by Table 601 prohibits its classification as Type I or II construction. Similarly, the lack of required fire resistance in any element required by Table 601 to be protected will result in a fully nonrated building.

Table 601 identifies the required fire-resistance ratings of building elements based upon the specified type of construction. Exterior walls are further regulated by Table 602 based upon the building's location in relation to adjoining lot lines and public ways. Reference is made to Section 703.2 for those building elements required to have a fire-resistance rating by Table 601. Section 703.2 establishes the appropriate test procedures for building elements, components and assemblies that are required to have a fire-resistance rating.

The provisions of Chapter 6 in regard to fire-resistance are intended to address the structural integrity of the building elements under fire conditions. Unlike those fire-resistance-rated assemblies, such as fire walls and fire barriers, whose intent is to safeguard against the spread of fire, the protection afforded by the provisions of Chapter 6 is solely that of structural integrity. As such, the protection of door and window openings, ducts and air transfer openings is not required for building elements required to be fire-resistance rated by Table 601 unless mandated by other provisions of the IBC.

The IBC intends that the provisions of the code are minimum standards. Thus, Section 602.1.1 directs that buildings not be required to conform to the requirements for a type of construction higher than the type that meets the minimum requirements of the IBC based on occupancy. A fairly common case in this regard is where a developer may construct an industrial building that complies in most respects to the requirements of the code for a Type IIIB building, but the occupancy provisions are such that a Type VB building would meet the requirements of the code. In this latter case, it would be clearly inappropriate and, in fact, a violation of the code for the building official to require full compliance with requirements for a Type IIIB building. However, where the building does comply in all respects to Type IIIB, the building official may so classify it.

Types I and II. Buildings classified as Type I and II are to be constructed of noncombustible **602.2** materials unless otherwise modified by the code. The various building elements in these noncombustible buildings are regulated by Table 601. Although Type I and II buildings are defined as noncombustible, it is evidenced by Section 603 that combustible materials are permitted in limited quantities. Wood doors and frames, trim and wall finish are permitted, as well as combustible partitions, insulation and roofing materials. Where these combustibles are properly controlled, they have proven, over the years, to not add significantly to the fire hazard.

Furthermore, Type I buildings are to be of the highest levels of fire-resistance-rated construction. The fire-resistance ratings required for Type I buildings historically have provided about the same protection over the years and, thus, have proved to be satisfactory for occupancies with low-to-moderate fire loadings, such as office buildings, hotels and retail stores. Type IB construction is very similar to Type IA construction except for a reduction of 1 hour in the required ratings for interior and exterior bearing walls, and the structural frame, while providing a $1/_2$-hour reduction for roof construction. Thus, and particularly because of the reduction in the fire-resistance rating required for the structural frame, the Type IB building does not enjoy all of the unlimited height and areas that accrue to the Type IA building. It will be noted from Table 503 that Type IB construction typically has height and, to some degree, area limits placed on it.

Buildings of Type II construction, although noncombustible, may be of either protected (Type IIA) or unprotected (Type IIB) construction. The building elements of a Type IIA building are typically required to be protected to a minimum fire-resistance rating of 1 hour. Such elements in a Type IIB structure may be nonrated.

602.3 **Type III.** The Type III building grew out of the necessity to prevent conflagrations in heavily built-up areas where buildings were erected side-by-side in congested downtown business districts. After the severe conflagrations of years past in Chicago and Baltimore, it became apparent that some control must be made to prevent the spread of fire from one building to another. As a result, the Type III building was defined. The Type III building is, in essence, a wood-frame building (Type V) with fire-resistance-rated noncombustible exterior walls.

Around the turn of the 20th century, and prior to the promulgation of modern building codes, Type III buildings were known as ordinary construction. They later became known in some circles as ordinary masonry construction. However, as stated previously, the intent behind the creation of this type of construction was to prevent the spread of fire from one combustible building to another. Thus, the early requirements for these buildings were for a certain thickness of masonry walls, such as 13 inches (330 mm) of brick for one-story and 17 inches (432 mm) for two-story buildings of bearing-wall construction. Later, the required fire endurance was specified in hours. Thus, any approved noncombustible construction that would successfully pass the standard fire test for the prescribed number of hours was permitted.

In spite of the requirement for noncombustible exterior walls, Type III buildings are considered combustible structures and are either protected (Type IIIA) or unprotected (Type IIIB). Interior building elements are permitted to be either combustible or noncombustible. There is an allowance for the use of fire-retardant-treated wood as a portion of the exterior wall assembly, provided such wall assemblies have a fire-resistance rating of 2 hours or less.

602.4 **Type IV.** Type IV buildings are designated as heavy-timber buildings. In the eastern United States during the 1800s, a type of construction evolved that was known as mill construction. Mill construction was developed by insurance companies to reduce the heavy losses they were facing in the heavy industrialized areas of the Northeast.

This type of construction has also been known as slow burning. Wood under the action of fire loses its surface moisture, and when the surface temperature reaches about 400°F (204°C), flaming and charring begin. Under a continued application of the heat, charring continues, but at an increasingly slower rate, as the charred wood insulates the inner portion of the wood member. There is quite often enough sound wood remaining during and after a fire to prevent sudden structural collapse. In recognition of these characteristics, the insurance interests reasoned that replacement of light-wood framing on the interior of factory buildings with heavy-timber construction would substantially decrease their fire losses.

The Type IV building is essentially a Type III building with a heavy-timber interior. It is of interest to see how the 1943 edition of the *National Building Code*, developed by the National Board of Fire Underwriters, defined heavy-timber construction:

> "Heavy-timber construction," as applied to buildings, means that in which walls are of approved masonry or reinforced concrete; and in which the interior structural elements, including columns, floors and roof construction, consist of heavy timbers with smooth, flat surfaces assembled to avoid thin sections, sharp projections and concealed or inaccessible spaces; and in which all structural members which support masonry walls shall have a fire-resistance rating of not less than 3 hours; and other structural members of steel or reinforced concrete, if used in lieu of timber construction, shall have a fire-resistance rating of not less than 1 hour.

From this definition, it can be seen that in the early development of heavy-timber construction, not only did the heavy-timber members have large cross sections to achieve the slow-burning characteristic, but, furthermore, surfaces were required to be smooth and flat. Sharp projections were to be avoided, as well as concealed and inaccessible spaces. Thus, the intent of the concept is to provide open structural framing without concealed spaces and without sharp projections or rough surfaces, which are more easily ignitable. In this case, flame spread along the surface of heavy-timber members is reduced; and without concealed blind spaces, there is no opportunity for fire to smolder and spread undetected.

In accordance with Table 601 and Section 602.4.6, modern-day heavy-timber construction can be a mixture of heavy-timber floor and roof construction and 1-hour fire-resistance-rated bearing walls and partitions. Although heavy-timber construction is not generally recognized as equivalent to 1-hour fire-resistance-rated construction, the code considers heavy timber to provide equivalent protection in Type IV buildings.

In keeping with the concept of slow-burning construction by means of wood members with large cross sections, the IBC specifies minimum nominal dimensions for wood members used in heavy-timber construction. As the code specifies the size of members as nominal sizes, the actual net surfaced sizes may be used. For example, an 8-inch by 8-inch (203 mm by 203 mm) member nominally will actually be a net size of $7^1/_2$ inches by $7^1/_2$ inches (191 mm by 191 mm). Therefore, even though the code calls for a nominal 8-inch by 8-inch (203 mm by 203 mm) member, the net $7^1/_2$-inch by $7^1/_2$-inch-size (191 mm by 191 mm) member meets the intent of the code. As indicated earlier, the minimum sizes for heavy-timber construction as listed in this section are based on experience and the good behavior in fire of heavy-timber construction.

Wherever framing lumber or sawn timber is specified, structural glued-laminated timber may also be used, as all have the same inherent fire-resistive capability. However, because solid sawn wood members and glued-laminated timbers are manufactured with different methods and procedures, they do not have the same dimensions. Table 602.4 compares the solid sawn sizes with those of glued-laminated members to indicate equivalency in regard to compliance with the Type IV construction criteria.

Section 602.4.6 specifies that partitions shall be of either solid-wood construction or 1-hour fire-resistance-rated construction. However, various provisions of the code address the use of fire partitions and fire barriers. In these cases, the fire-resistant-rated fire partitions or fire barriers in heavy-timber buildings should be constructed as required by the code for the required rating. For example, where there is a requirement for fire-resistance-rated corridors in heavy-timber buildings, 1-hour fire-resistance-rated construction must be used rather than solid-wood construction for the partitions.

It is highly unusual for any building designed and constructed today to be considered compliant as a Type IV structure. As previously addressed, in order for a building to be properly classified, all portions must be in conformance with the established criteria. Many buildings may have some heavy-timber elements that qualify as Type IV, however, the floor construction and/or roof construction does not fully comply with the prescriptive requirements of Sections 602.4.4 and 602.4.5, respectively. In such cases, the building cannot be classified as Type IV. Such buildings are most likely Type III or V construction. However, even if the building as a whole is not considered a Type IV structure, the recognition of individual heavy-timber elements is very important. For example, the provisions of Section 705.2.3 recognize heavy-timber projections for use in locations where unprotected combustible construction is not permitted. For this and other reasons, the requirements for Type IV heavy-timber must be fully understood.

Type V. Type V buildings are essentially construction systems that will not fit into any of **602.5** the other higher types of construction and may be constructed of any materials permitted by the code. The usual example of Type V construction is the light wood-frame building consisting of walls and partitions of 2-inch by 4-inch (51 mm by 102 mm) or 2-inch by 6-inch (51 mm by 152 mm) wood studs. The floor and ceiling framing are usually of light wood joists of 2-inch by 6-inch (51 mm by 152 mm) size or deeper. Roofs may also be framed with light wood rafters of 2-inch by 4-inch (51 mm by 102 mm) or deeper cross sections or, as is now quite prevalent, framed with pre-engineered wood trusses of light-frame construction. Wood-frame Type V buildings may be constructed with larger framing members than just described, and these members may actually conform to heavy-timber sizes. Such structures sometimes have a limited number of noncombustible building elements. However, unless the building complies in all respects to one of the other four basic types of construction, it is still a Type V building.

Type V construction is divided into two subtypes:

1. Type VA. This is protected construction and required to be of 1-hour fire-resistance-rated construction throughout.

2. Type VB. This type of construction has no general requirements for fire resistance and may be of unprotected construction, except where Section 602.1 and Table 602 require exterior wall protection because of proximity to a lot line.

Section 603 *Combustible Materials in Type I and II Construction*

Buildings of Type I and II construction are considered noncombustible structures. As such, all of the building elements, including walls, floors and roofs, are to be constructed of noncombustible materials. There are, however, a variety of exceptions to the general rule that allow a limited amount of combustibles to be used in the building's construction. It has been determined that the level of combustibles permitted by Section 603.1, as well as their control, does not adversely impact the fire-severity potential caused by the materials of construction.

The following listing provides an overview of some of those combustible materials permitted in Type I and II buildings:

1. Fire-retardant-treated (FRT) wood may be used in the construction of interior nonbearing partitions where the required fire-resistance rating of the partitions does not exceed 2 hours. In nonbearing exterior walls, FRT wood is permitted, provided no fire rating of the exterior walls is mandated. Roofs constructed of fire-retardant-treated wood are also acceptable in most buildings. This would include roof girders, trusses, beams, joists or decking, as well as blocking, nailers or similar components that may be a part of the roof system. Where the building is classified as other than Type IA construction, the use of fire-retardant-treated wood roof elements is permitted in all cases, regardless of building height. The same allowance is permitted in one- and two-story buildings of Type IA construction. For Type IA buildings exceeding two stories in height, the use of FRT wood in the roof construction is only allowed if the uppermost story has a height of at least 20 feet (6096 mm). Logically, the 20-foot measurement would be taken in a manner consistent with that described in Footnote b of Table 601, from the floor to the lowest point of the roof construction above.

   The allowances provided in Section 603.1, Item 1.3 do not reduce any required level of fire resistance mandated for wall or roof construction as established by Table 601. Rather, they simply allow the use of fire-retardant-treated wood in the locations listed where noncombustible construction is otherwise required. In reviewing the permitted use of FRT wood, there are two obvious building elements where such materials are not permitted in Type I or II construction. In Type I or II buildings, fire-retardant-treated wood is not permitted to be used in the floor construction and any bearing wall assemblies.

2. Combustible insulation used for thermal or acoustical purposes is acceptable, provided the flame-spread index is limited. Additional regulations addressing the use of thermal- and sound-insulating materials within buildings are found in Section 719.

3. Foam plastics installed under the limitations of Chapter 26 are permitted, as are roof coverings having an A, B or C classification as specified in Section 1505.

4. Wood doors, door frames, window sashes and frames, trim, and other combustible millwork and interior surface finishes are acceptable, as is blocking for handrails,

grab bars, cabinets, window and door frames, wall-mounted fixtures and similar items. Combustible stages and platforms are also permitted when complying with Section 410, and wood-finish flooring may be used when applied directly to the floor slab or installed over wood sleepers and fireblocked in accordance with Section 805.1.2.

A common allowable use of combustible elements in noncombustible buildings, detailed in Item 11, addresses the situation where nonbearing partitions divide portions of stores, offices or similar spaces occupied by one tenant only. The key words in this item are "occupied by one tenant only." It is the intent of the IBC that this expression applies to an area or building that is under the complete control of one person, organization or other occupant. This would be contrasted to multitenant occupancies, where the various tenant spaces in the building would be under the control of two or more individuals, companies or occupants. In such a multitenant space, the walls common to the public areas and to other tenants would not be regulated under this allowance. However, within each of the tenant spaces, those nonbearing walls and partitions not common with other tenants or public areas could utilize the optional construction methods of Item 11.

A multistory building owned by a large company would also qualify as being occupied by one tenant only if the large company that owned the building also completely occupied the building. A government office building owned by a city or county occupied by several departments of the government would also be considered occupied by one tenant only. If the government office building contains an assembly room, the assembly room itself would not qualify for the special provisions of Section 603.1 unless accessory to the office use. These exceptions typically apply only to stores, offices and similar uses. The intent of these provisions is to provide exceptions to the construction requirements of Chapter 6. Thus, one of the three types of partitions addressed in Item 11 of Section 603 could be constructed regardless of other requirements of Chapter 6 regulating the construction of partitions. For example, a 1-hour fire-resistance-rated combustible partition constructed of ordinary wood studs would be permitted to be installed in a Type IA building in accordance with the provisions of this section.

These provisions are based on the common practice in offices and stores to create large, open areas. When subdivided, low-height partitions are often utilized. Sometimes a few areas are completely partitioned off with full ceiling-height partitions in order to provide privacy or security for storage, etc. As these partitions are nonbearing and are subject to being moved to create various space configurations, the code permits the modification. Except for the partial-height panels of light construction, the other permitted partitions do provide some type of barrier to the spread of fire. They are to be constructed of fire-retardant-treated wood or be of a minimum 1-hour fire-resistance-rated construction. In the case of the partial-height partitions, the concept is that being only a portion of the room height, persons in one portion of the area are aware of what is going on in the other portions, and if a fire develops, the occupants would be aware of that fact and take appropriate action.

It should also be noted that combustible partitions permitted by the code under this provision are not to be used in the construction of corridor walls where the corridor serves an occupant load of 30 or more. The allowance for combustible materials is intended to address types of construction concerns; therefore, construction elements of a corridor must comply with the general requirements.

Reference is also made under Item 24 to Section 717.5 for the allowance of specific combustible elements within concealed spaces. The allowance for combustible items in concealed spaces is limited because of the increased potential for fire spread. Therefore, the flame spread index and smoke-developed index of the permitted items is often highly regulated. Combustible piping is permitted to be installed within partitions, shaft enclosures and concealed ceiling spaces. Various combustible materials are also permitted in plenums of Type I and II buildings, including wiring, fire sprinkler piping, pneumatic tubing and foam plastic insulation under the limitations imposed by Section 602 of the *International Mechanical Code*® (IMC®).

Table 601—Fire-resistance Rating Requirements for Building Elements. This table provides the basic fire-resistance rating requirements for the various types of construction. It also delineates those fire-resistance ratings required to qualify for a particular type of construction. As previously discussed, even though a building may have some features that conform to a higher type of construction, the building shall not be required to conform to that higher type of construction as long as a lower type will meet the minimum requirements of the code based on occupancy. Nevertheless, any building must comply with all the basic fire-resistance requirements in this table if it is indeed the intent to classify it for that particular type of construction. For example, in order for a building to be classified as Type IIA noncombustible construction, a minimum 1-hour fire-resistance rating is required for any and all structural frame members, bearing walls, floor construction and roof construction. In addition, and with limited exceptions, all of these building elements must be recognized as noncombustible.

Footnote a. Limited to buildings of Type I construction, the fire-resistance ratings of structural frame elements and interior-bearing walls supporting only a roof may be reduced by 1 hour. In other words, primary structural-frame members or interior-bearing walls providing only roof support shall have a minimum fire-resistance rating of 2 hours in Type IA buildings and 1 hour in Type IB construction. Additional provisions addressing the protection of certain structural frame members are found in Section 704.

Footnote b. This footnote, an exception to the general rule for roof construction, addresses those situations where the roof and its components are 20 feet (6096 mm) or more above any floor immediately below. Under these circumstances, the roof and its components, including structural frame members, roof framing and decking, may be of unprotected construction. The reduction of the fire-resistance rating would apply to buildings of Type IA, IB, IIA, IIIA and VA construction. The footnote mandates that all portions of the roof construction must be located at or above the 20-foot (6096 mm) height requirement. For example, in a sloped roof condition, it is not acceptable to merely protect those portions below the 20-foot (6096 mm) point and leave the remainder unprotected. See Figure 601-1.

The reduction in rating applies to all occupancies other than Groups F-1, H, M and S-1, where fire loading is typically higher. In all occupancies other than those just listed, the relaxation of the requirements is based on the fact that where the roof is at least 20 feet (6096 mm) above the nearest floor, the temperatures at this elevation during most fire incidents are quite low. As a result, fire protection of the roof and its members, including the structural frame, is not necessary. For those occupancies where the fire loading and the consequent

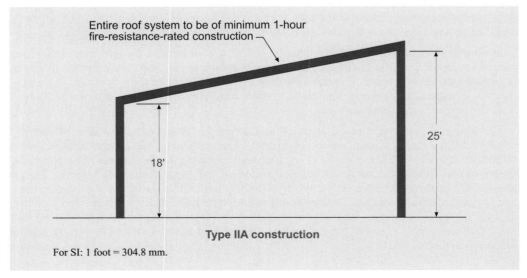

Figure 601-1 Fire-resistive ratings for roof construction

For SI: 1 foot = 304.8 mm.

potential fire severity is relatively high, such as factory-industrial, hazardous, mercantile or storage uses, the code does not permit a reduction in roof protection. It is also quite common in these occupancies for combustible or hazardous materials to be located in close proximity to the roof structure, as in the case of high-piled storage.

Footnote c. Applicable to Types IB, IIA, IIB, IIIA, and VA construction, the code permits heavy-timber members complying with Section 602.4 to be used in the roof construction without any fire-resistance rating as required by the table. It is assumed that roof members sized and constructed in compliance with the details of heavy-timber construction are equivalent to roof construction having a 1-hour fire-resistance rating. In addition, heavy-timber members are permitted to be utilized in the roof construction of an otherwise noncombustible Type IB, IIA or IIB building.

Footnote d. In this footnote, the code permits, for Type IIA, IIIA and VA buildings, the substitution of an automatic fire-sprinkler system for the 1-hour fire-resistance-rated construction, provided the sprinkler system is not otherwise required by other provisions of the IBC. The footnote cannot be applied where the sprinkler system has been utilized for allowable height or area increases under Sections 504.2 and 506.3.

It should be noted that the substitution of an automatic fire-sprinkler system under this provision does not waive or reduce the required fire-resistance rating of exterior walls under the provisions of this footnote. Such a component represents a specific fire-resistive requirement to counter specific hazards, whereas, on the other hand, the 1-hour fire-resistance-rated construction required for Type IIA, IIIA and VA buildings applies generally to the entire building.

In the final analysis, the IBC intends through this footnote that a building utilizing the automatic fire sprinkler system as a substitution for 1-hour fire-resistance-rated construction may still be classified as a Type IIA, IIIA or VA building.

It is not entirely clear what the phrase "not otherwise required by other provisions of the code" is intended to address. The most conservative approach would be based on Section 901.2, which states that "any fire protection system for which an exception or reduction to the provisions of this code has been granted shall be considered to be a required system." Therefore, the fire-resistive-substitution provisions of Footnote d would not be applicable where the sprinkler is used to modify or eliminate a code requirement. With the wide variety of allowances in the IBC for a sprinklered building, the use of this footnote would seem to be quite limited.

It is also clarified that when fire sprinklers are used for 1-hour substitution, any increase in area or height that is due to the presence of an automatic sprinkler system is not permitted. An automatic sprinkler system may not be used to increase both the area and the height, and still be used to eliminate the 1-hour fire-resistance-rated construction—but one or the other use may be selected, at the designer's option. Where fire sprinklers are installed, it is almost always more advantageous to classify the building as nonrated and use the benefits and allowances permitted by the IBC in Chapter 5. In other words, the result of this footnote is typically not the most advantageous use of a sprinkler system for type of construction purposes. See Application Example 601-1.

Footnote f. In addition to any required fire-resistance rating based upon the type of construction of the building per Table 601, it is also necessary that such rating requirements for exterior walls, both bearing and nonbearing, be in compliance with Table 602. The table regulates the hourly fire-resistance ratings for exterior walls based on fire separation distance. This footnote specifically indicates that exterior bearing walls have a fire-resistance rating based upon Table 601 or 602, whichever provides for the highest hourly rating. Exterior nonbearing walls are totally regulated by the rating requirements found in Table 602. The provisions of Section 704.10 must also be consulted where load-bearing structural members are located within the exterior walls or on the outside of the building. See Application Example 601-2 for the appropriate use of these provisions. In addition, applicable provisions in Section 603 regulating combustible material in Type I and Type II construction may apply to exterior walls.

Application Example 601-1

GIVEN: Fully sprinklered Group B office building of nonrated combustible construction.

DETERMINE: Maximum allowable height and area.

- Constructed as VB, using permitted sprinkler increases of Chapter 5
- Classified as Type VB
- Determination based on:
  - Table 503
  - Section 504.2 (height increase)
  - Section 506.3 (area increase)
- 9,000 sq ft, 2 stories (T503)
  - 18,000 sq ft increase (200%)
  - 1 story height increase

∴ Maximum height – 3 stories

   Maximum area – 27,000 sq ft/story

- Constructed as VB, using sprinkler system to classify as Type VA
- Classified as Type VA
- Determination based on:
  - Table 503
  - Table 601, Note d (increases of 504.2 and 506.3 not applicable
- 18,000 sq ft, 3 stories (T503)
  - No further increases permitted

∴ Maximum height – 3 stories

   Maximum area – 18,000 sq ft /story

For SI: 1 square foot = 0.093 m².

Thus, more advantageous to ignore use of Footnote d

TYPE OF CONSTRUCTION DETERMINATION

Application Example 601-2

GIVEN: A Group A-1 auditorium of Type IB construction. A portion of the exterior wall is located 8 feet from an interior lot line.

DETERMINE: The minimum required fire-resistance rating of the wall if it were (a) a bearing wall, and (b) a nonbearing wall.

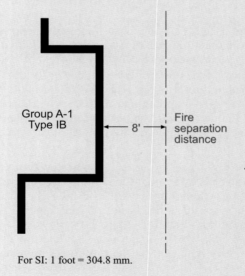

Group A-1
Type IB
← 8' →
Fire separation distance

	Bearing	Nonbearing
Table 601	2	0, see Table 602
Table 602	1	1

∴ Therefore, if wall is bearing, the minimum required rating is 2 hours. A minimum 1- hour rating is required if the wall is nonbearing.

Note: Per Section 704.10, load-bearing structural members located within the wall are to have minimum 2-hour rating.

For SI: 1 foot = 304.8 mm.

FIRE-RESISTANCE OF EXTERIOR WALLS

Table 602—Fire-resistance Rating Requirements for Exterior Walls Based on Fire Separation Distances. The IBC, as far as exterior wall protection is concerned, operates on the philosophy that an owner can have no control over what occurs on an adjacent lot and, therefore, the location of buildings on the owner's lot must be regulated relative to the lot line. In fact, the location of all buildings and structures on a given piece of property is addressed in relation to the real lot lines as well as any assumed or imaginary lines between buildings on the same lot. The assumption of imaginary lines is discussed with other exterior wall provisions in Section 705.

The lot-line concept provides a convenient means of protecting one building from another insofar as exposure is concerned. Exposure is the potential for heat to be transmitted from one building to another under conditions in the exposing building. Radiation is the primary means of heat transfer.

The code specifically provides that the fire separation distance be measured to the center line of a street, alley or public way. As the code refers to public way, this would also be applicable to appropriate open spaces other than streets or alleys that the building official may determine are reasonably likely to remain unobstructed through the years.

The regulations for exterior wall protection based on proximity to the lot line are contained in Table 602. The IBC indicates that the distances are measured at right angles to the face of the exterior wall (see definition of "Fire separation distance" in Section 702.1), which would result in the fire-resistive requirements for exterior walls not applying to walls that are at right angles to the lot line. See Figure 602-1.

In order to properly utilize Table 602, it is necessary to identify the fire separation distance, the occupancies involved and the building's type of construction. As the fire separation distance increases, the fire-resistance rating requirements are reduced, based upon the occupancy group under consideration. Figure 602-2 illustrates the application of exterior wall protection where the exterior walls of the building are parallel and perpendicular to the lot line. In this case, the illustration assumes that the building is one story of Type VB construction and used for offices (Group B). Referring to Table 602, it is noted that exterior walls less than 10 feet (3048 mm) from the lot line must be of minimum 1-hour fire-resistance-rated construction. Figure 602-3 depicts a similar building located such that the exterior walls are not parallel and perpendicular to the lot line, but are at some angle other than 90 degrees (1.57 rad). The regulation of openings in exterior walls is set

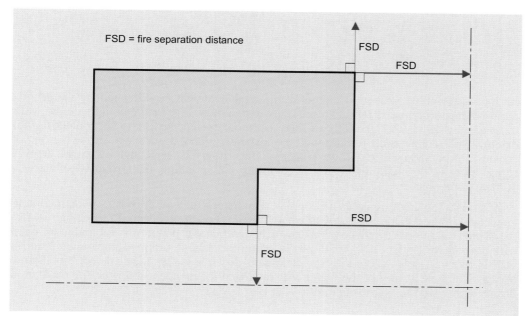

Figure 602-1
Fire separation distance

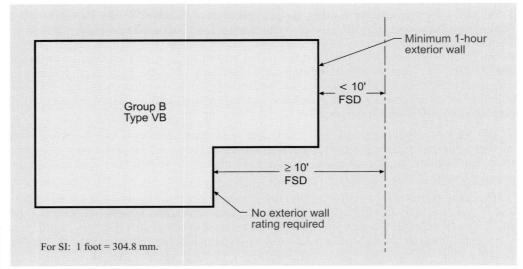

Figure 602-2
**Exterior wall
rating**

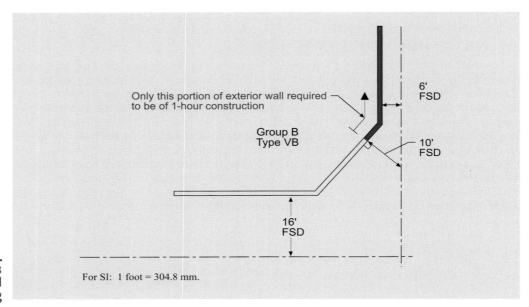

Figure 602-3
**Exterior wall
rating**

forth in Section 704.8. Several footnotes to the table address modifications to the general requirements. Footnote a repeats a previous requirement that load-bearing exterior walls must comply with both Tables 601 and 602. Footnote b refers the user to Section 406.1.2 for exterior wall requirements for Group U occupancies. Where used solely for the parking of private motor vehicles, a Group U garage or carport is regulated by either Item 1 (mixed-occupancy building) or Item 2 (building only contains the Group U) of Section 406.1.2 rather than by Table 602.

Although Table 602 requires a Group S-2 occupancy of Type I, II or IV construction to have a minimum 1-hour exterior wall where the fire separation distance is less than 30 feet (9144 mm), Footnote d reduces that distance significantly where it is a complying open parking garage. Under such conditions, a minimum 1-hour fire-resistance-rated exterior wall is required only where the fire separation distance is less than 10 feet. Footnote e indicates that each story of the building is regulated independently for the fire separation distance provisions as shown in Figure 602-4.

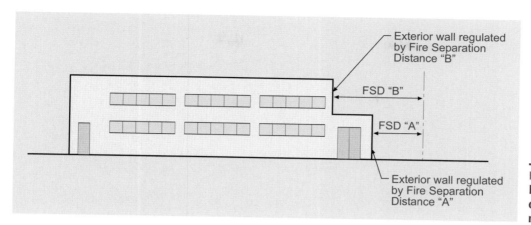

Figure 602.4
Fire separation distance measurement

KEY POINTS

- Buildings are classified in general terms as combustible or noncombustible, as well as protected or unprotected.

- Table 601 identifies the required fire-resistance ratings of building elements based upon the specified type of construction.

- Unless a fire wall is utilized, structures can be classified into only one type of construction.

- The structural frame is regulated in a manner apart from that of walls, floors and roofs.

- Type I and II buildings are considered noncombustible (required), whereas Type III, IV and V are viewed as combustible (permitted) construction.

- Very few structures fully comply with the provisions for heavy-timber construction.

- Type V buildings are by far the most common type of construction.

- Various reductions in fire resistance are permitted for nonbearing partitions.

- Table 602 regulates the protection of exterior walls insofar as exposure to an adjacent building is concerned.

- Combustible materials identified in Section 603 are permitted in otherwise noncombustible construction (Types I and II).

FIRE AND SMOKE PROTECTION FEATURES

The types of construction and the fire-resistance requirements of the *International Building Code*® (IBC®) are based on the concept of fire endurance. Fire endurance is the length of time during which a fire-resistive construction assembly will confine a fire to a given area, or continue to perform structurally once exposed to fire, or both. In the IBC, the fire endurance of an assembly is usually expressed as a "___-hour fire-resistance-rated assembly." Chapter 7 prescribes test criteria for the determination of the fire-resistance rating of construction assemblies and components, details of construction of many assemblies and components that have already been tested, and other information necessary to secure the intent of the code as far as the fire resistance and the fire endurance of construction assemblies and components are concerned. Additionally, Chapter 7 addresses other construction items that must be incorporated into a building's design in order to safeguard against the spread of fire and smoke.

Section 702 *Definitions*

Quite a few definitions that pertain to Chapter 7 are presented in this section.

ANNULAR SPACE. The open space created around the outside of a pipe, conduit or similar penetrating item where it passes through a vertical or horizontal assembly is considered the annular space. The code addresses methods to maintain the integrity of a fire-resistance-rated assembly, including methods to protect any annular space around a penetration.

BUILDING ELEMENT. Primary structural frame members, bearing walls, nonbearing walls and partitions, floor construction including secondary members, and roof construction including secondary members are considered to be building elements for the purposes of the IBC. Such elements are primarily regulated based on two criteria: fire-resistance and combustibility. In determining a building's type of construction, the building elements are evaluated based upon the criteria previously mentioned.

CEILING RADIATION DAMPER. Designed to protect air openings that occur in fire-resistance-rated roof/ceiling or floor/ceiling assemblies, ceiling radiation dampers are listed devices that automatically limit the radiative heat transfer from a room or space into the cavity above the ceiling. The damper, in conjunction with the fire-resistive ceiling membrane, protects the structural system within the floor/ceiling or roof/ceiling assembly from failure that is due to excessive heat.

DRAFTSTOP. Required by the code only in buildings of combustible construction, draftstops are utilized in large concealed spaces to limit air movement, accomplished through the subdivision of such spaces. Draftstops are to be constructed of those materials or construction identified by the IBC that effectively create smaller compartments within attics and similar areas.

F RATING. Penetration firestop systems are provided with F ratings to indicate the time periods in which they resist the spread of fire through the penetration. Tested to the requirements of ASTM E 814 or UL 1479, penetration firestop systems for fire-resistance-rated assemblies will typically have an F rating of at least 1 hour.

FIRE BARRIER. Fire-resistance-rated walls are considered fire barriers if constructed under the provisions of Section 707. The purpose of such assemblies is to create a barrier that will restrict fire spread to and from other portions of the building. All openings within a fire barrier must be protected with a fire-protective assembly. Fire barriers are often utilized to separate incompatible uses within the building, to create smaller fire areas containing the same uses or to provide for egress through a protected exit system.

FIREBLOCKING. Fireblocking is mandated by the code to address the spread of fire through concealed spaces of combustible buildings. Experience has shown that the greatest damage occurs to conventional wood-frame buildings during a fire when the fire travels unimpeded through concealed draft openings. Materials identified by the IBC as effective in resisting fire spread through concealed spaces include 2-inch (51 mm) nominal lumber, gypsum board and glass-fiber batts.

FIRE DAMPER. Regulated by test standard UL 555, fire dampers are devices located in ducts and air-transfer openings to restrict the passage of flames. Fire dampers close automatically upon the detection of heat, maintaining the integrity of the fire-resistance-rated assembly that is penetrated. The actuation of fire dampers creates some restriction to airflow from migrating throughout the duct system or through transfer openings, although not to the level of that required for a smoke damper.

FIRE DOOR ASSEMBLY. Where openings occur in fire-resistance-rated assemblies, fire door assemblies are permitted as a method to protect the openings. Addressed in Section 715, fire door assemblies include not only the door but also the door frame, the door hardware and any other components needed to provide the necessary fire-protective rating required for the specific application.

FIRE PARTITION. A fire partition is a wall or similar vertical element that is utilized by the code to provide fire-resistive protection under specified conditions. Typically required to be of fire-resistance-rated construction for 1 hour, fire partitions are commonly utilized for corridor walls, as well as walls separating dwelling units in apartment buildings. Openings that occur in fire partitions must be protected in accordance with Section 715. Regulated in Section 709, a fire partition is considered a lower type of fire-resistance-rated assembly than a fire barrier; thus, it is not permitted as an enclosure element for defining a fire area.

FIRE-PROTECTION RATING. Opening protectives, such as fire doors, fire windows and fire dampers, are assigned a fire-protection rating in order to identify the time period in which the protective is expected to confine fire spread. The specific rating, in either hours or minutes, varies based upon the details of the fire-resistance-rated assembly in which the opening protective is located.

FIRE-RESISTANCE RATING. Identified by a specific time period, a fire-resistance rating is assigned to a tested component or assembly based on its ability to perform under fire conditions. A fire-resistance-rated component or assembly is intended to restrict the spread of fire from a specified area or provide the necessary protection for the continued performance of a structural member. This performance is based on fire resistance, defined as the property of materials or assemblies that prevents or retards the passage of excessive heat, hot gases or flames.

FIRE-RESISTANT JOINT SYSTEM. Where a linear opening is placed in or between adjacent fire-resistance-rated assemblies to allow independent movement of the assemblies, it is considered by the IBC as a joint. One or more joints may be provided to address movement caused by thermal, seismic, wind or any other similar loading method. Designed to protect a potential breach in the integrity of a fire-resistance-rated horizontal or vertical assembly, a fire-resistant joint system is a tested assembly of specific materials designed to restrict the passage of fire through joints. The fire-resistance ratings required for the joint systems, as well as other requirements, are addressed in Section 714.

FIRE SEPARATION DISTANCE. The fire separation distance describes that distance between the exterior surface of a building and one of three locations—the nearest interior lot line; the centerline of a street, alley or other public way; or an imaginary line placed between two buildings on the same lot. The method of measurement is based on the distance as measured perpendicular to the face of the building. See Figure 702-1. The fire separation distance is important in the determination of exterior wall and opening protection based on the proximity to the lot lines. See the discussion on Table 602 for a further analysis of this subject.

FIRE WALL. Fully addressed in Section 706, one or more fire walls are building elements used to divide a single building into two or more buildings for the purpose of applying the

IBC. Starting at the foundation and continuing vertically to or through the roof, a fire wall is intended to fully restrict the spread of fire from one side of the wall to the other. Fire walls are higher level fire-resistance-rated elements than both fire barriers and fire partitions. Because the concept of fire walls is to create smaller buildings within one larger structure, with the code regulating each small building individually rather than collectively, it is critical that a fire wall be capable of maintaining structural stability under fire conditions. If construction on either side of a fire wall should collapse, such a failure should not cause the fire wall to collapse for the prescribed time period of the wall.

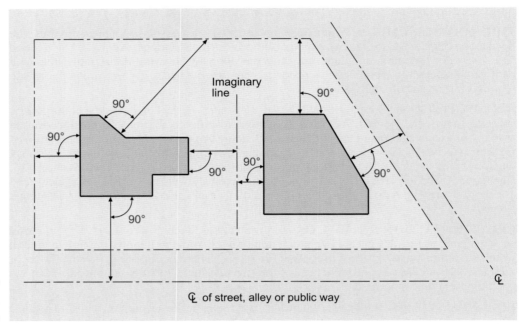

Figure 702-1
Fire-separation distance

FIRE WINDOW ASSEMBLY. Consistent with the purpose of a fire door assembly, a fire window assembly provides protection against the spread of fire through a glazed opening.

FLOOR FIRE DOOR ASSEMBLY. Although a fire door is typically viewed as an element protecting an opening in a vertical building element such as a wall, it is possible that such doors can be effective if installed horizontally for the protection of an opening in a fire-resistance-rated floor. The floor fire door assembly, like other fire door assemblies, includes the door, frame, hardware and other accessories that make up the assembly, and provides a specified level of fire protection for the opening.

HORIZONTAL ASSEMBLY. A horizontal assembly is the horizontal equivalent of a fire barrier. It is utilized to restrict vertical fire spread through an established degree of fire resistance as mandated through various provisions in the IBC. The specifics for horizontal assemblies, which include both floor and roof assemblies, are found in Section 712.

MEMBRANE PENETRATION. Similar to a through penetration in its performance requirements, a membrane penetration is an opening through only one membrane of a wall, ceiling or floor. An example of a very common membrane penetration is an electrical box.

MEMBRANE-PENETRATION FIRESTOP. Where a penetrating item such as a pipe or conduit passes through a single membrane of a fire-resistance-rated wall, floor/ceiling or roof/ceiling assembly, a membrane-penetration firestop may be required in order to adequately protect the penetration. Such a firestop consists of a device or construction that would effectively resist the passage of flame and heat through the opening in the membrane created by the penetrating item. A fire-resistance rating is assigned to a membrane-penetration firestop to indicate the time period for which the firestop is listed.

SELF-CLOSING. In order to eliminate a portion of the human element in maintaining the integrity of fire-resistance-rated assemblies, doors in the assemblies are typically required to be provided with self-closing devices that will ensure closing of the doors after having been opened. Occasionally, automatic closing fire doors are installed in specific locations on account of the nature of the situation. Under such conditions, the automatic closing fire assemblies are to be regulated by NFPA 80.

SHAFT. A shaft is considered the enclosed space that extends through one or more stories of a building. Its function is to connect vertical openings in successive floors that have been created to accommodate elevators, dumbwaiters, mechanical equipment or similar devices, as well as for the transmission of natural light or ventilation air.

SHAFT ENCLOSURE. A shaft enclosure is the building element defined by the boundaries of a shaft, which typically includes its surrounding walls and other forms of construction. Regulated by Section 708, it is required to be of fire-resistance-rated construction.

SMOKE BARRIER. Required under various circumstances identified by the code, smoke barriers are either vertical or horizontal membranes, or a combination of both, intended to restrict the movement of smoke. Walls, floors and ceiling assemblies may be considered smoke barriers where they are designed and constructed in accordance with the provisions of Section 710.

SMOKE COMPARTMENT. Where smoke barriers totally enclose a portion of a building, the enclosed area is considered a smoke compartment. By completely isolating the compartment from the remainder of the building by walls, floors and similar elements, smoke can either be contained within the originating area or prevented from entering other areas of the building. The use of smoke compartments is predominant in Group I-2 and I-3 occupancies.

SMOKE DAMPER. Test standard UL 555S states that leakage-rated dampers (smoke dampers in the terminology of the IBC) are intended to restrict the spread of smoke in heating, ventilating and air-conditioning (HVAC) systems that are designed to automatically shut down in the event of a fire, or to control the movement of smoke within a building when the HVAC system is operational in engineered smoke-control systems. The IBC simply identifies smoke dampers as listed devices designed to resist the passage of air and smoke through ducts and air-transfer openings. Smoke dampers must operate automatically unless manual control is desired from a remote command station.

T RATING. The T rating is defined as the time required for a specific temperature rise on the unexposed side of a penetration firestop system. More specifically, the penetration firestop system as well as the penetrating item must provide for a maximum increase in temperature of 325°F (163°C) above its initial temperature for the time period reflected in the fire-resistance rating. The establishment of the rating of the through-penetration firestop is determined by tests in accordance with ASTM E 814 or UL 1479. A more detailed discussion of the subject is found later in this chapter.

THROUGH PENETRATION. A through penetration is considered an opening that passes through an entire assembly, accommodating various penetrating items such as cables, conduit and piping. Where membrane construction is provided, such as gypsum board applied to both sides of a stud wall, a through penetration would pass entirely through both membranes and the cavity of the wall.

THROUGH-PENETRATION FIRESTOP SYSTEM. In order to adequately protect the penetration of a fire-resistance-rated assembly by conduit, tubing, piping and similar items, a through-penetration firestop system is sometimes required. Such a system may selectively include various materials or products that have been designed and tested to resist the spread of fire through the penetration. Through-penetration firestop systems are fire-resistance rated based on the criteria of ASTM E 814 or UL 1479, and provided with an hourly rating for both fire spread (F rating) and temperature rise (T rating).

Fire-resistance Ratings and Fire Tests

It is the intent of the IBC that materials and methods used for fire-resistance purposes are limited to those specified in this chapter. Materials and assemblies tested in accordance with ASTM E 119 or UL 263 are considered to be in full compliance with the code, as are building components whose fire-resistance rating has been achieved by one of the alternate methods specified in Section 703.3.

703.2 **Fire-resistance ratings.** This section indicates that building elements are considered to have a fire-resistance rating when tested in accordance with the procedures of ASTM E 119 or UL 263. Figures 703-1 through 703-5 depict the fundamental testing requirements of the two standards. The intent of the IBC is that any material or assembly that successfully passes the end-point criteria depicted for the specified time period shall have its fire-endurance rating accepted and the assembly classified in accordance with the time during which the assembly successfully withstood the test.

Although early fire testing in the United States began as long ago as the 1890s, the standard fire-endurance test procedure using a standard time-temperature curve and specifying fire-endurance ratings in hours was developed in 1918. The significance of 1918 and later standards is the fact that they were and are intended to be reproducible so that the test conducted at Underwriters Laboratories (UL) can be compared with the test of the same assembly conducted at the University of California, Ohio State University or other testing facility. An often-expressed criticism of a standard such as ASTM E 119 or UL 263 is that "it does not represent the real world." This is true in many cases, and for that reason it should not be thought of as representing the absolute behavior of a fire-resistance-rated assembly under most actual fires in buildings. There are too many variables that affect the fire endurance of an assembly during an actual fire, such as fuel load, room size, rate of oxygen supply and restraint, to consider that the test establishes absolute values of the real world fire endurance of an assembly. However, it is a severe test of the fire-resistive qualities of a material or an assembly, and because of its reproducibility, it provides a means of comparing assemblies.

In addition to the fire-endurance fire ratings obtained from the standard fire tests of ASTM E 119 and UL 263, it is also possible to obtain, as expressed in the standard, the protective membrane performance for walls, partitions, and floor or roof assemblies. In the case of combustible walls or floor or roof assemblies, it is also referred to as the finish rating. Although the test standard does not limit the determination of the protective membrane performance to combustible assemblies, its greatest significance is with combustible assemblies.

The end-point criteria for determining the finish rating are that the average temperature at the surface of the protected materials shall not be greater than 250°F (121°C) above the beginning temperature. Furthermore, the maximum temperature at any measured point shall not be greater than 325°F (163°C) above the beginning temperature. These temperatures relate to the lower limit of ignition temperatures for wood. Figure 703-4 illustrates the determination of the finish rating for a wall assembly, which is usually determined during a fire-endurance test of the assembly.

The condition of acceptance, also referred to as failure criteria and end-point criteria, of fire-resistance-rated assemblies are as follows:

1. For load-bearing assemblies, the applied load must be successfully sustained during the time period for which classification is desired. There shall be no passage of flame or gases hot enough to ignite cotton waste on the unexposed surfaces.

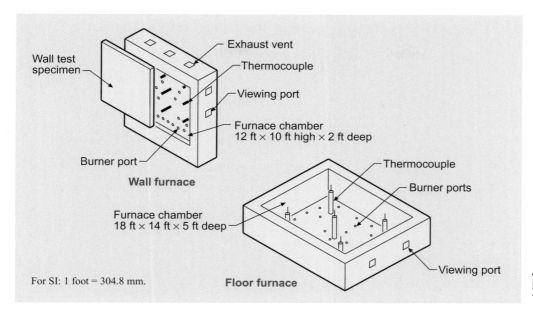

For SI: 1 foot = 304.8 mm.

**Figure 703-1
Test furnaces**

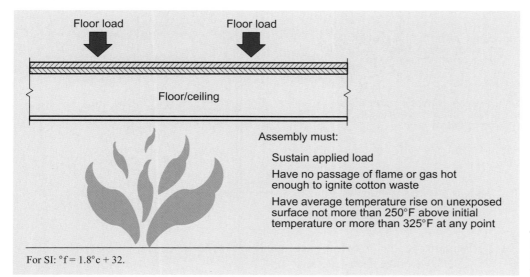

For SI: °f = 1.8°c + 32.

**Figure 703-2
Floor assembly
fire test**

2. The average temperature rise on the unexposed surface shall be not more than 250°F (121°C) above the initial temperature during the time period of the test.

3. The maximum temperature on the unexposed surface shall not be more than 325°F (163°C) above the initial temperature during the time period of the test.

4. Walls or partitions shall withstand the hose-stream test without passage of flame or gases hot enough to ignite cotton waste on the unexposed side or the projection of water from the hose stream beyond the unexposed surface.

In addition to the conditions of acceptance just described, load-carrying structural members in roof and floor assemblies are subject to special end-point temperatures for:

1. Structural steel beams and girders—1,100°F (593°C) average at any cross section and 1,300°F (704°C) for any individual thermocouple, for unrestrained assemblies.

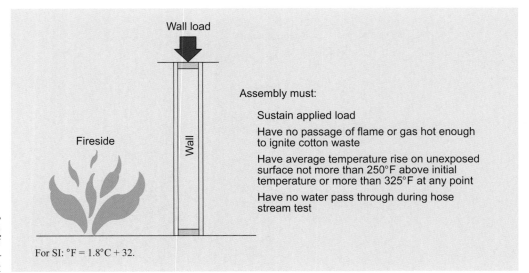

Wall load

Fireside

Wall

Assembly must:

Sustain applied load

Have no passage of flame or gas hot enough to ignite cotton waste

Have average temperature rise on unexposed surface not more than 250°F above initial temperature or more than 325°F at any point

Have no water pass through during hose stream test

For SI: °F = 1.8°C + 32.

Figure 703-3
Conditions of acceptance— wall fire test

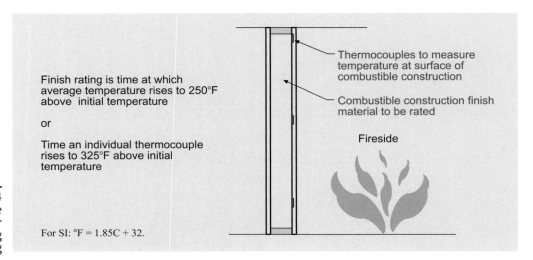

Finish rating is time at which average temperature rises to 250°F above initial temperature

or

Time an individual thermocouple rises to 325°F above initial temperature

Thermocouples to measure temperature at surface of combustible construction

Combustible construction finish material to be rated

Fireside

For SI: °F = 1.85C + 32.

Figure 703-4
Combustible assembly for determining finish rating

2. Reinforcing steel in cast-in-place reinforced concrete beams and girders—1,100°F (593°C) average at any section.

3. Prestressing steel in prestressed concrete beams and girders—800°F (427°C) average at any section.

4. Steel deck floor and roof units—1,100°F (593°C) average on any one span.

As columns are exposed to fire on all surfaces, the standard has special temperature and testing criteria for these members:

1. The column is loaded so as to develop (as nearly as practicable) the working stresses contemplated by the structural design. The condition of acceptance is simply that the column sustain the load for the duration of the test period for which a classification is desired.

2. Alternatively, a steel column may be tested without load, and the column will be tested in the furnace to determine the adequacy of the protection on the steel column. The test and end points are depicted in Figure 703-5.

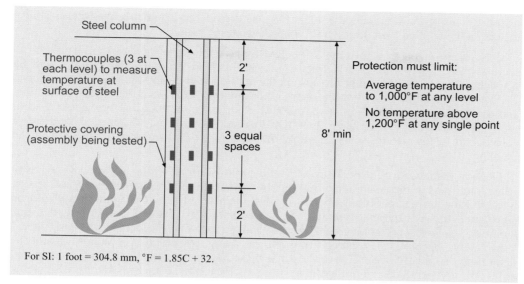

For SI: 1 foot = 304.8 mm, °F = 1.85C + 32.

Figure 703-5
Alternate fire test of steel column protection

The exception to this section is intended to modify the acceptance criteria for exterior bearing walls so that the walls will receive a rating based on which of the two following sets of criteria occurs first during the test:

1. Heat transmission or flame and hot gases transmission for nonbearing walls.

2. Structural failure or hose-stream application failure.

The first set of end points measures the wall's ability to prevent the spread of fire from one side to the opposite side. It is considered overly restrictive to require that exterior bearing walls comply with this first set of end points for a longer time than would be required for a nonbearing wall located at the same distance from the lot line if it is still structurally capable of carrying the superimposed loads.

Nonsymmetrical wall construction. At times, an interior wall or partition is constructed **703.2.1** nonsymmetrical as far as its fire protection is concerned, with the membrane on one side of the wall differing from that on the opposing side. Where the wall is to be fire-resistance rated, it must be tested from both sides in order to determine the fire-resistance rating to be assigned to the assembly. Based on the two tests, the shortest time period is determined to be the wall's rating. An assembly tested from only one side may be approved by the building official, provided there is adequate evidence furnished to show that the wall was tested with the least fire-resistive side exposed to the furnace. The provisions for exterior walls of nonsymmetrical construction differ somewhat from those addressing interior walls and are regulated by Section 705.5.

Restrained classification. A dual classification system is utilized in ASTM E 119 and UL **703.2.3** 263 for roof and floor assemblies, including their structural members. This dual classification system involves the use of the terms *restrained* and *unrestrained*. The use of the word *restrained* entails the concept of thermal restraint (restrained against thermal expansion as well as against rotation at the ends of an assembly or structural member).

For example, if a structural beam of a uniform cross section is subjected to heat on its bottom surface, such as would be the case in the standard test furnace, it will attempt to expand in all directions with the longitudinal expansion being the primary component. If the beam is restrained at the ends so that it cannot expand, compressive stresses will build up within the beam, and it will in effect behave in similar fashion as a prestressed beam. As a result, the thermal restraint will be beneficial in terms of improving the beam's ability to sustain the applied load during the fire test. If the same beam is restrained only for the lower one-half of its cross section, it will tend to deflect upward owing to the conditions of restraint. This upward deflection tendency is also considered to enhance the beam's ability to sustain the applied load during a fire-endurance test.

Conversely, if the end restraint is applied only to the upper half of the beam's cross section, the beam will tend to deflect downward and, in this case, the restraint will be detrimental to the beam's ability to sustain the applied load during the fire-endurance test. As the heat is applied to the bottom surface during a fire, it creates a downward deflection, and the two downward deflections are additive. In an actual building this could lead to premature failure. It can be seen, then, that thermal restraint may be either beneficial or detrimental to the fire-resistant assembly, depending on its means of application in the building.

General guidance for the building official is provided in ASTM E 119 and UL 263 as to what conditions in the constructed building provide restraint. It is generally agreed that an interior panel of a monolithically cast-in-place reinforced-concrete floor slab would be considered to have thermal restraint. Also, Footnote k to Table 720.1(1) provides that "interior spans of continuous slabs, beams and girders may be considered restrained." Conversely, because the restraint present in many construction systems cannot be determined so neatly, the IBC requires that these assemblies be considered unrestrained unless the registered design professional shows by the requisite analysis and details that the system qualifies for a restrained classification. Furthermore, the code requires that any construction assembly that is to be considered restrained be identified as such on the drawings.

703.3 **Alternative methods for determining fire resistance.** In addition to those assemblies and materials considered fire-resistance-rated construction based upon compliance with ASTM E 119 or UL 263, a number of alternative methods for determining fire resistance are set forth in this section. Where it can be determined that the fire-resistance rating of a building element is in conformance with one of the five listed methods or procedures, such a rating is considered acceptable.

703.4 **Noncombustibility tests.** Throughout the IBC, particularly in Chapter 6, the terms *combustible* and *noncombustible* are utilized. Under many different conditions, limits are placed on the use of combustible building materials, particularly in buildings of Type I or II construction. This section sets forth the two methods for determining if a material is noncombustible.

For most materials, ASTM E 136 is the test standard used to determine if a material is noncombustible. Composite materials such as gypsum board are also considered noncombustible if they comply with the criteria of Section 703.4.2. Such materials must have a structural base of noncombustible materials with a surfacing limited in thickness and flame spread.

Note that the term *noncombustible* does not apply to surface finish materials.

703.5 **Fire-resistance-rated glazing.** The use of fire-resistance-rated glazing typically only occurs where the limitations placed on fire-protection-rated glazing make it undesirable or impractical. Fire-resistance-rated glazing is subjected to the ASTM E 119 or UL 263 testing criteria, which includes stringent limitations on temperature rise through the assembly. Because the glazing is regulated as a wall assembly rather than an opening protective, its use is not limited by any of the provisions of Section 715. It is only regulated under the appropriate code requirements for a fire-resistance-rated wall assembly.

The labeling requirements are similar to those of Section 715 for glazing used as opening protectives with the exception of the identifier "W-XXX." The "W" indicates that the glazing meets the requirements of ASTM E 119 or UL 263, thus qualifying the glazing to be used as part of a wall assembly. It also indicates that the glazing meets the fire-resistance, hose-stream and temperature-rise requirements of the test standard. The fire-resistance rating of the glazing will then follow the "W" designation. See Figure 703-6.

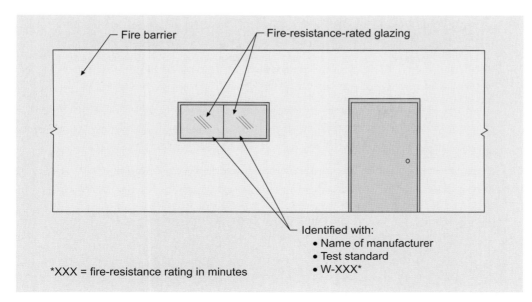

Figure 703-6
Fire-resistance-rated glazing in fire barriers

Marking and identification. The integrity of fire and/or smoke separation walls is subject to compromise during the life of a building. During maintenance and remodel activities, it is not uncommon for new openings and penetrations to be installed in a fire or smoke separation without the recognition that the integrity of the construction must be maintained or that some type of fire or smoke protective is required. The reduction or elimination of protection that occurs is typically not malicious. Rather, the installation of an inappropriate door or window, or the penetration of the separation without the proper firestopping, is often done due to the lack of information regarding the wall assembly's function and required fire rating.

703.6

Through the identification of fire and smoke separation elements, it is possible for tradespeople, maintenance workers and inspectors to recognize the required level of protection that must be maintained. The requirements apply to all wall assemblies where openings or penetrations are required to be protected. This would include exterior fire-resistance-rated walls as well as fire walls, fire barriers, fire partitions, smoke barriers and smoke partitions. The identifying markings must be located at maximum 30-foot (9144 mm) intervals to increase the possibility that they would be visible during any work on the wall assemblies. A minimum letter height is also prescribed along with sample language for the marking. See Figure 703-7.

It is intended that the identification marks be located in areas not visible to the general public. Specific locations set forth in the provisions indicate that the identification is to be provided within those concealed spaces that are accessible, such as above suspended ceilings or in attic and under-floor areas. The requirement for markings does not apply to fire or smoke separations in Group R-2 occupancies unless a decorative ceiling system is installed. There is an expectation that the separation walls in these types of uses are seldom altered and thus the identification markings are not necessary.

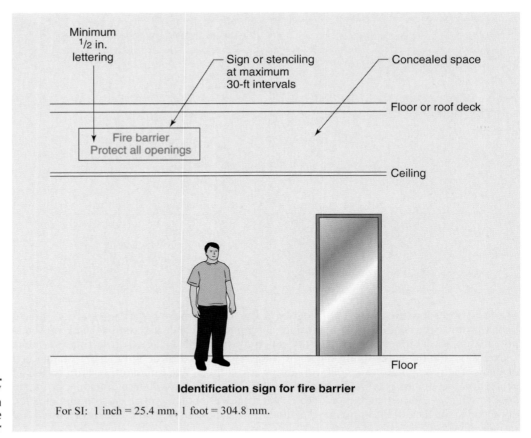

Figure 703-7
**Identification
sign for fire
barrier**

Identification sign for fire barrier

For SI: 1 inch = 25.4 mm, 1 foot = 304.8 mm.

Section 704 *Fire-resistance Rating of Structural Members*

Structural frame members such as columns, trusses, beams and girders are regulated for fire resistance based on a building's type of construction. The higher types of construction mandate a higher level of fire endurance for structural members and assemblies on account of the critical nature of their function. Type of construction considerations is based primarily on the potential for building collapse when subjected to fire. Therefore, the structural frame is specifically addressed in Table 601 as to the required fire-resistance ratings. This section provides further details for the protection of structural members.

Figure 704-1 provides simple details of fire protection of structural members that indicate the principle of *mass effect*. Mass effect is beneficial to the protection requirements for structural members of a heavy cross section. In the case of steel members, the amount of protection depends on the weight of the structural steel member. A heavy, massive structural steel cross section behaves such that the heat applied to the surface during a fire is absorbed away from the surface, resulting in lower steel surface temperatures. Thus, the insulating thicknesses indicated by tests or in Table 720.1(1) should not be used for members with a smaller weight than that specified in the test or table.

704.2 **Column protection.** Where a part of the structural frame system, columns required to have a fire-resistance rating shall be protected for their full height. The fire protection required for the column shall also be provided at the connections between the column and any beams or girders. Under all conditions, the columns must be protected by individual encasement an example of which is shown in Figure 704-2. Where a ceiling is provided, the fire resistance of the column is to be continuous from the top of the foundation or floor/ceiling assembly below through the ceiling space to the top of the column. Therefore, the columns shown individually protected in Figure 704-2 and in the fire-resistance-rated wall in Figure 704-3 require protection as shown for their full heights.

Protection of the primary structural frame other than columns. The code intends that the fire-resistive protection for primary structural frame members be applied to the individual structural member. This is based on the differences in both the testing procedure and the conditions of acceptance that were discussed in Section 703. See Figure 704-2. In other words, the code does not intend that the structural member be protected by a wall assembly or fire-resistance-rated horizontal assembly, except as permitted by this section.

Under certain restrictions, the code allows the use of a fire-resistance-rated wall, floor/ceiling or roof/ceiling assembly to provide protection for structural members, rather than requiring that they be individually protected. The criteria for use of alternate membrane protection in lieu of individual encasement are depicted in Figure 704-3 and are as follows:

1. The use of the ceiling protection applies only to horizontal structural members, such as girders, trusses, beams or lintels. (See Section 704.2 for column protection.)

2. The structural members shall not support directly applied loads from more than two floors or one floor and roof, or support a load-bearing wall or a nonload-bearing wall more than two stories in height.

3. The required fire-resistance rating of the assembly shall be at least equal to that required by the code for the individual protection of the structural members.

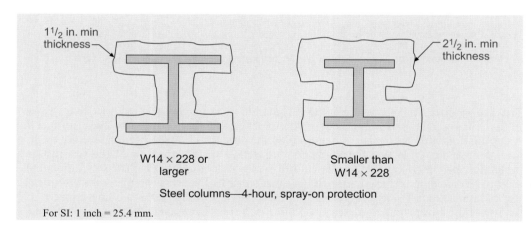

1¹/₂ in. min thickness

2¹/₂ in. min thickness

W14 × 228 or larger

Smaller than W14 × 228

Steel columns—4-hour, spray-on protection

For SI: 1 inch = 25.4 mm.

Figure 704-1
Mass effect

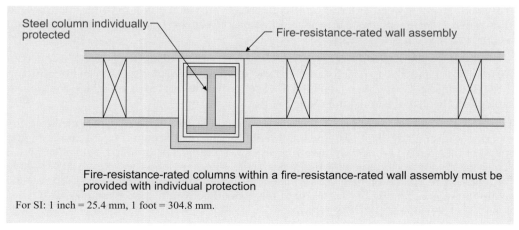

Steel column individually protected

Fire-resistance-rated wall assembly

Fire-resistance-rated columns within a fire-resistance-rated wall assembly must be provided with individual protection

For SI: 1 inch = 25.4 mm, 1 foot = 304.8 mm.

Figure 704-2
Individual protection of structural columns

704.4 **Protection of secondary members.** Secondary members, as defined in Section 202, may be protected in the same manner as primary structural frame members where a fire-resistance rating is required. Such elements can be individually encased or protected by a membrane or ceiling of a horizontal assembly. Floor joists and roof joists are examples of secondary members that are permitted to be protected by the horizontal assembly in which they are located.

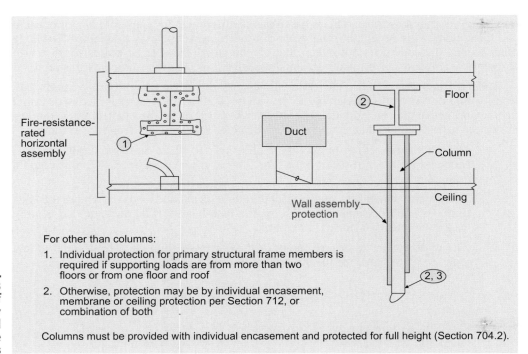

For other than columns:

1. Individual protection for primary structural frame members is required if supporting loads are from more than two floors or from one floor and roof

2. Otherwise, protection may be by individual encasement, membrane or ceiling protection per Section 712, or combination of both

Columns must be provided with individual encasement and protected for full height (Section 704.2).

Figure 704-3
Protection of primary structural frame members

704.5 **Truss protection.** It is the intent of the IBC that this item be applied to trusses that are a part of the primary structural frame as defined in Section 202. In this case, the code permits the encapsulation with fire-resistive materials of the entire truss assembly for its entirety. It is the intent of the code that the thickness and details of construction of the fire-resistive protection be based on the results of full-scale tests or of tests on truss components. Approved calculations based on such tests that show that the truss components provide the fire endurance required by the code are also acceptable. One application of this concept is in the use of the encapsulated trusses as dividing partitions between hotel rooms in multistory steel-frame buildings. Because the truss becomes part of the primary structural frame where it used to span between exterior wall columns, it provides a column-free interior. The fire-resistive design of the encapsulated protection can be based either on tests or on analogies derived from fire tests.

Additional criteria for the protection of primary structural members are illustrated in Figures 704-4 and 704-5, which depict details for attached metal members and reinforcing discussed in Sections 704.6 and 704.7. The provisions of Section 704.9 for impact protection are also illustrated in Figure 704-6.

704.10 **Exterior structural members.** The code provides that structural frame elements in the exterior wall or along the outer lines of a building must be protected based on the higher rating of three criteria. The minimum fire-resistance rating is determined by evaluating the requirements for 1) the structural frame per Table 601, 2) exterior bearing walls per Table 601 and 3) fire separation distance per Table 602. The highest of these three ratings is the minimum required rating of the structural members. See Application Example 704-1.

The intent of the provisions is that the structural frame should never have a lower fire rating than that required to protect the frame from internal fires. Nevertheless, if the exposure hazard from an external source is so great as to require exterior wall protection, a higher rating may be required.

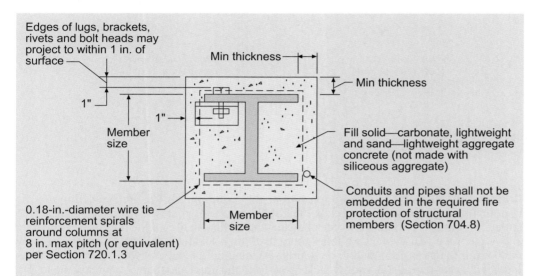

Edges of lugs, brackets, rivets and bolt heads may project to within 1 in. of surface

Min thickness

Min thickness

1"

1"

Member size

Fill solid—carbonate, lightweight and sand—lightweight aggregate concrete (not made with siliceous aggregate)

0.18-in.-diameter wire tie reinforcement spirals around columns at 8 in. max pitch (or equivalent) per Section 720.1.3

Member size

Conduits and pipes shall not be embedded in the required fire protection of structural members (Section 704.8)

For SI: 1 inch = 25.4 mm.

Table 720.1(1)					
Item Number	Member size	Minimum thickness (inches)			
		4 hour	3 hour	2 hour	1 hour
1-1.1	6 in. × 6 in. or greater	$2^{1}/_{2}$	2	$1^{1}/_{2}$	1
1-1.2	8 in. × 8 in. or greater	2	$1^{1}/_{2}$	1	1
1-1.3	12 in. × 12 in. or greater	$1^{1}/_{2}$	1	1	1

Figure 704-4
Protection of structural steel column

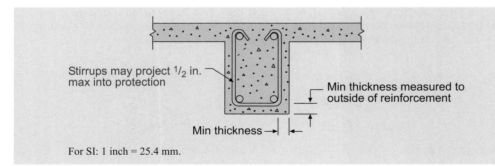

Stirrups may project $^{1}/_{2}$ in. max into protection

Min thickness measured to outside of reinforcement

Min thickness

For SI: 1 inch = 25.4 mm.

Figure 704-5
Reinforcing steel in concrete joists

Application Example 704-1

GIVEN: An exterior nonbearing wall in a Type IIIB building housing a Group M occupancy. The wall has a fire separation distance of 15 feet to an interior lot line.

DETERMINE: The minimum required fire-resistance rating for structural columns located within the exterior wall.

SOLUTION:

Per Table 601 for structural frame members, a minimum of 0 hours

Per Table 601 for exterior bearing walls, a minimum of 2 hours

Per Table 602 for a FSD of 15 feet, a minimum of 1 hour

∴ The columns shall have a minimum fire-resistance rating of 2 hours

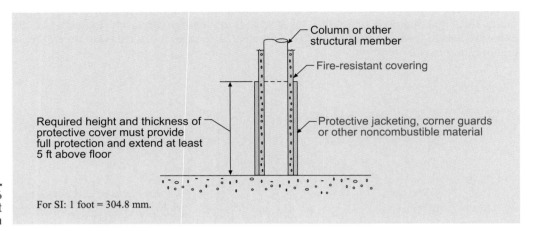

Figure 704-6
Impact
protection

For SI: 1 foot = 304.8 mm.

704.11 **Bottom flange protection.** Exempted from the requirements for fire protection are the bottom flanges of short-span lintels, and shelf angles or plates that are not part of the structural frame. It is assumed by the code that the arching action of the masonry or concrete above the lintel will prevent anything more than just a localized failure. Furthermore, only the bottom flange is permitted to be unprotected and, as a result, the wall supported by the lintel will act as a heat sink to draw heat away from the lintel and thereby increase the length of time until failure that is due to heat. This latter rationale also applies to shelf angles and plates that are not a part of the structural frame.

Section 705 *Exterior Walls*

Because of the potential for radiant heat exposure from one building to another, either on adjoining sites or on the same site, the IBC regulates the construction of exterior walls for fire resistance. Opening protection in such walls may also be required based upon the fire separation distances involved. In addition to the regulation of exterior walls and openings in such walls, the code addresses associated projections, parapets and joints.

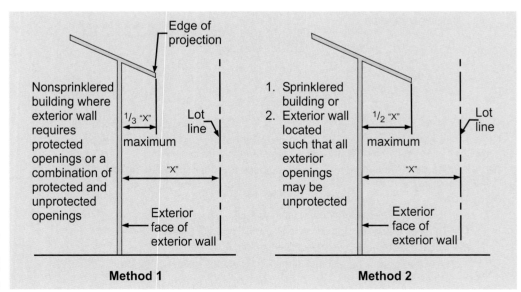

Figure 705-1
Limitations on
extension of
projections

Projections. Architectural considerations quite often call for projections from exterior **705.2** walls such as cornices, eave overhangs and balconies. Where these projections are from walls that are in close proximity to a lot line, they create problems that are due to trapping the convected heat from a fire in an adjacent building. As this trapped heat increases the hazard for the building under consideration, the code describes the location of an imaginary vertical plane beyond which any projection may not extend. In order to provide the necessary separation between the projection and the adjacent lot line, limits are placed upon the extent of the projection. If the building is provided with automatic sprinkler protection, a projection is permitted to extend to a point one-half the distance between the exterior face of the exterior wall and the lot line. This allowance is also applicable where the exterior wall is located with a sufficient fire separation distance such that Section 705.8 would allow unlimited unprotected openings. Where the exterior wall from which the projection extends is located such that Section 705.8 would require either protected openings or a combination of both protected and unprotected openings, the projection cannot extend beyond a point one-third the distance from the exterior wall to the lot line. The two methods are illustrated in Figure 705-1.

The use of either the one-half limitation or the one-third limitation must then be compared with the limitation based upon the point at which openings are first prohibited. The resulting projection limit is determined by the method which results in the lesser projection, in other words, the method which maintains the greatest distance between the leading edge of the projection and the adjacent lot line. It is important to note that the comparison of methods is limited to the pairings of Methods 1 and 3, or depending upon the sprinkler and fire separation conditions, Methods 2 and 3.

The reference to multiple buildings on the same lot is intended to address only those projections that extend beyond the opposing exterior walls of the adjacent buildings. For those exterior walls that directly oppose each other, the limits on projecting elements are not applicable. However, those projections that occur at exterior walls not located in opposition to those exterior walls of an adjacent building are to be regulated by the provisions of Section 705.2.

Projections from buildings are further regulated in order to prevent a fire hazard from inappropriate use of combustible materials attached to exterior walls. Thus, the IBC requires that projections from walls of Type I or II buildings be of noncombustible materials. However, it should be noted that certain combustible materials are permitted for balconies and similar projections as well as bay windows and oriel windows in accordance with Sections 1406.3 and 1406.4.

For buildings that the code considers to be of combustible construction (Type III, IV or V construction), both combustible and noncombustible materials are permitted. Where combustible projections are utilized and they extend into an area where openings are either not permitted or where they are required to be protected, the code requires that they be of at least 1-hour fire-resistance-rated construction, of heavy-timber construction, constructed of fire-retardant-treated wood, or as required in Section 1406.3 for balconies and similar projections. This requirement is based on a potential for a severe exposure hazard and, consequently, the code intends that combustible materials be protected or, alternatively, be of heavy-timber construction, which has comparable performance when exposed to fire. An example is shown in Figure 704-2, based on an evaluation of the criteria as established in Table 705.8.

Because projections are typically regulated independent of the roof construction, it is entirely possible that their construction types may be inconsistent. For example, Figure 705-3 shows two situations where the roof construction and resulting projections may differ in their required protection. Figure A relates a Type VA building with a 1-hour fire-resistance-rated roof system, but a nonrated projection. On the other hand, Figure B indicates a Type VB building with nonrated roof construction, but a minimum 1-hour-protected projection. In each case, the roof construction and its projection are regulated differently because of the concept of fire resistance being applied.

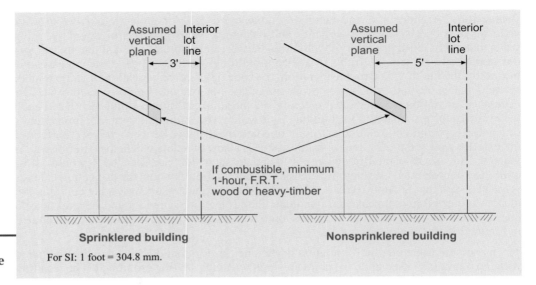

Figure 705-2
Combustible projections

For SI: 1 foot = 304.8 mm.

705.3   Buildings on the same lot. The IBC regulates exterior wall construction and protection based on the proximity of the walls to lot lines, either real or assumed. This section provides the code requirements for the establishment of imaginary lines between buildings on the same lot. Where two or more buildings are to be erected upon the same site, the determination of the code requirements for protection of the exterior walls is based on placing an assumed imaginary line between buildings. Figure 705-4 illustrates an example of two nonsprinklered Type IIIB buildings housing Group S-2 occupancies sharing a 32-foot-wide (9760 mm) yard, and it is noted that the imaginary line can be located anywhere between the two buildings so that the best advantage can be taken of wall and opening protection, depending on the use and architectural considerations for the exterior walls of the buildings. For example, if unprotected openings amounting to 25 percent of the area of the exterior walls of each nonsprinklered building were desired, the imaginary line would be located so that the distance between it and each building would permit such an amount of unprotected openings. Thus, the code would require that each building be placed slightly more than 15 feet (4572 mm) from the imaginary line in order to have unprotected openings totaling 25 percent of each opposing wall area. If one of the buildings were to have no openings in the exterior wall, the imaginary line could be placed at the exterior wall of the building without openings. The other building would be located at a distance of 32 feet (9760 mm) or more from the imaginary line and the other building. In the first case described, the opposing nonbearing exterior walls would both be required to be of minimum 1-hour fire-resistance-rated construction as they are each located less than 30 feet (9144 mm) from the imaginary line. However, the wall located 32 feet (9760 mm) from the imaginary line would not require any fire rating. Also, in the last example, Section 705.11 could possibly require that the exterior wall on the assumed lot line be provided with a parapet. See discussion of Section 705.11.

In the case where a new building is to be erected on the same lot as an existing building, the same rationale applies as depicted in Figure 705-4 except that the wall and opening protection of the existing building determines the location of the assumed imaginary line. As shown in Figure 705-5, the exterior wall and opening protection of the existing building must remain in compliance with the provisions of the IBC. In any case, where two or more buildings are located on the same lot, they may be considered to be a single building subject to the limitations depicted in Figure 705-6. For further discussion on this condition, see the commentary on Section 503.1.2.

705.5   Fire-resistance ratings. The IBC requires that exterior walls conform to the required fire-resistance ratings of Tables 601 and 602. Bearing walls must comply with the more restrictive requirements of both tables, whereas nonbearing exterior walls need only comply with Table 602. Table 601 is intended to address the fire endurance of bearing walls necessary to prevent building collapse that is due to fire for a designated time period. Table 602 is used to determine the required fire-resistance ratings that are due to exterior fire exposure from adjacent buildings, as well as the interior fire exposure that adjacent buildings are exposed to on account of the uses sheltered by the exterior walls. Where structural frame members are located within exterior walls, reference to Section 704.10 is required. Examples of the use of these provisions are shown in Application Examples 601-2 and 704-1.

Section 703.2.1 addresses nonsymmetrical interior wall construction, whereas this section of the code addresses nonsymmetrical construction for exterior walls. This method of construction, which provides for a different membrane on each side of the supporting elements, is much more typical for exterior applications. As an example, a nonsymmetrical exterior wall may consist of wood studs covered with gypsum board on the inside, with sheathing and siding on the exterior side. See Figure 705-7. Where exterior walls have a fire separation distance of more than 10 feet (3048 mm), the fire-resistance rating is allowed to be determined based only on interior fire exposure. This recognizes the reduced risk that is due to the setback from the lot line. For fire separation distances greater than 10 feet (3048 mm), the hazard is considered to be predominately from inside the building. See Figure 705-8. Thus, fire-resistance-rated construction whose tests are limited to interior fire exposure is considered sufficient evidence of adequate fire resistance under these circumstances. However, at a distance of 10 feet (3048 mm) or less, there is the additional hazard of direct fire exposure from a building on the adjacent lot and the possibility that it may lead to self-ignition at the exterior face of the exposed building. Therefore, exterior walls located very close to any lot line must be rated for exposure to fire from both sides. The listings of various fire-resistance-rated exterior walls will indicate if they were only tested for exposure from the inside, usually by a designation of "FIRE SIDE" or similar terminology. Where so listed, their use is limited to those applications where the wall need only be rated from the interior side. It should be noted that this allowance is applicable regardless of why the wall requires a rating.

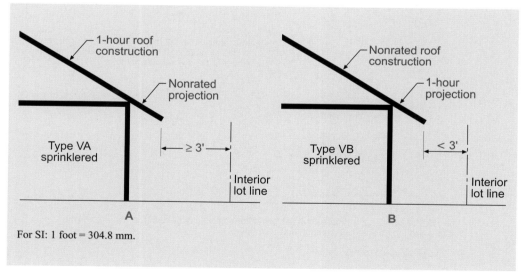

For SI: 1 foot = 304.8 mm.

Figure 705-3
Projection vs roof protection

Structural stability. This section refers the code user to Section 705.11 for parapets in determining the required height of exterior walls. It also requires the wall to have sufficient structural stability to remain in place for the duration of time indicated by the fire-resistance rating. The intent is that the wall must be provided with required fire-resistive materials for its entire height and be designed to support all applicable forces during the fire-resistive time period.

705.6

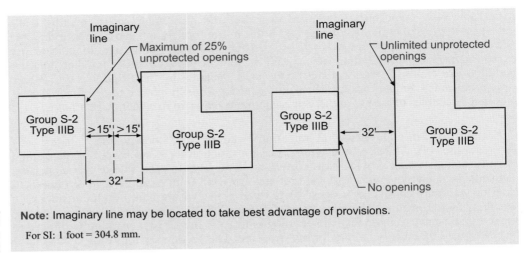

Figure 705-4
Buildings on the same lot

Note: Imaginary line may be located to take best advantage of provisions.

For SI: 1 foot = 304.8 mm.

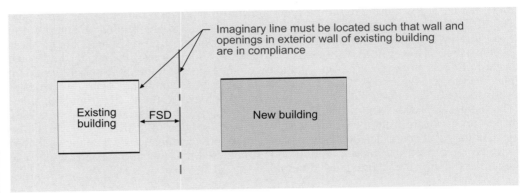

Figure 705-5
Buildings on the same lot

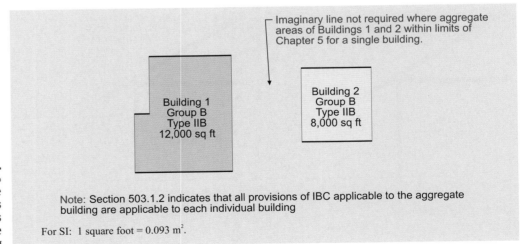

Figure 705-6
Multiple buildings considered as a single building

Note: Section 503.1.2 indicates that all provisions of IBC applicable to the aggregate building are applicable to each individual building

For SI: 1 square foot = 0.093 m².

705.7   **Unexposed surface temperature.** The provisions of this section provide a reduction of the prescriptive fire-resistance requirements for exterior walls under certain conditions. A fire-resistance-rated wall is generally required to meet the conditions of acceptance of ASTM E 119 or UL 263 for fire endurance and hose-stream tests on the surface exposed to the test fire, and heat-transmission limits on the unexposed surface. At fire separation distances beyond the point where no openings are allowed, typically 3 feet (914 mm), two more options are available:

1. Where opening protection is required, but the percentage of opening protection is not limited [typically a fire-separation distance of more than 20 feet (6096 mm)], compliance with the heat-transmission limits of ASTM E 119 or UL 263 is not required. This recognizes that, although heat transmission is an important consideration for interior walls, the fire hazard that the limit addresses is substantially reduced once the exterior wall of a building is set back far enough that the fire hazard it presents to (and receives from) a building on an adjacent lot does not warrant a limit on the percentage of opening protection to limit the hazard. It has the effect of compliance with the conditions of acceptance for fire assemblies of the same hourly rating.

   According to NFPA 252 for fire door assemblies and NFPA 257 for fire window assemblies, nearly identical conditions of acceptance for fire endurance and hose-stream tests are required, but not limits on heat transmission. Because an unlimited percentage of opening protection is allowed, the lack of a heat transmission limit for exterior walls is consistent with that for fire door and fire window assemblies. An exterior wall that does not meet the heat transmission limits is considered equivalent to an opening protective of the same hourly rating in its reduced ability to limit heat transmission.

2. Where the percentage of opening protection is limited [typically having a fire separation distance between 3 feet (914 mm) and 20 feet (6096 mm)], a similar reduction is possible, provided a correction is made according to the formula presented in this section. The formula converts the actual proposed area of protected openings to an increased equivalent area in proportion to the area of exterior wall surface under consideration that lacks adequate control of heat transmission. It places additional limits on the allowable percentage of opening protection.

   The formula increases the required percentage of opening protection, whereas Section 705.8 sets limits on the percentage. Relative to the limitations of Section 705.8, this method allows for a smaller percentage of opening protection at the same fire separation distance. Thus, a greater fire separation distance is required to maintain the same percentage of opening protection. The reduction of the heat transmission capacity of the exterior walls is compensated by a reduction in the allowable percentage of opening protection. Without this provision, a fire-resistance-rated exterior wall that does not meet the heat transmission limits would not be allowed.

If actual test results or other substantiating data are available, they may be used in the computations. In their absence, the standard time-temperature curve of ASTM E 119 and UL 263 would be used, which results in an equivalent area of protected openings equal to the actual area of protected openings plus the exterior wall area without adequate control of heat transmission per ASTM E 119 and UL 263. This is converted into a percentage of opening protection and compared to the limits of Section 705.8. The use of actual test results may reduce this effect. See Section 705.8 for the basic limits prior to modification by this provision.

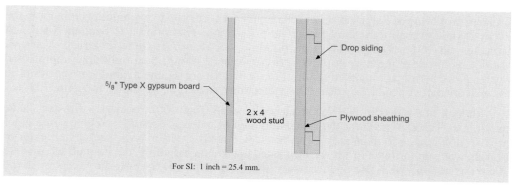

5/8" Type X gypsum board

2 x 4 wood stud

Drop siding

Plywood sheathing

For SI: 1 inch = 25.4 mm.

Figure 705-7
Nonsymmetrical exterior wall construction

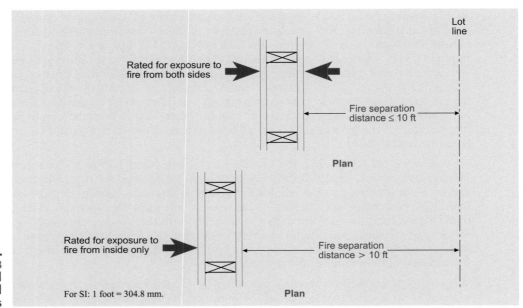

Lot
line

Rated for exposure to
fire from both sides

Fire separation
distance ≤ 10 ft

Plan

Rated for exposure to
fire from inside only

Fire separation
distance > 10 ft

For SI: 1 foot = 304.8 mm.

Plan

Figure 705-8
Nonsymmetrical
exterior wall
ratings

705.8.1   **Allowable area of openings.** Openings in an exterior wall typically consist of windows and doors. Occasionally, air openings such as vents are also present. The maximum area of either protected or unprotected openings permitted in each story of an exterior wall is regulated by this section. In addition, both unprotected and protected openings are permitted in the same exterior wall based upon a unity formula. The term *protected* in this section refers to those elements such as fire doors, fire windows and fire shutters regulated in Section 715. Protected openings have the mandated fire-protection rating necessary to perform their function. *Unprotected* openings are simply those exterior openings that do not qualify as protected openings. Opening protection presents a higher fire risk than fire-resistance-rated construction insofar as it does not meet the heat transmission limits of ASTM E 119 or UL 263, as previously discussed. At increasing distances from where openings are no longer prohibited, the hazard from heat radiation decreases, allowing the percentage of openings, both protected and unprotected, to increase. The high hazard of heat exposure at small fire separation distances justifies the prohibition of openings in order to limit the percentage of wall area without adequate heat transmission limits. As the fire separation distance increases, the percentage of openings is allowed to increase in compensation. At greater distances, the limit on the percentage of opening protection is eliminated. This recognizes that, at greater distances, the lack of adequate control of heat transmission does not pose a significant hazard to adjacent buildings, but containment of the fire to its origin inside the exposed building is still important.

There is a distance from lot lines where the hazard is reduced to such a degree that all opening limitations are no longer warranted. At this point, exposure to and from adjacent buildings is not significant and the need for fire resistance at exterior walls is reduced to fire protection of bearing walls and structural members in order to delay building collapse in the event of fire. Arguably the most important provision is Exception 2 to Section 705.8.1. It indicates that if the exterior wall of the building, and its primary exterior structural frame, are not required by the code to have a fire-resistance rating, then unlimited unprotected openings are permitted. In other words, if the wall does not require a rating, any openings in the wall are unregulated for area and fire protection. An example is shown in Figure 705-9.

Although not stated in the exception, only when Table 602 requires a fire-resistance rating does Table 705.8 limit the maximum area of exterior openings. Where some other provision of the code mandates a fire-resistance-rated exterior wall, such as an exterior bearing wall supporting a fire-resistance-rated horizontal assembly, the limitations of Table 705.8 do not apply. It is also not necessary to review Table 601 to apply the exception as the conditions are established in such a manner that Table 602 provides all of the necessary information. Directly stated, if Table 602 does not mandate a fire-resistance-rated exterior wall, an unlimited amount of unprotected openings are permitted.

The limitation on exterior openings is also not applicable for first-story openings in buildings of other than Group H as indicated in Exception 1 to Section 705.8. Limited in application, this exception allows an unlimited amount of unprotected openings at the first story under specified circumstances. Applicable only to those buildings that require a fire-resistance-rated exterior wall where the fire separation distance equals or exceeds 10 feet, the provisions are often utilized for opposing buildings having storefront systems. See Figure 705-10.

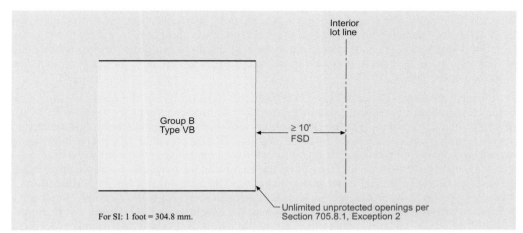

Interior lot line

Group B
Type VB

≥ 10'
FSD

For SI: 1 foot = 304.8 mm.

Unlimited unprotected openings per
Section 705.8.1, Exception 2

Figure 705-9
Unlimited
unprotected
openings

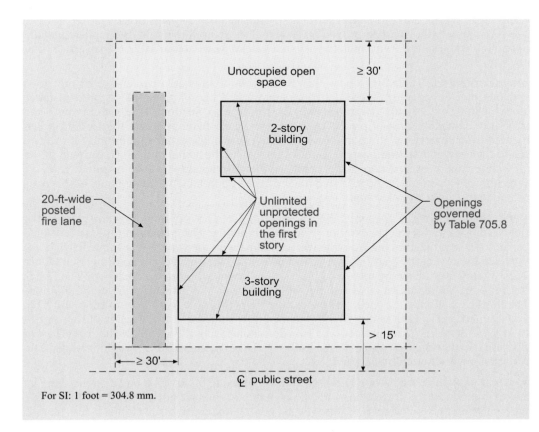

Unoccupied open
space

≥ 30'

2-story
building

20-ft-wide
posted
fire lane

Unlimited
unprotected
openings in
the first
story

Openings
governed
by Table 705.8

3-story
building

> 15'

≥ 30'

℄ public street

For SI: 1 foot = 304.8 mm.

Figure 705-10
Unlimited
openings in
the first story
exterior wall

How are exterior openings regulated in fully sprinklered buildings? Table 705.8 also recognizes an increase in the allowable area of unprotected exterior openings for those buildings that are provided with an automatic sprinkler system throughout. In other than higher level Group H occupancies, the maximum permitted area of unprotected openings in an exterior wall is allowed to be the same as the tabulated limitations for protected openings, provided the building is protected throughout with an NFPA 13 automatic sprinkler system. For example, the exterior wall of a fully sprinklered building having a fire separation distance of 15 feet (4572 mm) may have 75 percent of its surface area consist of unprotected openings. If the building is not sprinklered, the limit on unprotected openings is only 25 percent. The increased areas permitted due to sprinkler protection are all incorporated directly into Table 705.8. It is important to note that the presence of an automatic sprinkler system does not increase the maximum allowable opening area for protected openings. Whereas the benefits of such an increase would seem justifiable because of the increased level of protection, such an allowance is not addressed in the code. In addition, the unity formula (Equation 7-2) is not applicable to fully sprinklered buildings insofar as the code provides an increased allowance for unprotected openings to the amount permitted for protected openings.

705.8.2 **Protected openings.** Section 715.5 is referenced for identifying the level of protection required for windows needing opening protection. It further references Section 715.4 for fire doors and fire shutters. The use of sprinklers and water curtains to eliminate the required opening protection is addressed in the exception. It indicates that where the building is sprinklered throughout, those openings protected by an approved water curtain do not need to be fire-protective assemblies. However, the exception has virtually no application when the provisions of Table 705.8 are implemented, as the table allows for the elimination of protected openings in sprinklered buildings without the need for water curtains. There are a number of provisions throughout the IBC where the exception could be used. For example, Section 1027.5.2 typically mandates $^3/_4$-hour fire-protected openings in walls of egress courts less than 10 feet (3048 mm) in width. In a fully sprinklered building, the use of a complying water curtain would eliminate the need for such openings to have a fire-protection rating. Other provisions where the exception might be applied include Section 1007.7 for exterior areas for assisted rescue and Section 1022.6 for vertical exit enclosures.

705.8.4 **Mixed openings.** Table 705.8 specifies the maximum allowable percentage of protected and unprotected openings, considered separately and based on fire separation distance alone. The unity formula (Equation 7-2) as set forth in Section 705.8.4 determines the maximum allowable area of protected and unprotected openings where they are proposed together in an exterior wall at an individual story of a nonsprinklered building. It offers a traditional interaction relationship, namely, the sum of the actual divided by the sum of the allowable cannot exceed one. An example of the determination of the maximum area of exterior wall openings, where both protected and unprotected openings are utilized, is provided in Application Example 705-1.

705.8.5 **Vertical separation of openings.** The intent of this section is to limit the vertical spread of fire from floor to floor at the exterior wall of the building. The code requires exterior flame barriers either projecting out from the wall or in line with the wall. These flame barriers are intended to prevent the leap-frogging effect of a fire at the outside of a building. See Figure 705-11. However, there are three exceptions that eliminate the required barriers. The first is for buildings three stories or less in height. The second is for fully sprinklered buildings. The third exception is for open parking garages. It is probable that this provision will have very limited application, as it is doubtful there will be much new construction of four stories or more without sprinkler protection. Provisions addressing the spread of fire from floor to floor on the interior side of an exterior wall, such as at the intersection of a floor and curtain wall system, are found in Section 714.4.

705.8.6 **Vertical exposure.** The scope of this section is limited to buildings located on the same lot and to the issue of the protection of openings in the exterior wall of a higher building above the roof of a lower building. It requires each opening in the exterior wall that is less than 15 feet (4572 mm) above the roof of the lower building to be protected if the horizontal fire separation distance of the opening in the exterior wall of the higher building is less than 15 feet (4572 mm) from the exterior wall of the lower building. See Figure 705-12. There is an exception where the roof construction has at least a 1-hour fire-resistance rating, also illustrated in Figure 705-12. Application of this provision potentially mandates a higher level of protection than that required by the code for two buildings on separate adjoining lots. The presence of a lot line between two buildings institutes the concept of fire separation distance in the regulation of the opposing exterior walls and any openings in such walls. On a single lot with two buildings, the same concept is applied owing to the requirement for the placement of an assumed imaginary line between the buildings. This line is also the basis for regulating exterior wall and opening protection that is due to fire separation distance. The provisions of Section 705.8.6 introduce additional requirements that may not be mandated on account of the fire separation distance concept. In addition, where two buildings are located on the same lot, the provisions of Sections 705.3 and 503.1.2 permit them to be considered a single building if the aggregate area of the buildings is within the limits of Chapter 5 for a single building. For consistent application of the fire separation distance

Application Example 705-1

GIVEN: A nonsprinklered Group S-1 building of Type III B construction. The exterior wall shown is located 12 feet from an interior lot line.

DETERMINE: The maximum area permitted for unprotected openings.

SOLUTION:

$$\frac{A_a}{a} + \frac{A_u}{a_u} \leq 1.0$$

$$\frac{83}{(45\%)(18 \times 40)} + \frac{A_u}{(15\%)(18 \times 40)} = 1.0$$

$$\frac{83}{324} + \frac{A_u}{108} = 1.0$$

$$0.25 + \frac{81}{108} = 1.0$$

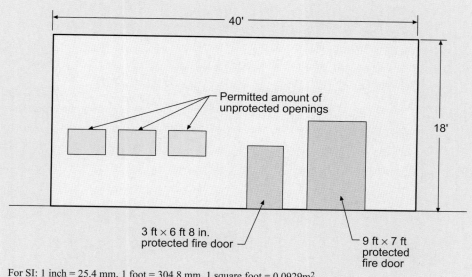

Permitted amount of unprotected openings

40'

18'

3 ft × 6 ft 8 in. protected fire door

9 ft × 7 ft protected fire door

For SI: 1 inch = 25.4 mm, 1 foot = 304.8 mm, 1 square foot = 0.0929m².

81 square feet of unprotected openings are permitted

concept, it would appear that the methodology for buildings on the same lot could be permitted to be utilized rather than the vertical exposure provisions of Section 705.8.6.

It appears that the provision is intended to mirror the termination requirements and allowances for fire walls where located in stepped buildings as established in Section 706.6.1. It would be logical to assume that if a fire wall is not necessary to obtain code compliance, resulting in no required application of the fire wall termination requirements, the same concept should be considered when applying this provision to buildings on the same lot that can be regulated as a single building.

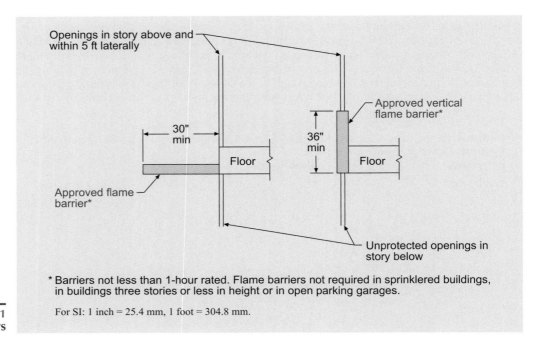

Openings in story above and within 5 ft laterally

30" min

36" min

Floor

Approved vertical flame barrier*

Floor

Approved flame barrier*

Unprotected openings in story below

* Barriers not less than 1-hour rated. Flame barriers not required in sprinklered buildings, in buildings three stories or less in height or in open parking garages.

For SI: 1 inch = 25.4 mm, 1 foot = 304.8 mm.

Figure 705-11
Flame barriers

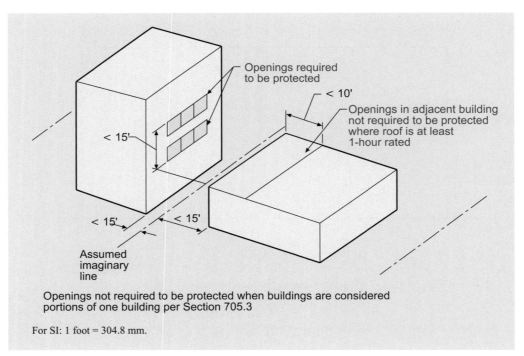

Openings required to be protected

< 10'

< 15'

Openings in adjacent building not required to be protected where roof is at least 1-hour rated

< 15'   < 15'

Assumed imaginary line

Openings not required to be protected when buildings are considered portions of one building per Section 705.3

For SI: 1 foot = 304.8 mm.

Figure 705-12
Vertical exposure

Parapets. This section intends that the exterior walls of buildings shall extend a minimum **705.11** of 30 inches (762 mm) above the roof to form a parapet. There are two reasons for the parapet:

1. To prevent the spread of fire from the roof of the subject building to a nearby adjacent building.

2. To protect the roof of a building from exposure that is due to a fire in an adjacent nearby building.

Most buildings do not have complying parapets, and those that do utilize them to hide the roof slope or roof-top equipment. Therefore, the exceptions to this section tend to become the general rule. Three of the six exceptions listed in the code—1, 3 and 6—involve cases where the parapet would serve no useful purpose. In Exception 2, a concession is made to the small-floor area building, and in Exceptions 4 and 5, an alternate method for providing equivalent protection is delineated. It is not necessary that all of the exceptions listed apply. Compliance with only one of the exceptions is all that is necessary for the elimination of a parapet.

Certainly, walls not required to be of fire-resistance-rated construction would not benefit from a parapet. In the case of walls that terminate at 2-hour fire-resistance-rated roofs or roofs constructed entirely of noncombustible materials, the parapet would be of little benefit, as the construction of the roof would prevent the spread of fire from or into the building. The exception for noncombustible roof construction is not intended to preclude the use of a classified roof covering.

In the case of walls permitted to have unprotected openings in conformance with Exception 6, the code assumes that the exterior wall will be far enough away from either an exposing building or an exposed building so that the protection provided by the parapet will not be necessary. This distance will vary based on the presence of a sprinkler system in the building as shown in Figure 705-13.

The fourth exception makes a provision for 1-hour fire-resistance-rated exterior walls that are constructed similar to 2-hour fire walls that terminate at the underside of the roof sheathing, deck or slab. This provides designers with an alternate to the use of parapets while recognizing that these walls provide adequate protection of the structure and its occupants as well as consistency with Section 706.6 for fire walls. See Figure 705-14. Exception 5 applies only to Group R-2 and R-3 occupancies and is intended to protect at the roof line through the use of a noncombustible roof deck, fire-retardant-wood sheathing or a gypsum-board underlayment.

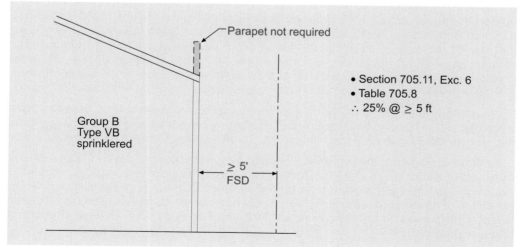

Parapet not required

Group B
Type VB
sprinklered

≥ 5'
FSD

• Section 705.11, Exc. 6
• Table 705.8
∴ 25% @ ≥ 5 ft

Figure 705-13
Parapet exception

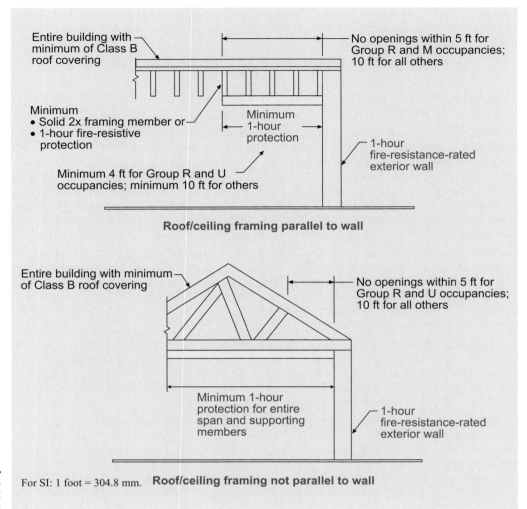

Entire building with minimum of Class B roof covering

No openings within 5 ft for Group R and M occupancies; 10 ft for all others

Minimum
• Solid 2x framing member or
• 1-hour fire-resistive protection

Minimum 1-hour protection

1-hour fire-resistance-rated exterior wall

Minimum 4 ft for Group R and U occupancies; minimum 10 ft for others

Roof/ceiling framing parallel to wall

Entire building with minimum of Class B roof covering

No openings within 5 ft for Group R and U occupancies; 10 ft for all others

Minimum 1-hour protection for entire span and supporting members

1-hour fire-resistance-rated exterior wall

Figure 705-14
Parapet alternative

For SI: 1 foot = 304.8 mm.   **Roof/ceiling framing not parallel to wall**

705.11.1   **Parapet construction.** In addition to having the same degree of fire resistance as required for the wall, the code also requires that the surface of the parapet that faces the roof be of noncombustible materials for the upper 18 inches (457 mm). Thus, a fire that might be traveling along the roof and reaching the parapet will not be able to continue upward along the face of the parapet and over the top and expose a nearby adjacent building. The requirement only applies to the upper 18 inches (457 mm) of the parapet to allow for extending the roof covering up the base of the parapet so that it can be effectively flashed. The 18-inch (457 mm) figure is based on a parapet height of at least 30 inches (762 mm).

As stated in the code, the 30-inch (762 mm) requirement is measured from the point where the roof surface and wall intersect. Therefore, when a cricket is installed adjacent to the parapet, the 30-inch (762 mm) dimension would be taken from the top of the cricket.

In those cases where the roof slopes upward away from the parapet and slopes greater than 2 units vertical in 12 units horizontal (16.7-percent slope), the parapet is required to extend to the same height as any portion of the roof that is within the distance where protection of openings in the exterior wall would be required. However, in no case shall the height of the parapet be less than 30 inches (762 mm). See Figure 705-15 for an illustration of this requirement.

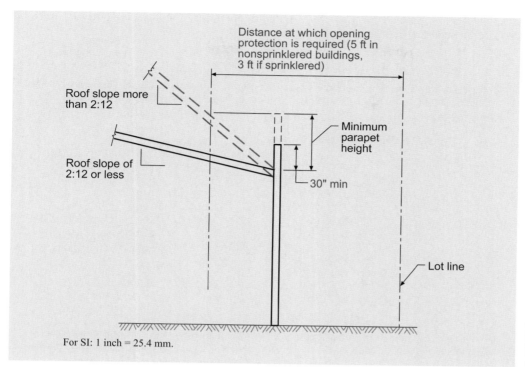

Distance at which opening protection is required (5 ft in nonsprinklered buildings, 3 ft if sprinklered)

Roof slope more than 2:12

Roof slope of 2:12 or less

Minimum parapet height

30" min

Lot line

For SI: 1 inch = 25.4 mm.

Figure 705-15
Parapet requirements

Section 706 *Fire Walls*

The IBC permits fire walls to be installed within a building, thereby creating one or more smaller-area buildings. It further intends that each portion of the structure so separated may be considered a separate building for all purposes of the code. The concept is based upon buildings on adjoining lots having a common party wall or two separate fire-resistance-rated walls located on the lot line. The high level of fire-resistance-rated construction between the two buildings, along with other controls, is deemed adequate for the protection of one building from its neighboring building. The use of one or more fire walls within a building is optional, based upon a decision by the designer. The code never mandates a fire wall be utilized, but rather offers it as an alternative to other mandated provisions. There are various reasons for utilizing fire walls within buildings; however, there are three such reasons that are quite common:

1. Allowable building area. The installation of one or more fire walls reduces the floor area in each of the separated buildings. Smaller floor areas can result in a reduction in the type of construction for one or more of the smaller buildings.

2. Multiple construction types. By separating a structure into separate buildings, they each are regulated independently for type of construction. Thus, not all of the structure would need to be classified based upon the lowest construction type involved.

3. Automatic sprinkler systems. Fire walls can be used to reduce building size for the purpose of eliminating a requirement for the installation of an automatic sprinkler system.

Examples of these various uses of fire walls are shown in Application Examples 706-1 through 706-3. A fourth application of the fire wall concept is found in Appendix B of the *International Fire Code*® (IFC®) relating to fire-flow requirements for buildings. Where structures are separated by fire walls without openings, the divided portions may be considered separate fire-flow calculation areas.

Application Example 706-1

GIVEN: A single-story Group E high school with a total floor area of 135,000 square feet. The building is fully sprinklered. Assume no frontage increase for allowable area purposes.

DETERMINE: How fire walls can be used to allow for Type IIB construction.

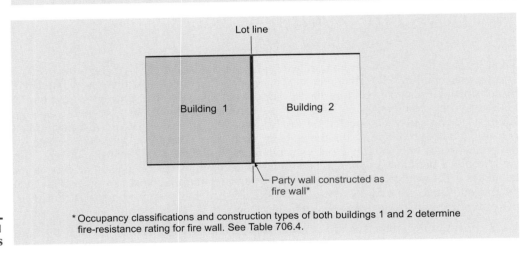

Type IIB construction

Minimum 2-hour fire wall per Table 706.4, Note a

≤ 58,000 sq ft | ≤ 58,000 sq ft

≤ 58,000 sq ft

T503	14,500 sq ft
Sprinkler	43,500 sq ft
Total	58,000 sq ft

For SI: 1 square foot = 0.093 m².

By creating a minimum of three separate buildings under one roof, Type IIB construction is acceptable. Slight increases in allowable area are possible by including available frontage increases; however, each of the three buildings must be evaluated individually for open space.

Lot line

Building 1   Building 2

Party wall constructed as fire wall*

* Occupancy classifications and construction types of both buildings 1 and 2 determine fire-resistance rating for fire wall. See Table 706.4.

Figure 706-1
Party walls

706.1 **General.** As previously mentioned, one or more fire walls may be constructed in a manner such that the code considers the portions separated by the fire walls to be separate buildings. Because a fire wall is such a critical element in the prevention of the spread of fire from one separated building to another, it is of great importance that the wall be situated and constructed properly. It must provide a complete separation. It should be noted that when a wall serves both as a fire barrier separating occupancies and a fire wall, the most restrictive requirements of each separation shall apply. The code also prohibits any openings in fire walls that are constructed on lot lines (defined as party walls).

706.1.1 **Party walls.** A common wall located on the lot line between two adjacent buildings is considered a party wall under this provision of the code. Regulated as a fire wall in accordance with the provisions of Section 706, a party wall can be used in lieu of separate and distinct exterior walls adjacent to the lot line as depicted in Figure 706-1. The hazard created by neighboring buildings adjacent to each other is further addressed through the requirement that no openings be permitted in a party wall. For purposes of this section, and consistent with the general provisions of Section 503.1 for structures containing fire walls, separate buildings are created.

GIVEN: An existing one-story sales building of Type VB construction.

DETERMINE: How a two-story office addition of Type IIB construction can be provided without causing the building to be considered Type VB throughout.

Application Example 706-2

Group M
Type VB

Group B
Type IIB

Fire wall creates two buildings, permitting independent construction types in each building

Minimum 3-hour fire wall per Table 706.4

GIVEN: A neighborhood retail center of 30,000 square feet total, divided into three 10,000 square foot tenants. The building is Type IIA construction.

DETERMINE: How the use of a fire wall will eliminate the requirement for a sprinkler system.

Application Example 706-3

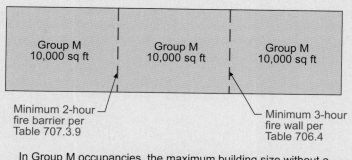

Group M
10,000 sq ft

Group M
10,000 sq ft

Group M
10,000 sq ft

Minimum 2-hour fire barrier per Table 707.3.9

Minimum 3-hour fire wall per Table 706.4

In Group M occupancies, the maximum building size without a sprinkler is 24,000 sq ft (Section 903.2.7, Item 3). Using a fire wall creates two separate buildings, each of which does not exceed the limit.

Structural stability. The objective of a fire wall is that a complete burnout can occur on one side of the wall without any effects of the fire being felt on the opposite side. Furthermore, the only damage to the wall will be the effects of fire and the shock of hose-stream application on the fire side. The code is very clear that fire walls should remain in place for the expected time period. Therefore, structural failure on either side of the wall shall not cause the collapse of the wall, nor can the required fire-resistance rating be diminished. In addition, structural members (especially members that conduct heat) that penetrate fire walls could limit their effectiveness and do not comply with this provision. Any structural member that passes through a fire wall could also adversely affect the integrity of the required fire-resistance-rated construction.

706.2

The intent of this section can be partially traced back to Section 101.3, which states that one of the goals of the code is to provide safety to fire fighters and emergency responders during emergency operations. During a fire, a fire wall provides a safe haven on the nonfire side for fire fighters to stage and fight a fire. It is critical that the fire wall does not pose a threat of collapse to the fire department personnel. This is more easily achieved where the fire wall is a nonbearing wall and is not penetrated by load-bearing elements. However, where a fire wall is proposed as a bearing wall, the building official should ensure that those structural members that frame into the wall will not cause the premature collapse of the fire wall prior to the hourly rating established for the wall. The structural engineer of record should provide evidence to this fact. If all structural elements framing into the fire wall, as well as their supporting members, have the same fire-resistance rating as the fire wall, it is reasonable to assume that the intent of the provision has been met.

706.3   **Materials.** In buildings of other than Type V construction, fire walls shall be constructed of noncombustible materials. The high degree of protection expected from a fire wall mandates that noncombustible construction be utilized for all but the lowest type of construction.

706.4   **Fire-resistance rating.** It is obvious that a fire wall performs the very important function of acting as a barrier to fire spread so that a fire on one side of the wall will not be transmitted to the other. On this basis, the fire wall must have a fire-resistance rating commensurate with the occupancy and type of construction in which it is constructed. The IBC provides that fire walls be of either 2-hour, 3-hour or 4-hour fire-resistance-rated construction as specified in Table 706.4. Permitted openings in fire walls are addressed in Section 706.8.

706.5   **Horizontal continuity.** A fire wall must not only separate the interior portions of the building but must also extend at least 18 inches (457 mm) beyond the exterior surfaces of exterior walls. See Figure 706-2. A number of exceptions permit the fire wall to terminate at the interior surface of the exterior finish material, with Exception 1 illustrated in Figure 706-3. Where combustible sheathing or siding materials are used, the wall must be protected for at least 4 feet (1220 mm) on both sides of the fire wall by minimum 1-hour construction with any openings protected at least 45 minutes. If the sheathing, siding or other finish material is noncombustible, such noncombustible materials shall extend at least 4 feet (1220 mm) on both sides of the fire wall; however, unlike the previous exception, no opening protection is required. As an option, where the separate buildings created by the fire wall are sprinklered, the fire wall may simply terminate at the interior surface of noncombustible exterior sheathing.

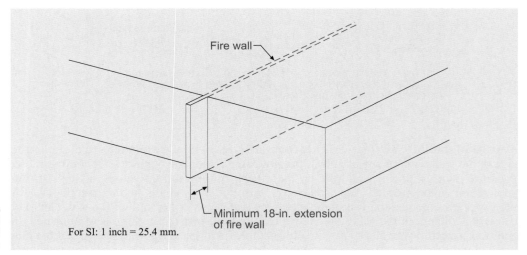

Fire wall

Minimum 18-in. extension of fire wall

For SI: 1 inch = 25.4 mm.

Figure 706-2
Horizontal
continuity

706.5.1   **Exterior walls.** Where a fire wall creating separate buildings intersects with the exterior wall, there is the potential for direct fire exposure between the buildings at the exterior. Unless the intersection of the exterior wall and the fire wall forms an angle of at least 180 degrees (3.14 rad), such as a straight exterior wall with no offsets, a condition occurs similar to that of two buildings located on the same site. The proximity of the two buildings may be such that the distance between them would allow for direct fire or substantial radiant heat to be transferred from one building to the other. This condition is also possible where the two buildings on the lot are portions of a larger structure with fire wall separations.

Where the fire wall intersects the exterior wall to form an angle of less than 180 degrees (3.14 rad), the exterior wall for at least 4 feet (1220 mm) on both sides of the fire wall shall be of minimum 1-hour fire-resistance-rated construction, and all openings within the 4-foot (1220 mm) portions of the exterior wall are to be protected with 45-minute fire assemblies. See Figure 706-4. As an option, an imaginary lot line may be assumed between the two buildings created by the fire wall and the exterior wall and opening protection would be based upon the fire separation distances to the imaginary lot line. This method is consistent with the provisions of Section 705.3 for addressing two buildings on the same lot. An example is shown in Figure 706-5.

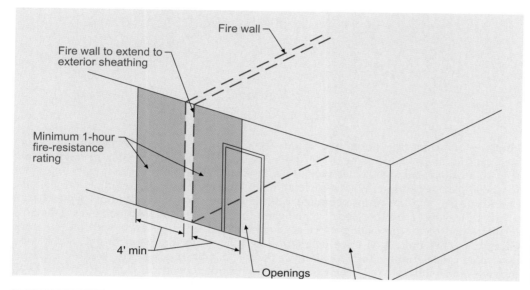

Figure 706-3
Horizontal continuity

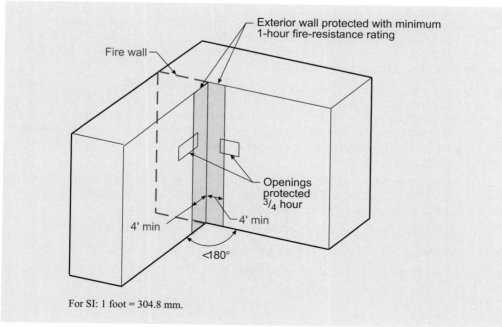

For SI: 1 foot = 304.8 mm.

Figure 706-4
Fire wall intersection with exterior walls

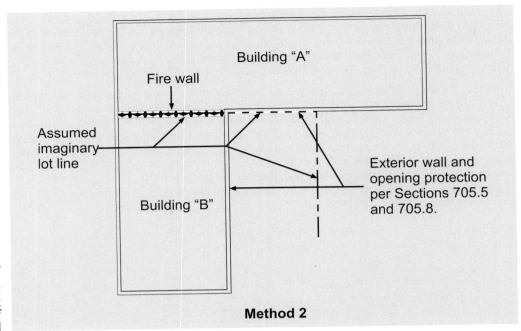

Figure 706-5
Imaginary lot
line at
extension of
fire wall

Building "A"

Fire wall

Assumed
imaginary
lot line

Exterior wall and
opening protection
per Sections 705.5
and 705.8.

Building "B"

Method 2

706.5.2 Horizontal projecting elements. Under the conditions where a horizontal projecting element such as a roof overhang or balcony is located within 4 feet (1220 mm) of a fire wall, the wall must extend to the outer edge of the projection. This general requirement provides for a complete separation by totally isolating all building elements, including projections, on either side of a fire-resistance-rated wall. However, such a condition is typically not visually pleasing. Therefore, the code indicates the fire wall is not required to extend to the leading edge of the projecting element if constructed in compliance with one of three exceptions. The protection must extend through the projecting element unless it has no concealed spaces. Where the projecting element is combustible, the fire wall shall extend through the concealed area, whereas in noncombustible construction, the extension need only be 1-hour fire-resistance-rated construction. Under all of the exceptions,, the exterior wall behind and below the projecting element is to be of 1-hour fire-resistance-rated construction for a distance not less than the depth of the projecting element on both sides of the wall. All openings within the rated exterior wall are to be protected by fire assemblies having a minimum fire-protection rating of 45 minutes. Figure 706-6 depicts these various conditions addressed in the exceptions.

706.6 Vertical continuity. Having established the intent of the IBC that fire walls prevent the spread of fire around or through the wall to the other side, the IBC further ensures the separate building concept by specifying that the wall shall extend continuously from the foundation to (and through) the roof to a point 30 inches (762 mm) or more above the roof. The 30-inch (762 mm) parapet prevents the spread of fire along the roof surface from the fire side to the other side of the wall.

Several exceptions, some of which are illustrated in Figure 706-7, allow the fire wall to terminate at the underside of the roof sheathing, deck or slab, rather than terminate in a parapet. The basis for such exceptions include:

1. Equivalent protection being provided by an alternate construction method;

2. Aesthetic considerations, as parapets disrupt the appearance of the roof; or

3. A combination of the two previous reasons.

It is emphasized that the term *fire wall* also limits its use to vertical walls. Therefore, there can be no horizontal offsets nor can the plan view of the wall change from level to level. See Figure 706-8.

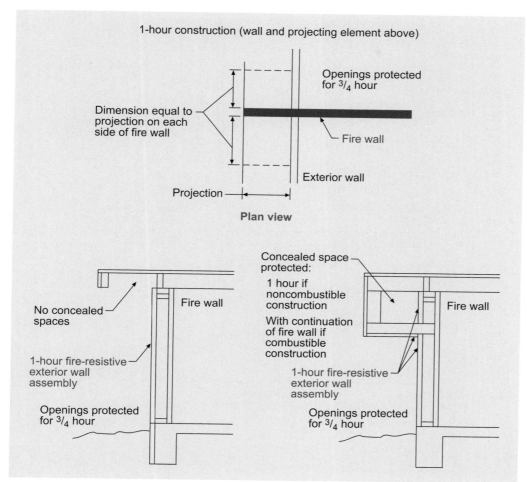

Figure 706-6
Horizontal projecting elements

Stepped buildings. Quite often, a fire wall is provided at a point in the building where the roof changes height. Under such conditions, the fire wall must extend above the lower roof for a minimum height of 30 inches (762 mm). In addition, the exterior wall shall be of at least 1-hour fire-resistance-rated construction for a total height of 15 feet (4572 mm) above the lower roof. The exterior wall shall be of fire-resistance-rated construction from both sides. Any opening that is located in the lower 30 inches (762 mm) of the wall shall be regulated based upon the rating of the fire wall. Openings above the 30-inch (762 mm) height, but not located above a height of 15 feet (4572 mm), shall have a minimum fire-protection rating of 45 minutes. An illustration of this provision is depicted in Figure 706-9.

706.6.1

An alternative is described in the exception that allows the fire wall to terminate at the underside of the roof sheathing, deck or slab of the lower roof. It is very similar to Exception 2 in Section 706.6. Because the greatest exposure occurs from a fire penetrating the lower roof and exposing the adjacent exterior portion of the fire wall, the code mandates that all protection be applied to the roof assembly of the lower roof. As shown in Figure 706-10, the lower roof assembly within 10 feet (3048 mm) of the wall shall be of minimum 1-hour fire-resistance-rated construction. In addition, no openings are permitted in the lower roof within 10 feet (3048 mm) of the fire wall.

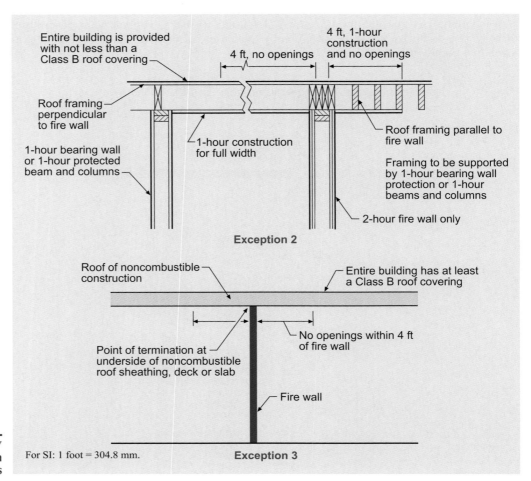

Figure 706-7
**Termination
of fire walls**

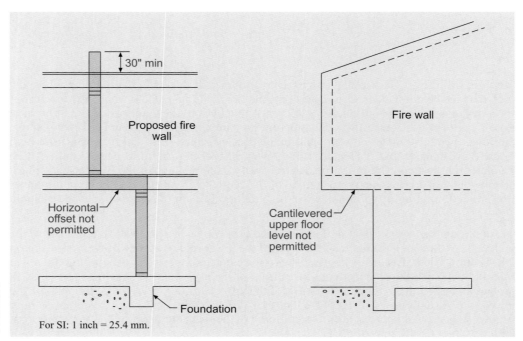

Figure 706-8
**Fire wall
vertical
continuity**

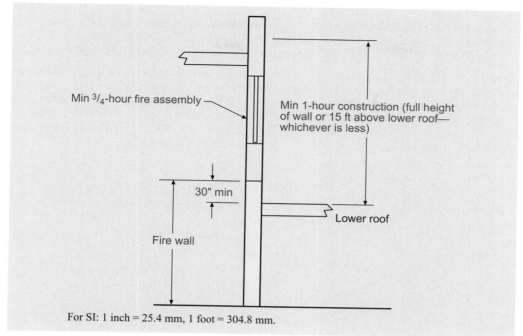

For SI: 1 inch = 25.4 mm, 1 foot = 304.8 mm.

Figure 706-9
Stepped buildings

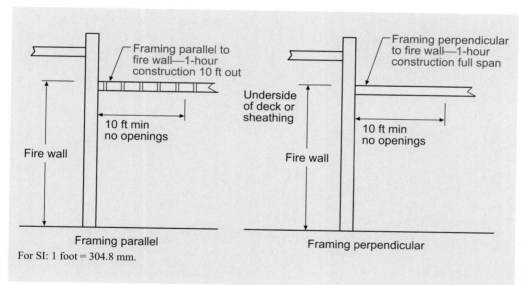

For SI: 1 foot = 304.8 mm.

Figure 706-10
Stepped buildings

Combustible framing in fire walls. This section defines the limitations placed on the penetration of combustible framing into concrete or masonry fire walls. Adjacent combustible members framing into a concrete or masonry fire wall from opposite sides require a minimum distance of 4 inches (102 mm) between embedded ends as shown in Figure 706-11. All hollow spaces at this location shall be solidly filled for the full thickness of the wall and for a distance not less than 4 inches (102 mm) above, below and between the structural members with noncombustible materials approved for fireblocking. **706.7**

Openings. In order to provide for the efficient use of a structure containing one or more fire walls, the code permits their penetration with protected openings. The fire-protective rating **706.8**

required for openings is addressed in Table 715.4. Each opening is limited to 156 square feet (15 m²), with multiple openings permitted. However, the aggregate width of all openings at any floor level is limited to 25 percent of the length of the fire wall. See Figure 706-12. The limit of 156 square feet (15 m²) per opening does not apply in fully sprinklered buildings.

A party wall, defined as a fire wall located on a lot line between two adjacent buildings and used for joint service of the buildings, is prohibited by Section 706.1.1 from having any door, window or other openings. This would include any openings above the lower roof normally permitted under Section 706.6.1 for stepped buildings. This provision is consistent with the requirements of Section 705.8 for exterior walls located on an interior lot line. Penetrations by ducts or air-transfer openings are also prohibited.

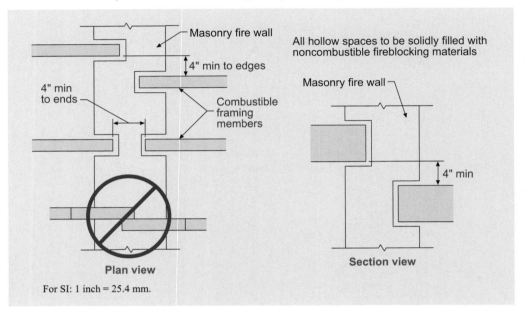

Figure 706-11
Combustible framing in fire walls

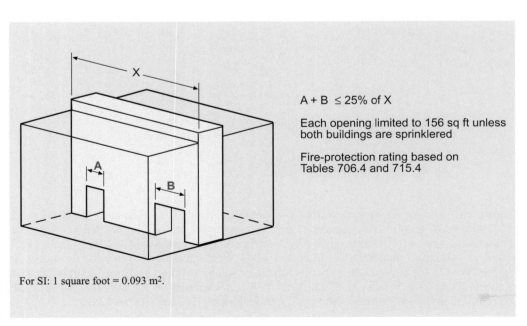

Figure 706-12
Openings in fire walls

Section 707 *Fire Barriers*

A common function of a fire barrier is to totally isolate one portion of a floor level from another through the use of fire-resistance-rated walls and opening protectives. Fire-resistance-rated horizontal assemblies are also often used in conjunction with fire barriers in multistory buildings in order to isolate areas vertically. This section identifies the different uses for fire barriers, as well as the method in which fire barriers are to be constructed.

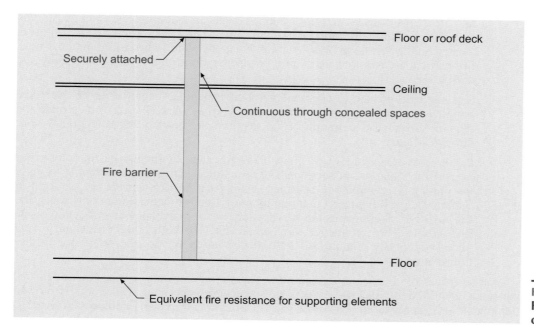

Figure 707-1
Fire barrier continuity

Fire-resistance rating. A fire barrier shall be used to provide the necessary separation for the following building elements or conditions: **707.3**

1. Shaft enclosure. The minimum required degree of fire-resistance for fire barriers used to create a shaft enclosure is based primarily on the number of stories connected by the enclosure. A minimum 2-hour fire-resistance rating is mandated where four or more stories are connected, with only a 1-hour rating required where connecting only two or three stories. In all cases, the rating of the fire barriers creating a shaft enclosure must equal or exceed that of the floor assembly that is penetrated by the enclosure.

2. Exit enclosure. The separation between an exit enclosure (stair enclosure) and the remainder of the building shall be accomplished with fire barriers having either a 1- or 2-hour fire-resistance rating as required by Section 1022.1.

3. Exit passageway. An exit passageway must be isolated from the remainder of the building by minimum 1-hour fire-resistance-rated fire-barrier walls. In multistory buildings, minimum 1-hour fire-resistance-rated horizontal assemblies must also be used to totally isolate the exit passageway. Where an exit passageway is a continuation of an exit enclosure, it must, at minimum, maintain the fire-resistance rating of the enclosure.

4. Horizontal exit. A minimum 2-hour fire-resistance-rated fire barrier may be utilized to create a horizontal exit when in compliance with all of the other

provisions of Section 1025. The fire barrier creates protected compartments where occupants of the building can travel to escape the fire incident.

5.   Atrium. Unless a complying glazing system or $^3/_4$-hour glass block construction is used, minimum 1-hour fire barriers are required when isolating an atrium from surrounding spaces.

6.   Incidental accessory occupancies. Table 508.2.5 indicates the required separation or protection required for special hazard areas such as storage rooms, laboratories and furnace rooms. Where a 1- or 2-hour fire-resistance-rated wall is required, it shall be a fire barrier.

7.   Control area. Table 414.2.4 identifies the minimum required fire-resistance rating for fire barriers used to create control areas in buildings housing hazardous materials. A minimum rating of 1 hour is mandated for separating control areas located on the first three floor levels above grade plane, whereas minimum 2-hour fire barriers are required for control area separations on all floor levels above the third level.

8.   Separated occupancies. The separation of dissimilar occupancies in the same building is accomplished by fire barriers. Table 508.4 is used to determine the required fire-resistance rating of the required fire barriers, ranging from 1 hour through 4 hours.

9.   Fire area. Where a building is divided into fire areas by fire barriers in order to not exceed the limitations of Section 903.2 for requiring an automatic sprinkler system, the minimum required fire-resistance ratings of the fire barriers are set forth in Table 707.3.9. Ranging from a minimum of 1 hour to a maximum of 4 hours, the fire-resistive requirements are based solely on the occupancy classification of the fire areas. The provisions are applicable to both single-occupancy and mixed-occupancy conditions. See the discussion on Section 901.7 for further information.

Note also that fire barriers are required as separation elements in other miscellaneous locations identified by the code, such as stage accessory areas (Section 410.5) and flammable finish spray rooms (Section 416.2). Throughout the code, references are made to fire barriers as the method of providing the appropriate fire-resistance-rated separation intended. In addition, many of the other *International Codes* also address the use of fire barriers to create protected areas.

707.5 Continuity. Fire barriers must begin at the floor and extend uninterrupted to the floor or roof deck above. Where there is a concealed space above a ceiling, the fire barrier must continue through the above-ceiling space. See Figure 707-1. Fireblocking, usually only required in combustible construction, is also required at every floor level if the fire barrier contains hollow vertical spaces. The intent of a fire barrier is to provide a continuous separation so as to completely isolate one area from another. As with many other fire-resistance-rated elements, the supporting construction must be of an equivalent rating to the fire barrier supported. A reduction relates to 1-hour incidental accessory occupancy separations in nonrated construction.

707.6 Openings. The provisions of Section 715 regulate the protection of openings in fire barriers. The fire-protection ratings mandated for fire-barrier openings in Tables 715.4 and 715.5 vary depending on the fire-resistance rating of the fire barrier as well as its purpose. The required rating may be as little as $^3/_4$ hour to as much as 3 hours.

The code limits the amount of openings that can occur in a fire barrier to an aggregate width of 25 percent of the wall length. In addition, any single opening is limited to 156 square feet (15 m²) in area. Where the fire door serves an exit enclosure, or where the opening protective assembly was tested in accordance with ASTM E 119 or UL 263 and rated equivalent to the fire barrier, the limitations are not applicable; in a fully sprinklered building there is no limit to the size of the openings.

Section 708 *Shaft Enclosures*

It is well known that one of the primary means for the spread of fire in multistory buildings, particularly older buildings, has been the transmission of hot gases and fire upward through unprotected or improperly protected vertical openings. The primary cause of death in the hotel portion of the MGM Grand Hotel in Las Vegas, Nevada, as a result of the fire in November 1980, was the upward transmission of smoke through inadequately protected elevator shafts, stair shafts, and heating and ventilating shafts. It is because of this potential for fire spread vertically through buildings that this section requires that vertical openings be protected with fire-resistance-rated shaft enclosures.

Shaft enclosure required. Applicable to both fire-resistance-rated and nonfire-resistance-rated horizontal assemblies, this section is the charging language requiring shaft protection for vertical openings. It refers to other provisions of the section regarding such issues as fire-resistance ratings, continuity, openings, penetrations and extent of enclosure. Additionally, this section is the starting point in the evaluation of openings that occur in a floor/ceiling assembly. Wherever an opening occurs, it is first regulated by this section as needing protection through the use of a shaft enclosure. If a shaft enclosure is not desired, a variety of other protective measures are set forth. There are a number of exceptions that modify the general enclosure requirements by eliminating the need for shaft protection: **708.2**

1. Within individual dwelling units not exceeding four stories in height, shaft enclosures are not required.

2. In fully sprinklered buildings, the vertical openings created for an escalator or stairway not required as a means of egress need not be protected by a shaft enclosure when protected by one of the following two methods:

    2.1. By limiting the size of the openings, along with the installation of draft curtains and closely-spaced sprinklers, the code assumes that the vertical openings in sprinklered buildings do not present an untenable condition. The purpose of the required draft curtain is to trap heat so that the sprinklers will operate and cool the gases that are rising. The curtain is not a fabric; it is constructed of materials consistent with the type of construction of the building. In other than mercantile and business occupancies, the use of this method is limited to openings connecting four or fewer stories.

    2.2. Approved power-operated automatic shutters may also be used to cut off the openings between floors. Required to have a minimum fire-protection rating of at least $1^1/_2$ hours, the shutters shall close immediately upon activation of a smoke detector. Obviously, operation of an escalator must stop once the shutter begins to close.

3. Penetrations of cables, cable trays, conduit, tubing or similar penetrating items are permitted without complying with the requirements for shaft construction, provided the openings are protected in conformance with the provisions of Section 713.4. It is important to note that vents are also identified in this exception, but only those vents that convey products of combustion as defined in the International Mechanical Code® (IMC®). The exception is not intended to apply to exhaust ducts.

4. As an alternate to the required shaft enclosures for ducts penetrating floor systems, the provisions of Section 716 also regulate the penetration of ducts through horizontal assemblies. If the provisions of Section 716.6 do not mandate the installation of a damper, the provisions of Sections 713.1.1 and Sections 713.4 through 713.4.2.2 are applicable. Where Section 716.6 mandates a damper in the duct or air transfer opening, then Section 716 applies. If the installation is in full compliance with these sections as applicable, ducts need not be protected by shaft enclosures. Otherwise, a shaft enclosure is required.

5. Atriums are intended to be open vertically. Where such special building features are designed and constructed in compliance with their own unique provisions, shaft enclosures are not required. The use of an atrium is a design option, voluntarily applied by the designer, and thus provided as an alternative to the shaft enclosure requirements. For further information, see the discussion of Section 404.

    It is a widely held belief that the allowance for atriums is also applicable to covered mall buildings that meet the special requirements of Section 402. However, there is no specific exception to Section 708.2 that exempts multistory covered mall buildings from the shaft enclosure requirements. The mall portion of the building could be exempted under the provisions of Exceptions 5 or 7, but a general exemption for the entire building is not available.

    The use of Exception 5 is typically limited to those buildings where the floor openings connect three or more stories. Where a floor opening connects only two stories, Exception 7 is commonly applied to all occupancies other than Groups I-2 and I-3. In such institutional buildings, the use of an atrium is permitted as an alternative to shaft enclosure protection.

6. A masonry chimney extending through one or more floor levels does not require shaft protection where the annular space around the chimney is protected in the manner specified by Section 717.2.5.

7. This exception permits two adjacent stories to intercommunicate with each other without protection of the openings between the two stories, except in the case of Group I-2 and I-3 occupancies. As long as these intercommunicating openings serve only the one adjacent floor, shaft protection is not required. This provision is commonly used where office buildings have a lobby that extends up through the second story so that individuals on the second floor may look down over a guard into the lobby below. It is important that the unprotected floor opening be appropriately separated from floor openings serving other floors.

    In addition, the opening between floor levels cannot be concealed within the construction of a wall or floor-ceiling assembly, cannot be open to a fire-resistance-rated corridor in Group I-1 and R occupancies, cannot open to a fire-resistance-rated corridor on nonsprinklered floors in any occupancy, and cannot be a part of the required means of egress (unless permitted by Exceptions 3 or 4 to Section 1016.1, or Exception 1 to Section 1022.1. The limitation on concealment is intended to prevent unprotected openings that are completely enclosed by walls partitions, chases or floor/ceiling assemblies. Where the openings are concealed in this manner, they permit a fire within the concealed space to burn undetected and distribute products of combustion to the upper floor.

8. Automobile ramps in both enclosed and open parking garages, when in compliance with the provisions of Sections 406.3 and 406.4, respectively, are not required to be enclosed by shaft construction. Because the nature of these uses makes it impractical for shaft enclosures, other safeguards are provided by the code in Section 406.

9. By definition, a complying mezzanine is intended to be open into the room below. As such, floor openings between the mezzanine and the lower floor need not be enclosed by shaft construction.

10. Joints protected by a fire-resistant joint system, like other building elements protected by approved methods, do not create any additional hazard that needs to be addressed by enclosure in a shaft.

11. Interior stairways permitted under the conditions of Exceptions 3 and 4 to Section 1016.1 do not require enclosure. In both cases, they may not connect more than two stories and are not permitted to be open to any other stories in the building. It would also seem logical that Exception 1 to Section 1022.1 be included, as it also addresses a stairway that connects no more than two stories.

12. Where a floor fire door assembly complying with Section 712.8 is installed, a shaft enclosure is not required.

13. In a Group I-3 occupancy, openings in floors within a housing unit are permitted without a shaft enclosure provided four specific conditions are met.

14. The enclosure of elevator hoistways in open parking garages and enclosed parking garages is not required for those hoistways that only serve the parking garage.

15. Shaft protection is not mandated for the enclosure of mechanical exhaust and supply duct systems in both types of garage facilities. The protection of vertical openings provided to accommodate elevators, as addressed in Item 14, as well as exhaust ducts and supply ducts, is unnecessary since the vehicle ramps of open and enclosed parking garages are permitted to be open at all levels.

16. Throughout the code, there may be other allowances for floor penetrations or openings that are adequately regulated without the need for shaft enclosures. Where permitted, these openings must comply with the specifics of their use.

Fire-resistance rating. To provide an acceptable level of protection for vertical openings between floors, this section mandates that all shaft enclosures have a fire-resistance rating at least equivalent to the rating of the floor being penetrated, but never less than 1 hour. Therefore, in Type I construction, or where the shaft enclosure connects four or more stories, a minimum 2-hour enclosure is mandated. A shaft enclosure is never required to have a higher fire-resistance rating than 2 hours. **708.4**

Continuity. Shaft enclosures are required to be constructed as fire barriers, extending from the top of the floor/ceiling assembly below to the underside of the floor or roof deck above, except as permitted by Sections 708.11 and 708.12. It is important that the walls continue through any concealed spaces such as the area above a ceiling, and that any hollow vertical spaces within the shaft wall be fireblocked at each floor level. In addition, the supporting elements of any shaft enclosure construction must be of fire-resistance-rated construction equivalent to that of the shaft construction. The enclosure, fire-resistance-wise, should be continuous from the lowest floor opening through to its termination. **708.5**

Exterior walls. Unless required to be fire-resistance rated because of the proximity to an exterior exit balcony, vertical exit enclosure, or an exterior exit stairway or ramp, the exterior walls of a shaft enclosure need only be protected because of their location on the lot as regulated by Section 705. **708.6**

Enclosure at the bottom. Many shafts do not extend to the bottom of the building or structure. Therefore, it is necessary to provide an approved method for maintaining the integrity of the shaft enclosure at its lowest point. This section identifies three methods for enclosing the bottom of a shaft enclosure. First, the shaft can be enclosed with fire-resistance-rated construction equivalent to that for the lowest floor penetrated, with a minimum rating consistent with that of the shaft enclosure. Second, a termination room related to the purpose of the shaft can be considered to be the enclosure at the bottom, provided the room is separated from the remainder of the building with fire-resistance-rated construction and opening protectives equivalent to those of the shaft enclosure. Third, approved horizontal fire dampers can be used to protect openings at the lowest floor level in lieu of the enclosure at the bottom of the shaft enclosure. See Figures 708-1a, 1b and 1c. **708.11**

The first of the three exceptions eliminates the fire-resistance-rated room separation where there are no other openings into the shaft enclosure other than at the bottom. All portions of the enclosure bottom must be closed off except for the penetrating items, unless the room is provided with an automatic fire-suppression system. An example of this concept would be a vent enclosure. The second exception requires that a shaft enclosure containing a refuse or laundry chute be used for no other purpose and shall end in a termination room per Section 708.13.4. Exception 3 applies where the shaft enclosure contains no combustible materials. In this situation, there is no need for either the fire-resistance-rated room separation or protection at the bottom of the enclosure. An example would be a light well that extends through several floor levels to the roof as illustrated in Figure 708-2. It would be considered an extension of the floor below the level of the floor opening.

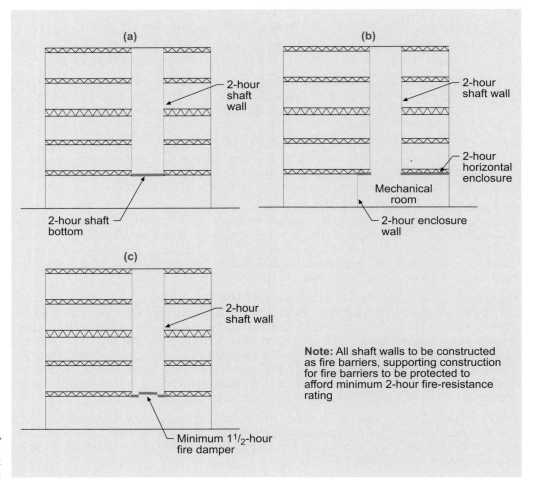

Figure 708-1
Enclosure at shaft bottom

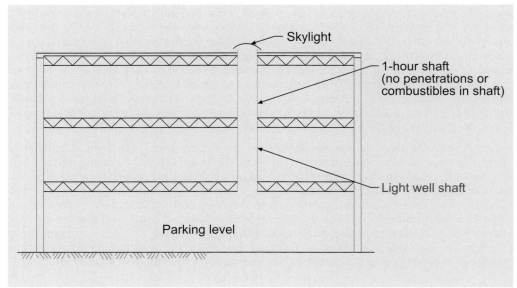

Figure 708-2
Vertical shafts— bottom enclosure

Enclosure at the top. Most shafts extend to or through the roof deck at the exterior, where there is no requirement to maintain the fire-resistance rating of the shaft enclosure construction. However, where the enclosure does not extend to the roof, the top of the shaft must be enclosed. The required fire-resistance rating of the shaft lid shall be equivalent to the rating of the topmost floor penetrated by the shaft, but in no case less than the fire-resistance rating required for the shaft. See Figure 708-3. **708.12**

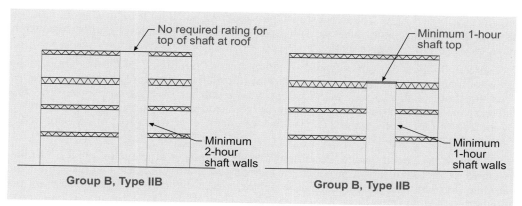

Figure 708-3
Top enclosure of shaft

Refuse and laundry chutes. The requirements of this section are intended to further strengthen the shaft-enclosure provisions where chutes and termination rooms for refuse or laundry are constructed. Refuse and laundry areas are often poorly maintained, with a greater potential for a fire incident than most other areas of a building. Coupled with the shaft conditions that are created by the chutes, these types of areas pose hazards that exceed those typically encountered. See Figure 708-4. To further secure the intent, Section 903.2.11.2 requires sprinkler protection for the chutes and termination rooms. **708.13**

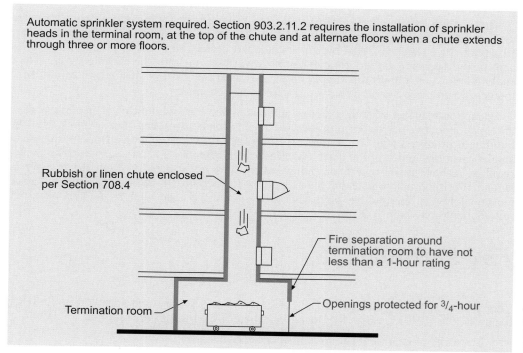

Figure 708-4
Rubbish and linen chutes

708.14.1 Elevator lobby. To reduce the potential for smoke to travel from the floor of fire origin to any other floor of the building by way of an elevator shaft enclosure, elevator lobbies are to be provided at each floor and must provide a complete separation utilizing fire partitions and the required opening protection. This separation essentially isolates the elevator shaft from the remainder of the building through the use of construction that provides both fire and smoke protection. Typical elevator hoistway doors, although fire-rated, cannot provide the necessary barrier required to keep smoke from passing from floor to floor through an elevator shaft enclosure. To restrict smoke passage from one floor level to other floor levels within the building, there must be some point where the floor level can be adequately separated from the elevator shaft.

Elevator lobbies are not required in all multistory buildings, as the requirement for elevator lobbies is not applicable where the elevator shaft connects only two or three stories. The threshold of four stories is consistent with a number of other code provisions that increase the level of protection where four or more stories are involved. In application, the scope of four or more stories is just one of many exceptions to the elevator lobby requirement.

There are also seven specific exceptions to the requirement for elevator lobbies. The first exempts a street floor lobby, provided the entire street floor is protected with an automatic sprinkler system. The second exception applies where elevator-shaft enclosures are not required to meet the provisions of Section 708.2. Third, additional doors may be utilized in lieu of an elevator lobby, provided they are installed in accordance with Section 3002.6. In this scenario, the lobby is not necessary, because the smoke-infiltration problem is addressed by the additional door. The fourth exception exempts fully sprinklered buildings, provided the building is not a Group I-2 or I-3 occupancy and is not by definition a high-rise building. Conversely, where the building is considered a high-rise, elevator lobbies are mandated. Thus, elevator lobbies are not required in fully sprinklered buildings, other than those housing Group I-2 or I-3 occupancies, that are not defined by Section 202 as high-rise buildings. The sixth exception permits pressurization of the elevator shaft enclosure as an alternative to creating an elevator lobby. The criteria for the pressurization method are established in Section 707.14.2.

The seventh exception eliminates the lobby requirement for elevators that serve open parking garages. This exception should also extend to enclosed parking garages since Exception 14 of Section 708.2 indicates that shaft protection is not required in both open and enclosed parking garages for elevator hoistways serving only the parking garage. Since enclosed elevator lobbies are simply an extension of the shaft enclosure protection, such lobbies are not necessary if a shaft enclosure is not required.

Exception 5 differs from the other exceptions in that it does not provide an alternative to an elevator lobby, but rather modifies the means for constructing the lobby. Where the building is protected with an automatic sprinkler system, the exception allows for the use of smoke partitions in lieu of fire partitions. Described in Section 711, smoke partitions are not required to have a fire-resistance rating but must be constructed to limit the transfer of smoke. It is important to note that the exceptions to Section 708.14.1 do not apply where an elevator lobby is required by some other provision of the code, such as the area of refuge requirement in Section 1007.4.

Section 709 *Fire Partitions*

This section regulates the design and construction of fire partitions installed in the listed locations. The IBC identifies five locations where fire partitions are required:

1. Walls separating dwelling units per Section 420.2
2. Walls separating sleeping units per Section 420.2

3. Walls separating tenant spaces in covered mall buildings as required by Section 402.7.2

4. Walls of fire-resistance-rated corridors per Section 1018.1

5. Elevator lobby separation as required by Section 708.14.1

Fire-resistance rating. The minimum fire-resistance rating of fire partitions is to be 1 hour, unless a reduction is permitted by one of two exceptions. Exception 1 refers to Table 1018.1, which identifies the required fire-resistance rating of a corridor based upon three factors—the occupancy classification of the area served by the corridor, the occupant load the corridor serves, and whether or not the building is sprinklered. Where conditions warrant, the table indicates that the corridor needs only a $^1/_2$-hour fire-resistance rating. If no fire-resistance rating is mandated, fire partitions are not required and none of the provisions of Section 709 are applicable. The second exception applies to walls separating dwelling units and sleeping units in buildings of nonrated construction. The presence of an automatic sprinkler system complying with NFPA 13 reduces the required fire-partition rating to 30 minutes. It should be noted that the exception does not permit this reduction where an NFPA 13R system is installed. **709.3**

Continuity. Consistent with the required continuity of fire barriers, the general requirement for fire partitions is that they must extend from the floor to the floor or roof deck above. However, unlike the provisions for fire barriers, fire partitions may terminate short of the floor or roof deck under various conditions. Where a fire-resistance-rated floor/ceiling or roof/ceiling is provided, a fire partition need only extend to, and be securely attached to, the ceiling membrane. For an example of this provision as it relates to corridor construction, refer to Figure 709-1. Under this condition in combustible construction, fireblocking or draftstopping must be installed at the partition line in the concealed space above the ceiling. Any supporting construction is to be at least 1-hour fire-resistance rated, except for tenant and sleeping unit separation walls and corridor walls in buildings of Type IIB, IIIB and VB construction. **709.4**

The following exceptions modify the continuity provisions of this section:

1. Where a crawl space exists below a floor assembly of at least 1-hour fire-resistance-rated construction, the fire partition does not need to extend into the underfloor space. See Figure 709-2.

2. The arrangement shown in Figure 709-3 would meet the code requirement for adequately enclosing a corridor. The corridor walls are protected on the side of the occupied use spaces by a fire-resistance-rated membrane extending from the floor to the floor or roof above. In this case, the ceiling over the corridor may be considered part of a fire-resistance-rated floor or roof assembly, and the corridor side of the ceiling protected by appropriate ceiling materials would satisfy the fire-resistance rating for the assembly.

3. The code provides that the corridor ceiling may be of the same construction as permitted for corridor walls as shown in Figure 709-4. In all probability, typical wall construction might not pass the 1-hour test when tested in a horizontal position. However, this arrangement, generally referred to as tunnel construction, is considered to be adequate protection for the corridor separating it from the spaces above.

    By establishing various methods for the enclosure of fire-resistance-rated corridors, the code is essentially attempting to get a minimal separation between the exit corridor and the occupied-use spaces. Any arrangement of the 1-hour fire-resistance-rated construction that effectively intervenes between these use spaces and the corridor would satisfy this requirement.

4. In covered mall buildings, fire partitions separating tenant spaces may terminate at the underside of a ceiling, even if the ceiling is not part of a fire-resistance-rated assembly. No type of extension of the fire partition is required by this section for attics and similar spaces above the ceiling.

5. In attic areas of Group R-2 occupancies less than five stories in height above grade plane, the draftstopping or fireblocking required by this section may be omitted where the attic area is subdivided by draftstopping into areas not exceeding two dwelling units or 3,000 square feet (279 m²), whichever is less. This exception is also found in Section 717.4.2.

6. In combustible buildings where the fire partitions stop at the fire-resistance-rated ceiling membrane, fireblocking or draftstopping is not required at the partition line if the building is fully sprinklered. Under this exception, sprinklers must be installed in the combustible floor/ceiling and roof/ceiling spaces. Exceptions in Sections 717.3.2 and 717.4.2 provide the same criteria.

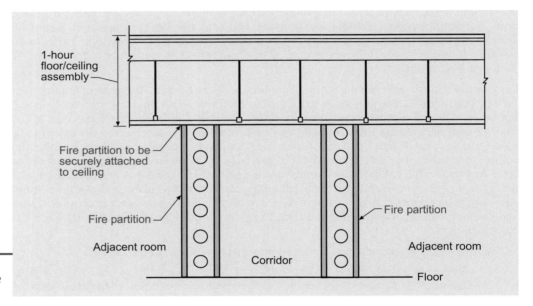

Figure 709-1
Corridor fire partitions

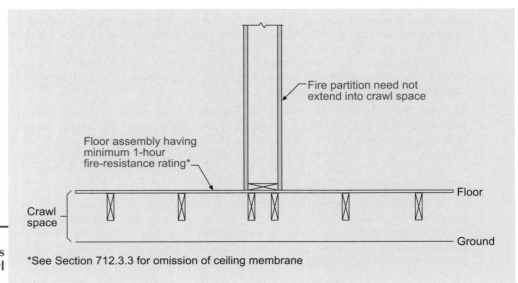

Figure 709-2
Fire partitions above a crawl space

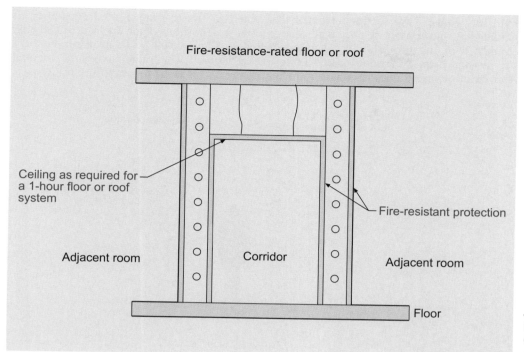

Fire-resistance-rated floor or roof

Ceiling as required for
a 1-hour floor or roof
system

Fire-resistant protection

Adjacent room

Corridor

Adjacent room

Floor

Figure 709-3
Corridor
construction

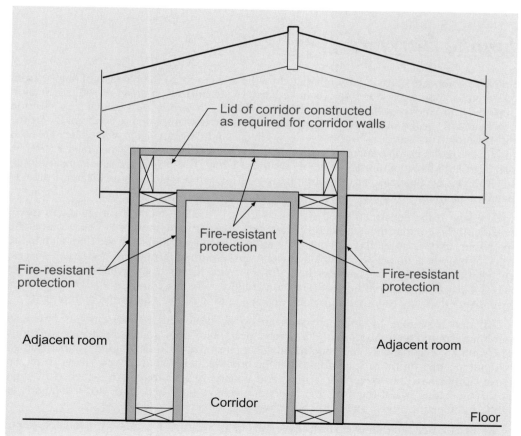

Lid of corridor constructed
as required for corridor walls

Fire-resistant
protection

Fire-resistant
protection

Fire-resistant
protection

Adjacent room

Adjacent room

Corridor

Floor

Figure 709-4
Tunnel
corridor

709.5   **Exterior walls.** This section clarifies that where a fire-resistance-rated separation is bounded by one or more exterior walls, the exterior wall portions of the enclosure need only comply with the provisions of Section 705. See Figure 709-5 for an example of a fire-resistance-rated corridor located along an exterior wall. The limited exposure potential from exterior spaces does not warrant the rating of exterior walls for enclosure purposes.

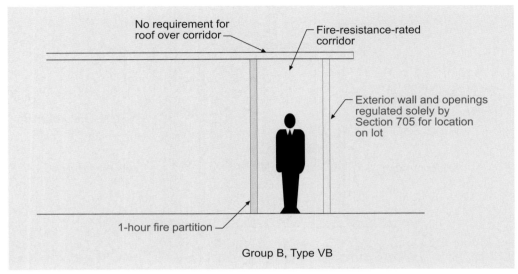

Figure 709-5
Corridor protection at exterior walls

Section 710   *Smoke Barriers*

Smoke barriers are occasionally mandated by the code to resist the passage of smoke from one area to another. For example, smoke barriers are utilized in areas of refuge (Section 1007.6.2), in smoke-control systems (Section 909.5), in Group I-3 occupancies (Section 408.6) and in various other building areas where smoke transmission is a concern. By far the most common utilization of smoke barriers is in Group I-2 occupancies, where they are used to create smoke compartments (Section 407.4). Smoke barriers must not only resist the passage of smoke, they must also be of minimum 1-hour fire-resistance-rated construction. In Group I-3 occupancies, an exception permits the use of 0.10-inch-thick (2.5 mm) steel in lieu of 1-hour construction.

The key to the construction of a smoke barrier is that all avenues for smoke to travel outside of the compartment created by the smoke barrier are eliminated. This requires the membrane to be continuous from outside wall to outside wall, and from the floor slab to the floor or roof deck above. The smoke barriers must continue through all concealed spaces, such as those above ceilings, unless the ceilings provide the necessary resistance against fire and smoke passage. In buildings of rated construction, all smoke barriers shall be supported by construction consistent with the fire-resistance rating of the wall or floor supported.

All door openings in smoke barriers are to be protected with assemblies having a minimum fire-protection rating of 20 minutes per Table 715.4. In cross-corridor situations in Group I-2 occupancies, the code mandates a pair of opposite-swinging doors installed without a center mullion. Such doors shall be provided with an approved vision panel; be close fitting; have no louvers or grilles; and undercuts are limited to $^3/_4$ inch. Although positive latching is not required, the doors are to have head and jamb stops, astragals or rabbets at meeting edges, and automatic-closing devices.

Smoke barriers, like smoke partitions regulated by Section 711, are only mandated where specifically identified by the code. As an example, Section 508.2.5.2 requires the use of "construction capable of resisting the passage of smoke" as a potential physical separation for incidental accessory occupancies. Thus, only construction that will perform the intended function is required, and not necessarily a smoke barrier or smoke partition.

Section 711 *Smoke Partitions*

Unlike the other separation elements used in the code, such as fire walls and smoke barriers, smoke partitions are not specifically defined. Their definition is simply a function of the requirements of Section 711. The purpose of a smoke partition is limited to the concerns of smoke movement under fire conditions, with no intent to regulate for the resistance to flame and heat.

Smoke partitions are mandated by the code in limited applications, most commonly in the construction of corridors in Group I-2 occupancies. As such, corridors in hospitals, nursing homes and similar I-2 occupancies are regulated by the provisions of both Sections 711 and 407.3. It requires a comparison of the two sections to determine the requirements for corridors, particularly corridor doors and air openings.

Openings. This section prohibits the installation of louvers in doors in smoke partitions. This is consistent with the provisions of Section 407.3.1 mandating an effective barrier against the transfer of smoke. The provisions of this section also require that doors in smoke partitions be tested in accordance with UL 1784 and be self-closing or automatic closing, but only where required elsewhere in the code. A review of Section 407.3.1 does not require the UL test, and it specifically states that self-closing or automatic-closing devices are not required. In this case, the provisions in Section 407.3.1 take precedence. **711.5**

Ducts and air transfer openings. The provisions of both Sections 716.5.7 and 711.7 mandate the need for smoke dampers in air transfer openings that occur in smoke partitions. Smoke dampers are not required at duct penetrations of smoke partitions, but only at unducted air openings. **711.7**

Section 712 *Horizontal Assemblies*

This section is applicable where floor and roof assemblies are required to have a fire-resistance rating. This will occur where the type of construction mandates protected floor and roof assemblies, such as in Type I, IIA, IIIA and VA construction, and where the floor assembly is used to separate occupancies or create separate fire areas. For example, in a building of Type IIA construction, Table 601 requires minimum 1-hour fire-resistance-rated floor construction. Where the floor separates a Group A-2 occupancy from a Group B, Table 508.4 addressing separated occupancies mandates a 1- or 2-hour separation.

As referenced in Section 420.3, complimentary to the provisions of Section 709 for fire partitions, floor assemblies separating dwelling units or sleeping units are required to be of at least 1-hour fire-resistance-rated construction. An exception reduces the required level of protection for the floor assembly to $^1/_2$ hour in buildings of Type IIB, IIIB or VB construction, provided the building is protected by an automatic sprinkler system.

Ceiling panels. The protection of a ceiling membrane also includes the adequacy of the panelized ceiling system to withstand forces generated by a fire and other forces that may try to displace the panels. These forces can generate positive pressures in a fire compartment that need to be counteracted. As a result, lay-in ceiling panels that provide a portion of the fire resistance of the floor/ceiling or roof/ceiling assembly should be capable of resisting this upward or positive pressure so that the panels stay in position and continue to maintain the integrity of the system. The code defines the pressure to be resisted as 1 pound per square **712.3.1**

foot (48 Pa). Section 712.3.2 permits the installation of access doors in ceilings where they are tested in accordance with ASTM E 119 or UL 263 as horizontal assemblies.

712.3.3 **Unusable space.** Figure 712-1 illustrates how this provision is applied in regard to unusable spaces such as crawl spaces and attics. For 1-hour fire-resistance-rated floor construction over a crawl space, the ceiling membrane is not necessary in the crawl-space area. Similarly, in 1-hour fire-resistance-rated roof construction, the floor membrane is not required in the attic. Note that the elimination of the membranes in the attic and crawl space is only applicable where the required rating of the floor or roof assembly is a maximum of 1 hour.

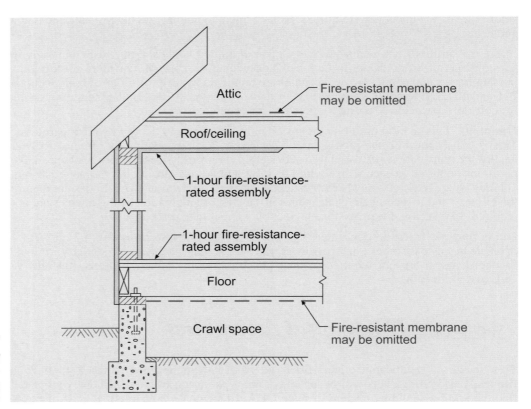

Figure 712-1
Omission of ceiling or flooring membrane

712.4 **Continuity.** Unless otherwise permitted by the provisions for shaft enclosures, penetrations or fire-resistive joint systems, horizontal fire-resistance-rated assemblies are to be continuous without any openings. The installation of unprotected skylights and other penetrations through the roof deck of a fire-resistance-rated roof assembly is permitted, provided the structural integrity of the roof construction is maintained. Section 705.8.6 should be consulted where vertical exposure to an adjacent building is possible.

This section also mandates that the horizontal assembly be supported by structural members or walls having at least the equivalent fire rating as that for the horizontal assembly. For example, in a Type IIA school building of two stories where the floor construction is required to be a 2-hour fire-resistance-rated assembly in order to separate fire areas, any walls or structural members in the first story supporting the second floor would be required to also be of 2-hour fire-resistance-rated construction. This would be the case even though the building generally is required to be only of 1-hour fire-resistance-rated construction. Obviously, if the horizontal assembly is not supported by equivalent fire-resistance-rated construction, the intent and function of the separation are negated if its supports fail prematurely.

Smoke barrier. Horizontal assemblies utilized for smoke barrier purposes must have **712.9** penetrations and joints protected in a manner similar to smoke barrier walls as established in Sections 713.5 and 714.6. Both of these code sections mandate compliance with the appropriate test standards for air leakage purposes with the maximum air leakage rate established in the code. Penetrations shall meet the requirements of UL 1479 and joint systems shall be tested in accordance with UL 2079. In addition, openings through horizontal fire barrier assemblies must comply with the shaft enclosure provisions of Section 708, modified in a manner such that no unprotected vertical openings are permitted.

Where a shaft enclosure housing an elevator passes through a horizontal smoke barrier assembly, the hoistway opening must be protected in accordance with the provisions of Section 708.14.1 addressing elevator lobbies. Since the purpose of the provisions in 708.14.1 is to limit the spread of smoke from floor to floor, the protection afforded by an elevator lobby is necessary to protect the opening created by the elevator shaft. It is important to note that the mandate for an elevator lobby applies to all multistory buildings regardless of the number of stories, not just those four or more stories in height. See Figure 712-2.

A companion provision, Section 407.4.3, specifically addresses the issue of horizontal smoke barrier assemblies in Group I-2 occupancies. Such institutional buildings must typically be provided with smoke barriers in order to create mandated smoke compartments. Where the building is multistory, it is important that both smoke barrier walls and horizontal smoke barrier assemblies be utilized in order to create the necessary smoke compartmentation. The floor/ceiling assemblies between stories provide the horizontal limits of each smoke compartment when constructed in accordance with Section 712.9.

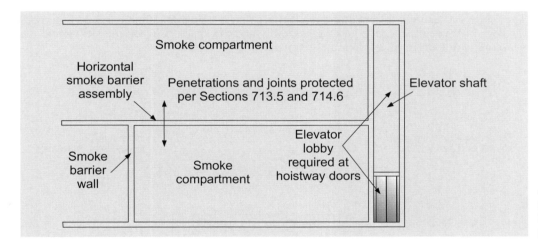

Figure 712-2
Horizontal smoke barrier

Section 713 *Penetrations*

The integrity of fire-resistance-rated horizontal and vertical assemblies is jeopardized where penetrations of such assemblies are not properly addressed. Cables, cable trays, conduit, tubing, vents, pipes and similar items are those types of penetrating items regulated by the code. This section of the IBC identifies the appropriate materials and methods of construction used to protect both membrane penetrations and through penetrations.

Ducts and air transfer openings. Section 716.5 identifies the various conditions under **713.1.1** which fire-resistance-rated wall assemblies penetrated by ducts or air transfer openings must be provided with fire and/or smoke dampers. There are a limited number of locations where a damper is not required, such as that permitted by Exception 3 of Section 716.5.2 for fire barrier penetrations. See Figure 713-1. In such situations, it is necessary that the penetrations be protected in accordance with the appropriate provisions of Section 713 in order to maintain the integrity of the fire-resistive assembly.

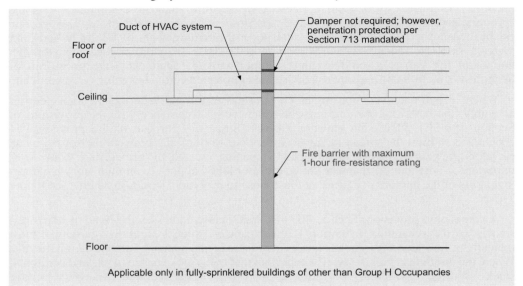

Figure 713-1
**Duct
penetration of
a 1-hour fire
barrier**

713.2 **Installation details.** As illustrated in Figure 713-2, sleeves used in the process of creating a through-penetration of a fire-resistance-rated building element must be properly installed. They must be securely fastened to the assembly that is being penetrated. In addition, both the space between the sleeve and the assembly and the space between the sleeve and the penetrating item must be appropriately protected.

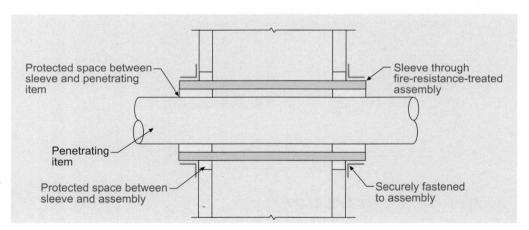

Figure 713-2
**Penetration
sleeve**

713.3 **Fire-resistance-rated walls.** This section regulates the penetration into or through fire walls, fire barriers, fire partitions and smoke barrier walls. The protection of penetrations in fire-resistance-rated exterior walls is not addressed; however, where such exterior walls are bearing walls it is necessary to consider penetration protection in order to maintain the structural integrity of the walls during fire conditions. Fire-resistance-rated interior bearing walls are also not specifically identified as elements regulated for penetrating items; however, any penetrations of such walls should be addressed in order to maintain the necessary structural fire resistance. For the most part, membrane penetrations are addressed in the same manner as through penetrations.

Through-penetrations. As a general rule, through penetrations (where the penetrating **713.3.1** items pass through the entire assembly) are required to be firestopped with approved through-penetration firestop systems when the penetrations pass through fire-resistance-rated walls, unless the approved wall assembly is tested with the penetrations as a part of the assembly. The firestop system is required to have an F rating at least equivalent to that of the fire-resistance rating of the wall penetrated as shown in Figure 713-3. There is no requirement for a T rating on a wall penetration, justified on the basis that there is no need for such a restrictive temperature rating for the penetration of wall assemblies.

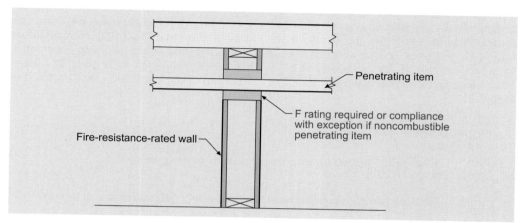

Figure 713-3
**F rating
required**

The IBC contains an exception to the general rule for firestopped wall penetrations allowing small noncombustible penetrating items no larger than 6-inch (152 mm) nominal diameter to penetrate concrete or masonry walls, provided the full thickness of the wall, or the thickness required to maintain the fire resistance, is filled with concrete, grout or mortar. The size of the opening is limited to 144 square inches (0.0929 m²). A second exception that is used extensively will allow the annular space around the same type of noncombustible penetrating item to be filled with a material that prevents the passage of flame or hot gases sufficient to ignite cotton waste when tested under the time-temperature fire conditions of ASTM E 119 or UL 263, and under a positive pressure differential of 0.01-inch (0.25 mm) water column. When properly installed around the penetrations of noncombustible items, these materials provide adequate firestopping between the penetrating item and the fire-resistive membrane of the wall. See Figure 713-4.

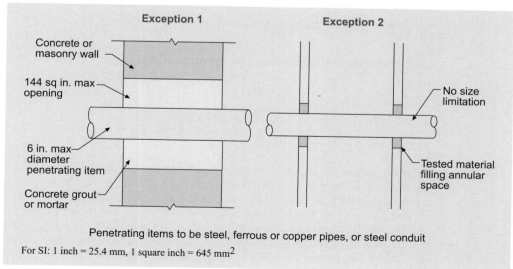

For SI: 1 inch = 25.4 mm, 1 square inch = 645 mm²

Figure 713-4
**Through-
penetrations
of wall**

713.3.2   **Membrane penetrations.** This section addresses penetrations through a single membrane of fire-resistance-rated walls. For the most part, a membrane penetration is to be protected by one of the methods established for through-penetrations as previously described. However, there are some membrance penetrations that are allowed without a specific firestopping material in the annular space around such penetrations. Openings for steel electrical boxes are specifically addressed where located in walls with a maximum 2-hour rating, provided that they are no more than 16 square inches (0.0103 m²) in area and the aggregate area of the boxes does not exceed 100 square inches (0.0645 m²) for any 100 square feet (9.29 m²) of wall area. The annular space between the wall membrane and any edge of the electrical box is limited to ¹/₈ inch (3.1 mm). Also, to prevent an indirect through-penetration, electrical boxes on opposing sides of a fire-resistance-rated wall shall be horizontally separated by no less than 24 inches (610 mm). As an alternate, boxes may be separated horizontally by the depth of the cavity if the cavity is filled with cellulose loose-fill, rockwool or slag mineral wool insulation, by solid fireblocking in accordance with Section 717.2.1, by protection of both outlet boxes with listed putty pads, or by any other listed methods and materials. Examples of several of these methods are illustrated in Figure 713-5.

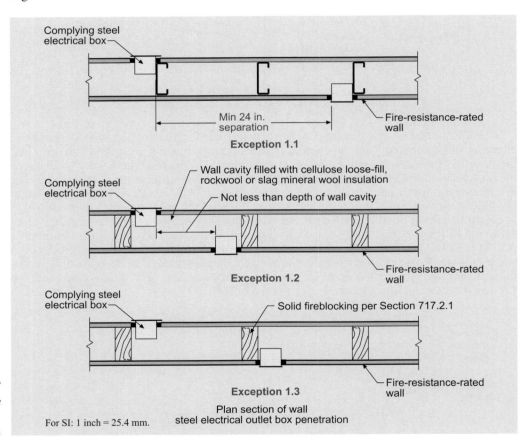

Figure 713-5
Penetration of fire-resistance-rated walls

A second exception for membrane penetrations of electrical-outlet boxes allows outlet boxes of any material, provided they are tested for use in fire-resistance-rated assemblies and installed in accordance with the instructions for the listing. Limitations are also placed on the annular space surrounding the box and conditions where the boxes are placed on opposite sides of the wall. Exception 3 allows for penetrations by electrical boxes of any size or type provided they are listed as a part of a wall opening protective material system, while Exception 4 addresses boxes, other than electrical boxes, that have annular space protection provided by an approved membrane penetration firestop system. The fifth exception permits the annular space created by the penetration of a fire sprinkler to be unprotected, provided that such a space is covered by a metal escutcheon plate. Because the escutcheon is a part of the listed sprinkler, it is inappropriate to require firestopping at this location. It should be noted that this exception applies to the penetration of sprinklers, not sprinkler piping or cross mains that might be penetrating fire-resistance-rated construction. See Figure 713-6.

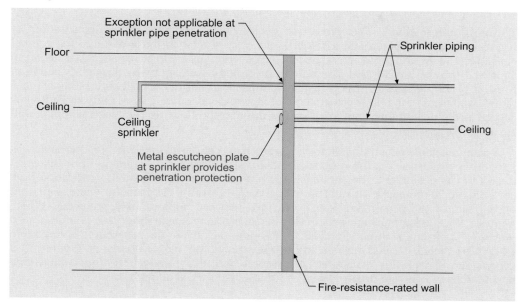

Figure 713-6
Membrane penetration protection

Dissimilar materials. This provision is intended to limit the occasional practice of using a noncombustible penetrating item (such as a short metal coupling) to penetrate a fire-resistance-rated wall, then connect to a combustible item (such as plastic piping or conduit) on the room side of the wall. The building official can accept such a condition where it is demonstrated that the fire-resistive integrity of the wall will be maintained. See Figure 713-7.

713.3.3

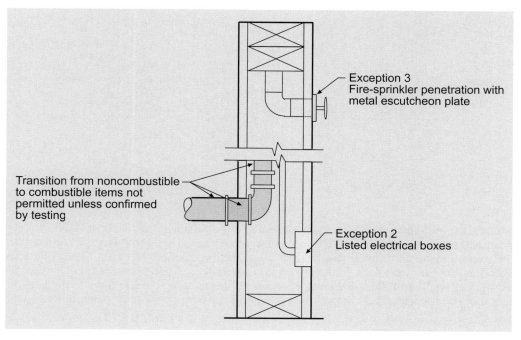

Figure 713-7
Membrane penetrations of walls

713.4 **Horizontal assemblies.** The shaft enclosure provisions of Section 708 intend to maintain a level of protection that is compromised when one or more openings occur in a floor or floor/ceiling assembly. However, penetrations by pipes, tubes, conduit, wire, cable and vents are permitted without shaft enclosure protection where in compliance with this section. In addition, this section addresses penetrations that occur in the ceiling of a roof/ceiling assembly. Penetrations occurring in both fire-resistance-rated horizontal assemblies and nonfire-resistance-rated assemblies are addressed.

713.4.1.1 **Through-penetrations.** The protection requirements for the through-penetration of fire-resistance-rated horizontal assemblies are very similar to those required for vertical elements. The general provisions state that the penetrations are to be installed as tested in an approved fire-resistance-rated assembly or protected by an approved through-penetration firestop system. Where a firestop system is utilized, it must have both an F rating and a T rating equivalent to the floor penetrated, but in no case less than 1 hour. Only an F rating is needed if the penetrating item, as it passes through the floor, is contained within a wall cavity above or below the floor. See Figure 713-8.

Noncombustible penetration items are granted exceptions to the general requirements as previously discussed for fire-resistance-rated walls. Where only a single fire-resistance-rated floor is penetrated, the annular space around the noncombustible penetration item need only to be protected with an approved material that fills the opening. There is no limit on the size of the penetrating items, provided they are appropriately protected. Where multiple floor assemblies are penetrated, the size of any penetrating item is limited to 6 inches (152 mm) in nominal diameter. In addition, the area of the penetration is limited to 144 square inches (92 900 mm²) in any 100 square feet (93 m²) of floor area. Figure 713-9 depicts the use of this exception. Allowances are also provided for noncombustible penetrations of concrete floors as well as for tested electrical outlet boxes.

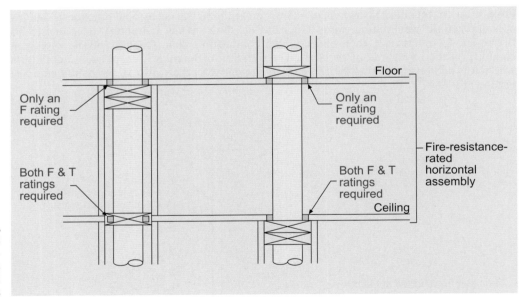

Only an F rating required

Only an F rating required

Floor

Fire-resistance-rated horizontal assembly

Both F & T ratings required

Both F & T ratings required

Ceiling

Figure 713-8
Through-penetrations of horizontal assemblies

713.4.1.2 **Membrane penetrations.** Fire-resistance-rated horizontal assemblies must be adequately protected at penetrations of the floor or ceiling membrane. Therefore, they are regulated in the same manner as through-penetrations addressed in Section 713.4.1.1. The code also specifies that any recessed fixtures that are installed in fire-resistance-rated horizontal assemblies shall not reduce the level of required fire resistance. Exceptions to the general requirement for approved firestop systems apply to noncombustible penetrations, steel electrical boxes, boxes listed as part of an opening protective material system, listed electrical-outlet boxes and fire sprinklers. See Figure 713-10.

Nonfire-resistance-rated assemblies. Figure 713-11 illustrates the provisions for penetrations of those horizontal assemblies not required to have a fire-resistance rating. Section 708 for shaft enclosures will regulate such penetrations where this section is not applicable. Where penetrations connect only two stories, the annular space around the penetrating items must simply be protected with a material that resists the free passage of fire and smoke. If the penetration items are noncombustible, up to three stories may be connected, provided the annular space is filled appropriately.

713.4.2

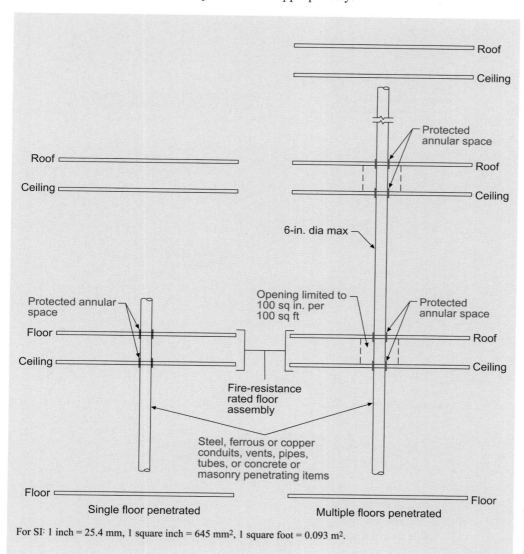

For SI: 1 inch = 25.4 mm, 1 square inch = 645 mm², 1 square foot = 0.093 m².

Figure 713-9
Penetrations of horizontal assemblies

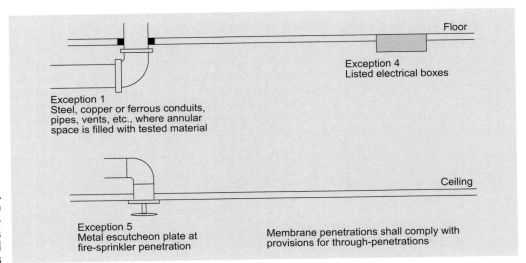

Figure 713-10
Membrane penetrations of horizontal assemblies

Floor

Exception 4
Listed electrical boxes

Exception 1
Steel, copper or ferrous conduits, pipes, vents, etc., where annular space is filled with tested material

Ceiling

Exception 5
Metal escutcheon plate at fire-sprinkler penetration

Membrane penetrations shall comply with provisions for through-penetrations

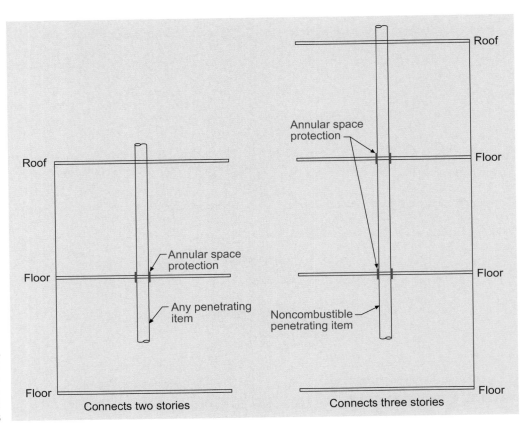

Figure 713-11
Nonfire-resistance-rated assemblies

Roof

Annular space protection

Floor

Annular space protection

Any penetrating item

Noncombustible penetrating item

Floor

Floor

Roof

Floor

Floor

Connects two stories

Connects three stories

Section 714 *Fire-resistant Joint Systems*

A fire-resistant joint system is defined in Section 702.1 as "an assemblage of specific materials or products that are designed, tested, and fire-resistance rated in accordance with either ASTM E 1966 or UL 2079 to resist, for a prescribed period of time, the passage of fire through joints made in or between fire-resistance-rated assemblies." The term *joint* is also defined in Section 702.1 as "the linear opening in or between adjacent fire-resistance-rated assemblies that is designed to allow independent movement of the building, in any plane, caused by thermal, seismic, wind or any other loading." The approved joint system should be designed to resist the passage of fire for a time period not less than the required fire-resistance rating of the floor, roof or wall in or between which it is installed. See Figure 714-1.

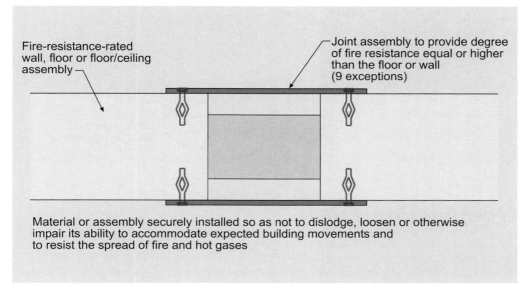

Fire-resistance-rated wall, floor or floor/ceiling assembly

Joint assembly to provide degree of fire resistance equal or higher than the floor or wall (9 exceptions)

Material or assembly securely installed so as not to dislodge, loosen or otherwise impair its ability to accommodate expected building movements and to resist the spread of fire and hot gases

Figure 714-1
Fire resistant joint systems

The code lists nine locations where it is not necessary to provide fire-resistant joint systems. For most of the applications listed, they are also locations where fire assemblies are not required to protect openings in the horizontal or vertical assemblies. Item 9 references maximum $^5/_8$-inch (15.9 mm) control joints when tested in accordance with ASTM E 119 or UL 263.

Exterior curtain wall/floor intersection. Vertical passages without barriers allow fire and hot gases to circumvent the protection for occupants in the floors above. When floors or floor/ceiling assemblies do not extend to the exterior face of a building, this section requires an approved barrier at the intersection at least equal to the fire resistance of the floor or floor/ceiling assembly. See Figure 714-2.

714.4

ASTM E 2307 is identified for the specification and testing methods used to determine the necessary fire resistance. This test method measures the performance of a perimeter fire-barrier system and its ability to maintain a seal to prevent fire spread during the deflection and deformation of the exterior wall assembly and floor assembly expected during a fire condition, while resisting fire exposure from both an interior compartment and the flame plume emitted from a window burner below.

A minimum level of protection is also mandated at any voids created at the intersection of an exterior curtain wall and a nonrated floor or floor assembly. The required method is consistent with that required in the code for the penetration of ducts and other items through nonfire-resistance-rated floor systems. The protection of the annular space is provided through the installation of an approved noncombustible material that resists the free passage of flame and the products of combustion.

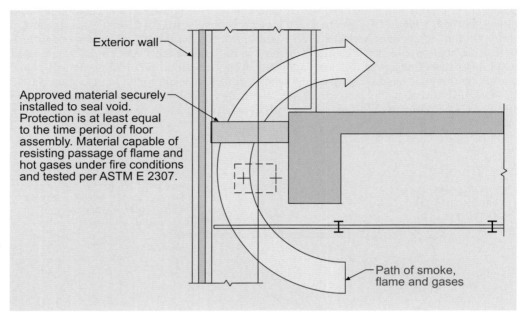

Exterior wall

Approved material securely installed to seal void. Protection is at least equal to the time period of floor assembly. Material capable of resisting passage of flame and hot gases under fire conditions and tested per ASTM E 2307.

Path of smoke, flame and gases

Figure 714-2
Exterior curtain wall/floor intersection

Section 715 *Opening Protectives*

In the context of the IBC, an opening protective refers to a fire door, fire shutter or fire-protection-rated glazing, including the required frames, sills, anchorage and hardware for its proper operation. Generally, whenever any fire door, fire shutter or fire-protection-rated glazing is referred to, it is the intent of the code that the entire fire assembly be included.

715.4 **Fire door and shutter assemblies.** This section sets forth the test standards and additional criteria necessary for the acceptance of fire door and fire shutter assemblies. In addition, Table 715.4 identifies the minimum fire-protection rating for an opening protective based on the type of assembly in which it is installed. For example, a door assembly in a 1-hour fire barrier wall separating hazardous material control areas would need to have a minimum $^3/_4$-hour fire-protective rating, whereas a 1-hour fire-resistance-rated exit enclosure would require a 1-hour door assembly.

Side-hinged or swinging doors are to be tested for conformance with NFPA 252 or UL 10C. It is important that the NFPA 252 test provides for positive pressure in the furnace as established by this section. See Figure 715-1. For other types of doors, the pressure level need only be maintained as nearly equal to the atmosphere's pressure as possible.

715.4.3 **Door assemblies in corridors and smoke barriers.** Fire door assemblies located in fire-resistance-rated corridor walls or smoke barrier walls are further regulated where required by Table 715.4 to have a 20-minute fire-protection rating. They are commonly referred to as smoke- and draft-control assemblies. Their primary purpose is to minimize smoke leakage around the door and through the opening. For this reason, these doors shall not contain louvers and must be installed in accordance with NFPA 105.

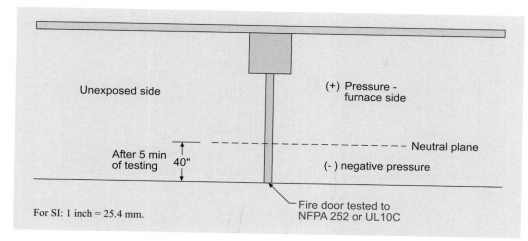

For SI: 1 inch = 25.4 mm.

Figure 715-1
Fire test of door assemblies

The protection of fire-rated corridors is intended to be a two-way protection. Although the general intent is to protect the corridor from smoke that might be generated by a fire occurring within the adjacent use spaces, there are occasions where it is just as important to protect the occupied use spaces from smoke in the corridor. The fire test for corridor and smoke-barrier doors is essentially the same test of the door as for other fire-door assemblies, except that the fire test for the 20-minute assembly does not include the hose-stream test. In addition, Section 715.4.3.1 requires the door assembly to be tested for smoke infiltration through the UL 1784 air leakage test and identifies the criteria for acceptance. Note that glazing other than in the door itself, such as in sidelites or transoms, must be tested with the hose-stream test as set forth in NFPA 257 or UL 9.

An exception permits the installation of a viewport through the door for purposes of observation. These viewports must be installed under the limitations of, and in accordance with, the conditions specified in the exception. Corridor door provisions are modified in Section 407.3.1 for Group I-2 occupancies and in multitheater complexes as shown in Figure 715-2. In addition, where horizontal sliding doors are used in smoke barriers of Group I-3 occupancies as specified, the 20-minute fire-protection rating not required.

Doors in exit enclosures and exit passageways. In addition to the normal requirement for fire doors, the IBC is concerned that fire doors installed in vertical exit enclosures and exit passageways shall be capable of limiting the temperature transmission through the door. It specifies that the temperature rise above ambient temperature shall be limited to a maximum of 450°F (232°C) at the end of 30 minutes of the normal fire test. However, in buildings equipped with an automatic sprinkler system, the temperature limitation is not applicable.

715.4.4

The purpose of these highly protected exit elements and their openings is to protect the building occupants while they are exiting the building. It is intended that in a properly enclosed and protected vertical exit enclosure, building occupants from the floors above the fire floor will be able to pass through the fire floor inside the enclosure and eventually pass down and out of the building. The end-point limitation on temperature transmission through the fire door, then, is literally to protect the person inside the enclosure from excessive heat radiation from the fire door as he or she passes through the fire floor. In sprinklered buildings, the maximum transmitted temperature end point is not required. It is expected that a sprinkler system will limit the fire growth to the point where such extra care is unnecessary.

Where glazing is provided in exit enclosure doors, it shall be fire tested as a component of the door assembly (including the temperature-transmission criteria where required) unless it is no greater than 100 square inches (0.065 m^2) in size. Where limited in size, the glazing is merely regulated as a glass light in a fire door.

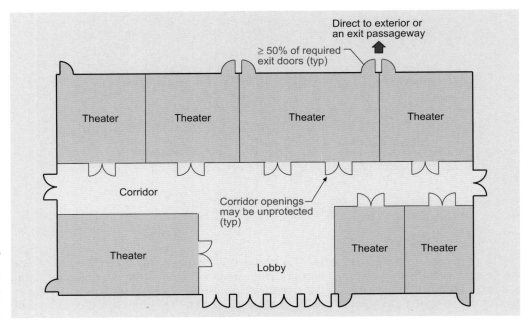

Figure 715-2
Corridor doors in a multitheater complex

715.4.6 **Labeled protective assemblies.** Fire doors are required to have an approved label or listing mark permanently affixed at the factory. The label must contain information that identifies the manufacturer, the third-party inspection agency and the fire-protection rating. Where applicable, the maximum transmitted temperature end point or the smoke- and draft-control designation must be identified.

Listing agencies will typically only label door assemblies that have been tested. However, some door assemblies are too large to be tested in available furnaces. As a result, the code permits the installation of oversized fire doors under the conditions of this section. As oversized fire doors are not subjected to the standard fire test, an approved testing agency must provide a certificate of inspection from them certifying that, except for the fact that the doors are oversized, they comply with the requirements for materials, design and construction for a fire door of the specified fire-endurance rating. An approved agency may also provide a label on the door indicating it is oversized. Where the certificate or label of an approved agency has been provided, there is assurance that the fire door will protect the opening as required by the code.

The letter "S" on a fire door indicates that it is in compliance with UL 1784, the air leakage testing. Through this identifying mark, it is possible to quickly identify the door as appropriate where smoke and draft control doors are mandated.

715.4.7.1 **Size limitations.** Fire-protection-rated glazing installed in fire doors is permitted where conforming to the size limitations of NFPA 80. The installation of these types, as well as wired glass, are limited by two exceptions. First, the only time a fire door in a fire wall is permitted to contain fire-protection-rated glazing is where the fire wall serves as a horizontal exit. Owing to the use of a horizontal exit as a required means of egress, it is often beneficial to provide a glass light of limited size so that occupants may view the egress path ahead of them. Otherwise, the high integrity of a fire wall should not be compromised by fire doors having glazed openings. Second, the maximum size of all types of fire-protection-rated glazing in 1^1/$_2$-hour fire doors is limited to 100 square inches (0.065 m^2) when installed in a fire barrier.

Identification. Glazing utilized in fire door assemblies must be identified for verification of **715.4.7.3.1** its appropriate application. The "D" designation indicates the glazing can be used in a fire door assembly, with the remaining identifiers providing specific information as to the glazing's capability to meet the hose-stream test and temperature limits. See Figure 715-3.

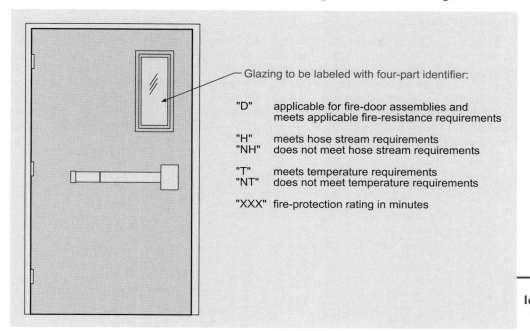

Glazing to be labeled with four-part identifier:

"D" applicable for fire-door assemblies and
meets applicable fire-resistance requirements

"H" meets hose stream requirements
"NH" does not meet hose stream requirements

"T" meets temperature requirements
"NT" does not meet temperature requirements

"XXX" fire-protection rating in minutes

Figure 715-3
**Identification
of glazing in
fire doors**

Door closing. The code mandates that fire doors be provided with closers to allow them to **715.4.8** shut and protect the opening without manual operation. One exception to this broad-based requirement applies to those fire doors located in the common walls between sleeping units of hotels and motels. These doors are so seldom open that it is unreasonable to require door-closing hardware.

Automatic-closing fire door assemblies. Where automatic-closing devices are used **715.4.8.2** instead of self-closing devices on fire doors, they must also comply with the provisions of NFPA 80 for self-closing action. Although they are generally held in an open position, doors equipped with automatic-closing devices become self-closing when actuated. The use of automatic-closing devices is typically a design decision; however, the code does mandate such devices in two applications. Automatic-closing devices are required by the exception to Section 710.5 on cross-corridor doors located in smoke barriers of Group I-2 occupancies. They must also be installed on cross-corridor doors located in a horizontal exit as set forth in Section 1025.3.

Smoke-activated doors. This section identifies eleven locations where a smoke detector is **715.4.8.3** to be utilized to actuate the closing operation for an automatic-closing fire door where such a closing device is provided. The detectors must be installed in accordance with the provisions of Section 907.3 and, furthermore, they must be of an approved type that will release the door in the event of a power failure. Automatic-closing fire-door assemblies are often used to increase the reliability of the opening protection. Swinging fire doors with self-closers are all too often propped open with wood blocks or wedges. Although this section regulates the method for activating automatic-closing fire doors, it does not identify where automatic-closing doors are mandated.

Fire-protection-rated glazing. In many situations, it is necessary to provide glazed **715.5** openings in fire-resistance-rated walls. The provisions of this section address fire window assemblies installed as opening protectives in fire partitions and exterior walls, as well as 1-hour fire barriers. Wired glass in steel frames is considered equivalent to $^3/_4$-hour fire window assemblies. Otherwise, in order to provide a fire-protection rating of 45 minutes, window assemblies must be tested in accordance with NFPA 257 or UL 9. An individual light of wired glass cannot exceed 9 square feet (0.84 m²) in surface area, nor be more than 54 inches (1372 mm) in either dimension. Other types of fire-protection-rated glazing are also permitted, provided they are installed and sized in accordance with NFPA 80. In all cases, a fire-window assembly must include an approved frame, fixed in position or automatic-closing.

In interior applications, fire-protection-rated glazing is limited to fire partitions, smoke barriers and fire barriers having a maximum fire-resistance rating of 1 hour. The total aggregate area of fire windows cannot exceed 25 percent of the area of the common wall between areas as shown in Figure 715-4. In making this 25-percent calculation, it is permissible to assume the entire area of the common wall even though a portion of that area might be taken up by doors. This gross area is usable in calculating the maximum percentage of area for windows. Where the ceilings are of different heights, the lower ceiling establishes the gross area.

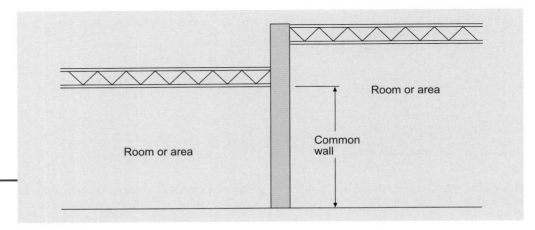

Figure 715-4
Glazing limitations

Fire-protection-rated glazing is not permitted to be located in any fire wall or fire barrier having a fire-resistance rating exceeding 1 hour; however, Section 715.2 recognizes that glazing tested as a part of a wall assembly is permitted in all applications and, therefore, not regulated by this section. This type of glazing is referred to as fire-resistance-rated glazing, rather than fire-protection-rated glazing, and is addressed in Section 703.5. In addition, the 25-percent area limitation is not applicable. Glazing that has a fire-resistance rating under ASTM E 119 or UL 263 that meets the fire-resistance rating of the wall may be used in more applications than fire-protection-rated glazing complying with NFPA 257 or UL 9. Also see the discussion of Section 703.5.

There are various conditions under which exterior fire window assemblies are required to be located in exterior walls. The specifics for exterior wall protection are found in Section 705. An overview of the requirements for exterior glazed openings is provided in Figure 715-5.

Exterior glazed openings			
	Wall having rating exceeding 1 hour per Table 602	Wall having rating of 1 hour per Table 602	Protection of exterior openings per Sections 705.8.5 and 705.8.6
Allowable area of openings Section 705.8	1¹/₂ hours	³/₄ hour	
Vertical separation Section 705.8.5			³/₄ hour
Vertical exposure Section 705.8.6			³/₄ hour

Figure 715-5
Exterior glazed openings

Section 716 *Ducts and Air Transfer Openings*

Where a duct or air transfer opening penetrates a fire-resistance-rated assembly, it is often necessary that a method of protection be provided to maintain the integrity of the assembly. Many times, dampers are utilized to protect the opening created by the duct or transfer opening. If dampers are not required to be provided under the provisions of this section, it is still necessary to protect the penetration under the provisions of Section 712.

Installation. This section states that fire and smoke dampers shall be installed in **716.2** accordance with their listing. The test standards for each of the types of damper carry specific requirements that manufacturers provide installation and operating instructions, and that a reference to these instructions shall be a part of the required marking information on the damper.

Damper testing. Dampers must not only be listed but also bear a label of an approved **716.3.1** testing agency. The label shall indicate that the damper is in compliance with the appropriate standard as identified by this section. For example, for fire dampers the required information on the damper includes the hourly rating; the words "Fire Damper"; whether or not the damper is to be in a dynamic or static system (or both); maximum rated airflow and pressure differential across the closed damper for dampers intended for use in dynamic systems; an arrow showing direction of airflow for dampers intended for use in dynamic systems; the intended mounting position (vertical, horizontal or both); top of damper; and of course, the manufacturer's name and model number. UL 555 (which applies to fire dampers) requires that all of this information shall be available on the damper label, which is installed at the factory, and that all labels shall be located on the internal surface of the damper and be readily visible after the damper is installed. UL 555 indicates that fire dampers tested under that standard are intended for use in HVAC duct systems passing through fire-resistive walls, partitions or floor assemblies.

Just as fire dampers are tested for different hourly ratings, they are also tested for different installation positions. A damper listed for vertical installation cannot arbitrarily be installed in the horizontal position.

This section also states that only fire dampers labeled for use in dynamic systems shall be installed in systems intended to operate with fans on during a fire. The test standard for fire dampers states that fire dampers are intended for use in either static systems that are automatically shut down in the event of a fire, or in dynamic systems that are operational in the event of a fire. If the HVAC system has not been designed and constructed to shut down in case of a fire, then dynamically listed fire dampers are necessary. Special attention should be paid to damper listings when smoke-control systems are installed under the provisions of Section 909.

Test standard UL 555S is to be used to determine the compliance of smoke dampers. This standard states that leakage-rated dampers (smoke dampers in the IBC) are intended to restrict the spread of smoke in HVAC systems that are designed to automatically shut down in the event of a fire, or to control the movement of smoke within a building when the HVAC system is operational in engineered smoke-control systems.

In addition to fire dampers and smoke dampers, two other types of dampers are referenced in the IBC. Where combination fire/smoke dampers are required, they must comply with the requirements of both UL 555 for fire dampers and UL 555S for smoke dampers. Figure 716-1 illustrates the installation of an automatic-closing combination fire and smoke damper. Ceiling radiation dampers, intended for installation in air-handling openings penetrating the ceiling membranes of fire-resistance-rated floor/ceiling and roof/ceiling assemblies, are to meet the conditions of UL 555C.

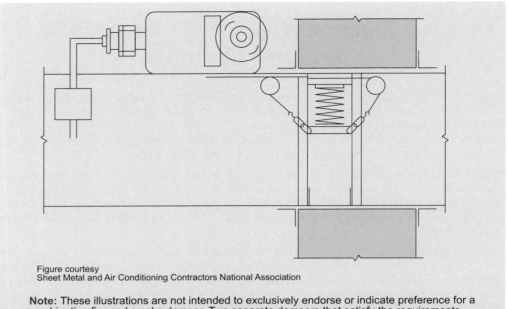

Figure courtesy
Sheet Metal and Air Conditioning Contractors National Association

Figure 716-1
Combination fire and smoke dampers

Note: These illustrations are not intended to exclusively endorse or indicate preference for a combination fire and smoke damper. Two separate dampers that satisfy the requirements for the respective functions may also be used for fire and smoke control.

716.3.2.1 **Fire-protection rating.** Test standard UL 555 covers fire dampers ranging from $^1/_2$ hour to 3 hours. Because fire dampers carry an hourly rating, plans should reflect the rating required at a particular location if more than one rating is required within a building. Table 716.3.2.1 indicates whether a $1^1/_2$-hour or 3-hour rating is required for a fire damper, based upon the fire-resistance rating of the assembly in which it is installed. In all applications rated at 3 hours or greater, a 3-hour-rated damper is mandated; otherwise, a $1^1/_2$-hour damper is acceptable.

716.3.2.2 **Smoke damper ratings.** A minimum Class II leakage rating is required for smoke dampers, which also must have an elevated temperature rating of at least 250°F (121°C). The class designation indicates the maximum leakage permitted in cubic feet per minute per square foot (cubic mm per minute per mm²) for the particular class. The four classes progress from Class I (least leakage or best performance) through Class IV (greatest leakage or poorest performance). The IBC requires conformance with Class II, so a damper rated as Class I would also be acceptable. These leakage ratings are determined at ambient temperature after exposing the damper to temperature degradation at an elevated temperature, with 250° F (121°C) being the lowest elevated temperature allowed by the code. Dampers can be tested using higher degradation temperatures [one as high as 850°F (454°C)], but most listed dampers seem to have been tested at either 250°F (121°C) or 350°F (177°C).

The provisions of Section 716.3.3.2 specifically instruct the designer or installer on how to control smoke dampers. Smoke dampers are required to be closed by activation of smoke detectors installed in accordance with Section 907.3 for fire-detection systems and any of the five specified methods of control listed in this provision. These methods of control, each having benefits and drawbacks, were proposed by those individuals involved with damper installation and should provide consistent and logical control methods for the dampers.

716.4 **Access and identification.** Both fire dampers and smoke dampers shall be installed so that they are accessible for inspection and servicing. It is important that any access openings in a fire-resistance-rated assembly be adequately protected in order to maintain the integrity of the assembly. This will typically involve the use of an access door having the required fire-protection rating. Permanent identification of the access points to fire-damper and smoke-damper locations is also mandated.

Where required. This section lists those specific locations where the various dampers are required. Dampers need only be installed in ducts and transfer openings where specifically identified by this section. In some locations, both a fire damper and a smoke damper are required. This means that either two dampers must be installed or a damper listed for both heat and smoke control must be used. See Figure 716-2 for an overview of the required locations for smoke and fire dampers. **716.5**

Fire walls. Because of the importance of maintaining the separation provided by fire walls used to divide a structure into two or more separate buildings, the code requires the use of approved fire dampers under all conditions. Such dampers are to be installed at all permitted duct penetrations and air-transfer openings of fire walls. Where the fire wall serves as a party wall as addressed in Section 706.1.1, ducts and air transfer openings are prohibited. There is no requirement for smoke dampers at duct penetrations and air openings through fire walls except for those fire walls serving as horizontal exits. **716.5.1**

Fire barriers. Much like fire walls, fire barriers are designed to totally isolate one area of a building from another. Therefore, the general requirement is that all duct penetrations and air transfer openings of fire barriers be protected by complying fire dampers. There are, however, several exceptions that may eliminate the need for dampers. Of special note is the elimination of fire dampers in certain sprinklered buildings. Fire dampers are not required for duct penetrations and air openings in fire barriers where all of the following conditions exist: **716.5.2**

1. The penetration consists of a duct that is a portion of a ducted HVAC system.

2. The fire-resistance rating of the fire barrier is 1 hour or less.

3. The area is not a Group H occupancy.

4. The building is fully protected by an automatic fire-sprinkler system.

A fire barrier that serves as a horizontal exit must also be provided with a listed smoke damper at each point a duct or air transfer opening penetrates the fire barrier.

Shaft enclosures. Section 1022.4 limits the penetration of exit enclosures by ductwork to those situations where independent pressurization is provided. Three methods of ventilating an exit enclosure are also addressed in Section 1022.5. Otherwise, the fire-resistance rating of shaft enclosures where ducts and air transfer openings penetrate the shaft must be maintained as mandated by this section. **716.5.3**

Both a fire damper and a smoke damper, or a combination fire/smoke damper, must be installed where a duct or air transfer opening penetrates a shaft enclosure. Four exceptions identify conditions under which fire dampers are not required. As shown in Figure 716-3, the first exception permits fire dampers to be eliminated where steel exhaust subducts enter an exhaust shaft. The subducts must extend vertically at least 22 inches (559 mm), and there must be continuous air flow upward to the outside through the shaft. Smoke dampers are also not required under similar conditions, but the exception is limited to fully sprinklered Group B and R occupancies where the fan providing continuous airflow is on standby power.

Fire partitions. Fire partitions are not regulated by the code as highly as fire barriers or fire walls, so it is consistent that the requirements for dampers through such partitions are not as restrictive. The general rule is that a fire damper is required in any duct or air transfer opening that penetrates a fire partition. However, where the building is fully sprinklered, ducts penetrating tenant separation walls in covered mall buildings or fire-resistance-rated corridor walls need not be fire dampered. In addition, fire dampers are not necessary for small steel ducts installed above a ceiling, provided the duct does not communicate between a corridor and adjacent rooms and does not terminate at a register in a fire-resistance-rated wall. **716.5.4**

Location		Fire damper	Smoke damper
Fire walls		Required	Not Required[26]
Fire barriers		Required[1,2,3]	Not Required[26]
Shaft enclosure[7]		Required[1,2,4,5,29]	Required[2,5,6,29]
Fire partitions		Required[8,9,22]	Not Required[2,27]
Fire-resistance-rated corridors[23]		Not Required[28]	Required[10,18]
Smoke partitions		Not Required	Required[2,21]
Smoke barriers		Not Required	Required[11]
Horizontal assemblies[12]	Through penetrations	Required[13,19]	Not Required
	Membrane penetrations[24]	Required[14,20,25]	Not Required
	Nonfire-resistance-rated assemblies	Required[15,16,17]	Not Required

[1] Not required for penetrations tested in accordance with ASTM E 119 or UL 263 as part of the rated assembly.

[2] Not required for ducts used as a part of an approved smoke control system in accordance with Section 909.

[3] Not required in sprinklered building of other than Group H for 1-hour walls penetrated by ducted HVAC systems.

[4] Not required for steel exhaust subducts extending at least 22 inches vertically in exhaust shafts having continuous airflow upward to the outside.

[5] Not required in parking garage supply or exhaust shafts that are separated from other building shafts by a minimum of 2-hour fire-resistance-rated construction.

[6] Not required in fully sprinklered Group B and R occupancies for kitchen, clothes dryer, bathroom and toilet room exhaust openings with steel exhaust subducts that extend at least 22 inches vertically and an exhaust fan installed at upper terminus of shaft is powered continuously per Section 909.11 with a continuous upward airflow to the outside.

[7] See Sections 1022.4 and 1022.5 for permitted penetrations of exit enclosures.

[8] Not required in sprinklered buildings of other than Group H for tenant separation or corridor walls, provided duct protected per Section 713 as a penetration.

[9] Not required in buildings of other than Group H where duct penetration is limited to 100 square inches; is of minimum 0.0217-inch steel; does not have communicating openings between a corridor and adjacent spaces; is installed above a ceiling; does not terminate at a wall register of the fire-resistance-rated wall; and minimum 12-inch-long steel sleeve centered and secured in opening.

[10] Not required for corridor penetrations of minimum 0.019-inch steel ducts with no openings into corridor.

[11] Not required where openings in steel ducts are limited to a single smoke compartment.

[12] General requirement of Section 708.2 mandates shaft enclosures for openings in floor and roof systems.

[13] In other than Group I-2 and I-3, fire dampers are permitted in lieu of shaft enclosures for penetration of fire-resistance-rated horizontal assembly that connects two floors.

[14] Where shaft enclosure is not provided, an approved ceiling damper is required at the ceiling line of a fire-resistance-rated floor/ceiling or roof/ceiling assembly where duct penetrates ceiling or diffuser is installed without a duct.

[15] Not required where duct does not connect more than two stories and the annular space around the duct is filled with noncombustible material.

[16] Limited to three connected stories without shaft enclosure, provided fire dampers are installed at each floor line and annular space is filled.

[17] Not required in ducts within individual dwelling units.

[18] Not required in building with a smoke control system if not necessary for operation and control of system.

[19] Not required at each floor where 1) penetrating three floors or less, 2) duct of steel construction and within wall cavity, 3) duct opens into only one dwelling unit or sleeping unit and is continuous from unit to building exterior, 4) duct limited to 4 inches in diameter and total area limited to 100 square inches per 100 square feet, 5) annular space around duct protected with materials that prevent passage of flame and hot gases, and 6) grille openings in ceiling of fire-resistance-rated floor/ceiling or roof/ceiling assembly protected with ceiling radiation damper.

[20] Ceiling radiation dampers not required for exhaust duct penetrations protected per Section 713.4.1.2, where ducts located within wall cavity and do not pass through another dwelling unit or tenant space.

[21] Only required to protect air transfer openings per Section 711.7.

[22] Not required for tenant partitions in covered mall buildings where walls not required to extend to floor or roof deck above.

[23] Fire partition provisions also applicable to fire-resistance-rated corridors.

[24] Applicable to penetrations of ceiling membrane of fire-resistance-rated floor/ceiling or roof/ceiling assembly.

[25] Ceiling radiation dampers not required where tests per ASTM E 119 or UL 263 have shown that dampers are not necessary in order to maintain the fire-resistive rating of the assembly.

[26] Only required where duct or air transfer opening penetrates a horizontal exit wall.

[27] Only required where penetration is an air transfer opening.

[28] Only required for duct or air transfer openings in fire-resistance-rated corridor walls in a nonsprinklered building.

[29] Not required in kitchen and clothes dryer exhaust systems when installed per IMC.

Figure 716-2
Fire and smoke damper location

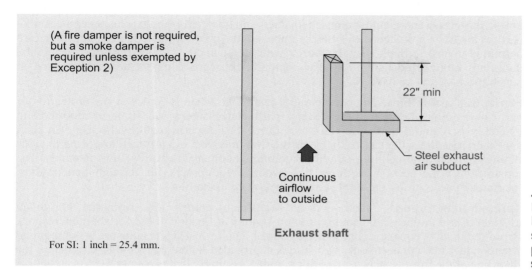

(A fire damper is not required, but a smoke damper is required unless exempted by Exception 2)

22" min

Steel exhaust air subduct

Continuous airflow to outside

Exhaust shaft

For SI: 1 inch = 25.4 mm.

Figure 716-3
Exhaust subducts penetrating shafts

Corridors. Because a fire-resistance-rated corridor is intended to be an exit access component providing a limited degree of occupant protection during egress activities, it is logical that air openings into the corridor be addressed. This section mandates, with exceptions, that all corridors required to be protected with smoke- and draft-control doors shall also be provided with smoke dampers where ducts or air transfer openings penetrate the corridor enclosures. As illustrated in Figure 716-4, an important exception eliminates the need for smoke dampers where steel ducts pass through, but do not serve, the corridor.

716.5.4.1

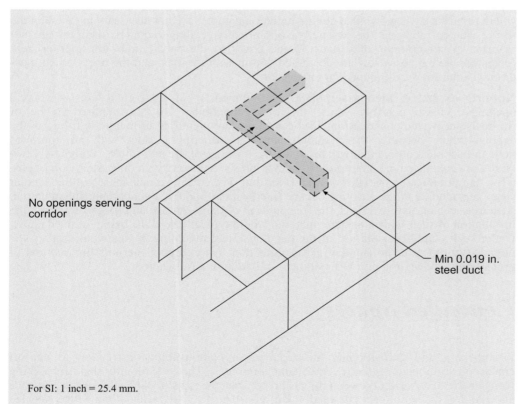

No openings serving corridor

Min 0.019 in. steel duct

For SI: 1 inch = 25.4 mm.

Figure 716-4
Ducts crossing corridor

716.5.5 **Smoke barriers.** Those air openings, both ducts and transfer openings, that penetrate smoke barriers are to be provided with smoke dampers at the points of penetration. Steel ducts are permitted to pass through a smoke barrier without a damper, provided the openings in the ducts are limited to a single smoke compartment. Fire dampers are not required at penetrations of smoke barriers.

716.6 **Horizontal assemblies.** The code is quite restrictive when it comes to the protection of vertical openings between floor levels, particularly where the floor or floor/ceiling assembly is required to be fire-resistance rated. This section requires the use of a shaft enclosure to address the hazard that is created where a duct or air transfer opening extends through a floor, floor/ceiling assembly or ceiling membrane of a roof/ceiling assembly. The remainder of the section modifies this general requirement for through penetrations, membrane penetrations and nonfire-resistance-rated assemblies.

716.6.1 **Through-penetrations.** Ducts and air transfer openings that penetrate horizontal assemblies are initially regulated by the provisions of Section 708.2 for shaft enclosures. Permitted in all occupancies other than Groups I-2 and I-3, a shaft enclosure is not required where a duct that connects only two stories is provided with a fire damper installed at the floor line of the fire-resistance-rated floor/ceiling assembly that is penetrated. As an option, the duct may be protected in a manner prescribed in Section 713.4 for the penetration of horizontal assemblies. The code's intent to limit fire and smoke migration between smoke compartments vertically in Group I-2 and I-3 occupancies is maintained through the limitation imposed in this section. The exception goes on to eliminate the fire damper requirement in specified residential occupancies for dwelling units and sleeping units, as well as sleeping units in Group I-1 occupancies.

716.6.2 **Membrane penetrations.** A shaft enclosure need not be provided where an approved ceiling radiation damper is installed at the ceiling line of a fire-resistance-rated floor/ceiling or roof/ceiling assembly penetrated by a duct or air-transfer opening. Designed to protect the construction elements of the floor or roof assembly, the ceiling damper is not required where fire tests have shown that ceiling radiation dampers are not necessary to maintain the fire-resistance rating of the assembly. Additionally, ceiling radiation dampers are not required at penetrations of exhaust ducts, provided the penetrations are appropriately protected, the exhaust ducts are contained within wall cavities, and the ducts do not pass through adjacent dwelling units or tenant spaces.

716.6.3 **Nonfire-resistance-rated assemblies.** The elimination of shaft enclosures at vertical openings is also possible where the floor assemblies are not required to be of fire-resistance-rated construction. Two conditions are identified utilizing the filling of the annular space between the assembly and the penetrating duct with an approved noncombustible material that will resist the free passage of fire and smoke. Where only two stories are connected, no other protective measures are necessary. In three-story conditions, a fire damper must be installed at each floor line. The code also mandates that the annular space surrounding the penetrating duct be filled with an approved noncombustible material, such as a sealant, that will resist the free passage of flame, smoke and gases. However, the installation of such sealant or other material would typically void the listing of the damper. Under such conditions, the use of the damper's steel mounting angles would satisfy the intended purpose of the annular space protection. An exception permits this method of protection in a dwelling unit without the installation of a fire damper.

Section 717 *Concealed Spaces*

Fireblocking and draftstopping are required in combustible construction to cut off concealed draft openings (both vertical and horizontal). The code requires that fireblocking form an effective barrier between floors and between the top story and attic space. The code also requires that attic spaces be subdivided, as will be discussed later, along with concealed spaces within roof/ceiling and floor/ceiling assemblies. Figures 717-1 through 717-4 depict IBC requirements for fireblocking.

Experience has shown that some of the greatest damage occurs to conventional wood-framed buildings during a fire when the fire travels unimpeded through concealed draft openings. This often occurs before the fire department has an opportunity to control the fire, and greater damage is created as a result of the lack of fireblocking.

For these reasons, the code requires fireblocking and draftstopping to prevent the spread of fire through concealed combustible draft passageways. Virtually any concealed air space within a building will provide an open channel through which high-temperature air and gases will spread. Fire and hot gases will spread through concealed spaces between joists, between studs, within furred spaces and through any other hidden channel that is not fireblocked.

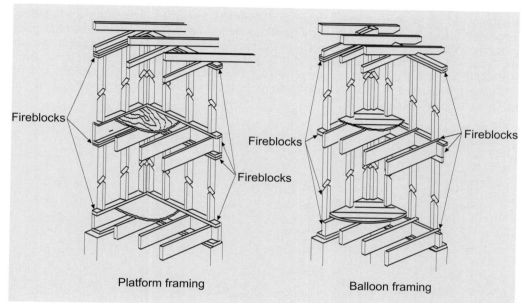

Platform framing

Balloon framing

Figure 717-1
Fireblocking

Fireblocking. The platform framing method that is used most often today in wood-frame construction provides adequate fireblocking between stories in the stud walls, but care must be exercised to ensure that furred spaces are effectively fireblocked to prevent transmission of fire and hot gases between stories or along a wall. For this reason, the code requires that fireblocking be provided at 10-foot (3048 mm) intervals horizontally along walls that are either furred out, of double-wall construction, or of staggered-stud construction.

717.2

Fireblocking provisions for wood flooring used typically in gymnasiums, bowling alleys, dance floors and similar uses containing concealed sleeper spaces are found in Section 717.2.7. As long as the wood flooring described in this section is in direct contact with a concrete or masonry fire-resistance-rated floor, there is no significant hazard. However, if there is a void between the wood flooring and the fire-resistance-rated floor, a blind space is created that is enclosed with combustible materials and provides a route for the undetected spread of fire. Therefore, the code requires that where the wood flooring is not in contact with the fire-resistance-rated floor, the space shall be filled with noncombustible material or shall be fireblocked. Two exceptions to these fireblocking requirement are:

1. The first exception exempts slab-on-grade floors of gymnasiums. In this case, the code presumes a low hazard, as gymnasiums are usually only one story in height. If the floor is at or below grade, it is unlikely that any ignition sources would be

present to start a fire that would spread through the blind space under the wood flooring.

2. Bowling lanes are exempted from fireblocking except as described in the code, which provides for areas larger than 100 square feet (93 m²) between fire blocks. Fireblocking intermittently down a bowling lane would create problems for a consistent lane surface.

Fireblocking materials are required to consist of lumber or wood structural panels of the thicknesses specified, gypsum board, cement fiber board, mineral wool, glass fiber or any other approved materials securely fastened in place.

Batts or blankets of mineral wool and glass fiber materials are allowed to be used as fireblocking and work especially well where parallel or staggered stud walls are used. Loose-fill insulation should not be used as a fireblocking material unless specifically tested for such use. It must also be shown that it will remain in place under fire conditions. Even in the case where it fills an entire cavity, a hole knocked into the membrane enclosure for the cavity could allow the loose-fill insulation material to fall out, negating its function. Therefore, loose-fill insulation material shall not be used as a fire block unless it has been properly tested to show that it can perform the intended function. The main concern is that the loose-fill material, even though it may perform adequately in a fire test to show sufficient fire-retardant characteristics to meet the intent of this section, would not be adequately evaluated for various applications because of the physical instability of the material in certain orientations.

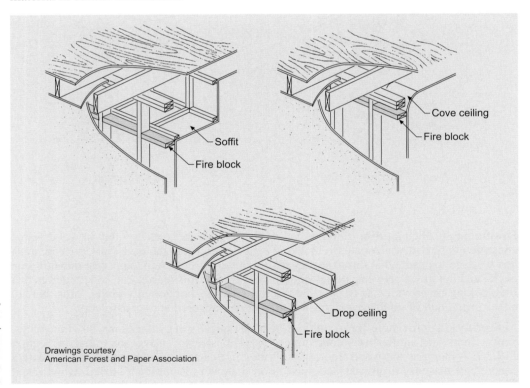

Figure 717-2
**Fire blocks—
vertical and
horizontal
space
connections**

Drawings courtesy
American Forest and Paper Association

717.3   **Draftstopping in floors.** Draftstops are often utilized to subdivide large concealed spaces within floor/ceiling assemblies of combustible construction. Figure 717-5 shows IBC requirements for draftstopping in these locations. Gypsum board, wood structural panels, particleboard, mineral wool or glass fiber batts and blankets, and other approved materials are considered satisfactory for the purpose of subdividing floor/ceiling areas, provided the materials are of adequate thickness, adequately supported and their integrity is maintained.

Draftstops are to be installed in floor/ceiling assemblies as follows:

1. Residential occupancies. The code requires that draftstops be installed in line with the wall separating tenants or dwelling units from each other and the remainder of the building, consistent with the provisions of Section 717.3.2 for dwelling unit and sleeping unit separations. In this case, a fire originating in a dwelling unit or hotel room will find draftstops in the concealed space blocking the transmission of fire and hot gases into another hotel room or apartment. Where the residential occupancy is fully sprinklered, draftstopping is not required. Where a residential-sprinkler system is utilized, automatic sprinklers must also be installed in the combustible concealed floor areas.

2. All other occupancies. For uses other than residential occupancies, the code intends that the concealed space within the floor/ceiling assembly be separated by draftstopping so that the area of any concealed space does not exceed 1,000 square feet (93 m²). An exception permits the elimination of draftstopping where automatic fire sprinklers are installed throughout the building.

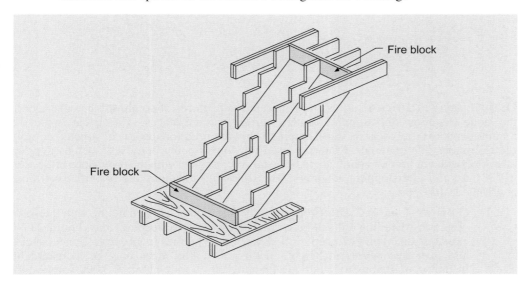

Figure 717-3
**Fire block—
stairs**

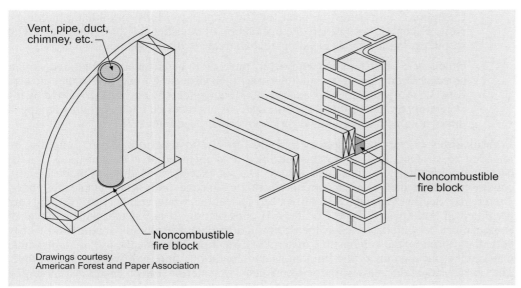

Drawings courtesy
American Forest and Paper Association

Figure 717-4
**Fire blocks—
pipes,
chimneys, etc.**

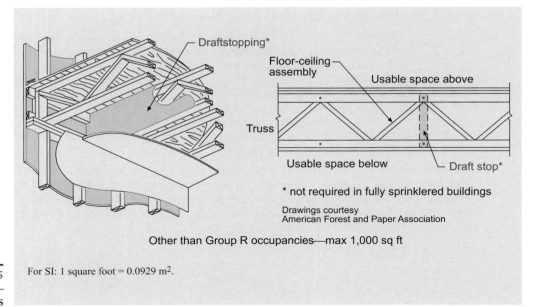

Figure 717-5
Draft stops—floors

Draftstopping*

Floor-ceiling assembly

Usable space above

Truss

Usable space below

Draft stop*

* not required in fully sprinklered buildings

Drawings courtesy
American Forest and Paper Association

Other than Group R occupancies—max 1,000 sq ft

For SI: 1 square foot = 0.0929 m².

717.4   Draftstopping in attics. In attics and concealed roof spaces of combustible construction, the code requires draftstopping under certain circumstances. Consistent with the requirements for fireblocking, draftstopping is not required for spaces constructed entirely of noncombustible materials. Materials used for draftstopping purposes, such as gypsum board, plywood or particleboard, are to be installed consistent with the provisions of Section 717.3.1 for the draftstopping of floors. The following locations are identified as those requiring draftstopping:

1. Groups R-1 and R-2. Draftstops are to be installed above and in line with the walls separating dwelling units or between walls separating sleeping units. Figure 717-6 explains the intent of Exception 1. Exception 3 applies to Group R-2 occupancies less than five stories in height. In this case, attic areas may be increased by installing draftstops to subdivide the attic into a maximum of 3,000-square-foot (279 m²) spaces, with no area to exceed the inclusion of two dwelling units. Exception 2 eliminates the need for draftstopping in fully sprinklered buildings, while Exception 4 considers the installation of a residential-sprinkler system with sprinklers in the attic as an acceptable alternate to draftstopping.

2. Other uses. Draftstops are required by the code to be installed in attics and similar concealed roof spaces of buildings other than Groups R-1 and R-2 so that the area between draftstops does not exceed 3,000 square feet (279 m²). As permitted by the exception, draftstopping of the attic space is not required in any building equipped throughout with an automatic sprinkler system. See Figure 717-7.

717.5   Combustibles in concealed spaces in Type I or II construction. Where buildings are intended to be classified as noncombustible, it is intended that combustibles not be permitted, particularly in concealed spaces. The six exceptions to this limitation identify conditions under which a limited amount of combustible materials is acceptable. Exception 1 references Section 603, which identifies 22 applications where combustible materials are permitted in buildings of Type I or Type II construction. It is felt that the low level of combustibles permitted, as well as their control, does not adversely impact the fire-severity potential caused by the combustible materials. Exception 2 permits the use of combustible materials in plenums under the limitations and conditions imposed by IMC Section 602. The third exception allows the concealment of interior finish materials having a flame-spread index of Class A. Exception 4 addresses combustible piping, provided it is located within partitions or enclosed shafts in a complying manner. As an example, the presence of plastic pipe within the wall construction of a Type I or II building does not cause the building to be considered combustible construction. Exception 5 allows for the installation of combustible piping within concealed ceiling areas of Type I and II buildings, while Exception 6 permits combustible insulation on pipe and tubing in all concealed spaces other than plenums.

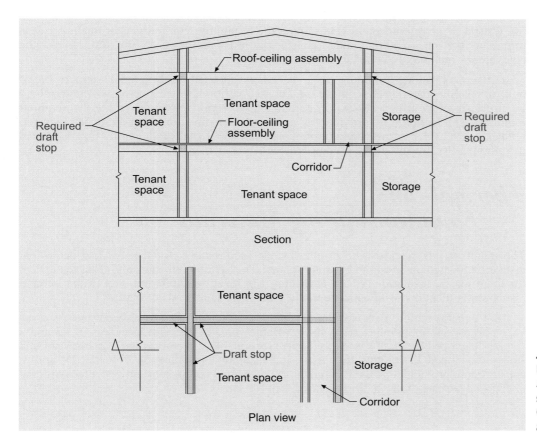

Figure 717-6
Attic draft stops—Groups R-1 and R-2

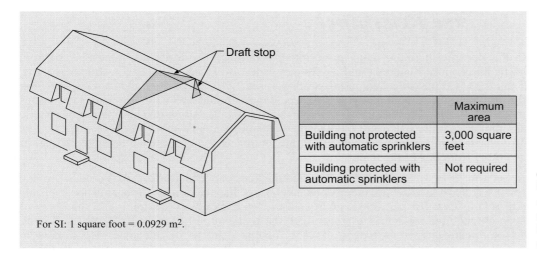

	Maximum area
Building not protected with automatic sprinklers	3,000 square feet
Building protected with automatic sprinklers	Not required

For SI: 1 square foot = 0.0929 m^2.

Figure 717-7
Attic draft stops—other than Group R occupancies

Fire-resistance Requirements for Plaster

Where gypsum plaster or portland cement plaster is considered a portion of the required fire-resistance rating of an assembly, it must be in compliance with this section. Appropriate fire tests shall be referenced in determining the minimum required plaster thickness. It is important that the material under consideration is addressed in the test, unless the equivalency method of Section 718.2 is utilized.

In noncombustible buildings, it is necessary that all backing and support be of noncombustible materials. Except for solid plaster partitions or where otherwise determined by fire tests, it is also necessary in certain plaster applications to double the required reinforcement in order to provide for additional bonding, particularly under elevated temperatures. Under specific conditions, it is permissible to substitute plaster for concrete in determining the fire-resistance rating of the concrete element.

Thermal- and Sound-insulating Materials

The intent of this section is to establish code requirements for thermal and acoustical insulation located on or within building spaces. This section regulates all insulation except for foam-plastic insulation, which is regulated by Section 2603, and duct insulation and insulation in plenums, which must comply with the requirements of the IMC.

As a general requirement, insulation, including facings used as vapor retarders or as vapor permeable membranes must have a flame spread index not in excess of 25 and a smoke-developed index not to exceed 450. Section 719.2.1 waives the flame-spread and smoke-developed limitations for facings on insulation installed in Type III, IV and V construction, provided that the facing is installed behind and in substantial contact with the unexposed surface of the ceiling, floor or wall finish.

Prescriptive Fire Resistance

In this section, there are many prescriptive details for fire-resistance-rated construction, particularly those materials and assemblies listed in Table 720.1(1) for structural parts, Table 720.1(2) for walls and partitions, and Table 720.1(3) for floor and roof systems. For the most part, the listed items have been tested in accordance with the fire-resistance ratings indicated. In addition, a similar footnote to all of the tables allows the acceptance of generic assemblies that are listed in GA 600, the Gypsum Association's *Fire-Resistance Design Manual*. It is important to review all of the applicable footnotes when using a material or assembly from one of the tables.

Section 720.1.1 intends that the required thickness of insulating material used to provide fire resistance to a structural member cannot be less than the dimension established by Table 720.1(1), except for permitted modifications. An example of the minimum thickness of concrete required for a structural-steel column is shown in Figure 720-1. Note that Figure 714-4 illustrates that the edges of such members are to be adequately reinforced in compliance with the provisions of Section 720.1.3. Figure 720-2 illustrates the minimum concrete-thickness requirements for protecting reinforcing steel in concrete columns, beams, girders and trusses. Refer to Section 704 for additional provisions regarding structural members.

As previously mentioned, the fire-resistance ratings for the fire-resistance-rated walls and partitions outlined in Table 720.1(2) are based on actual tests. Figure 720-3 shows two samples from the table. For reinforced concrete walls, it is important to note the type of aggregate as discussed earlier in this chapter. The difference in aggregates is quite significant for a 4-hour fire-resistance-rated wall, as it amounts to a difference in thickness of almost 2 inches (51 mm). For hollow-unit masonry walls, the thickness required for a particular fire-endurance rating is the equivalent thickness as defined in Section 721.3.1 for concrete masonry and Section 721.4.1.1 for clay masonry. Figure 720-4 outlines the manner in which the equivalent thickness is determined.

Table 720.1(3) of the IBC provides fire-resistance ratings for floor/ceiling and roof/ceiling assemblies, and Figure 720-5 depicts the construction of a 1-hour fire-resistance-rated wood floor or roof assembly. Of special note is Footnote n, which exempts unusable space from the flooring and ceiling requirements. See Figure 712-1.

Often, materials such as insulation are added to fire-resistance-rated assemblies. It is the intent of the IBC to require substantiating fire test data to show that when the materials are added, they do not reduce the required fire-endurance time period. As an example, adding insulation to a floor/ceiling assembly may change its capacity to dissipate heat and, particularly for noncombustible assemblies, the fire-resistance rating may be changed. Although the primary intent of the provision is to cover those cases where thermal insulation is added, the language is intentionally broad so that it applies to any material that might be added to the assembly.

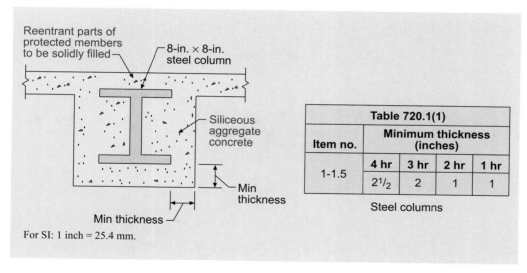

For SI: 1 inch = 25.4 mm.

Figure 720-1
Prescriptive fire resistance

Bonded prestressed concrete tendons. Figure 720-6 depicts the requirements specified in Items 1 and 2 for variable concrete cover for tendons. It must be noted that for all cases of variable concrete cover, the average concrete cover for the tendons must not be less than the cover specified in IBC Table 720.1(1).

720.1.5

As prestressed concrete members are designed in accordance with their ultimate moment capacity, as well as with their performance at service loads, Item 3 provides two sets of criteria for variable concrete cover for the multiple tendons:

1. Those tendons having less concrete cover than specified in Table 720.1(1) shall be considered to be furnishing only a reduced portion of the ultimate-moment capacity of the member, depending on the cross-sectional area of the member.

2. No reduction is necessary for those tendons having reduced cover for the design of the member at service loads.

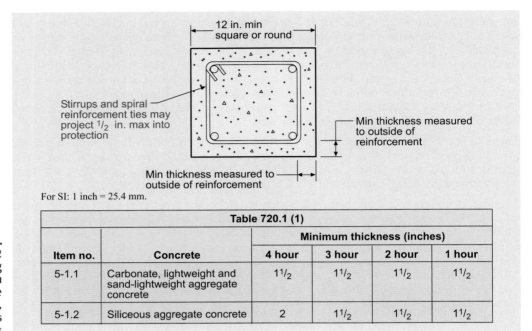

For SI: 1 inch = 25.4 mm.

		Minimum thickness (inches)			
		Table 720.1 (1)			
Item no.	Concrete	4 hour	3 hour	2 hour	1 hour
5-1.1	Carbonate, lightweight and sand-lightweight aggregate concrete	1$^1/_2$	1$^1/_2$	1$^1/_2$	1$^1/_2$
5-1.2	Siliceous aggregate concrete	2	1$^1/_2$	1$^1/_2$	1$^1/_2$

Figure 720-2
Reinforcing steel in concrete columns, beams, girders and trusses

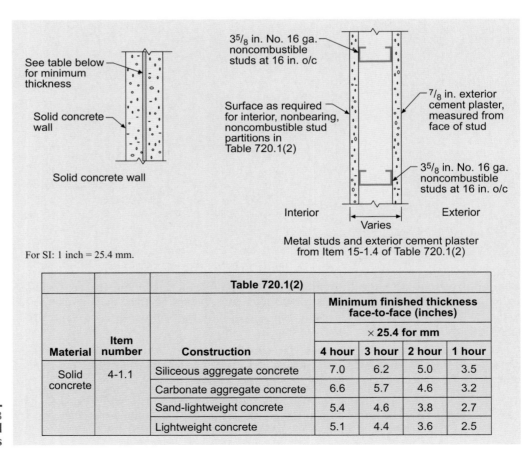

For SI: 1 inch = 25.4 mm.

			Minimum finished thickness face-to-face (inches)			
			× 25.4 for mm			
Material	Item number	Construction	4 hour	3 hour	2 hour	1 hour
Solid concrete	4-1.1	Siliceous aggregate concrete	7.0	6.2	5.0	3.5
		Carbonate aggregate concrete	6.6	5.7	4.6	3.2
		Sand-lightweight concrete	5.4	4.6	3.8	2.7
		Lightweight concrete	5.1	4.4	3.6	2.5

Table 720.1(2)

Figure 720-3
Walls and partitions

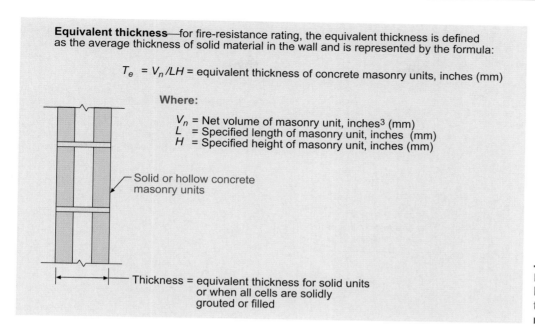

Equivalent thickness—for fire-resistance rating, the equivalent thickness is defined as the average thickness of solid material in the wall and is represented by the formula:

$$T_e = V_n/LH = \text{equivalent thickness of concrete masonry units, inches (mm)}$$

Where:

V_n = Net volume of masonry unit, inches^3 (mm)
L = Specified length of masonry unit, inches (mm)
H = Specified height of masonry unit, inches (mm)

Solid or hollow concrete masonry units

Thickness = equivalent thickness for solid units or when all cells are solidly grouted or filled

Figure 720-4
Equivalent thickness of masonry walls

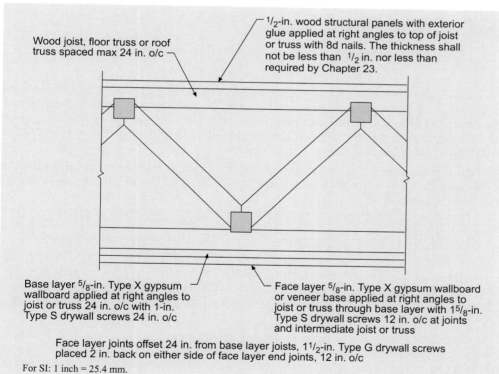

$1/_2$-in. wood structural panels with exterior glue applied at right angles to top of joist or truss with 8d nails. The thickness shall not be less than $1/_2$ in. nor less than required by Chapter 23.

Wood joist, floor truss or roof truss spaced max 24 in. o/c

Base layer $5/_8$-in. Type X gypsum wallboard applied at right angles to joist or truss 24 in. o/c with 1-in. Type S drywall screws 24 in. o/c

Face layer $5/_8$-in. Type X gypsum wallboard or veneer base applied at right angles to joist or truss through base layer with 1$5/_8$-in. Type S drywall screws 12 in. o/c at joints and intermediate joist or truss

Face layer joints offset 24 in. from base layer joists, 1$1/_2$-in. Type G drywall screws placed 2 in. back on either side of face layer end joints, 12 in. o/c

For SI: 1 inch = 25.4 mm.

Figure 720-5
1-hour wood floor or roof assembly item number 21-1.1

As the ultimate-moment capacity of the member is critical to the behavior of the member under fire conditions, the code requires the reduction for those tendons having cover less than that specified by the code. However, behavior at service loads is less affected by the heat of a fire; therefore, the code permits those tendons with reduced cover to be assumed as fully effective.

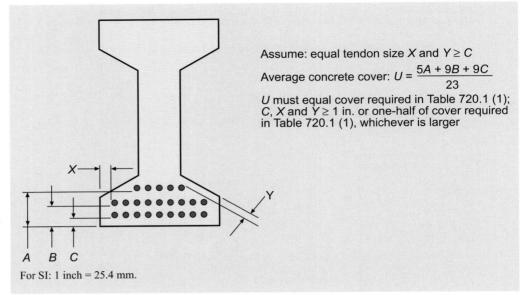

Assume: equal tendon size X and $Y \geq C$

Average concrete cover: $U = \dfrac{5A + 9B + 9C}{23}$

U must equal cover required in Table 720.1 (1);
C, X and $Y \geq 1$ in. or one-half of cover required
in Table 720.1 (1), whichever is larger

For SI: 1 inch = 25.4 mm.

Figure 720-6
Variable protection of bonded prestressed tendons, multiple tendons

Section 721 *Calculated Fire Resistance*

Fire research and the theory of heat transmission have combined to make it possible with the present state-of-the-art technology to calculate the fire endurance for certain materials and assemblies. As a result of this testing and research, this section permits the calculation of the fire-resistance rating for assemblies of structural steel, reinforced concrete, wood, concrete masonry and clay masonry.

At the present time, it is doubtful that the fire resistance of many buildings will be based on calculations. Even so, the code users should be aware of the useful information presented in this section, including:

Reference	Subject
Section 721.5.1.2	Attachment of gypsum wallboard around structural-steel columns
Section 721.2.1.3.1	Thickness of ceramic blanket joint material for precast concrete wall panels
Section 721.6.2	Wood wall, floor and roof assemblies
Section 721.6.3	Design of exposed wood members for 1-hour fire-resistance rating

The procedure set forth in Section 721.6.3 should be used when someone wishes to consider exposed heavy timber as 1-hour construction in something other than a Type IV building. One of the factors affecting a wood member's fire-resistance rating is the load on the member as a percentage of its allowable structural capacity. See Application Example 721-1.

721.1.1 **Definitions.** Several definitions that pertain to Section 721 are presented in this section. Of these, four definitions for concrete made from different aggregate types are important not only for structural considerations but also from a fire-endurance standpoint. Because the aggregate type bears on concrete performance, minimum concrete thickness listings in this section, as well as in Tables 720.1(1), 720.1(2) and 720.1(3), provide listings for different aggregates. Generally, the use of siliceous aggregate results in lower fire-resistance ratings, whereas structural lightweight concretes have better fire resistance than normal weight concrete. This is illustrated by the equivalent thicknesses for concrete walls as shown in Table 721.2.1.1. For a 4-hour wall, the following minimum thicknesses are required:

Concrete type	Thickness (inches)
Siliceous aggregate	7.0
Carbonate aggregate	6.6
Sand-lightweight	5.4
Lightweight	5.1

For SI: 1 inch = 25.4 mm.

Application Example 721-1

DETERMINE: 8-inch by 8-inch timber column's fire-resistance rating at 50 percent, 75 percent and 100 percent of its structural capacity.

SOLUTION: According to Section 721.6.3, the column fire-resistance rating is calculated from the formula, $2.54\, Z_d\, [3 - (d/b)]$ for columns that may be exposed to fire on four sides.

WHERE:

b = larger side of column [inches (mm)]

d = smaller side of column [inches (mm)]

Z = 0.9 + 30/r (load factor based on Figure 721.6.3 (1)

r = ratio of applied load to allowable load expressed as a percent of allowable

When r = 50% of less, Z = 1.5

Calculated Z =
    50% capacity Z = 1.5
    75% Z = 0.9 + 30/75 = 1.3
    100% Z = 0.9 + 30/100 = 1.2

Calculating rating:

$2.54Z\, (7.5)\, [3 - (7.5/7.5)] = 38.1Z$

50% capacity   38.1 (1.5) = 57.15 minutes
75% capacity   38.1 (1.3) = 49.53 minutes
100% capacity   38.1 (1.2) = 45.72 minutes

The same column loaded to 50 percent of capacity, but exposed on only three sides, would have a fire-resistance rating of 100 minutes.

For SI: 1 inch = 25.4 mm.

HEAVY-TIMBER CALCULATION

KEY POINTS

- Fire endurance is the basis for the fire-resistance requirements in the IBC.

- Materials and assemblies tested in accordance with ASTM E 119 or UL 263 are considered to be in full compliance with the code for fire-resistance purposes.

- Elements required to be fire-resistance-rated include structural frame members, walls and partitions, and floor/ceiling and roof/ceiling assemblies.

- The method for protecting fire-resistance-rated elements must be in exact compliance with the desired listing.

- In many cases, fire-resistance protection for structural members must be applied directly to each individual structural member.

- Exterior walls of buildings located on the same lot are regulated by the placement of an assumed line between the two buildings.

- Where an exterior wall is located an acceptable fire-separation distance from the lot line, the wall's fire-resistance rating is allowed to be determined based only on interior fire exposure.

- Opening protection presents a higher fire risk than fire-resistance-rated construction insofar as it does not need to meet the heat-transmission limits of ASTM E 119 or UL 263.

- The maximum area of both protected or unprotected openings permitted in each story of an exterior wall is regulated by the fire separation distance.

- The code intends that each portion of a structure separated by a fire wall be considered a separate building.

- The objective of fire walls is that a complete burnout can occur on one side of the wall without any effects of the fire being felt on the opposite side.

- The purpose of a fire barrier is to totally isolate one portion of a floor from another through the use of fire-resistance-rated walls and opening protectives as well as fire-resistance-rated horizontal assemblies.

- Fire barriers are used as the separating elements for vertical exit enclosures, exit passageways, horizontal exits, incidental accessory occupancies, occupancy separations and other areas where a complete separation is required.

- Fire barriers must begin at the floor and extend uninterrupted to the floor or roof deck above.

- The potential for fire spread vertically through buildings mandates that openings through floors be protected with fire-resistance-rated shaft enclosures.

- Various modifications and exemptions for the enclosure of horizontal openings are found in the IBC.

- Fire partitions are utilized to separate dwelling units, sleeping units, tenant spaces in covered mall buildings and fire-resistance-rated corridors from adjacent spaces.

- Fire partitions are permitted to extend to the membrane of a fire-resistance-rated floor/ceiling or roof/ceiling assembly.

- Smoke barriers are required in building areas where smoke transmission is a concern.

- The membrane of smoke barriers must be continuous from outside wall to outside wall and from floor slab to the floor roof deck above, to eliminate all avenues for smoke to travel outside of the compartment created by the smoke barriers.

- Smoke partitions are intended to solely restrict the passage of smoke.

- Horizontal assemblies are required to have a fire-resistance rating where the type of construction mandates protected floor and roof assemblies, and where the floor assembly is used to separate occupancies or create separate fire areas.

- Penetration firestop systems are approved methods of protecting openings created through fire-resistance-rated walls and floors for piping and conduits.

- A limited level of protection is permitted for penetrations of noncombustible items.

- Both through penetrations and membrane penetrations are regulated, typically in similar fashion.

- Joints, such as the division of the building designed for movement during a seismic event, must often be protected if they occur in a fire-resistance-rated vertical or horizontal element.

- An opening protective refers to a fire door, fire shutter or fire-protection-rated glazing, including the required frames, sills, anchorage and hardware for its proper operation.

- Table 715.4 identifies the minimum fire-protection rating for an opening protective based on the type of assembly in which it is installed.

- In interior applications, fire-protection-rated glazing is limited to fire partitions and fire barriers having a maximum fire-resistance rating of 1-hour.

- In addition to fire dampers and smoke dampers, ceiling radiation dampers and combination fire/smoke dampers are referenced in the IBC.

- Fireblocking and draftstopping are required in combustible construction to cut off concealed draft openings.

- Prescriptive methods for fire-resistance-rated construction are detailed for structural parts, walls and partitions, and floor and roof systems.

- The calculation of fire resistance is permitted for structural steel, reinforced concrete, wood, concrete masonry and clay masonry.

INTERIOR FINISHES

Unfortunately, a number of building code provisions are enacted only after a disaster (usually with a large loss of life) indicates the need to regulate in a specific area. This is true of the interior wall and ceiling finish requirements of Chapter 8. In this case, the 1942 Coconut Grove nightclub fire in Boston, with a loss of almost 500 lives, provided the impetus to develop code requirements for the regulation of interior finish. Based on fire statistics, lack of proper control over interior finish (and the consequent rapid spread of fire) is second only to vertical spread of fire through openings in floors as a cause of loss of life during fire in buildings.

The dangers of unregulated interior finish are as follows:

The rapid spread of fire. Rapid spread of fire presents a threat to the occupants of a building by either limiting or denying their use of exitways within and out of the building. This limitation on the use of exits can be created by:

1. The rapid spread of the fire itself so that it blocks the use of exitways.

2. The production of large quantities of dense, black smoke (such as smoke created by certain plastic materials), which obscures the exit path and exit signs.

The contribution of additional fuel to the fire. Unregulated finish materials have the potential for adding fuel to the fire, thereby increasing its intensity and shortening the time available to the occupants to exit safely. However, because ASTM E 84 or UL 723 does not require the determination of the amount of fuel contributed, the *International Building Code*® (IBC®) does not regulate interior finish materials on this basis.

Section 801 *General*

It is the intent of the IBC to regulate the interior finish materials on walls and ceilings, as well as coverings applied to the floor. In addition, limitations on the use of trim and decorative materials are found in Section 806, with the exception of foam plastics used as trim or finish material. These are addressed in Chapter 26. Combustible materials are permitted as finish materials in buildings of any type of construction, provided the wall, ceiling or floor finishes are in compliance with this chapter.

As established in Section 803.2, it is not the intent of the IBC to regulate thin materials such as wallpaper that are less than 0.036 inches (0.9 mm) thick. These thin materials behave essentially as the backing to which they are applied and, as a result, are not regulated. In some cases, however, repeated applications of wallpaper where the original materials are not removed can accumulate to a thickness of such magnitude that they must be regulated. The IBC, as stated in Section 803.3, also does not regulate the finish of exposed heavy-timber members complying with Section 602.4 insofar as this type of construction is not subject to rapid flame spread.

Section 803 *Wall and Ceiling Finishes*

803.1.1 **Interior wall and ceiling finish materials.** The standard test for the determination of flame spread and smoke-development characteristics is set forth in both ASTM E 84 and UL 723. This test is commonly known as the Steiner Tunnel Test. Based upon the results of this test, interior wall and ceiling materials are classified. These classifications are divided into three groups as Class A, Class B and Class C.

Room corner test for interior wall or ceiling finish materials. As an option to testing interior wall and ceiling finish materials under the criteria of ASTM E 84, such materials, other than textiles, can be tested per NFPA 286. Described as the "room corner" test, NFPA 286 describes the conditions of the test, whereas the code sets forth the minimum acceptance criteria.   **803.1.2**

Textiles. This section regulates carpet—as well as other textiles that are napped, tufted, looped, woven or nonwoven—where applied as wall finish materials. Textile wall coverings present a unique hazard because of the potential for extremely rapid fire spread. The code provides three options for the acceptance of textiles used as interior wall finish. One method of testing includes the surface burning characteristics test of ASTM E 84 or UL 723. The textile must have a flame-spread index of Class A and be protected by automatic sprinklers. A second option is based on the room corner test for textiles as established in NFPA 265, where the testing must be done in accordance with the Method B protocol. It is important that the testing be done in the same manner as the intended use of the textile materials. A third approach is the utilization of the ceiling and wall finish room corner test as set forth in NFPA 286. This test is also based on the intended application of the textile material and must include the product-mounting system. Where textiles are intended to be applied as ceiling finish materials, only the methods utilizing ASTM E 84 and NFPA 286 are to be used.   **803.5**

Interior finish requirements based on group. Table 803.9 is divided based upon the presence, or lack of, an automatic sprinkler system. Note that an extensive number of footnotes modify the general provisions of Table 803.9.   **803.9**

As a general rule, vertical exit enclosures and exit passageways are regulated at the highest level because of their importance as exit components in the means of egress. Both types of exits permit unlimited travel distance and are typically single-directional. Corridors are also highly regulated, but not to the extent that vertical exit enclosures and exit passageways are. Interior finish requirements for rooms and other enclosed areas are not as restrictive as for exitways; however, the wall and ceiling finishes are still regulated to some degree.

When it comes to occupancy groups, the high-hazard, institutional and assembly occupancies typically have the most restrictive flame-spread classifications. On the other hand, utility occupancies have no restrictions. Using a nonsprinklered office building as an example, the maximum flame spread classification of finish materials, based on occupancy group and location within the building, is shown in Figure 803-1.

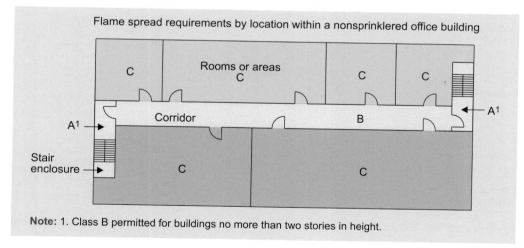

Flame spread requirements by location within a nonsprinklered office building

Note: 1. Class B permitted for buildings no more than two stories in height.

Figure 803-1
Flame-spread requirements by location within a nonsprinklered office building

Based on the cooling that an automatic sprinkler system provides, the code permits a reduction of one classification for many of the occupancies and locations. However, this is not a standard reduction. The table must be referenced to identify those reductions available for sprinklered applications. There is also no allowance for reducing a Class C requirement to a lower classification based on sprinkler protection. It should be noted that the required sprinkler system need only be provided in those exitways or rooms where the classification reduction is taken, and not throughout the entire building.

Figure 803-2 provides flame-spread classifications of woods commonly used in construction and finish work. A glance at the chart will show that most species of wood qualify for a Class C rating. There are few wood species that warrant a Class B rating, and no species is shown that qualifies for a Class A rating. However, there are many paints and varnishes on the market that manufacturers refer to as fire-retardant coatings. Because of intumescence, these paints or coatings bubble up or swell up under the action of flame and heat to provide an insulating coating on the surface of the material treated. Certain intumescent paints can reduce the flame spread of combustible finishes to as low as Class A and the smoke density to considerably below 450. These flame-spread-reducing intumescent coatings are particularly useful when correcting an existing nonconforming combustible interior finish.

Species of wood	Flame spread	Source*
Birch, yellow	105-110	UL
Cedar, eastern red	110	HUD/FHA
Cedar, Pacific Coast yellow	78	CWC
Cedar, western red	70	HPMA
Cottonwood	115	UL
Cypress	145-150	UL
Fir, Douglas	70-100	UL
Gum, red	140-155	UL
Hemlock, West Coast	60-75	CWC
Lodgepole	93	CWC
Maple flooring	104	CWC
Oak, red or white	100	UL
Pine, eastern white	85	UL
Pine, Idaho white	72	CWC
Pine, northern white	120-215	HPMA
Pine, ponderosa	105-200	UL
Pine, red	142	HUD/FHA
Pine, southern yellow	130-190	CWC
Pine, western white	75	HUD/FHA
Poplar	170-185	UL
Redwood	70	UL
Spruce, northern	65	UL
Spruce, western	100	UL
Spruce, white	65	UL
Walnut	130-140	CWC
Plywoods		UL
Douglas fir, $\frac{1}{4}$-inch	120	HUD/FHA
Lauan, three-ply urea glue, $\frac{1}{4}$-inch	110	HUD/FHA
Particleboard, $\frac{1}{2}$-inch	135	HPMA
Redwood, $\frac{3}{8}$-inch	95	CRA
Redwood, $\frac{5}{8}$-inch	75	CRA
Walnut, $\frac{3}{4}$-inch	130	HUD/FHA

*Source:

CRA:	California Redwood Association Association Data Sheet-2D2-7L (Lumber) Data Sheet -2D2-7P (Plywood)	HPMA: HUD/FHA:	Hardwood Plywood Manufactures Test No. 337, Test No. 592, Test No. 596 Flame-spread Rating for Various Material
CWC:	Canadian Wood Council Data File FP-6	UL:	Underwriters Laboratories UL 527, May, 1971

Figure 803-2
Flame-spread classification of woods

Stability. The IBC requires that the method of fastening the finished materials to the interior surfaces be capable of holding the material in place for 30 minutes under a room temperature of 200°F (93°C). If there is any question as to the adequacy of the fastening, appropriate tests should be required to determine compliance with this provision of the code. **803.10**

Application of interior finish materials to fire-resistance-rated structural elements. This section is applicable only where finish materials are applied on walls, ceilings or structural elements required to have a fire-resistance rating, or where such building elements must be of noncombustible construction (typically Type I or II construction). The greatest concern is where interior finish is not applied directly to a backing surface, creating concealed spaces that provide the opportunity for fire to originate and spread without detection until the interior finish material has burned through. Section 803.11.1 allows for the installation of furring strips, provided they are installed directly against the surface of the wall, ceiling or structural member. In addition, the concealed space created by the furring strips must be either fireblocked at maximum 8-foot (2438 mm) intervals, or filled completely with a Class A or organic material. The maximum depth of the concealed space is limited to $1^3/_4$ inches (44 mm). This section is also referenced by Section 803.11.3 for fireblocking in heavy-timber construction. **803.11**

Where interior finish materials are set out or suspended more than the $1^3/_4$ inches (44 mm) specified in Section 803.4.1, the potential exists for the fire to gain access to the space through joints or imperfections and to spread along the back surface as well. In this case, the flame spreads at a much faster rate than on one surface as the flame front will be able to feed on the material from two sides. Therefore, the provisions of Section 803.11.2 are intended to protect against this type of hazard. In the case where a wall is set out, the wall, including the portion that is set out, is required by the code to be of fire-resistance-rated construction as would be required by the code for the occupancy and type of construction.

It should be noted that the provisions of Sections 803.11.1 and 803.11.2 are applicable only where the walls and ceiling assemblies are required to be of either fire-resistance-rated or noncombustible construction. Where the walls and ceiling assemblies are of unprotected combustible construction, only the fireblocking provisions of Section 717.2 are applicable.

Section 803.11.4 requires that thin materials—no more than $1/_4$ inch (6.4 mm) thick—other than noncombustible materials, be applied directly against a noncombustible backing unless they are qualified by tests where the material is suspended from the noncombustible backing. The reason for this requirement is similar to that in Section 803.11.2. There are many buildings where thin paneling, such as luan plywood, is installed on walls and ceilings. When not installed against a noncombustible backing, these materials readily burn through and permit an almost uncontrolled rapid flame spread because the flame proceeds on both surfaces of the material.

Section 804 *Interior Floor Finish*

Floor finishes such as wood, terrazzo, marble, vinyl and linoleum present little, if any, hazard that is due to the spread of fire along the floor surface. However, other flooring materials such as carpeting are highly regulated by the IBC because of their potential for helping increase the growth of a fire.

Classification. For the purpose of regulating floor finishes based on the occupancy designation of the area where the finishes are installed, the code identifies two classes: Class I and Class II. Determined by test standard NFPA 253, the classifications are based on the critical radiant flux. The critical radiant flux is determined as that point where the heat flux level will no longer support the spread of fire. Class I is considered to have a critical radiant flux of 0.45 watts per square centimeter or greater, where Class II need only exceed 0.22 **804.2**

watts per square centimeter. Therefore, the Class I material is more resistant to flame spread because of the higher heat-flux level characteristics.

804.4 Interior floor finish requirements. Interior floor finishes are regulated differently by the code based upon two factors: the occupancies in the building and where the finish materials are located in relationship to the means of egress. The IBC selectively requires that exit enclosures, exit passageways and corridors be provided with a floor finish exceeding the critical radiant flux level established by the DOC (U.S. Department of Commerce) FF-1 "pill test." In addition, the floor finish materials of all rooms or spaces unseparated from a corridor by full-height partitions are regulated in a like manner, as there is evidence that corridor floor coverings can propagate flame when exposed to a fully developed fire in a room that opens into a corridor. Those rooms that have no direct connection with a corridor are simply regulated for the DOC FF-1 criteria, as are rooms that are separated from a corridor by full-height partitions.

The DOC FF-1 "pill test" requires a minimum radiant flux of 0.04 watts per square centimeter and is used to regulate all carpeting sold in the United States. Fire tests have demonstrated that carpet on the floor that passes the pill test is not likely to become involved in a room fire until the fire has reached or approached flashover.

Only those finish materials having a Class I or II classification may be installed in corridors, exit passageways and exit enclosures of fully-sprinklered Group I-1, I-2 and I-3 occupancies. The same limitation holds true for floors of exitways in areas of nonsprinklered buildings housing Group A, B, E, M or S occupancies. In other occupancies, the interior floor finish in the listed exitways need only comply with DOC FF-1 listing. The commentary above assumes the reduction in the classifications of floor finishes that is permitted where the building is fully sprinklered. Class II floor finish materials are permitted in lieu of Class I materials, whereas materials complying with the DOC FF-1 pill test may be used instead of Class II materials.

As a note, the entire building, and not just the area where the floor finish is located, must be provided with an automatic sprinkler system.

Section 805 *Combustible Materials in Type I and II Construction*

805.1 Application. Where combustible flooring materials are installed in or on floors in noncombustible buildings, they are regulated by this section. Combustible sleepers may only be installed where the space between the fire-resistance-rated floor deck and the sleepers is completely filled with approved noncombustible materials or is fireblocked in the manner described in Section 717. Finish flooring of wood shall be attached to sleepers, which if not imbedded, shall be appropriately fireblocked. As long as the wood flooring is in direct contact with a fire-resistance-rated floor, there is no significant hazard. However, if there is a space between the wood flooring and the fire-resistance-rated floor, a concealed space is created that is enclosed with combustible materials and provides a route for the undetected spread of fire. Therefore, the code requires that where wood flooring is not in direct contact with the fire-resistance-rated floor, the space be filled with noncombustible material or be fireblocked. Based upon the controls placed upon wood sleepers and finish flooring, it is also reasonable that combustible insulating boards be permitted where installed in a similar manner.

Section 806 *Decorative Materials and Trim*

Curtains, hangings and other decorative materials can potentially assist the spread of flame through a room or area. Therefore, for certain occupancies, the code regulates such materials as to their flame resistance and limits their use. In Groups A, E, I, R-1 and dormitories of Group R-2 occupancies, decorative materials hanging from the walls or ceilings must be flame resistant or noncombustible. In Group I-3 occupancies, only noncombustible materials are permitted. Test standard NFPA 701 is to be utilized in order to determine the effectiveness of the materials for flame resistance. In other than auditoriums of Group A, the total amount of flame-resistant decorative materials is limited to 10 percent of the total area of the walls and ceiling within the space. The amount of decorative material in auditoriums may be up to 75 percent of the aggregate wall and ceiling area, provided the building is fully sprinklered and the material is applied in a manner consistent with Section 803.11. As in all cases involving interior finish materials, documentation for such materials must be available for review and analysis by the building official.

Although typically not an issue, the amount of combustible trim in a room is limited to 10 percent of the aggregate wall or ceiling area of the room. By excluding handrails and guardrails, the code is only concerned with those rare situations where an extensive amount of decorative trim could present a problem. All trim must have a minimum flame-spread index and smoke-developed index of Class C.

KEY POINTS

- Regulation of finish materials by the IBC includes those on walls and ceilings, as well as floor coverings.

- Unregulated interior finish materials contribute to the rapid spread of fire, presenting a threat to the occupants by limiting or denying their use of exitways.

- Enclosed vertical exitways and exit passageways are the most highly regulated building elements for the application of interior finish materials, with corridors moderately controlled, and rooms or areas the least-regulated portions of the building.

- Installation of an automatic sprinkler system often allows a one-class reduction in the requirement for flame-spread classification.

- Textile wall and ceiling coverings are more highly regulated than other finish materials because of the potential for extremely rapid fire spread.

- Carpeting and similar floor covering materials are regulated in specific locations of specific occupancies.

- In certain occupancies, the code regulates curtains, hangings and other decorative materials as to their flame resistance and limits their use.

9

FIRE PROTECTION SYSTEMS

Chapter 9 provides requirements for three distinct systems considered vital for the creation of a safe building environment. The first of these systems is intended to control and limit fires and to provide building occupants and fire fighters with the means for fighting fire. Included are fire-extinguishing and standpipe systems. The detection and notification of a fire condition is addressed by the second system. Manual fire alarms, automatic fire detection and emergency alarm systems are included in this grouping. The third system is intended to control smoke migration. Included are design installation standards for smoke-control systems required by other chapters of the *International Building Code®* (IBC®) as well as smoke- and heat-venting systems. In addition to the provisions for fire protection systems, criteria are provided to increase the efficiency and safety of fire department personnel during emergency operations. Topics addressed include emergency responder safety features and radio coverage, the fire department command center, fire department connections and fire pump rooms.

Section 901 *General*

It is the intent of this chapter to require fire protection systems in those buildings and with those uses which through experience have been shown to present hazards requiring the additional protection provided by fire protection systems. The installation, repair, operation and maintenance of such systems are based upon the provisions of the IBC and the *International Fire Code®* (IFC®). Furthermore, it is the intent of the code to prescribe standards for those systems that are required. However, there are times when the installation of a fire-protection system is not based on a code mandate. In such situations, the nonrequired system must still meet the provisions of the code. Once fire protection is provided to some degree, it is expected that the system is properly installed.

An exception to this section permits a fire protection system or any portion of that system that is not required by the code to be installed for partial or complete protection, provided that the installation meets the code requirements. As an example, fire-sprinkler protection may be provided only in a specific area of a building, based on a request by the owner rather than on a requirement of the code. Although the sprinkler system must be installed in accordance with the proper design standard (in most cases NFPA 13), it is not necessary that the sprinkler system extend into other areas of the building.

More than likely, however, a fire protection system is utilized to gain exceptions to, or reductions in, other code requirements. Under these conditions, the fire protection system is considered a required system and is subject to all of the requirements imposed by the IBC and IFC.

901.7 **Fire areas.** The fire area concept is based upon a time-tested approach to limiting the spread of fire in a building. Through the use of fire-resistive elements, compartments can be created that are intended to contain a fire for a prescribed period of time. The floor area that occurs within each such compartment is considered to be the fire area. By definition as established in Section 902.1, a fire area is the aggregate floor area enclosed and bounded by fire walls, fire barriers, exterior walls or fire-resistance horizontal assemblies of a building. See Figure 901-1. In addition, any areas beyond the exterior wall that are covered with a floor or roof above, such as a canopy extending from the building, are considered part of the building for fire area purposes. This approach is consistent with the determination of building area in Section 502.1. An example is shown in Application Example 901-1. By isolating a fire condition to a single fire area through the use of fire separation elements, only a portion of the building is considered at risk because of a single fire incident.

The use of fire areas as a fire protection tool is limited only as an alternative to the requirements for automatic sprinkler systems. Other fire protection systems, such as fire alarm systems, are regulated by methods that are not based on the fire area concept. Even within the automatic sprinkler provisions of Section 903.2, only a portion of the requirements use the fire area approach as an alternate means of protection. The fire area methodology set forth in the IBC, applicable only in limited occupancy groups under limited conditions, allows for the omission of automatic sprinkler protection.

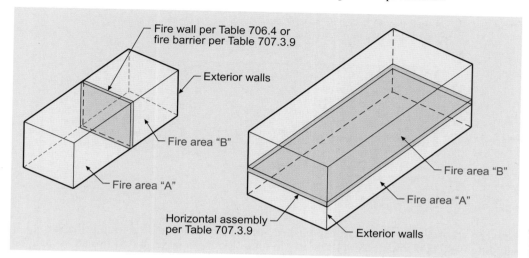

Figure 901-1
Fire areas

As an example, the provisions of Section 903.2.3, Item 1 require that a fire area containing a Group E occupancy that exceeds 12,000 square feet (1115 m²) in floor area be provided with an automatic sprinkler system. Conversely, where the fire area size does not exceed the established threshold of 12,000 square feet (1115 m²), a sprinkler system is not required unless mandated by another code provision. Where the building under consideration is limited to a maximum of 12,000 square feet (1115 m²), it can be viewed as a single fire area, and no sprinkler system is mandated. However, where the building exceeds 12,000 square feet (1115 m²) in floor area, two or more fire areas must be established to eliminate the sprinkler requirement. Table 707.3.9 is referenced because it sets forth the minimum required level of fire-resistance necessary to create an adequate fire separation

Application Example 901-1

GIVEN: A Group M retail sales building that includes a roofed exterior sales area of 3,000 sq.ft.
DETERMINE: If a sprinkler system is required.

Group M retail sales 11,000 sq ft

Covered exterior sales area 3,000 sq ft

SOLUTION: As a single Group M fire area of 14,000 sq. ft. a sprinkler system is required per Section 903.2.7, Item 1. However, no sprinkler system is required if a complying 2-hour fire barrier per Section 707.3.9 separates the interior and exterior sales areas.

FIRE AREA DEFINED

Application Example 901-2

GIVEN: A 16,000-square-foot building with two tenants: a 10,000-square-foot Group M retail store and a 6,000-square-foot Group F-1 fabrication shop.

DETERMINE: The requirements for using fire areas rather than sprinkler protection.

SOLUTION: No sprinkler system required for:
- Group M fire area ≤ 12,000 square feet
- Group F-1 area ≤ 12,000 square feet
- Total of Group M and F-1 fire areas ≤ 24,000 square feet

Group M 10,000 sq ft	Group F-1 6,000 sq ft

Fire area separation minimum of 3-hour fire barrier per Table 707.3.9

For SI: 1 square foot = 0.093 m².

between the fire areas that are established. In the example, and assuming the Group E building is 20,000 square feet (1858 m²) in total floor area, at least two fire areas must be created as an alternative to sprinkler protection. Neither of the two fire areas is allowed to exceed 12,000 square feet (1115 m²), and Table 707.3.9 indicates that the minimum fire separation between the two fire areas must be 2 hours. Therefore, a minimum 2-hour fire-resistance-rated fire wall, fire barrier or horizontal assembly, or a combination of these elements, would be required.

A similar approach is taken in a mixed occupancy building where the multiple fire areas are of different occupancy classifications. The minimum required fire-resistance rating for the separation between the fire areas would also be based on the requirements of Table 707.3.9. For example, where a building contains a 10,000-square-foot (929 m²) Group M occupancy and a 6,000-square-foot (558 m²) Group F-1 occupancy, the minimum fire-resistive separation between the Group M fire area and the Group F-1 fire area would be 3 hours. Although the Group M requirement in Table 707.3.9 only mandates a 2-hour separation, a minimum 3-hour fire separation is required for a Group F-1 occupancy. For further information, see Application Example 901-2 and the discussion of Table 707.3.9.

Section 903 *Automatic Sprinkler Systems*

In general, automatic sprinkler systems are required when:

1. Certain special features and hazards of specific buildings, areas and occupancies are such that the additional protection provided by sprinkler systems is warranted.

2. There are inadequate numbers and sizes of openings in the exterior walls from which a fire may be fought from the exterior of the building. The provisions requiring sprinklers in these so-called *windowless buildings* apply to all buildings, regardless of occupancy, except for Group R-3 and U occupancies.

There are three general situations in which sprinkler or other fire suppression systems are to be provided within a building. An automatic sprinkler may be required throughout the building, throughout a fire area, or only in the specific room or space where sprinkler protection is necessary. Examples are depicted in Figure 903-1.

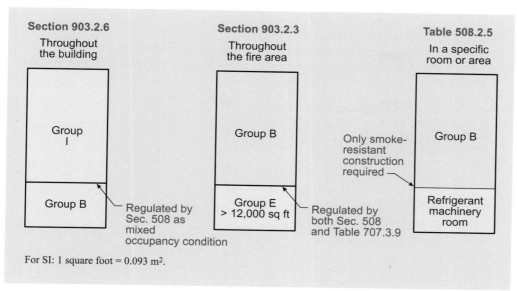

Figure 903-1
**Sprinkler
requirements**

1. Throughout the building. There are numerous applications of the code that require the entire building to be sprinklered, either mandated because of a code requirement or used as a substitute for other fire and life-safety features. Examples include the requirement for sprinklers throughout all buildings containing a Group I fire area per Section 903.2.6 and the elimination of corridor fire protection in some occupancies based on Table 1018.1. The extent of required sprinkler protection in Group A occupancies is described in a different manner, but quite often results in a fully-sprinklered condition. Initially, only the specific Group A occupancy must be provided with sprinklered protection where exceeding the limits established by Sections 903.2.1.1 through 903.2.1.4. However, as mandated by Section 903.2, the sprinkler system must also be provided throughout *the floor area* where the Group A occupancy is located. In application, it is appropriate that the sprinkler system be installed throughout the entire story or floor level containing the Group A occupancy. In addition, all floor levels between the Group A occupancy and the level of exit discharge must be sprinklered. This will commonly result in a requirement that the entire building be provided with an automatic sprinkler system. Additional commentary is provided in the discussion of Section 903.2.1.

2. Throughout a fire area. In Section 903.2, a variety of provisions require only those fire areas that exceed a certain size or occupant load, or are located in a specific portion of the structure to be sprinklered. The sprinkler requirements based on fire area include the provisions of Section 903.2.3 for Group E occupancies and Section 903.2.4.1 for woodworking operations.

    A variation of this requirement occurs in those mixed-occupancy buildings containing a Group H-2, H-3 or H-4 occupancy. The code only mandates that the sprinkler system be provided in the Group H portion, not the entire building. However, since all other occupancies in a mixed-occupancy building must be appropriately isolated from the Group H occupancy because of the *separated occupancies* provisions of Section 508.4, the result is basically a requirement to sprinkler the Group H compartment.

3. Specific rooms or areas. Occasionally, only a specific portion of the building requires the protection provided by a sprinkler system. The sprinkler addresses the particular hazard that occurs at possibly only a single location. For example, the allowance for a reduction in the flame-spread classification of interior finishes from Table 803.9 is based on sprinkler protection in the room, area or exitway where the finish under consideration is installed.

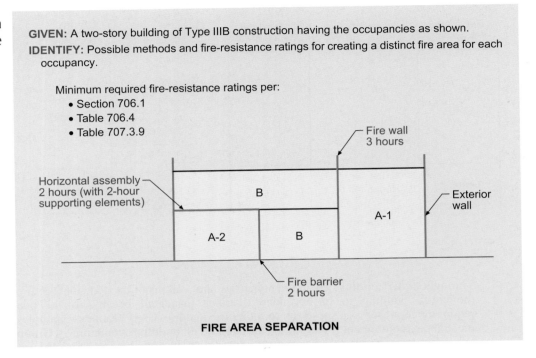

Application Example 903-1

GIVEN: A two-story building of Type IIIB construction having the occupancies as shown.

IDENTIFY: Possible methods and fire-resistance ratings for creating a distinct fire area for each occupancy.

Minimum required fire-resistance ratings per:
- Section 706.1
- Table 706.4
- Table 707.3.9

Fire wall 3 hours

Horizontal assembly 2 hours (with 2-hour supporting elements)

B

Exterior wall

A-1

A-2

B

Fire barrier 2 hours

FIRE AREA SEPARATION

903.1.1 **Alternative protection.** Where an automatic sprinkler system is addressed in the IBC, alternative automatic fire-extinguishing systems are acceptable, provided they are installed in accordance with approved standards. These systems would be special systems required by the IFC and other types of systems such as dry chemical, carbon dioxide or aqueous-foam systems. This is one of the few provisions in the IBC where approval must come from other than the building official. Although the building official is almost always charged with making any decisions regarding the building code, the fire code official is typically better able to evaluate and determine the appropriateness of an alternative fire-extinguishing system.

903.2 **Where required.** It is the intent of this section to specify those occupancies and locations where automatic sprinkler systems are required. A fire-extinguishing system is a system that discharges an approved fire-extinguishing agent such as water, dry chemicals, aqueous foams or carbon dioxide onto or in the area of a fire. A fire sprinkler system discharges water only. The code specifies a fire sprinkler system in this section, as it is the intent of the code that water be applied and not one of the other extinguishing agents. Generally, water is the most effective extinguishing agent for fires. Only where water creates problems, such as in magnesium or calcium carbide storage areas, would some other type of extinguishing agent be required. The allowance for the installation of a system other than an automatic sprinkler system is subject to approval by the fire code official. It is important to note that where an automatic fire-extinguishing system is installed as an alternative to the required automatic sprinkler system of Section 903, it cannot be utilized for the purposes of exceptions or reductions allowed by other code provisions. Alternative automatic fire-extinguishing systems are addressed in Section 904.

Fire areas. Most of the requirements of this section are based on the concept of fire areas. Where a fire area exceeds a specified size, is located in a certain portion of the building or exceeds a specified occupant load, the code often requires the installation of an automatic sprinkler system to address the increased hazards and concerns that exist. The provisions for fire areas can be found in various sections of the IBC.

The definition of "Fire area" is located in Section 902.1. A fire area is "the aggregate floor area enclosed and bounded by fire walls, fire barriers, exterior walls or fire-resistance-rated horizontal assemblies of a building." Complete isolation and separation of a portion of a building from all other interior areas is provided for a fire area through the use of fire-resistance-rated construction and opening protectives. The total floor area within the enclosed area, including the floor area of any mezzanines or basements, is considered the size of the fire area. See Application Example 903-1. It is also important to note that "areas of the building not provided with surrounding walls shall be included in the fire area if such areas are included within the horizontal projections of the roof or floor next above."

Where fire walls are utilized, Section 503.1 indicates that each portion of the structure included within the fire walls is considered a separate building. This concept would be consistent with that of separate and distinct fire areas being created through the use of fire walls. The fire-resistance rating of the wall and the fire-protective ratings of any openings in the fire wall are identified in Chapter 7. Also see the discussion of Section 706.1 where the fire wall separates different occupancies.

Fire barriers and fire-resistance-rated horizontal assemblies may also be used to create fire areas, provided the fire-resistance-rated construction totally separates one interior area from another. In order to determine the minimum fire-resistance rating of the vertical and horizontal elements, the occupancy classifications of the areas being separated must be identified. Table 707.3.9 is then referenced to determine the minimum fire-resistance rating of the separation. The use of this table is applicable to both single-occupancy and mixed-occupancy buildings as illustrated in Application Example 903-2. Where the fire area separation occurs between two fire areas of the same occupancy, the hourly rating established by Table 707.3.9 for that single occupancy classification is applied. If the fire areas are of different occupancy classifications, the controlling fire-resistance rating of the fire barrier or horizontal assembly separating the occupancies is based on the higher of the ratings as established by Table 707.3.9 for the occupancies involved. For further information, see the discussions of Section 901.7 and Table 707.3.9.

Because the majority of sprinkler provisions are based on the size of the fire area, it is sometimes possible for the designer to eliminate the requirement for sprinklers by reducing the floor area within the surrounding fire-resistance-rated construction. The use of fire walls, or fire barriers and fire-resistance-rated horizontal assemblies can subdivide a structure into smaller, less hazardous areas that are of such a size that sprinklers are not necessary. See Application Example 903-3. This concept of compartmentation has been used in building codes for decades as an effective method of reducing the loss of life and property in fires.

The exception to Section 903.2 eliminates the sprinkler requirement in telecommunications occupancies in those rooms or areas dedicated solely for essential telecommunications and power equipment. The alternative protection is provided through the required installation of an automatic fire alarm system, as well as fire-resistance-rated separation from other areas of the building.

Group A. Because of the potentially high occupant load and density anticipated in Group A **903.2.1** occupancies, coupled with the occupants' probable lack of familiarity with the means of egress system, various assembly uses must be protected by an automatic sprinkler system. Where an automatic sprinkler system is required for a Group A occupancy, the system must be installed throughout the entire floor level or story where the Group A occupancy is located. In addition, where the Group A occupancy requiring a sprinkler system is located on a floor level other than the level of exit discharge, all floor levels between the Group A occupancy and the nearest level of exit discharge must be sprinklered as well. By expanding the areas of the building required to be protected by an automatic sprinkler system beyond just the assembly areas, the code provides for protection adjacent to the Group A areas as well as throughout the means of egress. Figures 903-2 and 903-3 illustrate these fundamental provisions.

Application Example 903-2

GIVEN: The various occupancies housed in a building as shown below.

DETERMINE: The required fire-resistance ratings of the assemblies separating the occupancies in order to create different fire areas for the purpose of applying Section 903.2.

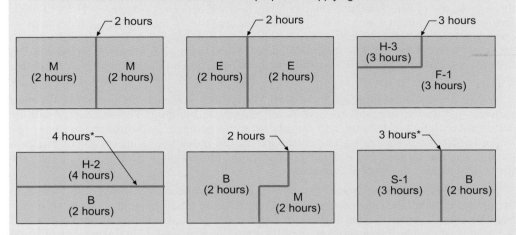

*Required minimum fire-resistance rating for fire barrier based upon higher of ratings as established by Table 707.3.9

Application Example 903-3

GIVEN: A large building is to be divided into various retail, business and assembly tenants having the floor areas indicated.

DETERMINE: A method in which the space can be subdivided into individual fire areas by fire barriers and not be required to be protected by an automatic sprinkler system.

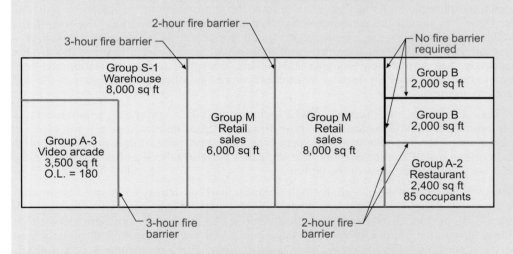

- Automatic sprinkler system not required by Section 903.2 based upon creation of complying fire areas

- Fire areas created with fire barriers rated in accordance with Table 707.3.9

For SI: 1 square foot = 0.093 m².

FIRE AREA DETERMINATIONS

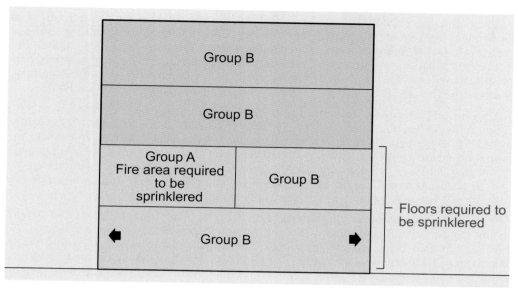

Figure 903-2
Group A sprinkler

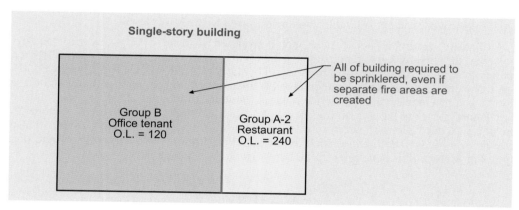

Figure 903-3
Group A fire area

Group A-1. The combination of highly concentrated occupant loads, high numbers of occupants, reduced lighting levels and potentially high fuel loads create a level of hazard that justifies the need for sprinkler protection. Therefore, fire areas containing theaters and similar assembly uses intended for the viewing of motion pictures or the performing arts shall be provided with an automatic sprinkler system where any one of the following conditions exist:

903.2.1.1

1. The fire area containing the Group A-1 occupancy exceeds 12,000 square feet (1115 m^2).

2. The occupant load of the fire area exceeds 299.

3. The fire area is located on any floor level other than that of the exit discharge.

It should also be noted that any fire area containing a multitheater complex, defined as two or more theaters served by a common lobby, shall be provided with a sprinkler system throughout the fire area.

Group A-2. Fire areas housing uses intended for food or drink consumption are regulated for sprinkler protection at a higher level than other enclosed assembly occupancies. Even where the occupant load is not excessive, the hazards associated with such uses warrant the protection provided by a sprinkler system. Oftentimes, the consumption of alcohol beverages by the building's occupants creates an environment more likely to be unsafe. The reduced lighting levels in some uses, along with the probability of loose chairs and tables, also increases the risk for obstructed egress. The record of casualties during fires in

903.2.1.2

buildings housing nightclubs, restaurants and similar types of uses demonstrates the need for the additional protection provided by fire sprinklers or, alternatively, the separation of the use into smaller compartments. The code intends that fire areas exceeding 5,000 square feet (465 m²) that contain Group A-2 uses be provided with an automatic sprinkler system, as well as such uses having an aggregate occupant load within the fire area of 100 or more, or where the fire area is located on a floor level other than the level of exit discharge.

903.2.1.3   **Group A-3.** The sprinkler threshold for a Group A-3 occupancy is identical to that for a Group A-1 occupancy. As such, where any fire area in a Group A-3 exceeds 12,000 square feet (1115 m²), where the fire area has an occupant load greater than 299 or where the assembly occupancy is located on any floor other than the exit discharge level, an automatic sprinkler system is required. In applying the provisions of this section, it is important to note that the occupant load threshold is based upon the number of people within the entire fire area, not just in each assembly room. See Application Example 903-4.

The code requirements for these types of uses, specifically for exhibition and display rooms, can be strongly attributed to the McCormick Place Fire in Chicago on January 16, 1967. McCormick Place was not sprinklered and consisted of three levels, including a main exhibit area of 320,000 square feet (29 728 m²) on the upper level. Both the upper and lower levels were in the final stages of readiness for a housewares exhibition and were heavily laden with combustibles when the fire broke out. The fire was reported to have originated in the storage area behind an exhibit booth on the upper level. The upper level was almost totally destroyed, and considerable damage occurred to the lower level.

Ordinarily, assembly occupancies are considered to have a very low fire loading; however, the need for built-in fire suppression for an assembly use that is used for exhibition or display purposes was clearly demonstrated by the McCormick Place fire. Display booths are most often constructed with combustible materials, and the storage area behind the booths is a receptacle for combustible materials and packing boxes. Thus, without built-in fire suppression, the large quantities of combustible materials and large areas combine to create an excessive hazard. Many other assembly occupancies classified as Group A-3 also present significant fire loading such as art galleries, libraries and museums. Therefore, sprinkler protection is beneficial for all large Group A-3 occupancies.

Application Example 903-4

GIVEN: A mixed-occupancy building containing a Group B office area and four Group A-3 conference rooms (each with an occupant load of 88)

DETERMINE: An appropriate method of fire area separation as an alternative to installation of a sprinkler system

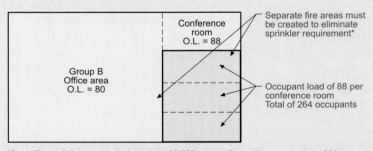

*Each Group A fire area to be less than 12,000 square feet with no more than 299 occupants

Note: Mixed occupancy conditions must also comply with Section 508.

For SI: 1 square foot = 0.093 m².

SOLUTION: A minimum of two fire areas must be created so as not to exceed the 299 occupant load and 12,000 square foot limitations

AGGREGATE GROUP A OCCUPANT LOADS

Group A-4. The fire-sprinkler requirements for Group A-4 occupancies (those assembly **903.2.1.4** uses provided with spectator seating for the viewing of indoor activities and sporting events) are identical to the provisions for Group A-3 uses. See Section 903.2.1.3 for a discussion of the sprinkler requirements.

Group A-5. The fire loading in stadiums and grandstands is typically quite low except for **903.2.1.5** specific accessory areas such as concession stands, storage and equipment rooms, press boxes and ticket offices. Therefore, assembly occupancies classified as Group A-5 do not require the installation of an automatic sprinkler system except for those support areas exceeding 1,000 square feet (92.9 m^2) in floor area. The limitation of 1,000 square feet (92.9 m^2) is based upon the floor area of each individual area and not on the aggregate area of all such spaces. Where such accessory spaces are of a considerable size, the hazards posed by the potentially large quantities of combustible materials can be reduced where such areas are sprinklered.

Group B ambulatory health care facilities. As a general rule, Group B occupancies do **903.2.2** not require a sprinkler system based solely on their occupancy classification. However, Section 903.2.2 mandates that a Group B ambulatory care facility be provided with an automatic sprinkler system when either of the following conditions exist at any time:

- Four or more care recipients are incapable of self-preservation, or
- One or more care recipients that are incapable of self-preservation are located at other than the level of exit discharge.

Although such facilities are generally regarded as moderate in hazard level due to their office-like conditions, additional hazards are typically created due to the presence of individuals who are temporarily rendered incapable of self-preservation due to the application of nerve blocks, sedation or anesthesia. While the occupants may walk in and walk out the same day with a quick recovery time after surgery, there is a period of time where a potentially large number of people could require physical assistance in case of an emergency that would require evacuation or relocation. The installation of an automatic sprinkler provides an important safeguard that enables the moderate-hazard classification of Group B.

Group E. History has shown that educational occupancies perform quite well when it comes **903.2.3** to fire- and life-safety concerns. Much of this can be attributed to the continuous control and supervision that takes place within schools, as well as the students' knowledge of egress responsibilities in case of a fire or other emergency. However, because of the potential for moderate to high combustible loading, fire areas in Group E occupancies that exceed 12,000 square feet (1115 m^2) in floor area must be provided with an automatic sprinkler system. In addition, fire sprinklers are required for those portions of educational buildings located below the level of exit discharge.

An exception to the general sprinkler provision is based on increased criteria for exiting. Where every classroom within an educational building has at least one exit door to the exterior and such exit doors are located at ground level, an automatic sprinkler system need not be provided. By providing direct egress for the students from their classrooms to the exit discharge, a very high level of occupant protection has been attained.

Group F-1. Without an automatic sprinkler system to limit the size of a fire, the fire can **903.2.4** spread very quickly to other portions of the structure. This is particularly true for large floor-area buildings containing combustible materials such as manufacturing facilities, warehouses and retail sales buildings. The IBC requires an automatic sprinkler system to be installed throughout any building containing a Group F-1 occupancy where the fire area containing the Group F-1 exceeds 12,000 square feet (1115 m^2). A fire sprinkler is also required where the building housing the Group F-1 occupancy is four stories or more in height or has an aggregate of Group F-1 fire areas in the building of more than 24,000 square feet (2230 m^2). The aggregate F-1 fire area would also include the floor area of any mezzanines involved. See Figure 903-4. The 24,000 square-foot (2230 m^2) limitation would also be applicable in single-story structures as shown in Figure 903-5.

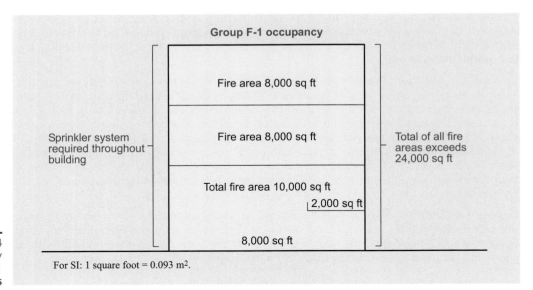

Group F-1 occupancy

Sprinkler system required throughout building

Fire area 8,000 sq ft

Fire area 8,000 sq ft

Total fire area 10,000 sq ft

2,000 sq ft

8,000 sq ft

Total of all fire areas exceeds 24,000 sq ft

Figure 903-4
Multistory
Group F-1
Occupancies

For SI: 1 square foot = 0.093 m².

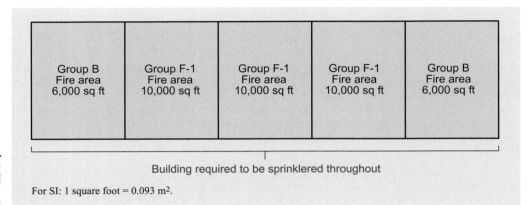

| Group B Fire area 6,000 sq ft | Group F-1 Fire area 10,000 sq ft | Group F-1 Fire area 10,000 sq ft | Group F-1 Fire area 10,000 sq ft | Group B Fire area 6,000 sq ft |

Building required to be sprinklered throughout

Figure 903.5
Sprinklered
Group F-1
Occupancies

For SI: 1 square foot = 0.093 m².

903.2.4.1 **Woodworking operations.** Because of the special hazards involving dusts created during woodworking operations such as sanding and sawing, this section requires that an automatic fire-sprinkler system be installed in fire areas of Group F-1 woodworking occupancies where the floor area of such operations exceeds 2,500 square feet (232 m²). Where equipment, machinery or appliances that generate finely divided combustible waste or that use finely divided combustible materials are a portion of a woodworking operation, the size of the operation is strictly limited unless sprinklers are installed. An example of this provision is shown in Figure 903-6. The provision is based on the size of the area where only the sanding, sawing and similar operations occur, not necessarily the floor area of the entire woodworking operation. However, because these types of operations occur quite often as an integral part of the overall woodworking activities, rather than isolated in their own room or area, some means of regulating and controlling the hazard should be provided.

903.2.5 **Group H occupancies.** Group H occupancies are high-hazard uses, and one special feature is that, in addition to presenting a local hazard within the building, it has a potential for presenting a high level of hazard to the surrounding properties. Therefore, the code requires sprinkler protection for all Group H occupancies. Note that the sprinkler system is not necessarily required throughout the entire building that contains a Group H-2, H-3 or H-4 occupancy. Only such Group H areas must be provided with a sprinkler system. In the case of a Group H-1 occupancy, no other occupancies are permitted in the same building. Therefore, a building containing a Group H-1 occupancy must be sprinklered throughout. In addition, buildings containing Group H-5 occupancies require sprinklers in other portions of the building in addition to the high-hazard area proper. This requirement is based on the original premise that the primary protection feature of this highly protected use is the automatic fire-extinguishing system. For the purpose of sprinkler-system design, all areas of a Group H-5 are considered Ordinary Hazard Group 2, except for storage rooms with dispensing operations, which are considered Extra Hazard Group 2.

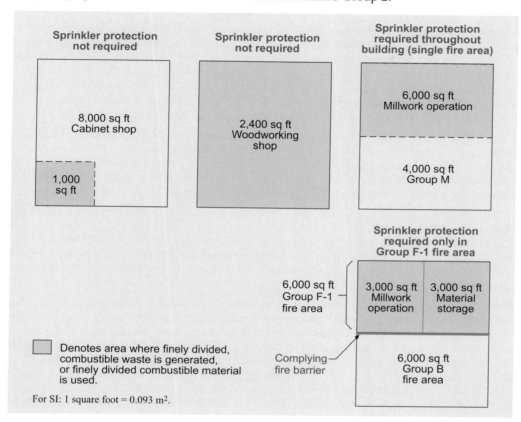

Sprinkler protection not required

8,000 sq ft
Cabinet shop

1,000 sq ft

Sprinkler protection not required

2,400 sq ft
Woodworking shop

Sprinkler protection required throughout building (single fire area)

6,000 sq ft
Millwork operation

4,000 sq ft
Group M

Sprinkler protection required only in Group F-1 fire area

6,000 sq ft
Group F-1
fire area

3,000 sq ft
Millwork operation

3,000 sq ft
Material storage

Complying fire barrier

6,000 sq ft
Group B
fire area

Denotes area where finely divided, combustible waste is generated, or finely divided combustible material is used.

For SI: 1 square foot = 0.093 m².

Figure 903-6
Woodworking operations

Group I. Because the mobility of the occupants of Group I occupancies is greatly diminished (in the case of hospitals and detention facilities, the self-mobility is essentially nonexistent), the code requires an NFPA 13 automatic sprinkler system throughout any building where a Group I fire area exists. For supervised residential facilities classified as Group I-1 occupancies, an allowance is made for the use of an approved NFPA 13D or 13R system. The similarities between this Group I use and those uses classified as Group R justify the reduction in sprinkler protection.

903.2.6

Group M. The typical American supermarket evolved during the construction boom that followed World War II. At that time, the typical supermarket consisted of a one-story building of moderately large area, e.g., 15,000 to 25,000 square feet (1394 m² to 2323 m²). During the 1950s, fire statistics indicated that large-area supermarkets without sprinkler protection were subject to a larger proportion of fires than were usually attributable to this use in the past. As a result, building codes began requiring sprinklers in larger retail sales occupancies. The present requirements, detailed in the discussion in Section 903.2.4, are based upon any of three factors; the size of the fire area, the number of stories or the combined fire area on all floors. In addition, reference to the IFC is made for sprinkler protection in mercantile buildings where merchandise is placed in high-piled or rack storage. The installation of an automatic sprinkler system is also mandated in any Group M occupancy that is used for the display and sale of upholstered furniture. This provision is not

903.2.7

based on the size of the fire area of the Group M occupancy – instead, it is based strictly on the specific contents of the mercantile occupancy. The requirement does not apply to retailers who sell only mattresses and box springs. It is also not applicable to the display and sale of furniture that is not upholstered, such as furniture constructed entirely of wood, plastic or metal. In addition, the provision is not applicable in Group S occupancies where upholstered furniture is stored.

903.2.8 Group R. In hotels, apartment buildings, dormitories and other Group R occupancies, occupants may be asleep at the time of a fire, and may experience delay and disorientation in trying to reach safety. In addition, fire hazards in residential uses are often unknown to most occupants of the building, as they are created within an individual dwelling unit or guestroom. This helps to explain why these occupancies have a poor fire record when it comes to injury and loss of life. Therefore, an automatic sprinkler system is required throughout any building containing a Group R use. The sprinkler requirement applies to the entire building and not just the fire area containing the Group R occupancy.

903.2.9 Group S-1. In a manner consistent with that for Group F-1 and M occupancies, buildings containing combustible storage and warehousing uses must be provided with an automatic sprinkler system where the floor area or height exceeds the specified threshold. The sprinkler requirement is based on the probable presence of large amounts of combustible materials, typically arranged in a highly concentrated manner.

Although the storage of commercial trucks and buses is typically regulated under the provisions of Section 903.2.10.1, there are situations where the parking of such vehicles occurs in the same area with other storage uses. These multipurpose spaces, such as fire station bays, are more appropriately classified as Group S-1 occupancies. In such cases, a more restrictive threshold of 5,000 square feet (464 m^2) is used to require sprinkler protection.

903.2.9.1 Repair garages. The unique hazards associated with vehicle repair garages may be addressed in part through the installation of an automatic sprinkler system. However, the requirement for sprinklers is limited only to those repair garages that present a high level of concern based on size or location. By locating the repair garage above grade in a building of one or two stories, the size of the fire area containing the garage becomes the controlling factor in the determination of whether or not a sprinkler system is required. Where there is vehicle parking in the basement of a building used for vehicle repair, the building must be sprinklered regardless of fire area size. The sprinkler requirement is applicable even where the repair activity occurs only above the basement level. In buildings where commercial trucks or buses are repaired, the threshold for sprinkler protection is consistent with that established in Section 903.2.10.1 for commercial parking garages.

903.2.10 Group S-2 enclosed parking garages. Because the bulk of the uses designated as Group S-2 occupancies present very low fire-load potential, there is generally no requirement for these low-hazard occupancies to be sprinklered. However, where the Group S-2 portion of a building is an enclosed parking garage, the hazard level is increased. There is a need to protect other uses housed above an enclosed parking garage; thus, a Group S-2 enclosed parking garage is required to be sprinklered where the garage is located below another occupancy. In fact, in such a situation the entire building must be sprinklered, regardless of the size of the garage itself. There is also an exception to the sprinkler requirement where an enclosed parking garage is located beneath an R-3 occupancy. Where the enclosed parking garage has no uses above, the required point at which an automatic sprinkler system is required is consistent with the threshold established for other moderate-hazard occupancies. The installation of an automatic fire sprinkler system for enclosed parking garages is required where the fire area containing the garage exceeds 12,000 square feet (1115 m^2) in floor area. The fire behavior in an enclosed parking garage, although similar to that in an open parking garage, is of greater concern since smoke ventilation will be more difficult due to the lack of sufficient exterior openings. This concern is addressed by the required installation of an automatic sprinkler system once the 12,000-square-foot (1115 m^2) fire area threshold is exceeded. The sprinkler requirement is not applicable to open parking garages.

Commercial parking garages. Where the vehicles stored within a building consist of commercial trucks and buses, the code mandates stringent floor areas when it comes to the requirement for an automatic sprinkler system. Where a fire area containing commercial parking exceeds 5,000 square feet (464 m^2) in floor area, the building housing the vehicles must be sprinklered throughout. The provision is intended to address those facilities housing larger vehicles. It is generally not applicable where pick-up trucks and similar-sized vehicles are being used for business activities. **903.2.10.1**

Stories without openings. The provisions of this section make specific the intent of the code to require automatic sprinkler protection in *windowless buildings*. A structure having inadequate openings on the exterior wall as determined by this section such that fire department access is insufficient is considered a *windowless building*. The requirements of this section apply to all occupancies except Groups R-3 and U. The provisions are applicable on a floor-by-floor basis and do not apply to any story or basement having a floor area of 1,500 square feet (139.4 m^2) or less: **903.2.11.1**

- **On the basis of each individual story above ground.** Each individual story is analyzed for the size and the number of exterior wall openings. Thus, in a multistory building, it is possible to have a requirement that a sprinkler system be installed in one story and not in another.

The code requires that the openings be:

1. **Installed entirely above the adjoining ground level.** This provision is necessary so that effective fire suppression and rescue can be accomplished from the exterior of the building.

   Where the openings cannot be located entirely above the adjoining ground level, the code permits the use of exterior stairways or ramps that lead directly to grade.

2. **Of adequate size and spacing.** Although it may be argued that the openings required by the code are not the equivalent of automatic fire-sprinkler protection, the access for fire fighting provided by the openings has proven satisfactory.

   Although not expressly stated in the code, there is an expectation that a below-grade opening used to satisfy this provision be simply a typical 3-foot by 6-foot, 8-inch (914 mm by 2032 mm) door leading directly to the exterior stairway or ramp. However, above-grade openings are more specifically addressed. A total of 20 square feet (1.86 m^2) of openings is mandated in each 50 lineal feet (15 240 mm) of exterior wall. It is not necessary to obtain all 20 square feet (1.86 m^2) from a single opening, as long as the minimum dimension requirement of 30 inches (762 mm) is met. Multiple 30-inch by 30-inch (762 mm by 762 mm) openings would comply; however, they may not be as effective as a larger single opening.

   The intent of the code is that there shall be at least one opening in each 50 linear feet (15 240 mm) of exterior wall. It may be better stated that *any* wall section of 50 feet in length be provided with complying openings. Thus, an exterior wall 100 feet (30 480 mm) long with 20-square-foot (1.86 m^2) openings located at third points along the wall would comply, as shown in Figure 903-7. There is no portion of the wall that is 50 feet (15 240 mm) in length that does not contain the necessary openings. However, the same wall with such openings located at each end, as depicted in Figure 903-8, will not comply with the intent of the code insofar as there is a length of wall that exceeds the 50-foot (15 240 mm) dimension without an opening. Certainly, the same wall with only one 40-square-foot (3.72 m^2) opening at one end also would not comply.

3. **Accessible to the fire department from the exterior.** Surely, the openings would be of no value for fire fighting if the fire-fighting forces could not gain access. The mere fact that the openings may be 30 or 40 feet (9144 mm or 12 192 mm) above grade does not mean the openings are inaccessible. However, if, with the resources available to the fire department, access cannot be obtained to the openings, they

would be considered inaccessible. The determination of accessibility rests with the building official. However, personnel in a fire department should be consulted for their professional opinions and also for their knowledge of the capabilities of their equipment.

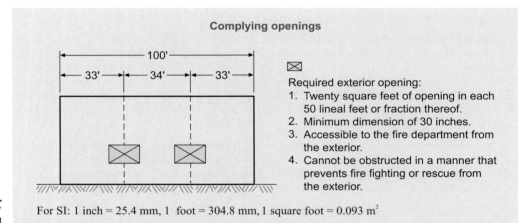

Complying openings

⊠
Required exterior opening:
1. Twenty square feet of opening in each 50 lineal feet or fraction thereof.
2. Minimum dimension of 30 inches.
3. Accessible to the fire department from the exterior.
4. Cannot be obstructed in a manner that prevents fire fighting or rescue from the exterior.

For SI: 1 inch = 25.4 mm, 1 foot = 304.8 mm, 1 square foot = 0.093 m^2

Figure 903-7
Required exterior openings

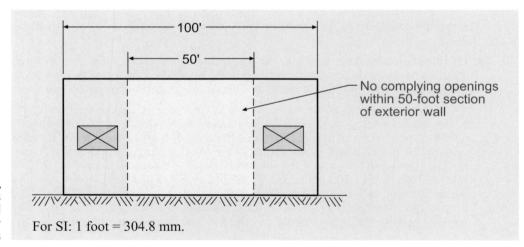

No complying openings within 50-foot section of exterior wall

Figure 903-8
Required exterior openings

For SI: 1 foot = 304.8 mm.

4. **Adequate to allow access for fire fighting to all portions of the interior of the building.** For this reason, the code requires that where openings are provided on only one side and the opposite exterior wall is more than 75 feet (22 860 mm) away, sprinklers shall be provided, or, as an alternative, openings shall be provided on at least two sides. The 75-foot (22 860 mm) distance is a straight-line measurement taken between the two opposing walls. Where complying openings are required in two exterior walls because of the 75-foot (22 860 mm) limitation, the openings are permitted on either two adjacent sides or opposite sides on the assumption that, with two exterior sides having openings, adequate access may be gained to effectively fight the fire.

The provisions requiring openings in exterior walls do not extend beyond the exterior wall line into the building. Thus, the code does not dictate specific openings for interior partition arrangements, because the normal openings provided through interior partitions provide adequate accessibility to all interior portions of the building.

5. **Applicability.** As previously noted, the provisions of Section 903.2.11.1 apply to every story or basement of all buildings where the floor area exceeds 1,500 square feet (139.4 m^2). Figures 903-9 and 903-10 provide additional graphic representations of the requirements of this section.

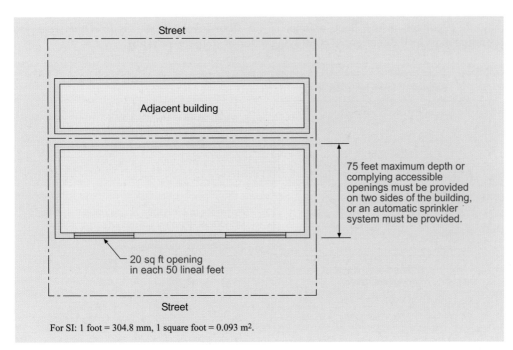

For SI: 1 foot = 304.8 mm, 1 square foot = 0.093 m^2.

Figure 903-9
Maximum distance between walls

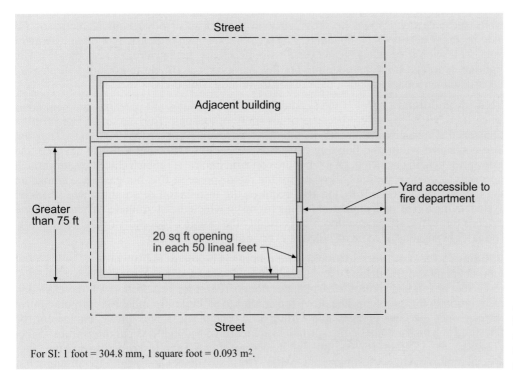

For SI: 1 foot = 304.8 mm, 1 square foot = 0.093 m^2.

Figure 903-10
Access to required openings

Basements are considered to be somewhat more difficult than stories above grade when it comes to fighting fires from the exterior of the building. Therefore, an additional requirement is imposed in addition to those of Section 903.2.11.1.2. The code provides that when any portion of a basement is located more than 75 feet (22 860 mm) from complying exterior wall openings, the basement is required to be provided with an automatic sprinkler system. The 75-foot (22 860 mm) measurement should be taken in a straight line, resulting in the use of the arc method as shown in Figure 903-11. The two methods of providing complying exterior openings set forth in Section 903.2.11.1 are both available for a basement condition. If the openings are available entirely above the adjoining ground level, they are regulated in the same manner as for floor levels above grade. Otherwise, the openings must lead directly to a complying exterior stairway or ramp.

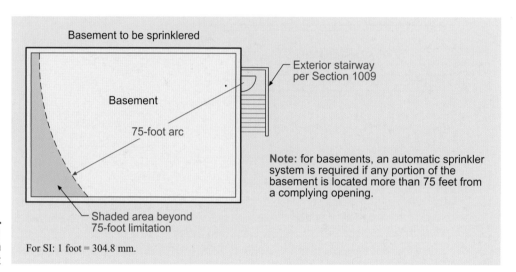

Basement to be sprinklered

Basement

75-foot arc

Exterior stairway per Section 1009

Note: for basements, an automatic sprinkler system is required if any portion of the basement is located more than 75 feet from a complying opening.

Shaded area beyond 75-foot limitation

For SI: 1 foot = 304.8 mm.

Figure 903-11
Openings in basement

With regard to the requirement that the exterior wall openings be entirely above grade, areaways and light wells may be considered to meet this requirement. However, the light wells and areaways should be provided with a stairway or other equivalent means for gaining ready access to the openings. Furthermore, the plan dimensions of the areaway or light well should be adequate to permit the necessary maneuvering to accomplish fire fighting or rescue from the opening. On this basis, it is advisable to consult with the fire department personnel to obtain their expertise in these situations.

903.2.11.2 **Rubbish and linen chutes.** Linen chutes and rubbish chutes are potential problem areas when it comes to fire safety because of a variety of reasons. They are often used for the transfer of combustible materials, including some levels of hazardous materials. They are also concealed within the building construction, possibly allowing a fire to smolder and grow prior to being detected. Of even more concern, linen and rubbish chutes create vertical openings through a building, allowing for the rapid spread of fire, hot gases and smoke up through the chute. Therefore, the IBC requires the installation of an automatic sprinkler system at the top of such chutes and in the rooms in which they terminate. Where the chute extends through at least three floors, as is usually the case in a four-story or higher building, sprinkler heads must also be installed at every other floor level.

903.2.11.3 **Buildings 55 feet or more in height.** Fire fighting in buildings that are over 55 feet (16 764 mm) in height is difficult, and many jurisdictions do not have the personnel or equipment to rescue occupants and control fires on upper floors. Therefore, this section was developed in recognition of this problem. The provision applies to all buildings other than airport control towers, open parking garages and Group F-2 occupancies. This provision lowers the threshold at which automatic fire sprinklers are required for high-rise buildings from 75 feet (22 860 mm) to 55 feet (16 764 mm), but it does not alter the applicability of any of the other special requirements set forth in Section 403. These still do not apply until the structure has occupiable floors more than 75 feet (22 860 mm) above the lowest level of fire-department vehicular access.

Other required suppression systems. A number of additional locations and uses require suppression systems based on requirements located throughout other portions of the IBC. Table 903.2.11.6 provides a cross reference to the sprinkler requirements for such subjects as atriums, stages, application of flammable finishes and unlimited area buildings. In addition, a reference is made to Section 903.2.13 of the IFC for many other specific buildings and areas where the installation of a fire-extinguishing system is mandated.

903.2.11.6

NFPA 13 sprinkler systems. Where the code requires the installation of an automatic sprinkler system in a building, it typically is referring to a system designed and installed in accordance with the criteria of NFPA 13. This standard is also applicable for those provisions that utilize a sprinkler system as an alternative to other code requirements. Throughout the code, the use of a sprinkler system "in accordance with Section 903.3.1.1" is referenced. In addition, where an automatic sprinkler system is required by the code with no direct reference to Section 903.3.1.1, the use of an NFPA 13 system is required.

903.3.1.1

Exempt locations. It is the intent of sprinkler protection that sprinklers be installed throughout the structure, including basements, attics and all other locations specified in the appropriate standard. It is also the intent of the IBC that when an automatic sprinkler system is required throughout, the same meaning is implied. One of the reasons for requiring protection throughout is the possibility of a fire in an unprotected area gaining such a foothold that the automatic sprinkler system would be overpowered. However, over the years, certain areas, locations or conditions have shown that they require special consideration, and the omission of sprinklers is permitted. In this section, the code itself provides the rationale for the omission of sprinklers.

903.3.1.1.1

NFPA 13R sprinkler systems. Although residential sprinkler systems installed in accordance with NFPA 13R may be used to satisfy the requirements of specific institutional and residential occupancies, they are not always recognized as full sprinkler protection for the purposes of exceptions or reductions permitted by other code requirements. However, where specifically mentioned through a reference to this section, such systems may be considered acceptable. Where the code indicates that a benefit can be derived from a sprinkler system installed "in accordance with Section 903.3.1.2," it intends that an NFPA 13R system can be utilized for the benefit. An important point is that an NFPA 13R sprinkler system is only permitted in residential-type buildings up to four stories in height.

903.3.1.2

Balconies and decks. Experience has shown that numerous fires in apartment buildings have started from grilling or similar activities on the balconies and patios. Because the NFPA 13R sprinkler standard does not mandate sprinklers in such locations, the code requires such sprinkler protection. The provision is applicable only to dwelling units in buildings of Type VA or VB construction. The automatic sprinkler protection is only required where there is a roof, deck or balcony directly above a balcony, deck or patio below. These areas will continue to typically be protected by sidewall orientation automatic sprinklers. If there is no horizontal element located directly above an exterior balcony, deck or ground floor patio, the additional sprinkler protection is not required.

903.3.1.2.1

Quick response and residential sprinklers. Based upon the timely performance of quick-response and residential automatic sprinklers, the code requires that they be installed in those occupancies where response or evacuation may not be immediate because of the condition of the occupants. Therefore, all spaces within a Group I-2 smoke compartment containing patient sleeping units and all dwelling units and sleeping units in Group R and I-1 occupancies are to be provided with these types of sprinklers. Such sprinklers are also required in light-hazard occupancies, where the quantity or combustibility of contents is low. Light-hazard occupancies included places of worship, education facilities, office buildings, museums, and seating areas of restaurants and theaters.

903.3.2

Section 904 *Alternative Automatic Fire-extinguishing Systems*

The code permits the use of automatic fire-extinguishing systems other than automatic sprinkler systems for those circumstances approved by the fire code official. However, the use of an alternative system in lieu of a sprinkler system does not gain the benefit of exceptions or reductions in code requirements. Only those buildings or areas protected by automatic sprinkler systems can take advantage of the allowances provided throughout the code.

The installation of automatic fire-extinguishing systems shall be in compliance with this section, which for the most part refers to the appropriate test standard and listing for each of the various types of systems. Those types addressed include wet-chemical, dry-chemical, foam, carbon dioxide, halon and clean-agent systems.

The inspection and testing of the system is emphasized because of the importance of a fully operating system. Specific items are identified for inspection, including the location, identification and testing of the audible and visible alarm devices. More specific criteria is also present for the installation and operation of a fire-extinguishing system for a commercial cooking system.

Section 905 *Standpipe Systems*

A standpipe system is a system of piping, valves and outlets that is installed exclusively for fire-fighting activities within a building. Standpipes are not considered a viable substitute for an automatic fire-sprinkler system. They are needed in buildings of moderate height and greater, and when used by trained personnel provide an effective means of fighting a fire.

905.3 **Required installations.** This section provides the scoping criteria for when a standpipe system must be provided. Figure 905-1 provides the basic requirements for when a standpipe system is required in a building. Building height is the primary consideration for the installation of a standpipe system. The general requirement calls for Class III standpipe systems where the vertical distance between the highest floor level in the building and the

	Class I	Class II	Class III
Buildings where highest floor level is located more than 30 feet above lowest point of fire-department vehicle access			x^1
Buildings where lowest floor level is located more than 30 feet below highest point of fire-department vehicle access			x^1
Nonsprinklered Group A buildings with an occupant load >1,000	x		
Covered mall buildings	x		
Stages greater than 1,000 square feet			x^2
Underground buildings	x		
Marinas and boatyards	See IFC Chapter 45		

For SI: 1 foot = 304.8 mm, 1 square foot = 0.093 m².
[1] Exceptions allow for Class I in sprinklered buildings, open parking garages and sprinklered basements.
[2] Only a 1¹/₂ in. hose connection is required where buildings or area sprinklered and connection installed per NFPA 13 or NFPA 14 for Class II or III Standpipes.

Figure 905-1
Standpipes required

lowest level of fire-department vehicle access exceeds 30 feet (9144 mm). See Figure 905-2. For measurement purposes, it is not necessary to consider any level of fire department vehicle access that, because of topographic features, makes access to the building from that point impractical or impossible. Such a condition typically occurs on steeply sloping sites and is illustrated in Figure 905-3. Several exceptions allow the use of Class I rather than Class III standpipes. Additional standpipe requirements may apply to assembly occupancies, covered mall buildings and stages.

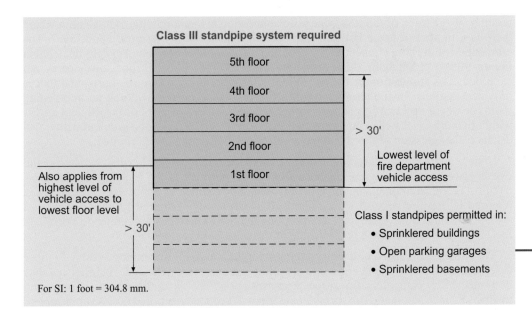

For SI: 1 foot = 304.8 mm.

Figure 905-2 Class III standpipe systems

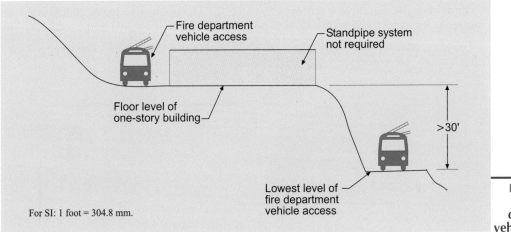

For SI: 1 foot = 304.8 mm.

Figure 905-3 Fire department vehicle access

Location of Class I standpipe hose connections. The code intends that Class I standpipes **905.4** are for the use of the fire department to fight fires within a building. Thus, the code requires that standpipe outlets be at every required stairway. The connections are to be located at every intermediate floor level landing of those stairways required by the code. As an alternative, the hose connections are permitted to be located at the floor level landing, but only where specifically allowed by the fire code official. However, the installation of hose connections at intermediate landings is typically preferred in order to avoid congestion at the stairway door. With these locations for standpipe outlets, the fire department personnel can bring a hose into the stair enclosure and make a hookup to outlets in a relatively protected area. For this reason, standpipe connections are typically not required for stairways permitted to be unenclosed by the exceptions to Section 1016.1 and 1022.1 as unenclosed stairways provide no protection for fire department personnel. However, it is very seldom that an unenclosed stairway is allowed to be utilized as a required means of egress in a building that requires a standpipe system.

Because a horizontal exit provides a barrier having a minimum fire-resistance rating of 2 hours, it is a logical location for Class I standpipe outlets. Such outlets are to be provided on both sides of the horizontal exit wall adjacent to the egress doorways through the horizontal exit wall regardless of whether or not egress is provided from both directions. An exception permits the omission of the standpipe hose connection at the horizontal exit opening where there is a limited distance between the opening and the stairway hose connection. The elimination of the hose connection is permitted on one side of horizontal exit, as depicted in Figure 905-4, or on both sides, provided the "100-foot plus 30-foot" distance is not exceeded. The application of the exception is most common where the horizontal exit is provided to allow for the termination of a fire-resistance-rated corridor at an intervening room, or where necessary to address an inadequate number of exits or insufficient egress width. In those cases where the horizontal exit is provided because of a problem with travel distance, the omission of the standpipe connections will seldom be permitted.

Under unique circumstances, standpipe outlets located in exit enclosures or at horizontal exit doorways may be a great distance from some portions of the building. Under these circumstances, additional standpipe outlets must be provided in approved locations when required by the fire code official. Figure 905-5 depicts the required locations of Class I standpipe connections. Standpipe connections required for stages exceeding 1,000 square feet ($93m^2$) are to be located on each side of the stage per Figure 905-6.

The exit enclosure also provides protection for the standpipe and piping system. In those cases where the risers and laterals are not within exit enclosures, the code requires that they be protected by equivalent fire-resistant construction. The exception to this requirement assumes that the automatic fire sprinkler system will keep the risers and laterals cool enough so that they will not be damaged by fire.

On those roofs that have a flat enough slope for fire fighters to move about, the code requires at least one roof outlet so that exposure fires can be fought from the roof.

The interconnection of the standpipe risers at the bottom for multiple standpipe systems is intended to increase the reliability.

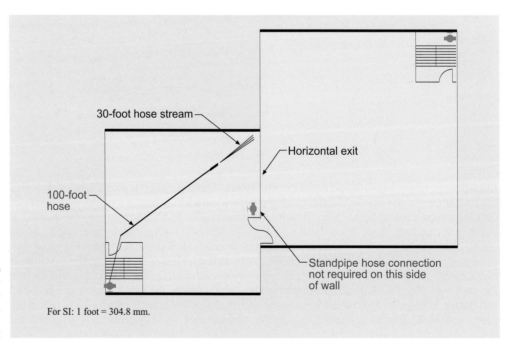

Figure 905-4
Connections at a horizontal exit

30-foot hose stream

Horizontal exit

100-foot hose

Standpipe hose connection not required on this side of wall

For SI: 1 foot = 304.8 mm.

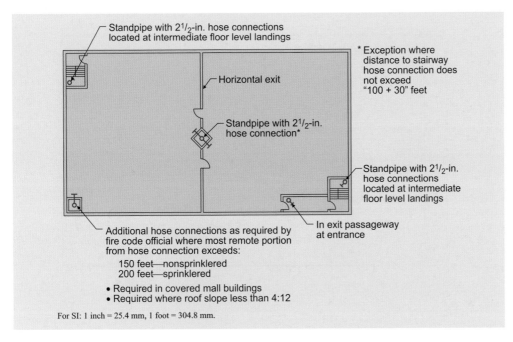

For SI: 1 inch = 25.4 mm, 1 foot = 304.8 mm.

Figure 905-5
Standpipe connection locations

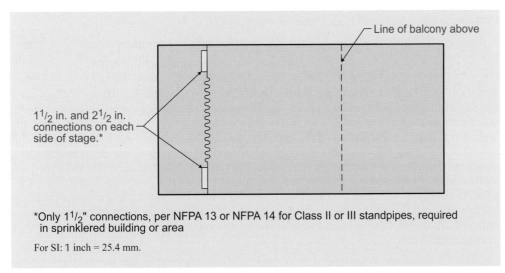

*Only 1¹/₂" connections, per NFPA 13 or NFPA 14 for Class II or III standpipes, required in sprinklered building or area

For SI: 1 inch = 25.4 mm.

Figure 905-6
Class II standpipe locations at stages > 1000 sq ft

Location of Class II standpipe hose connections. It is the intent of the code to require the location of hose cabinets for Class II standpipes at intervals ensuring that all portions of a building will be within 30 feet (9144 mm) of a nozzle attached to 100 feet (30 480 mm) of hose. In plan review, this would necessitate allowing for pulling the hose down corridors and through rooms such that several right-angle turns may be necessary before the hose stream can be placed on the fire. Therefore, judgment is necessary in the determination of standpipe locations. One method to account for this type of partitioning in a building where the future location of partitions is unknown is to subtract 30 feet (9144 mm) from the straight line distance between the hose cabinet and the remote location and then multiply the remainder by 1.4. If the result is more than 100 feet (30 480 mm), an additional standpipe connection will be required. Figure 905-7 illustrates the location of a Class II standpipe in a building where an office floor has a central corridor with offices on each side. In this particular arrangement, it is obvious from the layout that the one standpipe will suffice.

905.5

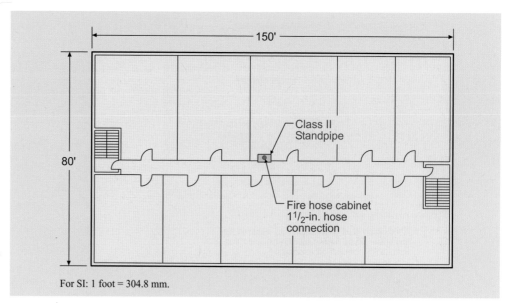

For SI: 1 foot = 304.8 mm.

Figure 905-7
Class II
connections

As there are no scoping provisions in Section 905.3 for a Class II or III standpipe system in a sprinklered Group A occupancy, other than the stage requirements of Section 905.3.4, the provisions of Section 905.5.1 have no application.

Because Class II standpipe systems are charged with water, the code does not require fire-resistive protection. The water within the system is considered adequate to keep the pipe cool enough to prevent damage.

905.6 **Location of Class III standpipe hose connections.** Because Class III standpipes are a combination of the benefits of Class I and II systems, containing both $1\frac{1}{2}$-inch (38 mm) outlets for occupant use and $2\frac{1}{2}$-inch (64 mm) outlets for fire department use, it is only logical that they be located so as to serve the building as required for both Class I and II standpipes. Figure 905-8 shows the typical arrangement for a Class III standpipe in a building. Usually, the hose rack for $1\frac{1}{2}$-inch (38 mm) outlets and $2\frac{1}{2}$-inch (64 mm) hose outlets are both located within the stair enclosure. Where the coverage requirements for Class II standpipes are such that stair enclosure locations will not cover the entire building, laterals are usually run to other locations from hose cabinets in order to provide for the required coverage. In this case, the laterals are not required to be protected, as they are charged with water.

Class III standpipe systems and their risers and laterals in sprinklered buildings are not required to have fire-resistive protection for the same reason as discussed for Class I standpipes.

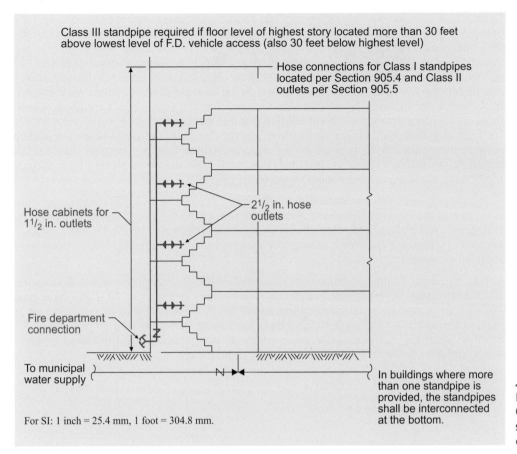

Class III standpipe required if floor level of highest story located more than 30 feet above lowest level of F.D. vehicle access (also 30 feet below highest level)

Hose connections for Class I standpipes located per Section 905.4 and Class II outlets per Section 905.5

2¹/₂ in. hose outlets

Hose cabinets for 1¹/₂ in. outlets

Fire department connection

To municipal water supply

In buildings where more than one standpipe is provided, the standpipes shall be interconnected at the bottom.

For SI: 1 inch = 25.4 mm, 1 foot = 304.8 mm.

Figure 905-8
Class III standpipe connections

Section 907 *Fire Alarm and Detection Systems*

One of the most effective means of occupant protection in case of a fire incident is the availability of a fire alarm system. An alarm system provides early notification to occupants of the building in the event of a fire, thereby providing a greater opportunity for everyone in the building to evacuate or relocate to a safe area. This section covers all aspects of fire alarm systems and their components.

Unlike most of the provisions of Section 903.2 addressing the required installation of an automatic sprinkler system, those requirements mandating a fire alarm system are not applied based on the fire area concept. Manual fire alarm systems are typically required based on the occupant load of the occupancy group under consideration, including any occupants of the same occupancy classification that may be identified in another fire area. Fire areas are not to be utilized in the application of this section regarding fire alarm systems.

Where required—new buildings and structures. Approved fire alarm systems, either **907.2** manual, automatic, or both manual and automatic, are mandated in those occupancies and areas identified by this section. Where automatic fire detectors are mandated, smoke detectors are to be provided unless normal operations would cause an inaccurate activation of the detector. All automatic fire-detection systems are to be installed in accordance with NFPA 72.

907.2.1 **Group A.** Where an assembly occupancy has a sizable occupant load, the safe egress of the occupants becomes an even more important consideration. When the occupant load reaches 300, the code mandates a manual alarm system to provide early notification to the occupants. The IBC also requires portions of a Group E educational occupancy that are occupied for accessory assembly purposes, such as a lunchroom or library, to have alarms as required for the Group E use, rather than based on the less restrictive requirements mandated for Group A occupancies. In addition, the Group E fire alarm provisions are applicable to those areas within a Group E school building that may be clasified as Group A occupancies, such as a gymnasium and/or auditorium. As is the case in many other occupancy classifications, manual alarm boxes are not required in those Group A occupancies where an automatic sprinkler system is installed that will immediately activate the occupant notification appliances (horn/strobes) upon water flow.

In assembly occupancies containing much larger occupant loads (1,000 or more people), activation of the required fire-alarm system must initiate a prerecorded announcement. The emergency voice/alarm communications system, upon approval of the building official, may also be used for live voice emergency announcements originating from a constantly attended location.

907.2.2 **Group B.** Larger business occupancies classified as Group B require the installation of a fire alarm system. Where the total occupant load exceeds 499 persons, or where more than 100 persons occupy Group B spaces above and/or below the lowest level of exit discharge, a manual fire alarm system shall be installed. See Figure 907-1. Similar to exceptions for other occupancies, the manual fire alarm boxes are not required in a sprinklered building where sprinkler water flow activates the notification devices.

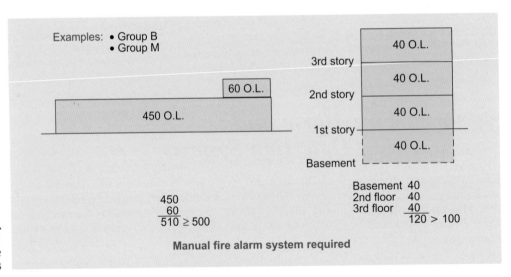

Figure 907-1
Required fire alarm systems

907.2.3 **Group E.** The IBC follows the philosophy of society in general that our children require special protection when they are not under parental control. Therefore, in addition to the other life-safety requirements for educational occupancies, manual fire alarm systems are required whenever the occupant load of any Group E occupancy is 50 or more. In addition, when the building is provided with smoke detection or a sprinkler system, such systems shall be connected to the building's fire alarm system. However, the more probable reason for such a low occupant load threshold being established is the exceptional value of such a system in an educational use. Students tend to react quickly and efficiently at the first notification of the alarm system, making safe egress possible. In addition, periodic fire drills reinforce the appropriate egress activity. Where the five conditions are met as listed in Exception 2, manual alarm boxes are not required. Exception 3 also exempts manual fire alarm boxes where, in a sprinklered building, the sprinkler water flow activates the notification appliances and manual activation is possible from a normally occupied location.

Group F. In multistory manufacturing occupancies, a fire alarm system is mandated where an aggregate occupant load of 500 or more is housed above and/or below the level of exit discharge. Similar to several other occupancies, the alarm system is necessary where there are very large occupant loads and where those occupants must travel vertically to exit. Application of the provision where occupants are located both above and below the lowest level of exit discharge is similar to that for Group B and M occupancies. In Group F occupancies, a manual fire alarm system is mandated where the aggregate occupant load of those levels, other than the exit discharge level, exceeds 499. Where the building is sprinklered, manual alarm boxes are not required if the water flow of the sprinkler system activates the notification appliances. **907.2.4**

Group H. Only those Group H occupancies associated with semiconductor fabrication or the manufacture of organic coatings need be provided with a manual alarm system by the IBC. Areas containing highly toxic gases, organic peroxides and oxidizers shall be protected by an automatic smoke detection system. **907.2.5**

Group I. As patients, residents or inmates of Group I occupancies are asleep during a large portion of the day, Section 907.2.6 provides for early warning of the occupants and staff, thus enhancing life safety. In addition, early response is beneficial because of the lack of mobility of many of the occupants. The general provisions require a manual fire alarm system be installed in all Group I occupancies. However, an automatic smoke detection system is only required in those occupancies classified as Group I-1, I-2 or I-3. Where a Group I-1 facility is provided with an automatic sprinkler system, only corridors and waiting areas open to such corridors need be equipped with the smoke detection system. The corridor smoke-detection system required in Group I-2 nursing homes and detoxification facilities may be omitted where the patient sleeping rooms have smoke detectors that provide a visual display on the corridor side of each patient room, as well as a visual and audible alarm at the appropriate nursing station. Another exception exempts the requirement for corridor smoke detection where patient room doors are equipped with automatic door closers having integral smoke detectors on the room side that performs the required alerting functions. **907.2.6**

Because of their special nature, in Group I-3 occupancies the provisions for fire alarm systems are greatly expanded. The manual and automatic fire alarm system is to be designed to alert the facility staff. Actuation of an automatic fire-extinguishing system, a manual fire alarm box or a fire detector must initiate an automatic fire alarm signal, which automatically notifies staff. For obvious reasons, manual fire alarm boxes need only be placed at staff-attended locations having direct supervision over the areas where boxes have been omitted. Under certain conditions, the installation of an approved smoke detection system in resident housing areas is required.

Group M. The threshold at which a manual fire alarm system is required for a Group M occupancy is the same as that for a Group B occupancy. During that portion of time when the building is occupied, the signal from the fire alarm box or water flow switch may be designed to only activate a signal at a constantly attended location, rather than provide the customary visual and audible notification. At this location, the use of an emergency voice/alarm communications system can be utilized to notify the customers of the emergency conditions. This provision is helpful in eliminating those nuisance alarms that may occur because of the presence of fire alarm boxes. **907.2.7**

Group R-1. When asleep, the occupants of residential buildings will usually be unaware of a fire, and it will have an opportunity to spread before being detected. As a result, a majority of fire deaths in residential buildings have occurred because of this delay in detection. It is for this reason that the IBC requires fire alarm systems in addition to smoke detectors in certain residential structures. **907.2.8**

In hotels and other buildings designated as Group R-1, the general provisions mandate that both a manual fire alarm system and an automatic fire detection system be installed.

There is an exception that eliminates the requirement for the required manual alarm system for such occupancies less than three stories in height where all guestrooms are completely separated by minimum 1-hour fire partitions and each unit has an exit directly to a yard, egress court or public way. This exception is based on the compartmentation provided by the separations between units and by the relatively rapid means of exiting available to the occupants. Where guestrooms are limited to egress directly to the exterior, early notification, although important, is not critical. A second exception requires the alarm system, but does not mandate the installation of fire alarm boxes throughout buildings that are protected throughout by an approved supervised fire sprinkler system. There is, however, a need for at least one manual fire alarm box installed in a location approved by the building official. In addition, sprinkler flow must activate the notification appliances. The automatic fire alarm system required by this section need only be provided within all corridors that serve guestrooms. An exception eliminates the requirement for the automatic fire detection system in buildings where egress does not occur through interior corridors or other interior spaces.

907.2.9 Group R-2. Group R-2 buildings such as apartment houses are to be provided with a fire alarm system based upon the number of dwelling units and sleeping units, as well as the location of any such units in relationship to the level of exit discharge. Where more than 16 dwelling units or sleeping units are located in a single structure, or where such units are placed a significant distance vertically from the egress point at ground level, it is beneficial that a detection and notification system be provided. If any one of the three listed conditions exist, the alarm system is required unless exempted or modified by one of the three exceptions. Exceptions similar to those permitted for Group R-1 occupancies apply to Group R-2 buildings as well.

907.2.11 Single- and multiple-station smoke alarms. As indicated in the introduction to the residential fire alarm provisions, residential fire deaths far exceed those of any other building classification. Furthermore, more than one-half of the fire deaths in residential buildings have occurred because of a delay in detection that is due to the occupants being asleep at the time of fire. Thus, the IBC requires smoke alarms in all residential buildings and in certain institutional occupancies. In Group R-1 occupancies, single- or multiple-station smoke alarms are to be installed in all sleeping areas, in any room along the path between the sleeping area and the egress door from the sleeping unit, and on each story within the sleeping unit. In all other residential occupancies, the code requires that smoke detectors be located in the sleeping rooms and on the ceiling or wall of the corridor or area giving access to the sleeping rooms. In addition, at least one smoke alarm shall be installed on each story of a dwelling unit, including basements. Where split levels occur in guestrooms or dwelling units, a smoke alarm need only be installed on the upper level, provided there is no intervening door between the adjacent levels. See Figure 907-2 for illustrations of these provisions.

Unless an automatic fire detection system is provided in compliance with Section 907.2.6, single- or multiple-station smoke alarms are to be installed in those sleeping areas of Group I-1 occupancies. It is not the intent of the exception to Section 907.2.11.2 to allow the omission of smoke alarms by referencing the provisions of Section 907.2.6, but rather to allow smoke detectors as a part of the fire alarm system to be installed in the sleeping rooms in lieu of single- or multistation smoke alarms. The hazards addressed by the residential smoke alarms also exist to some degree in institutional occupancies. In any new construction, the code requires that smoke alarms receive their power from the building wiring with a battery backup. See Figure 907-3.

In order to notify occupants throughout the dwelling unit or sleeping unit of a potential problem, multiple smoke alarms need to be interconnected. Therefore, when activation of one of the alarm devices takes place, activation of all the alarms must occur. The intent of the code is that the alarms be audible throughout the dwelling, particularly in all sleeping rooms.

907.2.12 Special amusement buildings. One of the factors contributing to the hazards within special amusement buildings has been the failure to detect a fire in its incipient stage.

Provisions for the detection of fires are required in order to protect the occupants in such structures. An automatic smoke detection system is required in all special amusement buildings unless conditions of the use of the building would result in nuisance alarms. In this case, an alternate type of detection method must be utilized. A provision of this section is that on the activation of the system as described, all confusing sounds and visual effects shall stop, an approved directional exit marking system shall activate and the exit path itself shall be illuminated. A public address system, which may also serve as the alarm system, is required so that occupants can be notified and given instructions in case of an emergency.

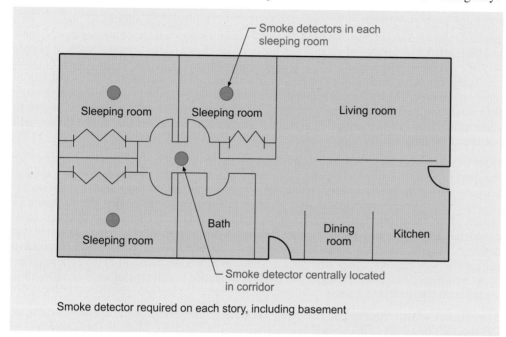

Figure 907-2
Location of smoke detectors

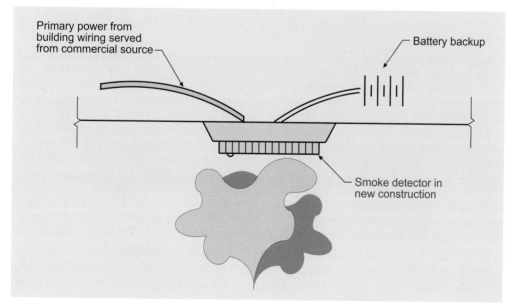

Figure 907-3
Smoke detectors

High-rise buildings. Among the more important life-safety features required by the code for high-rise buildings are the alarm and communication systems required by this section. Where it is expected that people will be unable to evacuate the building, it is imperative that

907.2.13

they be informed as to the nature of any emergency that may break out, as well as the proper action to take to exit to a safe place of refuge. Furthermore, a system is necessary in most cases to provide for communication between the fire officer in charge at the scene and the fire fighters throughout the building.

907.2.14 **Atriums connecting more than two stories.** Atriums connecting three or more stories shall be provided with an approved fire alarm system activated in accordance with Section 907.6. In addition, Group A, E or M occupancies containing an atrium must have an emergency voice/alarm communication system.

907.2.20 **Covered mall buildings.** It is common for several thousand people to be present in a covered mall building at any one time. Certainly, a large regional shopping center will have a sufficiently large number of tenants and anchor buildings to draw large numbers of people into the facility. Where such large crowds are gathered together in one facility, it is necessary to have a method for conveying instructions to the occupants in case of fire or other emergency. Thus, the code requires that covered mall buildings with a floor area more than 50,000 square feet (4645 m²) be provided with an emergency voice/alarm system to accomplish this. Furthermore, this system is required to be accessible to fire department personnel for their use.

907.3.1 **Duct smoke detectors.** Where required elsewhere in the code, duct smoke detectors are required to comply with this section. This usually occurs when duct detectors are used to activate smoke dampers, as part of a smoke control system, as permitted for Group I-3 sleeping units, or for mechanical unit shutdown. In order to determine the presence of smoke within the ducts of the mechanical system, smoke detectors located in ducts are to be connected to the control panel of the building's fire alarm system. Activation of the detectors should initiate a signal at a constantly attended location, or as an alternative activate the building's alarm notification devices. Where a fire alarm system is not required in the building, the duct smoke detectors must be designed to activate an audible and visible signal in an approved location.

907.4.2 **Manual fire alarm boxes.** The IBC often requires a manual fire alarm system because of the special occupants or hazards that exist within the building. Manual fire alarm boxes, defined as manually-operated devices used to initiate an alarm signal, and often referred to as pull stations, are utilized in many situations as a means for occupants to notify others of a potential fire emergency. This section identifies the proper locations for the installation of these alarm boxes.

In order that manual fire alarm boxes are readily available and accessible to all occupants of the building, they are to be located in close proximity to the point of entry to each exit. This would include placement within 5 feet (1524 mm) of exterior exit doors, as well as doors entering critical exit enclosures, exit passageways and horizontal exits. By placing the boxes adjacent to the exit doors, they will be available to occupants using any of the available exit paths. Additional boxes may be required in extremely large structures, as the maximum travel distance to the nearest alarm box cannot exceed 200 feet (60 960 mm). A manual fire alarm box, required to be red in color, must be located in a position so that it can be easily identified and accessed. The maximum height of 48 inches (1372 mm), measured from the floor to the activating lever or handle, is based on the high-end reach range limited by the accessibility provisions of ICC A117.1. The minimum height of 42 inches (1067 mm) keeps the activating mechanism in a position readily viewed and providing ease of manipulation. See Figure 907-4.

Unless the fire alarm system is monitored by a supervising station, a sign must be installed on or adjacent to each manual fire alarm box advising the occupants to notify the fire department. Where a supervised alarm system is in place, notification becomes automatic and the sign is not necessary.

Often it is necessary to develop a means of reducing the accidental or intentional damage or activation of the alarm initiating device. Therefore, the code gives the fire code official authority to accept protective covers placed over the listed boxes. The alarm box should remain easily identifiable when covered, with adequate instructions for operation.

Emergency voice/alarm communication system. This section requires that operation of any automatic fire detector, sprinkler water-flow device or manual fire alarm box result in an alert tone followed by appropriate voice instruction. The alert tone and voice instruction must be provided on the floor of the alarm, as well as the floors directly above and below the alarming floor as applicable. The tone and instruction may be directed on a general basis to the required floor levels, or selectively provided to only the following paging zones: **907.5.2.2**

1. Elevator groups

2. Exit stairways

3. Each floor

4. Areas of refuge as required by Section 1007

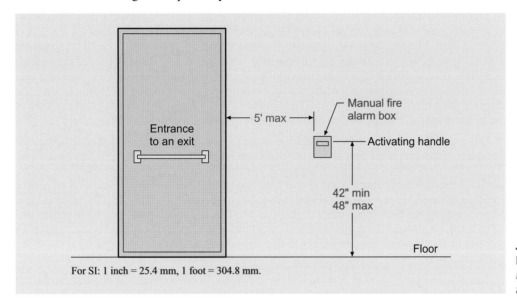

For SI: 1 inch = 25.4 mm, 1 foot = 304.8 mm.

Figure 907-4
Manual fire alarm boxes

Visible alarms. Visible alarms are intended to alert hearing-impaired individuals to a fire emergency. Of those locations where audible alarm systems are required, the IBC identifies the specific conditions under which visible alarm-notification appliances are also mandated. In those portions of the building deemed to be public areas or common areas, visual alarms shall be provided in addition to audible alarms. For example, in an office building, the lobby, public corridors and public restrooms would be considered public areas, whereas the corridors, toilet rooms, break areas and conference rooms inside an office suite would be considered common use. On the other hand, the individual offices of each employee would be considered private use. Although such offices would not require the installation of visible alarm notification appliances, wiring must be in place for future installation of the alarms as necessary. The potential of additional visible alarm notification appliances is taken into account by requiring at least 20 percent spare capacity for the appliance circuits. Keep in mind that visual alarms are only required in those occupancies where an alarm system is first required by the code. **907.5.2.3**

Table 907.5.2.3.3 is used to determine the number of sleeping units in Group R-1 or I-1 occupancies that must be provided with visible alarm notification appliances. In both occupancies, the appliances are to be activated by both the in-room smoke alarm and the building's fire alarm system. The number of units required to have both visual and audible alarms is based upon the total number of units in the building. Not strictly limited to sleeping units, the visible and audible alarm requirements are also applicable where dwelling units are located within a Group R-1 occupancy. In Group R-2 apartment houses and similar occupancies required by the code to have a fire alarm system, provisions must be made for the future installation of visible alarm notification devices as they become necessary.

907.6.3 **Zones.** In order to quickly identify the fire vicinity, zones must be created for alarm notification. The minimum requirement is to zone each floor separately. However, where a floor exceeds 22,500 square feet (1860 m²) in floor area or 300 feet (91 440 mm) in length in any direction, additional zones on the floor must be created with no single zone exceeding these limitations. An indicator panel installed in an approved location shall visually signal the zone location and maintain the identification until the system is reset.

<u>Section 908</u> Emergency Alarm
Systems

In addition to those alarms designed to provide early notification of a fire condition, emergency alarm systems are necessary in specific high-hazard-type occupancies to address other identifiable concerns. A reference is made to IBC Section 414.7, which requires a manual emergency alarm system in specific areas used for the storage of hazardous materials. The alarm system is intended to notify the other occupants of the building that there is an emergency situation involving hazardous materials. For the same reason, an alarm or communication method must also be provided where corridors and exit enclosures are utilized for the transportation of highly hazardous materials.

Other conditions call for the installation of a gas detection system designed to detect the presence of unacceptable levels of hazardous gas and perform additional safety functions. Areas identified in this section as requiring gas detection systems include HPM facilities, facilities involving the storage and use of toxic and highly toxic gases, ozone gas-generator rooms and repair garages for nonodorized-gas-fueled vehicles. The activation of a local alarm, audibly distinct from other alarms, is used to provide a warning both inside and outside the area where the gas is detected. In addition, the detection system must initiate the automatic shutdown of the gas supply to the area where the leak is detected.

<u>Section 909</u> Smoke Control
Systems

The provisions of this section are applicable to the design, construction, testing and operation of mechanical or passive smoke control systems only when they are required by other provisions of the IBC. Section 909 specifically exempts smoke- and heat-venting requirements that appear in Section 910 and are discussed in the next section of this handbook. Also, this section states that mechanical smoke control systems are not required to comply with Chapter 5 of the *International Mechanical Code®* (IMC®) for exhaust systems unless their normal use would otherwise require compliance.

The provisions of this section establish minimum requirements for the design, installation and acceptance testing of smoke control systems, but nothing within the section itself is intended to imply that a smoke control system is to be installed. Some sections that specifically reference Section 909 are the requirements for atriums (Section 404.4), and underground buildings (Section 405.5), and windowless buildings housing Group I-3 occupancies (Section 408.8). Smoke control systems are intended to provide a tenable environment in areas outside that of fire origination for the evacuation or relocation of occupants. The provisions are not intended for the preservation of contents or for assistance in fire suppression or overhaul activities.

Much of this section is based on the American Society of Heating, Refrigerating and Air-Conditioning Engineers publication, *Design of Smoke Control for Buildings*; NFPA publication 92-A, *Recommended Practice for Smoke Control Systems*; and the companion publication 92-B, *Technical Guide for Smoke Control Systems in Malls, Atriums and Large Areas.*

Although this section covers both passive and active smoke control systems, the majority of the material presented addresses the three mechanical methods—pressurization (Section 909.6), airflow (Section 909.7) and exhaust (Section 909.8)—with other sections addressing related subjects such as the design fire; equipment, including fans, ducts and dampers; power supply; detection and control systems; and the fire fighter's smoke-control panel.

An important segment of Section 909 addresses acceptance testing of the smoke control system. Smoke control system installation requires special inspection per Section 909.18.8 to be performed during erection of ductwork and prior to concealment. These inspections are intended for the purpose of testing for leaks, as well as for recording the specific device locations. The latter creates, in effect, as-built drawings for the system. Additional testing and verification prior to occupancy is also mandated. Section 909.18.8.3 requires the work of the special inspector to be documented in a final report. The report shall be reviewed by the responsible registered design professional who is required to certify the work. The final, designer-approved report, together with other information addressed in Section 909.18.9, shall be provided to the fire code official, and a copy shall be maintained on file at the building.

Smokeproof enclosures. The provisions of this section identify the methods for **909.20** complying with Section 1022.9 for the construction of smokeproof enclosures. Smokeproof enclosures are required by Sections 403 and 405 for high-rise buildings and underground buildings. There are two methods for construction of a ventilated smokeproof enclosure, both of which utilize an enclosed interior exit stairway. Either an exterior balcony or a ventilated vestibule can be used as the buffer between the floor of the building and the exit stairs. In addition, pressurization of the stair shaft is a permitted alternative.

Unless the pressurization provisions of Section 909.20.5 are utilized where a smokeproof enclosure is required, the exit path to the stair shall include a vestibule or an open exterior balcony. The minimum size of the vestibule is illustrated in Figure 909-1. A minimum 2-hour fire-resistance-rated fire barrier separates the smokeproof enclosure from the remainder of the building and also separates the stairway from the vestibule. The only openings permitted into the enclosure are the required means of egress doors. Construction of an open exterior balcony is based upon the required fire-resistive rating for the building's floor construction, which would typically be 2 hours.

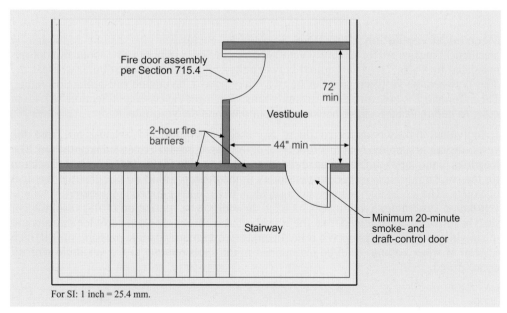

Fire door assembly per Section 715.4

72' min

Vestibule

2-hour fire barriers

44" min

Stairway

Minimum 20-minute smoke- and draft-control door

For SI: 1 inch = 25.4 mm.

Figure 909-1
Ventilated vestibule

909.20.3 **Natural ventilation alternative.** In this section, the code provides the details of construction where natural ventilation is utilized to comply with the concept of a smokeproof enclosure. Where an open exterior balcony is provided, fire doors into the stairway shall comply with Section 715.4. In a vestibule scenario, a similar complying fire door is required between the floor and the vestibule. Between the vestibule and the stairway, the door assembly need only have a 20-minute fire-protection rating. The necessary vestibule ventilation is to be provided by an opening in the exterior wall at each vestibule. Facing an outer court, yard or public way at least 20 feet (6096 mm) in width, the exterior wall opening must provide at least 16 square feet (1.5 m²) of net open area. See Figure 909-2.

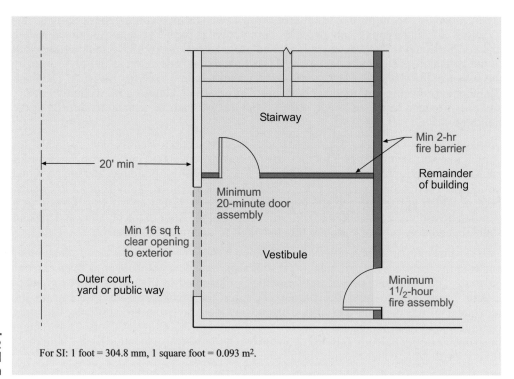

Figure 909-2
Natural ventilation

For SI: 1 foot = 304.8 mm, 1 square foot = 0.093 m².

909.20.4 **Mechanical ventilation alternative.** Smokeproof enclosures may also be ventilated by mechanical means. As for naturally ventilated vestibules, a minimum 1½-hour fire door as mandated by Table 715.4 is required between the building and the vestibule, whereas the door between the vestibule and the stairway may have a 20-minute fire-protection rating. The minimum 1½-hour fire door assembly must also meet the criteria of Section 715.4.3 in order to minimize air leakage between the building and the vestibule.

Individual tightly-constructed ducts are used to supply and exhaust air from the vestibule. Air is supplied near the floor level of the vestibule and exhausted near the top. The locations of the supply and exhaust registers are illustrated in Figure 909-3, as is the location for the smoke trap. It is important that doors in the open position do not obstruct the duct openings. The code also allows the use of a performance-based engineered vestibule ventilation system per Section 909.20.4.2.1. In addition to ventilation of the vestibule, air shall be provided and relieved from the stair shaft as well. By supplying an adequate amount of air while providing a dampered relief opening, a minimum positive pressure of 0.10 inch (29 Pa) of water column shall be maintained in the shaft relative to the vestibule with all doors closed.

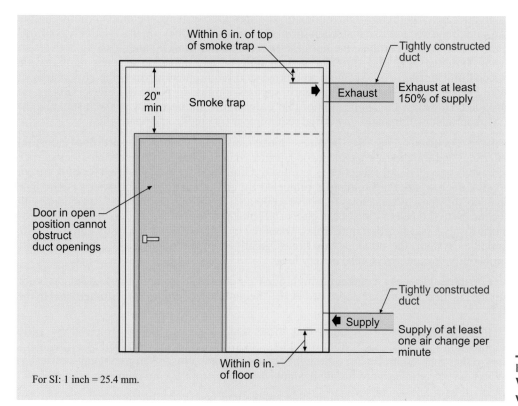

For SI: 1 inch = 25.4 mm.

Figure 909-3
Vestibule
ventilation

Stair pressurization alternative. The method addressed in this section applies only to those buildings equipped throughout with an automatic sprinkler system. Through the pressurization of the stair shaft to a prescribed level, the need for vestibules or open exterior balconies is eliminated. Pressurization levels for the interior exit stairways shall fall between 0.10 inch (25 Pa) and 0.35 (87 Pa) inch of water column in relationship to the building. With this extent of pressurization, ventilation methods are deemed to be unnecessary.

909.20.5

Section 910 *Smoke and Heat Vents*

Smoke and hot gases created by a fire rise to the underside of the roof structure above and then build up so as to cause reduced visibility to the point where fire fighting is relatively ineffective. Also, as the hot gases accumulate near the roof structure, the unburned products of combustion become superheated, and if a supply of air is introduced, these hot, unburned products of combustion will ignite violently. Thus, it has been found that it is imperative that industrial and warehouse-type occupancies be provided with smoke and heat vents in the roofs. Although the IBC only requires smoke and heat vents for one-story buildings, or one-story portions of multistory buildings, their use in other portions of multistory buildings is also sometimes utilized, particularly in the top story.

Smoke and heat vents must typically be installed in conjunction with draft curtains, which are sometimes referred to as curtain boards. Draft curtains installed at code-specified intervals confine the smoke and hot gases so they are not diluted and, thus, increase the effectiveness of automatic vents. Without the confinement of the smoke and hot gases by draft curtains, the vents would be relatively ineffective because of delay in operating, or even because of nonoperation.

In the case of buildings with automatic fire sprinkler systems, the draft curtains confine the smoke and hot gases so that when sprinklers are actuated they are in a relatively confined area. Without draft curtains, the hot gases would spread laterally and possibly activate so many sprinklers that the sprinkler system may be overcome on account of excessive water-flow demand. An exception to this general requirement permits the elimination of draft curtains within the area where early suppression fast-response (ESFR) sprinklers are installed. However, the application of this provision is not necessary as the installation of ESFR sprinklers eliminates any initial requirement for smoke and heat vents.

910.2 **Where required.** The IBC requires smoke and heat venting in specified industrial buildings and warehouses and in any occupancy where high-piled combustible stock or rack storage is provided. An exception is granted to Group S-2 occupancies used for bulk frozen food storage where the building is protected by an automatic fire sprinkler system. The intent is that those occupancies that have a potential to include large areas be provided with the means to rid the building of hot gases and smoke. In the case of frozen food warehouses storing only Class I or II commodities, when the building is protected by an automatic fire sprinkler system, the potential for fire is greatly reduced. Should a fire break out, the probability of rapid fire spread is very low.

For those buildings containing high-piled combustible storage, the requirements for smoke and heat venting are addressed in IBC Section 413 (Combustible Storage) and in the IFC.

It is becoming more common that these large areas are being protected by early suppression fast response (ESFR) sprinklers. Such sprinklers are designed to extinguish a fire, rather than control a fire, through the quick application of large amounts of water. Where ESFR sprinklers are installed, smoke and heat vents are not required.

910.3.2 **Vent operation.** The code requires that smoke and heat vents be approved, labeled and capable of manual and automatic operation. Required to be located in the roof, smoke and heat vents shall operate automatically in order to release the smoke and hot gases. Where the building is provided with an automatic sprinkler system, it is important to overall fire operations that the smoke and heat vents not operate automatically prior to activation of the sprinkler.

910.3.3 **Vent dimensions.** Aerodynamic studies of flow through rectangular openings have shown that the minimum cross-sectional area for smoke and heat vents should be 16 square feet (1.5 m^2), with a minimum dimension of 4 feet (1219 mm). Thus, the code has this same requirement. The code also permits projections such as ribs and rain gutters to project into the required 4 feet (1219 mm) as long as the total width for all projections does not exceed 6 inches (152 mm).

910.3.4 **Vent locations.** As smoke and heat vents are intended to release smoke and hot gases from a fire within the building, the code requires that they be placed a minimum of 20 feet (6096 mm) from fire walls and from any lot line in order to reduce exposure to adjacent property. Vents shall also be located at least 10 feet (3048 mm) from fire barriers. Such conditions as roof pitch, curtain location, sprinkler head location and structural members shall be considered in the location of vents.

910.3.5 **Draft curtains.** In order to be effective, draft curtains are required to be of the types of construction that prevent the passage of smoke through the draft curtain. Therefore, the code requires that they be constructed of sheet metal, lath and plaster, gypsum wallboard, or of another approved material that prevents the passage of smoke. Also, the joints and connections are required to be smoketight. Figure 910-1 illustrates the method for determining draft curtain depth and spacing utilizing the requirements of Table 910.3. In those areas where ESFR sprinklers are provided, draft curtains need only be installed at the point separating the ESFR sprinklers and the conventional sprinklers.

910.4 **Mechanical smoke exhaust.** An engineered mechanical smoke-exhaust system may be accepted by the fire code official in lieu of smoke and heat vents. Fans located within each draft curtain area can be utilized for exhaust purposes. Each fan shall be manually controlled, as well as automatically activated, by either an automatic sprinkler system or heat detectors. Supply air for the exhaust system shall be provided uniformly around the perimeter of the area served. It is the intent of the code that the engineered system provide results equivalent to the smoke- and heat-vent method.

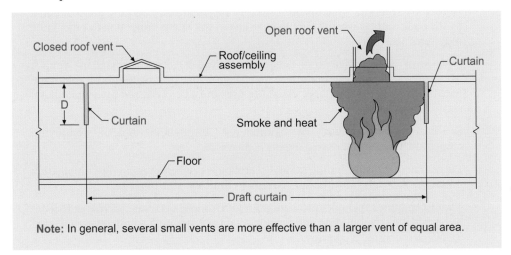

Note: In general, several small vents are more effective than a larger vent of equal area.

Figure 910-1
Roof vents and curtain boards

Section 911 *Fire Command Center*

The fire command center is the heart of the fire- and life-safety systems in a complex building. The IBC mandates a fire command center be provided in only very special structures such as high-rise buildings regulated by Section 403. The purpose of the command center is to provide a central location where fire personnel can operate during a fire incident or other emergency. Located as determined by the fire department, the fire command center shall be isolated from the remainder of the building by a minimum 1-hour fire-resistance-rated fire barrier. The code lists those system units, controls, display panels, indicators, devices, furnishings and plans that are to be contained in the command center.

Section 914 *Emergency Responder Safety Features*

Section 914 is intended to provide correlation to the current requirements in the IFC for the identification of shaftway hazards and the location of fire protection systems. These requirements are located in Sections 316.2 and 509.1 of the IFC. Section 101.3 of the IBC states the safety of emergency responders is part of its scope and intent. This new section reinforces the intent by specifying that interior and exterior shaftway hazards be identified as well as the location of fire protection systems, such as fire alarm control units or automatic sprinkler risers.

KEY POINTS

- Automatic sprinkler systems are typically installed because they are mandated by the code, or because they are to be used as equivalent protection to other code requirements.

- Because of the potentially high occupant load and density anticipated in Group A occupancies, coupled with the occupants' probable lack of familiarity with the means of egress system, large assembly uses must be protected by an automatic sprinkler system.

- School buildings must be sprinklered throughout unless complying compartmention or increased egress is provided.

- Large manufacturing buildings and warehouses, when containing combustible goods or materials, must be sprinklered to limit the size of a fire.

- The IBC requires sprinkler protection for all Group H occupancies owing to local hazards within the building and the potential for presenting a high level of hazard to the surrounding properties.

- Because the mobility of the occupants of Group I occupancies is greatly diminished, the code requires automatic fire suppression.

- On account of their fire record, hotels, apartment buildings, assisted-living facilities and all other residential occupancies must always be sprinklered.

- Adequate openings must be provided in exterior walls for fire department access, or a sprinkler system must be installed.

- Most buildings exceeding 55 feet (16 764 mm) in height are required to be equipped with an automatic sprinkler system throughout.

- Certain occupancies and uses are required to have standpipe protection.

- The locations of Class I standpipe connections are specifically identified in the code.

- The locations of some Class II standpipe hose cabinets are based on the distances that the fire hose can reach throughout the building.

- One of the most effective means of occupant protection in case of a fire incident is the availability of a fire alarm system.

- Pressurization, airflow and exhaust are the three methods of mechanical smoke control.

- A ventilated smokeproof enclosure utilizes either an exterior balcony or a ventilated vestibule, whereas pressurization of the stair shaft is a permitted alternative.

- Smoke and heat venting is required in large, open areas of manufacturing, warehouse and hazardous occupancies, as well as retail sales with high-piled stock.

- Curtain boards are used to divide the area below the roof into zones for smoke and heat venting.

- Special provisions intended to increase the efficiency and safety of fire department personnel during emergency operations are addressed, including emergency responder safety features, the fire department command center, fire department connections and fire pump rooms.

MEANS OF EGRESS

This chapter establishes the basic approach to determining a safe exiting system for all occupancies. It addresses all portions of the egress system and includes design requirements, as well as provisions regulating individual components, that may be used within the egress system. The chapter specifies the methods of calculating the occupant load that are used as the basis of designing the system and, thereafter, discusses the appropriate criteria for the number of exits, location of exits, width or capacity of the egress system, and the arrangement of the system. This arrangement is treated in terms of remoteness and accessibility of the egress system. The accessibility is handled both in terms of the system's being usable by building occupants and in terms of it being available within a certain maximum distance of travel. After having dealt with general issues that affect the overall system or multiple zones of the system defined as the exit access, exit and exit discharge, the chapter then establishes the design requirements and components that may be used to meet those requirements for each of the three separate zones.

In interpreting and applying the various provisions of this chapter, it would help to understand the four fundamental concepts on which safe exiting from buildings is based:

1. A safe egress system for all building occupants must be provided.

2. Throughout the system, every component and element that building occupants will encounter in seeking egress from the building must be under the control of the person wishing to exit.

3. Once a building occupant reaches a certain degree or level of safety, as that occupant proceeds through the exiting system, that level of safety is not thereafter reduced until the occupant has arrived at the exit discharge, public way or eventual safe place.

4. Once the exit system is subject to a certain maximum demand in terms of number of persons, that system must thereafter (throughout the remainder of the system) be capable of accommodating that maximum number of persons.

Egress for individuals with physical disabilities is to be provided under the provisions of this chapter, primarily through the design of an accessible means of egress system. Because many of the elements composing the egress system (doors, landings, ramps, etc.) may also form part of the accessible routes as required by Chapter 11, such requirements must be referenced where applicable.

This chapter includes the three-part definition of "Means of egress." The three-part system, or zonal approach as it is now used, was introduced by the National Fire Protection Association (NFPA) in 1956 and was incorporated over the years into all of the legacy model codes. This approach has established terms that are used throughout the design and enforcement communities to deal with the means of egress system. The three parts of the means of egress system are the exit access, the exit and the exit discharge. For conceptual ease, the exit access is generally considered any location within the building from where you would start your egress travel, and continues until you reach the door of an exit. The exit access would include all the rooms or spaces that you would pass through on your way to the exit. This may be the room you are in; an intervening room; a corridor; an exterior egress balcony; and any doors, ramps, stairs or aisles that you use along that path. An exit is the point where the code considers that you have obtained an adequate level of safety so that travel distance measurements are no longer a concern. Exits will generally consist of fire-resistance-rated construction and opening protection that will separate the occupants from any problem within the building. Elements that are considered exits include exterior exit doors at ground level, exit enclosures, exit passageways, exterior exit stairs, exterior exit ramps and horizontal exits. Exterior exit doors, exterior exit stairs and exterior exit ramps will not provide the fire-protection levels that the other elements provide, but insofar as the occupant will be outside the building, they will provide a level of safety by removing the occupants from the problem area. The last of the three parts is the exit discharge. The *International Building Code*® (IBC®) will generally view exterior areas at ground level as the exit discharge portion of the exit system. Therefore, the exit access will be the area within the building that gets the occupants to an exit, whereas the exit discharge will be the exterior areas at grade where the occupants go upon leaving the building in order to reach the public way.

Section 1001 *Administration*

This section requires that every building or portion thereof comply with provisions of Chapter 10. In dealing with portions of buildings, it is important to understand that the code intends this chapter to apply to all portions that are occupiable by people at any time. Therefore, areas such as storage rooms and equipment rooms, although often unoccupied, will still be regulated under the provisions of the chapter.

In order to provide an approved means of egress at all times, it is critical that the exiting system be maintained appropriately. Section 1030 of the *International Fire Code*® (IFC®) regulates maintenance of the means of egress for the life span of the building. Should there be alterations or modifications to any portion of the building, Section 1001.2 mandates that the number of existing exits not be reduced, nor the capacity of the means of egress be decreased, below that level required by the IBC. Section 4604 of the IFC also provides a limited number of specific provisions addressing the means of egress in existing buildings.

Section 1002 *Definitions*

Section 1002 contains a number of definitions that are especially important to this chapter. As with most other terms defined in the code, the definitions found in Section 1002.1 are specific to the IBC and typically differ from their ordinarily accepted meanings.

ACCESSIBLE MEANS OF EGRESS. In concert with efforts to make buildings accessible and usable for persons with disabilities, it is necessary that safe egress for physically disabled persons also be provided. Therefore, accessible exit paths of continuous and unobstructed travel are to be provided. Generally consistent with the provisions of other exiting systems, an accessible means of egress shall begin at any accessible point within the building and continue until reaching the public way. The primary provisions regulating accessible means of egress are located in Section 1007.

AISLE. Where furniture, fixtures and equipment limit the potential travel paths within the means of egress system, aisles and aisle accessways are created. Typically, aisles accept the contribution of occupant travel from adjoining aisle accessways. At times, multiple aisles may converge into a main aisle, which then may lead to exit access doorways or exits. Aisles are common throughout most buildings and are considered portions of the exit access.

AISLE ACCESSWAY. The path of travel from an occupiable point in a building to an aisle is considered to be an aisle accessway. Generally called a row in everyday language when adjacent to seats and fixtures, an aisle accessway is typically utilized by small numbers of people prior to converging with other persons at an aisle. An example of aisle accessways is shown in Figure 1002-1.

ALTERNATING TREAD DEVICE. By appearance more of a ladder than a stairway, an alternating tread device is shown in Figure 1002-2. Typically, steps are supported by a center rail placed at a severe angle from the floor. The key difference between this device and other forms of stairways addressed in the IBC is that the user, by nature of the design of the device, can never have both feet on the same level at the same time.

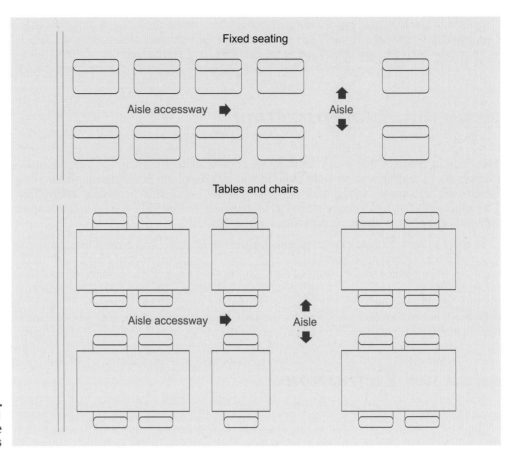

Figure 1002-1
Aisle accessways

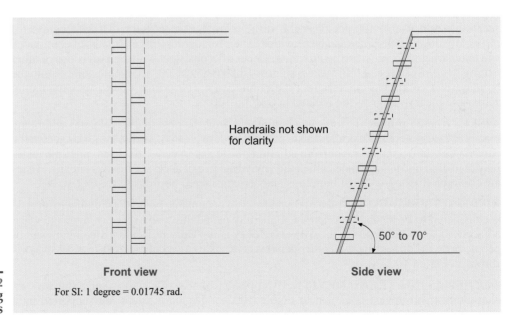

Figure 1002-2
Alternating tread devices

AREA OF REFUGE. It is common for an area of refuge to be included as a portion of the accessible means of egress. The intent of the refuge area is to provide a location where individuals unable to use stairways can gather, awaiting assistance or instructions during an emergency evacuation of the building. The size and construction requirements for areas of refuge are provided in Section 1007.6.

BLEACHERS. Structures designed for seating purposes containing tiered or stepped seating two or more rows high are considered bleachers. The definition of "Grandstand" is identical to that of bleachers, recognizing that the terms are interchangeable. Bleachers may, or may not, be provided with backrests. In addition, bleachers may be located either inside or outside a structure. The specific provisions for bleachers are found in ICC 300, *Bleachers, Folding and Telescopic Seating, and Grandstands.* Similar seating areas which are considered as building elements would not be defined as bleachers and are not regulated by ICC 300. A building element, as defined in Section 702.1, is deemed to be a fundamental component of the building's construction as listed in Table 601.

COMMON PATH OF EGRESS TRAVEL. It is important to limit the travel distance that occurs where only a single path is available to the user of the means of egress. For this reason, such limited travel is regulated as a common path of egress travel. Very similar to the concept addressed in the limitations of dead-end conditions in corridors, the intent of regulations regarding a common path of egress travel is based upon the lack of at least two separate and distinct paths of egress travel toward two or more remote exits. Included as the initial part of the permitted travel distance, a common path of egress travel occurs within the exit access portion of the means of egress. See Figure 1002-3.

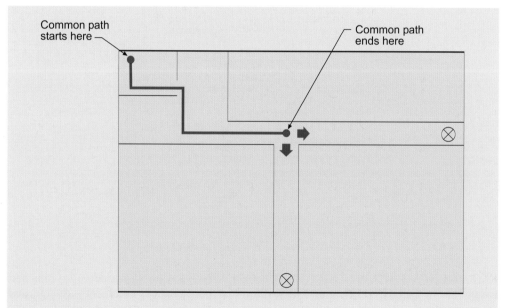

Figure 1002-3
Common path of egress travel

CORRIDOR. A corridor is considered a component of the exit access portion of the means of egress that provides an enclosed and directed path of egress travel to an exit. Regulated by Section 1018, corridors may or may not be required to be fire-resistance rated, but in all cases, an enclosure of some sort is anticipated. The determination as to whether or not a design element is to be regulated as a corridor is to be made by the building official. Many factors may be considered prior to making this determination, which could possibly include length, degree of enclosure, length-to-width ratio, adjacent spaces and other considerations. However, because of the reliance upon corridors as an important egress element, it is critical that they be appropriately regulated. A corridor's primary purpose is for the movement of occupants, both as a part of the building's circulation and its use as a means of egress. Although some spaces may have one or more of the characteristics of a corridor as previously mentioned (length, degree of enclosure, length to width ratio, etc.), their primary function is that of rooms, and they should not be considered corridors. A classic example can be found in many observation buildings in zoos, where a very long, narrow element is utilized as a means for occupants to view displays of wildlife. Although in plan view it may appear to be a corridor, it actually functions as a room and should be regulated as such. It should be noted that the placement of a few pieces of furniture or equipment within a corridor in an effort to consider the space a room rather than a corridor is not appropriate. It should be noted that corridors are never mandated by the IBC, but rather are utilized as design elements. Where provided, however, such enclosures must comply with the code.

DOOR, BALANCED. Balanced doors are a special type of double-pivoted door in which the pivot point is located some distance in from the door edge, thus creating a counter-balancing effect.

EGRESS COURT. That portion of the exit discharge at ground level that extends from a required exit to the public way is considered an egress court. An egress court may be a yard, a court, or a combination of a yard and a court, that extends from the end point of an exit, typically an exterior door at grade level, until it reaches a public way. An example is illustrated in Figure 1002-4.

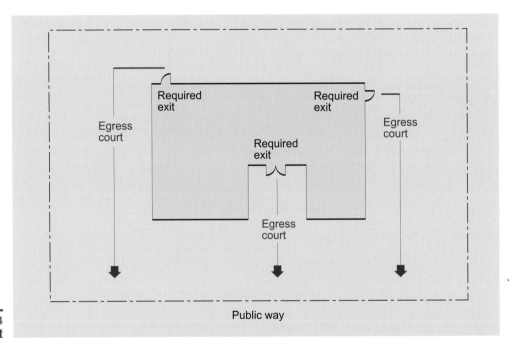

Figure 1002.4
Egress court

EMERGENCY ESCAPE AND RESCUE OPENING. Required in sleeping rooms and basements in specific residential and institutional occupancies, emergency escape and rescue openings are intended to allow for a secondary means of escape or rescue in the event of an emergency. Typically an operable window or door, such an opening is regulated by Section 1029 for minimum size, maximum height from the floor and operational constraints. Although the provisions are found in Chapter 10 of the code, the intent of the opening is for emergency escape or rescue access, and it is not intended to be considered an element of a complying means of egress.

EXIT. As stated in the introduction, an exit is the first portion of egress travel where the code believes that the occupant has obtained an adequate level of safety so that travel distance measurements are no longer a concern. In addition, an exit typically provides only single-direction egress travel. The adequate level of safety is provided by building elements that completely separate the means of egress from other interior spaces in the building. Inside the building, fire-resistance-rated construction and opening protectives are utilized to provide a protective path of egress travel between the exit access and exit discharge portions of the means of egress. Building elements that are considered exits include exterior exit doors at the level of exit discharge, exit enclosures, exit passageways, exterior exit stairs, exterior exit ramps and horizontal exits.

EXIT ACCESS. The exit access is identified as the initial component of the means of egress system, the portion between any occupied point in a building and the exit. Leading to one or more of the six exit components, the exit access makes up the vast majority of any building's floor area. Because the exit access begins at any point that may potentially be occupied, it is probable that only those concealed areas, such as penthouses, attics and under-floor spaces that are typically unoccupied, fall outside of the definition of the means of egress.

EXIT ACCESS DOORWAY. The term "exit access doorway" is commonly utilized in IBC Chapter 10 to establish a reference point within the exit access for applying various means of egress provisions, including those addressing arrangement, number, separation, opening protection and exit sign placement. Although one would expect the term "doorway" to be limited to those situations where an actual door opening, either with or without a door, is present, the IBC definition expands this traditional meaning by including certain access points that do not necessarily include doorways, such as unenclosed exit access stairs and ramps. In fact, any point at which the exit access is narrowed so as to create a single point of travel could potentially be considered as an exit access doorway as shown in Figure 1002-5.

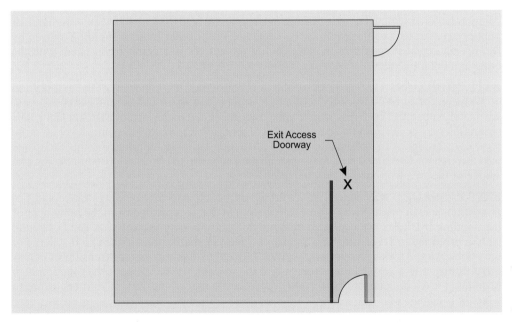

Figure 1002-5
Exit access doorway

EXIT DISCHARGE. Exit discharge is the last portion of the three-part means of egress system and is that portion between the point where occupants leave an exit and the point where they reach a public way. For conceptual ease, all exterior travel at ground level is considered a part of the exit discharge. An egress court is the primary component of exit discharge.

EXIT DISCHARGE, LEVEL OF. The story within a building where an exit terminates and the exit discharge begins creates a condition defined as the level of exit discharge. Because exit discharge occurs substantially at ground level, the level of exit discharge typically provides a horizontal path of travel toward the public way. At times, the code uses this term to define another floor level's relationship to egress at grade level. In most buildings, the first story above the level of exit discharge is typically the building's second story above grade plane as shown in Figure 1002-6.

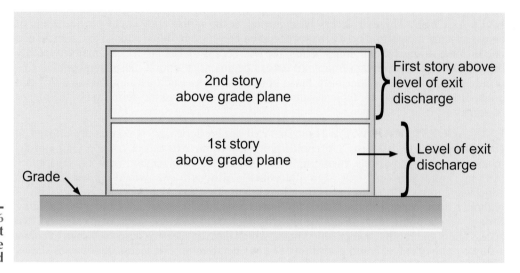

Figure 1002-6
Level of exit discharge defined

EXIT ENCLOSURE. An exit enclosure can provide a protected path of egress travel in either a vertical (vertical exit enclosure) or horizontal (exit passageway) direction. These components are regulated as exits, which must typically lead directly to the exit discharge or public way. An exit enclosure must be separated from other interior building spaces by fire-resistance-rated construction and opening protectives in a manner consistent with the other exit components.

EXIT, HORIZONTAL. The horizontal exit is considered a component of the exit portion of the means of egress. The concept of a horizontal exit is to provide a refuge compartment adequately separated from fire, smoke and gases generated in the area of a fire incident. The separation may occur from one building to another at approximately the same level, or more typically, through a fire-resistance-rated wall within a single building. As the occupants pass through the horizontal exit, they enter an area intended to afford safety from the fire and smoke of the area from which they departed.

EXIT PASSAGEWAY. Much like a vertical exit enclosure in providing smoke and fire protection, an exit passageway is an exit component that is separated from the remainder of the building by fire-resistance-rated construction and opening protectives. The horizontal path of protected travel afforded by an exit passageway shall extend to the exit discharge or the public way. An exit passageway is often utilized as the horizontal extension of egress travel connecting a vertical exit enclosure to the exterior of the building. Its function is to maintain a level of occupant protection equivalent to that of the vertical exit enclosure to which it is connected.

FIRE EXIT HARDWARE. Specifically listed for use on fire door assemblies, fire exit hardware is mandated for use where panic hardware is required on a door assembly that also requires a fire-protection rating.

FLIGHT. The use of the term "flight" is specifically defined in order to establish its use within the code. The definition addresses two separate issues. One, a flight is made up of the treads and risers that occur between landings. As an example, a stairway connecting two stories that includes an intermediate landing consists of two flights. Secondly, the inclusion of winders within a stairway does not create multiple flights. Winders are simply treads within a flight and are often combined with rectangular treads within the same flight.

FLOOR AREA, GROSS. As indicated in Table 1004.1.1, the determination of the occupant load in the design of the means of egress system for most building uses is typically based upon the gross floor area. This term describes the total floor area included within the surrounding exterior walls of a building, and the definition further states that the floor area of the building or portion thereof not provided with surrounding exterior walls shall be the usable area under the horizontal projection of the roof or floor above. The intent of this latter provision is to cover where a structure may not have exterior walls or may have one or more sides open without an enclosing exterior wall. Where buildings are composed of both enclosed and unenclosed areas, the gross floor area is typically determined as illustrated in Figure 1002-7. Projections extending beyond an exterior wall or column line that are not intended to create usable space below are not to be considered in the determination of gross floor area. Areas often considered accessory-type spaces, such as closets, corridors, elevator shafts and stairways, must also be considered a part of the gross floor area.

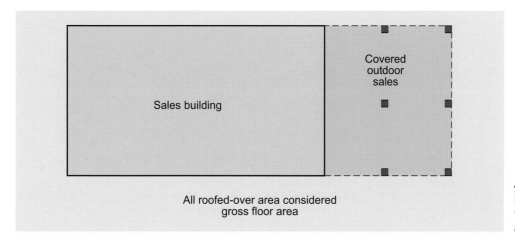

Figure 1002-7
Gross floor area

FLOOR AREA, NET. The net floor area is considered that portion of the gross floor area that is typically occupied. Normally unoccupied accessory areas such as corridors, stairways, closets, toilet rooms, equipment rooms and similar spaces are not to be included in the calculation of net floor area. In addition, the measurements are based upon clear floor space, allowing for the deduction of building construction features such as interior walls and columns, as well as elevator shafts and plumbing chases. The use of net floor area in the calculation of design occupant load is typically permitted only in assembly and educational uses as set forth in Table 1004.1.1. It is important to note that in calculating net floor area, as well as gross floor area, the floor space occupied by furniture, fixtures and equipment is not to be excluded in the calculation. The floor-area-per-occupant factor established in Table 1004.1.1 includes any such anticipated furnishings in the establishment of an appropriate density estimate.

FOLDING AND TELESCOPING SEATING. Folding and telescoping seats are structures that provide tiered seating, which can be reduced in size and moved without dismantling. Utilized quite often in school gymnasiums, such seating presents the same concerns and risks as permanently-installed bleacher seating when occupied.

GRANDSTAND. The definition of grandstand is also applicable to bleachers. Further information is provided in the discussion of the definition of bleachers.

GUARD. A component or system of components whose function is the minimization of falls from an elevated area is considered a guard. Placed adjacent to the elevation change, a guard must be of adequate height, strength and configuration to prevent someone from falling over or through the guard. Outside of the code, this element is more commonly described as a guardrail.

HANDRAIL. Typically utilized in conjunction with a ramp or stairway, a handrail is intended to provide support for the user along the travel path. A handrail may also be used as a guide to direct the user in a specified direction.

MEANS OF EGRESS. The means of egress describes the entire travel path a person encounters, beginning from any occupiable point in a building and not ending until the public way is reached. Often encompassing both horizontal and vertical travel, the means of egress should be direct, obvious, continuous, undiminished and unobstructed. It includes all components of the exiting system that might intervene between the most remote occupiable portion of the building and the eventual place of safety—typically the public way. Therefore, the means of egress includes all such intervening components as aisle accessways, aisles, doors, corridors, stairways and egress courts, as well as any other component that might be in the path of travel as depicted in Figure 1002-8. There are three distinct and separate portions of a means of egress—the exit access, the exit and exit discharge.

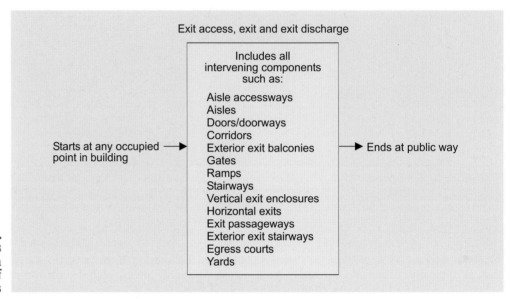

Figure 1002-8
Definition of a means of egress

MERCHANDISE PAD. A merchandise pad is created where racks, displays, shelving units and similar fixtures are grouped into a specific area. Aisle accessways are provided within the merchandise pad to allow for customer circulation and are utilized as exit access elements. Such aisle accessways connect to the aisles that partially or entirely surround the merchandise pad. An example is shown in Figure 1002-9.

NOSING. The leading edge portions of stair treads are considered nosings. Nosings may also be found where a landing is the top step of a stairway run.

OCCUPANT LOAD. Viewed as the basis for the design of the means of egress system, the occupant load is that number of persons considered for means of egress purposes both within and from any space, area, room or building. Variations in code requirements are often caused based upon the anticipated occupant load served by the specific building egress element. Although the establishment of an occupant load is critical in the application of many means of egress provisions located in Chapter 10, it is also often necessary when addressing minimum requirements for fire alarm systems, sprinkler systems in Group A occupancies and plumbing fixture counts.

PANIC HARDWARE. The term *panic hardware* describes an unlatching device that will operate even during panic situations, so that the force of individuals against the egress door will cause the door to unlatch without manual manipulation of the device.

PUBLIC WAY. The code defines a public way essentially as a street, alley or any parcel of land that is permanently appropriated to the public for public use. Therefore, the public's right to use such a parcel of land is guaranteed. The building occupants, having reached a public way, are literally free to go wherever they might choose. They are certainly free to go so far as to escape any fire threats in any building that they might have been occupying. There is an expectation that the public way be a continuation of the egress path, providing for continued exit travel away from the building until the necessary safety level has been achieved.

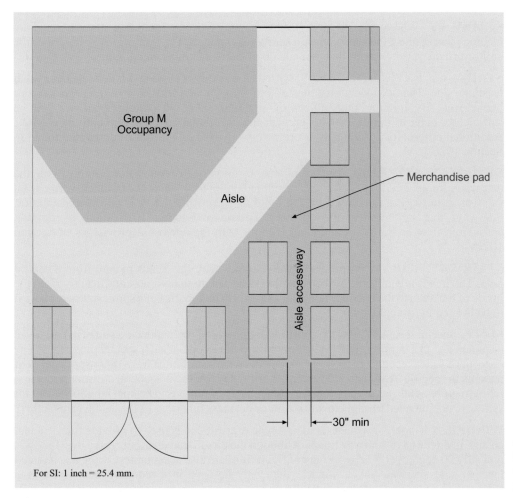

For SI: 1 inch = 25.4 mm.

Figure 1002-9
Merchandise pad

RAMP. Where the slope of a travel path exceeds 1 unit vertical to 20 units horizontal (1:20), it is considered a ramp. Where the travel path has a slope of 5 percent or less (less than or equal to 1:20), it is considered merely a walking surface.

SCISSOR STAIR. A unique design element, the scissor stair allows for two independent paths of travel within a single vertical exit enclosure. A scissor stair is considered a single means of egress because the exit paths are not isolated from each other with the required level of fire resistance.

SMOKE-PROTECTED ASSEMBLY SEATING. Where the means of egress for assembly seating areas is designed to be relatively free of the accumulation of smoke, the seating is considered to be smoke-protected. In order to qualify as smoke-protected, the seating area and its exiting system must comply with the provisions of Section 1028.6.2, which addresses the methods of smoke control, the minimum roof height, and the possible installation of sprinklers in adjacent enclosed spaces. Exterior seating facilities such as stadiums or amphitheaters are commonly considered to have smoke-protected assembly seating.

STAIR. Where one or more risers are provided to address a change in elevation, a stair is created. A stair may simply be a slight change in height from one floor level to another, commonly referred to as a step, or may be a series of treads and risers connecting one floor or landing to another. Also described in the code as a flight of stairs, a stair does not include the landings and floor levels that interrupt stairway travel.

STAIRWAY. Where one or more flights of stairs occur, including any intermediate landings that connect the stair flights, a stairway is created. The term *stairway* describes the entire vertical travel element that is made up of stairs, landings and platforms.

STAIRWAY, EXTERIOR. To be classified as an exterior stairway, it must be open on at least one side. The open side must then adjoin an open area such as a yard, egress court or public way. By limiting the number of enclosed sides, an exterior stairway will be sufficiently open to the exterior to prevent the accumulation of smoke and toxic gases. Additional criteria for defining an exterior stairway used as a means of egress are found in Section 1026.3.

STAIRWAY, INTERIOR. By definition, a stairway that does not comply with the definition for an exterior stairway is considered an interior stairway. In other words, if all sides of a stairway are enclosed by the building's construction, it is considered interior. Stairways that fail to meet the openness criteria of Section 1026.3 are, by default, considered interior stairways and shall comply with the enclosure provisions of Section 1022.

STAIRWAY, SPIRAL. A spiral stairway is one where the treads radiate from a central pole. The treads are uniform in shape, with a tread length that varies significantly from the inside of the tread to the outside. The dimensional characteristics of a spiral stairway cause it to be limited in its application.

SUITE. Special means of egress provisions are provided for health care suites in Group I-2 occupancies. The definition of "Suite" establishes the scope of such special provisions. The concept of suites recognizes those arrangements where staff must have more supervision of patients in specific treatment and sleeping rooms. Therefore, the general corridor width and rating requirements are not appropriate under such conditions. The special allowances for suites are not intended to apply to day rooms or business functions of the health care facility.

WINDER. A winder, or winder tread, is a type of tread that is used to provide for a gradual change in direction of stairway travel. Although the directional change created by winders is typically 90 degrees (1.57 rad), other configurations are also acceptable. Owing to a reduced level of safe stairway travel, winders may only be used in a required means of egress stairway when located within a dwelling unit.

Section 1003 *General Means of Egress*

The requirements and topics addressed in this section are used as basic provisions and are to be applied throughout the entire egress path. Examples of the types of general issues that are found here include ceiling height, protruding objects, floor surface, elevation change and egress continuity. For consistency purposes, the provisions for ceiling height and protruding objects are identical to the accessibility criteria of ICC A117.1.

1003.2 **Ceiling height.** In order to provide an exit path that maintains a reasonable amount of headroom clearance for the occupants, this section requires the means of egress to have a minimum ceiling height of 7 feet 6 inches (2286 mm). The intent of the provision is to address all potential paths of exit travel that can be created based upon multiple directions of egress and the layout of the room or space insofar as furniture, equipment and fixtures are concerned. Any portion of the floor area of the building that can reasonably be considered a possible exit path should be provided with a minimum 7-foot 6-inch (2286 mm) clear height, unless reduced by exceptions permitted for sloped ceilings, dwelling and sleeping units in residential occupancies, stairway and ramp headroom, door height and protruding objects. Additional exceptions reduce the minimum required clear height to 7 feet 0 inches (2134 mm) in parking garage vehicular and pedestrian traffic areas as well as above and below floors considered as mezzanines.

Protruding objects. Limitations are placed on the permitted projection of protruding objects for two purposes. First, to maintain an egress path that is essentially free of obstacles. Second, to provide a circulation path that is usable by all occupants, including those individuals with sight-related disabilities. For this reason, provisions regulate the accessibility concerns regarding protruding objects as well as the egress concerns. Note that projections into the required egress width and the minimum clear width of accessible routes are also limited by other provisions of the code.   **1003.3**

Headroom. Consistent with the allowance for stair headroom and doorway height to be reduced below the required egress height of 7 feet 6 inches (2286 mm), other portions of the egress system may likewise be reduced to a minimum height of 80 inches (2032 mm). The reduction for signage, sprinklers, decorative features, structural members and other protruding objects is limited to 50 percent of the ceiling area of the egress path. See Figure 1003-1. Though projections at an 80-inch (2032 mm) height are not unusual to building occupants, it is necessary to maintain a majority of the egress system at 7 feet 6 inches (2286 mm) or higher. Passage through a doorway may be further reduced in height to 78 inches (1981 mm) at the door closer or stop. This reduction at doors is also permitted for accessibility purposes by Section 307.2 of ICC A117.1.   **1003.3.1**

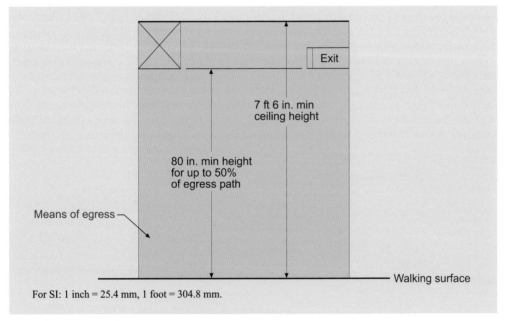

For SI: 1 inch = 25.4 mm, 1 foot = 304.8 mm.

Figure 1003-1
Means of egress headroom

Where a vertical clearance of 80 inches (2032 mm) cannot be achieved, the reduced-height portion of such floor area cannot be utilized as a portion of the means of egress system. It is also necessary to provide some type of barrier that will prohibit the occupant from approaching the area of reduced height. This is of particular importance where the occupant is sight impaired, with no method other than a barrier to identify the presence of an overhead protruding object. The mandated barrier is to be installed so that the leading edge is no more than 27 inches (686 mm) above the walking surface as shown in Figure 1003-2. By limiting the height of the barrier edge, it will be located in a manner so that a sight-impaired individual using a long cane will detect the presence of an obstruction and maneuver to avoid the hazard.

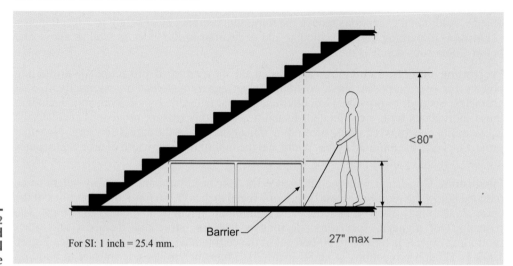

Figure 1003-2
Reduced vertical clearance

1003.3.2 **Post-mounted objects.** Free-standing objects mounted on a post or pylon that are located along or adjacent to the walking surface are potential hazards, particularly to a sight-impaired individual. Objects such as signs, directories or telephones that are mounted on posts or pylons are, therefore, limited to an overhang of 4 inches (102 mm) maximum if located more than 27 inches (686 mm), but not more than 80 inches (2032 mm), above the floor level. By limiting the overhang to 4 inches (102 mm), a cane will hit the post or pylon prior to the individual impacting the mounted object. See Figure 1003-3. Free-standing objects mounted at or below 27 inches (686 mm) will fall within the cane-detection zone, and objects mounted at 80 inches (2032 mm) or higher are sufficiently above the walking surface. Similar concerns are addressed where the obstruction is mounted between posts located more than 12 inches (305 mm) apart. Unless the lowest edge of the obstruction is at least 80 inches (2030 mm) above the walking surface, it must be located within the cane recognition area extending from the walking surface to a height of 27 inches (686 mm).

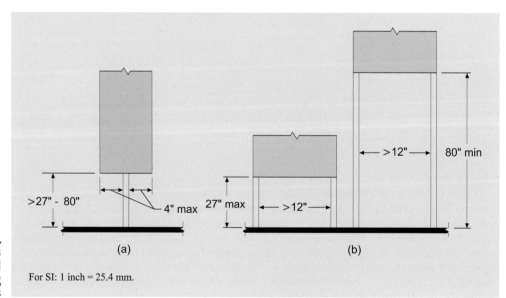

Figure 1003-3
Post-mounted protruding objects

Horizontal projections. Consistent with the other provisions for protruding objects, horizontal projections such as structural elements, fixtures, furnishings and equipment are considered hazardous where they fall outside of the area where cane detection can identify them. Visually-impaired individuals cannot detect overhanging objects when walking alongside them. Because proper cane techniques keep people some distance from the edge of a walking surface or from walls, a slight overhang of no more than 4 inches (102 mm) is not considered hazardous. An example of this provision is illustrated in Figure 1003-4. An exception permits handrails to protrude 4$^1/_2$ inches (114 mm).

1003.3.3

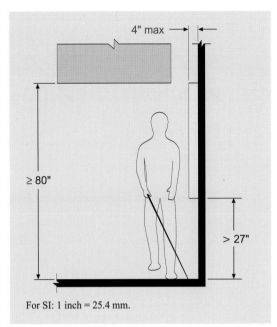

For SI: 1 inch = 25.4 mm.

Figure 1003-4
Limits of protruding objects

Although the provisions of this section, as well as those in Sections 1003.3.1 and 1003.3.2, are primarily based upon clearances established for accessibility purposes, their value to all users of the egress path is considerable. Projections into the means of egress potentially could result in a reduced travel flow, resulting in longer evacuation times during emergency conditions. In addition, injuries are possible to individuals who fail to pay proper attention to where they are going.

Floor surface. As evidenced by the requirements for ceiling height and protruding objects, the potential for exit travel to be impeded by obstructions is addressed throughout Chapter 10. Various provisions attempt to eliminate the opportunity for hazards along the exit path to slow travel. This section recognizes one area that is often taken for granted when it comes to egress—the walking surface of the means of egress. It is typically assumed that a walking surface that provides adequate circulation, and often accessibility, throughout a building will be acceptable for egress purposes as well. Although this is usually true, it is stated in the code that the egress path should have a walking surface that is slip-resistant and securely attached so there is no tripping or slipping hazard that would result in an obstruction of the exiting process. Although the regulation of floor surfacing materials is typically recognized for interior walking surfaces, the provision is also applicable to exterior egress paths. The performance criteria of this code section provide a basis for determining the appropriateness of any questionable exit discharge elements.

1003.4

Elevation change. The code is concerned that along the means of egress there is no change in elevation along the path of exit travel that is not readily apparent to persons seeking to exit under emergency conditions. Therefore, along the means of egress, any change in elevation of less than 12 inches (305 mm) must be accomplished by means of a ramp or other sloping surface. A single riser or a pair of risers is not permitted. See Figure 1003-5. Steps used to achieve minor differences in elevation frequently go unnoticed and as a consequence can cause missteps or accidents.

1003.5

This limitation on the method for a change of elevation, however, does not apply in certain locations. Where exterior doors are not required to be accessible by Chapter 11, a single step of 7 inches (179 mm) or less in height is permitted by Exception 1 at such exterior doors in Groups F, H, R-2, R-3, S and U. See Figure 1003-6. A second exception allows, under specific conditions, a stair with a single riser or with two risers and a tread at those locations not required to be accessible by Chapter 11. In this case, the risers and treads must comply with Section 1009.4, the tread depth must be at least 13 inches (330 mm), and a minimum of one complying handrail must be provided within 30 inches (762 mm) of the center line of the normal path of egress travel on the stair. See Figure 1003-7. A third exception applies to seating areas not required to be accessible. Risers and treads may be utilized on an aisle serving the seating where a complying handrail is provided.

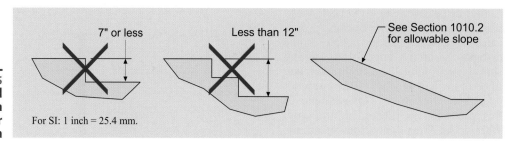

Figure 1003-5
Longitudinal section through corridor or other exit path

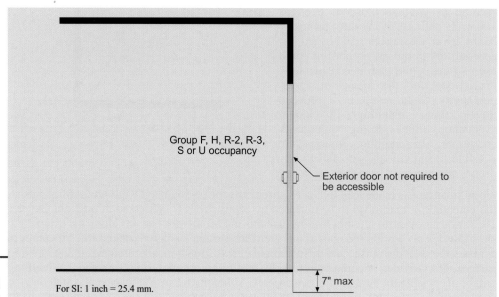

Figure 1003-6
Single step at exterior door

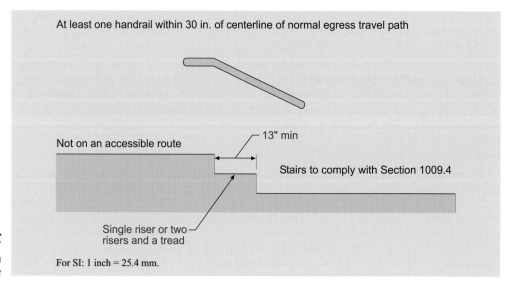

Figure 1003-7
Elevation change

Means of egress continuity. This section emphasizes that wherever the code imposes **1003.6**
minimum widths on components in an exiting system, such widths are to be clear, usable
and unobstructed. Nothing may project into these required widths so as to reduce the
usability of the full dimension, unless the code specifically and expressly states that a
projection is permitted. Two notable examples of permitted projections are doors, either
during the course of their swing or in the fully open position, and handrails. The limitations
on the amount of such projections are specified within the appropriate sections of the
chapter. Additionally, this section places into code language one of the four basic concepts
that were previously discussed—that once the exit system is subject to a certain maximum
demand in terms of number of persons, that system must thereafter be capable of
accommodating that maximum number of persons.

Elevators, escalators and moving walks. For a variety of reasons, elevators, escalators **1003.7**
and moving walks are not to be used to satisfy any of the means of egress for a building.
These building components are intended for circulation purposes and do not conform with
the detailed egress requirements found in Chapter 10. The only exception is for elevators
used as an accessible means of egress as addressed in Section 1007.4.

Section 1004 *Occupant Load*

Design occupant load. This section prescribes a series of methods for determining the **1004.1**
occupant load that will be used as the basis for the design of the egress system. The basic
concept is that the building must be provided with a safe exiting system for all persons
anticipated in the building. The process for determining an appropriate occupant load is
based on the anticipated density of the area under consideration. Because the density factor
is already established by the code for the expected use, variations in occupant load are
simply a function of the floor area assigned to that use. It is apparent that in many situations
the occupant load as calculated is conservative in nature. This is appropriate because of the
extent that the means of egress provides for life safety concerns. The egress system should
be designed to accommodate the worst-case scenario, based on a reasonable assumption of
the building's use.

This provision also mandates that the occupant loads are to be cumulative as the
occupants egress through intervening spaces. Under the conditions of Section 1014.2, the
path of travel through the intervening space must be discernable to allow for a continuous
and obvious egress path. Although the provision specifically addresses egress from spaces
through a primary area, it is also applicable where egress travel goes from an accessory area
through another accessory area, or from a primary area through an accessory area. See
Figure 1004-1. Another common application occurs as users of the means of egress merge at
aisles, corridors or stairways as shown in Figure 1004-2.

Areas without fixed seating. The vast majority of buildings contain uses that do not utilize **1004.1.1**
fixed seating. Unlike auditoriums, theaters and similar spaces, in most instances the
maximum probable number of occupants may not be known. Therefore, the code provides a
formula for determining an occupant load that constitutes the minimum number of persons
for which the exiting system must be designed. As a consequence, the code refers to the
number obtained by the formula as the design occupant load. Egress systems for all
buildings or building spaces must be designed to accommodate at least this minimum
number. Basic examples of the use of Table 1004.1.1 are illustrated in Figure 1004-3.

As the person responsible for interpreting and enforcing the code, the building official
will be called on to make decisions regarding the categories in Table 1004.1.1. Although
Table 1004.1.1 contains occupant load factors that will serve the code user under most
conditions, there will be occasions when either the table does not have an occupant load
factor appropriate for the intended use, or where the occupant load factor contained in the
table will not have a realistic application. In such instances, the building official has the

authority to establish an appropriate occupant load factor or an appropriate occupant load for those special circumstances and those special buildings.

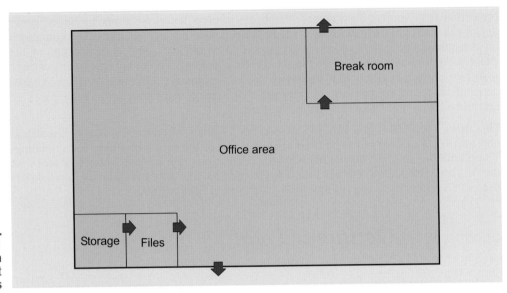

It may be meaningful to point out that the first column of Table 1004.1.1 is headed "Function of Space." The categories listed in the first column are not the specific groups identified in Chapter 3 for the purpose of assigning an occupancy classification but are the basic generic uses of building spaces. It has been pointed out in the discussion of the various occupancy groups that it is possible to have a classroom classified as a Group E occupancy, a Group B occupancy or, possibly, a Group A occupancy. In terms of occupant density, however, a classroom is a classroom, and it is reasonable to expect the same density of use in a classroom regardless of the occupancy group in which that classroom might be classified. Therefore, the table specifies that when considering classroom use, one must assume there is at least one person present for each 20 square feet (1.86 m²) of floor area.

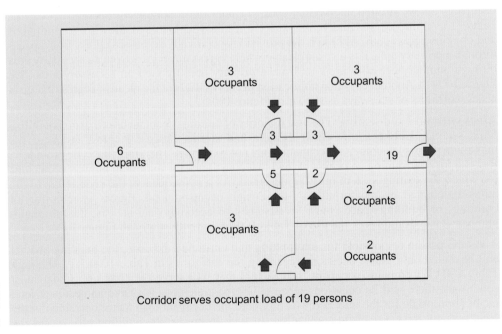

Corridor serves occupant load of 19 persons

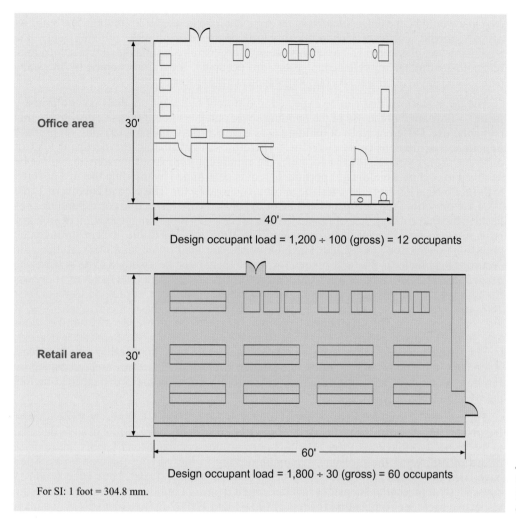

Office area   30'

◄——————— 40' ———————►

Design occupant load = 1,200 ÷ 100 (gross) = 12 occupants

Retail area   30'

◄——————————— 60' ———————————►

Design occupant load = 1,800 ÷ 30 (gross) = 60 occupants

For SI: 1 foot = 304.8 mm.

Figure 1004-3
Design
occupant load

In specifying how the occupant load is to be determined, the code intends that it is to be assumed that all portions of a building are fully occupied at the same time. It may be recognized, however, that in limited instances not all portions of the building are, in fact, fully occupied simultaneously. An example of this approach for accessory uses might include conference rooms in various occupancies or minor assembly areas such as lunch rooms in office buildings or break rooms in factories. It is important to note that the code does not provide for a method to address such conditions; thus, full occupancy should always be assumed. Only under rare and unusual circumstances should the building official ever consider reducing the design occupant load because of the simultaneous use concept. In such situations, he or she must determine that there are accessory spaces that ordinarily are used only by persons who at other times occupy the main areas of the building; therefore, it is not necessary to accumulate the occupant load of the separate spaces when calculating the total occupant load of the floor or building. It is always necessary, however, to provide each individual space of the building with egress as if that individual space were fully and completely occupied.

Another type of accessory-use area that must be considered in occupant-load calculation includes corridors, closets, toilet rooms and mechanical rooms. These uses are typical of most buildings and are to be included by definition in the gross floor area of the building. A quick review of Table 1004.1.1 will show that most of the uses listed are to be evaluated based on gross floor area, with no reduction for corridors and the like. However, a few of the listings indicate the use of the net floor area in the calculation of the occupant load. An example would be the determination of an occupant load in a school building. The building official should calculate the occupant load in such buildings utilizing only the administrative, classroom and assembly areas. It is generally assumed that when corridors, restrooms and other miscellaneous spaces are occupied, they are occupied by the same people who are at other times occupying the primary use spaces.

The occupant load that can be expected in different buildings depends on two primary factors—the nature of the use of the building space and the amount of space devoted to that particular use. Different types of building uses have a variety of characteristics. Of primary importance is the density characteristic. Therefore, in calculating the occupant load of different uses, by means of the formula, the minimum number of persons that must be assumed to occupy a building or portion thereof is determined by dividing the area devoted to the use by that density characteristic or occupant load factor. The second column of Table 1004.1.1 prescribes the occupant load factor to be used with respective corresponding uses listed in the first column. The occupant load factor does not represent the amount of area that is required to be afforded each occupant. The IBC does not limit, except through the provisions of Section 1004.2, the maximum occupant load on an area basis. Rather, the occupant load factor is that unit of area for which there must be assumed to be at least one person present. For example, when the code prescribes an occupant load factor of 100 gross for business use, it is not saying that each person in an office must be provided with at least 100 square feet (9.29 m²) of working space. Rather, it is saying that, for egress purposes, at least one person must be assumed to be present for each 100 square feet (9.29 m²) of floor area in the business use. It is important to note that the floor area, both net and gross, includes counters and showcases in retail stores, furniture in dwellings and offices, equipment in hospitals and factories, and similar furnishings. The floor areas occupied by furniture, equipment and furnishings are taken into account in the occupant load factors listed in the table.

The numbers contained in the second column of Table 1004.1.1 represent those density factors that approximate the probable densities that can usually be expected in areas devoted to the respective functions listed. For this purpose, the occupant load factors are really a means of estimating the probable maximum density in the varying function areas. They have been developed over a period of years and, for the most part, have been found to consistently represent the densities that one might expect in building spaces devoted to the respective uses. See Application Examples 1004-1 and 1004-2 for two methods of occupant load determination.

The exception to this section allows for a reduction in the calculated design occupant load on a very limited case-by-case basis. The building official is granted authority for the discretionary approval of lesser design occupant loads than those established by calculation. Although the provision allows the building official to be accommodating by recognizing the merits of the specific project, its use should be limited to very unique situations such as extremely large manufacturing or warehousing operations. See Application Example 1004-3. Where the exception is enacted in order to reduce the occupant load, the building official will typically impose specific conditions to help ensure compliance. It is critical that the reasoning for the occupant load reduction be justified and documented.

1004.2 **Increased occupant load.** The provisions of Section 1004.1.1 specify the method to be used in determining the anticipated occupant load for areas without fixed seating. The occupant load determined by this method is the minimum number of persons for which the exiting system must be designed. The provisions do not, as previously pointed out, intend that the maximum permitted occupant load be regulated or controlled on a floor area basis other than in the manner described by this section.

The provisions of this section specify how the maximum permitted occupant load in a building or portion of the building is to be determined. Here, the approach is taken that the occupant load determined as previously provided may be increased where the entire egress system is adequate, in all of its parts, to accommodate the increased number. In no case, however, shall the occupant load be established using an occupant load factor of less than 7 square feet (0.65 m²) of floor space per person. See Application Example 1004-4.

**Application
Example
1004-1**

GIVEN: A building assembly area and business areas as shown.

DETERMINE: The design occupant load of the building.

SOLUTION: The occupant load is simply 282, the combination of the assembly and business spaces. It is not necessary to consider the corridor, toilet rooms and other small accessory spaces that serve the entire building. Note that within the office area itself, such circulation and accessory areas would be included in the calculation.

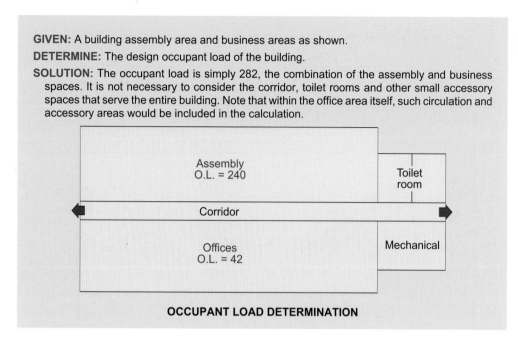

OCCUPANT LOAD DETERMINATION

**Application
Example
1004-2**

GIVEN: A 1600-square-foot conference room in a hotel.

DETERMINE: The design occupant load of the room.

SOLUTION: Because a variety of assembly activities can occur within the room, the use creating the largest occupant load would be evaluated.

  (1) Conference/seminar use with tables and chairs
    1 person per 15 sq ft = 106.67 = 106 occupants

  (2) Conference/seminar use with chairs only (auditorium-style seating)
    1 person per 7 sq ft for seating = 228.57 = 228

  Therefore, a design occupant load of 228 shall be designated. Note that other potential uses of the room (dining, receptions, dances, etc.) would also utilize these factors

OCCUPANT LOAD DETERMINATION

In order to analyze any increased occupant load, the building official must carefully review all aspects of the arrangement of space as well as the details of the total egress system, not only from the immediate space but continuously through all other building spaces that might intervene. In many cases, a diagram will be required indicating the approved furnishing and equipment layout.

Although it is critical that the building's means of egress system be designed to accommodate the increased occupant load, all other code requirements that are based upon the number of occupants must also be reviewed based upon the increased number. For example, if it is intended to increase the calculated occupant load of 258 in a Group A-3 conference facility to 340 occupants, all code requirements shall be applied based on the occupant load of 340. This would include the provisions of Section 903.2.1.3 that require an automatic sprinkler system, those of Section 907.2.1 mandating a manual fire alarm system, and the main exit requirements of Section 1028.2. Additional occupant-load based provisions that must be considered include those for plumbing fixtures and accessible parking spaces.

**Application
Example 1004-3**

GIVEN: A 210,000 square foot industrial building designed for final assembly of commercial aircraft.

DETERMINE: The design occupant load of the building.

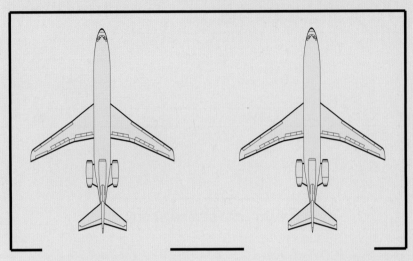

Based upon Table 1004.1.1, the design occupant load would be 2100.

$$\left(\frac{210{,}000 \text{ square feet}}{100 \text{ (factor for industrial areas)}} \right)$$

Where approved by the building official, a more realistic design occupant load is permitted based on the actual maximum number of occupants anticipated in the building.

OCCUPANT LOAD DETERMINATION

**Application
Example
1004-4**

GIVEN: A restaurant where the occupant load of the dining area is calculated at 135, based on Table 1004.1.1 (2025 sq ft/15). The restaurant's owner would like to establish a higher occupant load.

DETERMINE: The maximum permitted occupant load of the dining area.

SOLUTION: The absolute maximum occupant load per Section 1004.2 appears to be 289 (2025/7). However, it is obviously impossible for such an occupant load to safety occupy the space, even if adequate exit doors were provided. If tables and chairs were provided to seat 289 customers, there would be inadequate aisle accessways and aisles. If addition, the potential for egress obstruction would be significant. The appropriate maximum occupant load would be approved by the building official on a case-by-case basis, relying on the specific design of the space, the furniture and/or equipment layout, and the egress patterns created.

MAXIMUM OF OCCUPANT LOAD

1004.3   Posting of occupant load. Where a room or space is to be used as an assembly occupancy, this section requires the posting of a sign indicating the maximum permitted occupant load. This sign serves as a reminder to the occupants of the space, as well as building employees, that any larger occupant load would create an overcrowded condition. In order to be effective, the sign must be conspicuously located near the main exit of the room or space, and must be permanently maintained. An example of an occupant load sign is shown in Figure 1004-4. Where multiple uses causing varying occupant loads are anticipated, it is appropriate to designate the maximum occupant load for each use as shown in Figure 1004-5.

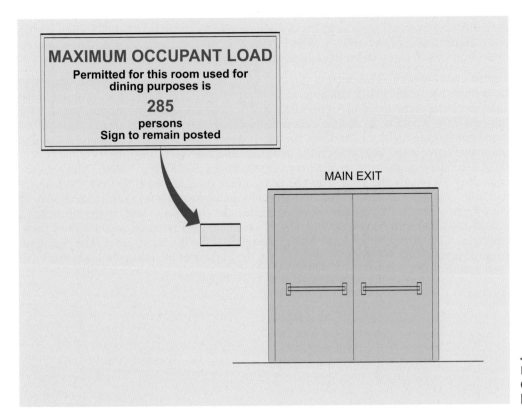

MAXIMUM OCCUPANT LOAD
Permitted for this room used for
dining purposes is
285
persons
Sign to remain posted

MAIN EXIT

Figure 1004-4
Occupant
load sign

MAXIMUM OCCUPANT LOAD

90 FOR CLASSROOM SEATING
WITH TABLES AND CHAIRS

192 FOR THEATER-STYLE SEATING
WITH CHAIRS ONLY

Figure 1004-5
Posting of
occupant load

Exiting from multiple levels. Where the means of egress includes multiple floors, the code **1004.4**
indicates that the capacity of the exitways be based upon the individual occupant loads of
each floor. In other words, the number of persons used to design the capacity of the stairway
is not based on any cumulative total number of persons, but rather on the required capacity
of the exits at each particular floor. In no case, however, shall exit capacity decrease along

the path of egress travel. A more in-depth discussion of this issue is found in Section 1005.1 and is illustrated in Figure 1005-2. Where exiting occurs from a mezzanine, the provisions of Section 1004.6 apply rather than those of this section.

1004.5 **Egress convergence.** This section directly addresses those situations where occupants from floors above and below converge at an intermediate level, rather than traveling in the same direction as discussed in Section 1004.4. The code states that the proper approach for this condition would be to add the occupant loads together; a method that is also utilized when converging aisles or corridors merge. In these cases, it can be assumed that the occupants arrive at the same point at the same time and, therefore, the capacity of the system must accommodate the sum of these converging floors. See Figure 1004-6. Although the code does not specify the approach to be taken where there are multiple floors both above and below the intermediate level of discharge, it is anticipated that the same methodology of convergence would be applied. For example, the occupant load assigned from a second-level basement would be added to the occupant load assigned from the third floor, resulting in a congregate occupant load converging at the discharge level. The minimum required egress capacity would be based upon the highest of the occupant loads that have been established.

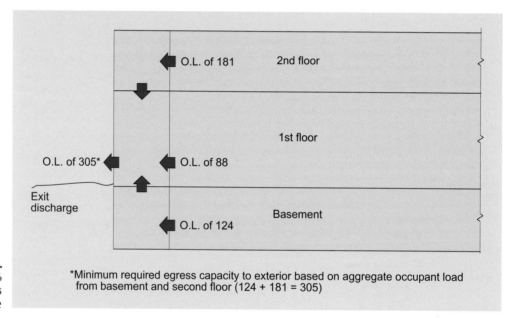

O.L. of 181   2nd floor

1st floor

O.L. of 305*   O.L. of 88

Exit discharge

Basement

O.L. of 124

Figure 1004-6
Egress convergence

*Minimum required egress capacity to exterior based on aggregate occupant load from basement and second floor (124 + 181 = 305)

1004.6 **Mezzanine levels.** Where a mezzanine level exits through a room on the level below, the occupant load of the mezzanine is to be added to the occupant load of the room below in order to determine the capacity. This is consistent with the concepts for outdoor areas in Section 1004.8 and exiting from accessory spaces through a primary space as addressed in Section 1004.1. It is important to note that the requirement for adding occupant loads is based solely on qualification of the upper level as a mezzanine. Where a floor level does not qualify as a mezzanine, the occupant load of that level is not to be added to the level below, even in those cases where an unenclosed stairway connects the two levels. This methodology is also unchanged where the unenclosed stairway serves as a required means of egress and passes through the level below.

1004.7 **Fixed seating.** The method of calculating occupant load discussed to this point—that is, the formula that divides an appropriate occupant load factor into the amount of space devoted to an occupancy—is used when dealing with building spaces without fixed seating. Where fixed seats are installed, the code specifies that the occupant load be determined simply by counting the number of seats. Although the code does not define the term *fixed seats*, it is intended by this term that the seats provided are, in fact, fastened in position, not easily movable, and maintained in those fixed positions on a more or less permanent basis. A primary example of a fixed-seat facility would be a performance theater. In determining the occupant load for this type of facility, only the number of fixed seats is utilized because the code also requires that the space occupied by aisles may not be used for any purpose other than aisles and, therefore, may not be used for accommodating additional persons. The aisle system through a fixed-seating facility is, in fact, the exiting system for those fixed seats and, as such, must remain unobstructed. Therefore, the code does not assume any occupancy in the areas that make up the aisles.

Under varying circumstances, fixed-seating assembly spaces may include other assembly areas capable of being occupied. Such areas could include wheelchair spaces, waiting areas, and/or standing room. Performance areas and similar spaces would also be evaluated and assigned an appropriate occupant load. The occupant load of all such areas must be added to that established for the fixed seating in the calculation of the total occupant load. An example is shown in Figure 1004-7. The inclusion of these additional occupiable areas provide for a more accurate determination of the potential number of persons who could occupy the room or space

In addition to those fixed-seating arrangements where the seating is provided by a chair-type seat, there will be those that utilize continuous seating surfaces such as benches and pews. When this type of seating is provided, it is necessary to assume at least one person present for each 18 inches (457 mm) of length of seating surface. Where seating is provided by use of booths, as is frequently done in restaurants, it must be assumed that there is a person present for each 24 inches (610 mm) of booth-seating surface. If the booth seating is curved, the code specifies that the booth-length be measured at the backrest of the seating booth. Where seating is provided without dividing arms, such as for benches and booths, it is reasonable to base the occupant load individually to each bench or booth. Similarly, it is appropriate to round the calculated occupant load down to the lower value, as this section only regulates each full 18 inches or 24 inches of width. See Application Example 1004-5.

Application Example 1004-5

GIVEN: A church sanctuary having pews as shown.

DETERMINE: The design occupant load of the sanctuary.

$$\frac{32 \text{ ft}}{1.5 \text{ ft}} = 21.33 = 21$$

(21 occupants/pew) × (24 pews) = 504 occupants

Total occupant load = 504 + occupant load of platform + occupant load of additional seating areas (wheelchair spaces, etc.)

Platform

12 pews at 32 ft   12 pews at 32 ft

Lobby

For SI: 1 inch = 25.4 mm.

MAXIMUM OF OCCUPANT LOAD

The method for determining occupant load in a small restaurant is depicted in Application Example 1004-6.

Application Example 1004-6

GIVEN: Information as shown in illustration.

DETERMINE: The occupant load for the small restaurant.

SOLUTION: Section 1004.7 states that where booths are used in dining areas, the occupant load shall be based on one person for each 24 inches (610 mm) of booth length. Based on this requirement, each 4-foot 6-inch booth would have an occupant load of four, or a total occupant load for the booth area of 16.

The fixed seats at the counter number eight, which, based on the first paragraph of Section 1004.7, would establish an occupant load of eight.

The open dining area (tables and chairs), having a floor area of 600 square feet (55.7 m^2) and using an occupant load factor of 15 square feet (1.39 m^2) per occupant, as set forth in Table 1004.1.1, would have an occupant load of 40.

The cooking area, having a floor area of 200 square feet (18.6 m^2) and using an occupant load factor of 200 square feet (18.6 m^2) per occupant, as set forth in Table 1004.1.1, would have an occupant load of one.

The total occupant load is as follows:

Booths 16
Counter 8
Dining area (tables and chairs). 40
Cooking area 1
Total Occupants 65

4 booths

Table and chair area 600 sq ft

Cooking area 200 sq ft

4' 6"

Counter

For SI: 1 inch = 25.4 mm, 1 foot = 304.8 mm, 1 square foot = 0.0929 m^2.

DETERMINATION OF OCCUPANT LOAD

1004.8 Outdoor areas. Occupiable yards, patios and courts that are used by occupants of the building must be provided with egress in a manner consistent with indoor areas. This provision is applicable to outdoor areas including building rooftops, that are occupied for a variety of uses, but primarily applied for outdoor dining at restaurants and cafés. For example, the occupant load of the dining area inside a café may require only one means of egress; however, the occupant load of the outdoor dining area would increase the café's total occupant load well over 50, thereby requiring two means of egress. The building official shall assign an occupant load in accordance with the anticipated use of outdoor areas. If an area's occupants need to pass through the building to exit, the cumulative total of the outdoor area and the building shall be used to determine the exiting requirements. This concept is consistent with the provisions of Section 1004.1.1. See Figure 1004-8.

Another example that is becoming more common is that of secured exterior areas serving nursing homes (Group I-2) or assisted living facilities (Groups I-1, R-3 or R-4). Such exterior spaces are often provided to enhance the livability of the facilities by providing outdoor spaces where the patients or residents are free to roam without individual supervision. When evaluating these spaces for egress purposes, there are several issues to consider. If the secured yard is provided with a means of egress independent of the facility, the gates must comply with all of the requirements for egress doors. Delayed egress devices installed in accordance with Section 1008.1.9.6 or 1008.1.9.7, as applicable, would be permitted as a means of addressing occupant safety for these areas. Without compliant gates, the means of egress must be designed for travel back through the facility. The facility must also egress independent of the secured yard unless all means of egress from the secured exterior area comply with the code.

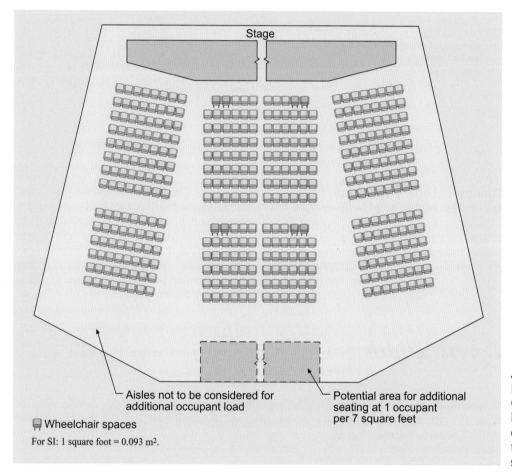

Aisles not to be considered for additional occupant load

Potential area for additional seating at 1 occupant per 7 square feet

Wheelchair spaces

For SI: 1 square foot = 0.093 m².

Figure 1004-7
Occupant load determination for fixed seating

The judgment of the building official is very important to the application of these provisions because the building official must determine exactly what occupant load should be considered, and to what degree the area is accessible and usable by the building occupants in order to establish the egress requirements. Some cases that will require judgment include large spaces that might have a very limited anticipated occupant load such as areas that are primarily for the service of the building. Where a portion of the required means of egress for the outdoor area is provided independent of travel back through the building, or where all of such required egress must pass through the building, the applicable provisions would be similar to those for travel through intervening spaces. The distribution of the occupant load from the outdoor area will depend on how many exits are required and how many means of egress paths are available.

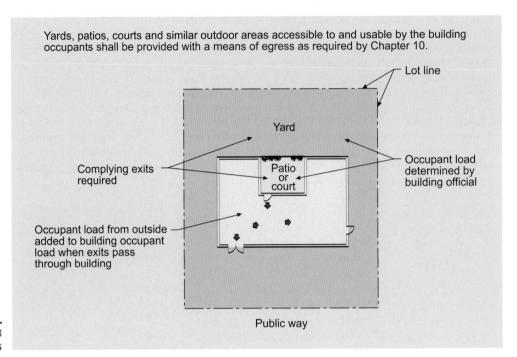

Yards, patios, courts and similar outdoor areas accessible to and usable by the building occupants shall be provided with a means of egress as required by Chapter 10.

Figure 1004-8
Outdoor areas

1004.9 **Multiple occupancies.** In many buildings there are two or more occupancies. Quite often, one or more of the egress paths from an individual occupancy will merge with egress paths from other occupancies. Within each individual occupancy, the means of egress shall be designed for that specific occupancy. However, where portions of the means of egress serve two or more different occupancies, the more restrictive requirements of the occupancies involved shall be met. An example might be where a sizable assembly occupancy shares an exit path with a business use. The more restrictive requirement for panic hardware would be applicable for any doors encountered along the shared egress route.

Section 1005 *Egress Width*

1005.1 **Minimum required egress width.** This section establishes the method for determining the capacity of the egress system, and more specifically, the minimum required capacity of each individual component in that system. It also establishes the method for distributing egress width to various egress paths where multiple means of egress are provided.

The formula for means of egress width is very succinct. It states that the total required width of the means of egress shall not be less than that obtained by multiplying the total occupant load served by an egress component by the appropriate factor as set forth in Section 1005.1, nor less than the minimum widths specified elsewhere in the code. It should be noted that the calculation of egress width in most assembly occupancies is not regulated by this section, but rather is governed by Section 1028.6. Where a Group A occupancy contains seats, tables, displays, equipment or other material, it must comply with the exiting provisions of Section 1028. In addition, where assembly rooms in an educational facility, such as a gymnasium and auditorium, are classified as a portion of the Group E occupancy, it is appropriate that the special egress provisions of Section 1028 also be applied to such rooms.

In designing the egress system, it is first necessary to determine the occupant load that must be accommodated through each individual portion of the system. Multiplying the occupant load by the appropriate factor will result in the minimum required width in inches (mm) necessary to accommodate the occupant load. An example is shown in Application Example 1005-1.

Application Example 1005-1

GIVEN: A home improvement center has an occupant load of 3,180. The building is fully sprinklered.

DETERMINE: The total required egress width from the building at the exit doors.

SOLUTION: For a Group M occupancy in a sprinklered building, Section 1005.1 indicates a width factor of 0.2 inches per occupant for egress components other than stairways.
3,180 (0.2) = 636 inches of clear door width to be distributed among available exit doors.

The formula stated in this section is the one used to determine if the egress system—and each individual component of that egress system—is adequate to accommodate the maximum permitted occupant load. To determine if the entire egress system is adequate or provides adequate capacity, it will be necessary to verify each individual element by the use of this formula.

In a given egress system, different components will afford different capacities. It is, of course, the capacity that is provided by the most restrictive component in the system that establishes the capacity of the overall system. It accomplishes nothing in increased exit capacity to arrange for a wide corridor that leads to a relatively narrow door. The capacity of this means of egress arrangement is established by the capacity of the doorway, and not by the apparent capacity of the wider corridor.

The occupant load served by the egress component is not the sole factor in determining the minimum required egress width. Egress components all have a minimum width established by other provisions in the code. This mandated width also cannot be reduced. It is the greater width established by these two methods (calculated width and component width) that becomes the minimum required egress width. See Application Example 1005-2. It should be noted that where the component width is the appropriate method for determining egress width, in many situations the required component widths may lessen along the egress path. For example, in an office building with an occupant load of 68 persons, a corridor required to be at least 44 inches (1118 mm) in width by Section 1018.2 may lead to an exit door with a minimum clear width of 32 inches (813 mm) as regulated by Section 1008.1.1. In this example, the minimum component widths for the corridor and the exit door provide for greater widths than required by Section 1005.1.

Application Example 1005-2

GIVEN: Various egress components and the occupant load served by each component.

DETERMINE: The minimum required width for each component in a Group B occupancy.

Egress Component	Occupant Load Served	Minimum Required Calculated Width	Minimum Required Component Width	Minimum Required Width
Aisle	64	64 (.2) = 12.8″	36″ Sec. 1017.2	36″
Corridor	130	130 (.2) = 26″	44″ Sec. 1018.2	44″
Stairway	200	200 (.3) = 60″	44″ Sec. 1009.1	60″
Door	180	180 (.2) = 36″	32″ Sec. 1008.1.1	36″

For SI: 1 inch = 25.4 mm.

It cannot be emphasized too strongly that when the code discusses width in terms of an egress system or component, it is referring to the clear unobstructed usable width afforded along the exit path by the individual components. Therefore, if it is determined, for example, that a means of egress must have a width of at least 3 feet (914 mm), it shall be arranged so that it is possible to pass a 36-inch-wide (914 mm) object through that egress path and each of its components. Unless the code specifically states that a projection is permitted into the required width, nothing may reduce the width of the component required to provide the necessary exit capacity.

Distribution of egress width. When the required egress capacity of the system has been determined by the use of the formula, and the required number of exit access doorways or exits has been determined in accordance with the provisions of Sections 1015.1 and 1021.1, the required egress capacity can be divided among the number of required means of egress. In fact, where additional complying means of egress are provided above the number required by the code, they too can be used for distribution purposes. The manner of distribution shall be such that the loss of any one means of egress will not reduce the available capacity to less than 50 percent of that required. Thus, after the loss of one means of egress at least one-half of the required capacity must be available. See Application Example 1005-3. It is the intent of this section that there be reasonable distribution of the egress capacity necessary to serve a given occupant load.

Application Example 1005-3

GIVEN: A retail store having 3 exits, with a total required exit width of 140 in.

DETERMINE: The manner in which the exit width may be distributed.

SOLUTION: Any distribution of egress width is acceptable provided that at least 50% of the required egress width (70 in.) is available after the width of the largest exit (66 in.) has been deducted from the total width provided (165 in.).

In this example, (165" − 66") ≥ 70", so it is an acceptable solution.

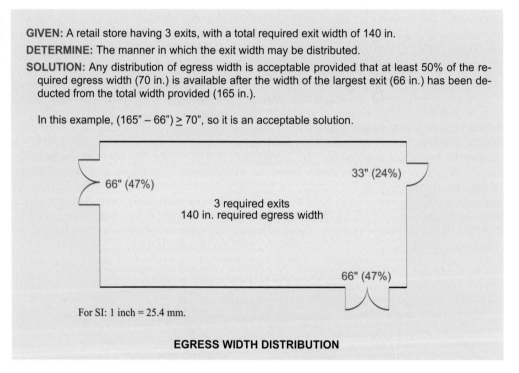

66" (47%)

33" (24%)

3 required exits
140 in. required egress width

66" (47%)

For SI: 1 inch = 25.4 mm.

EGRESS WIDTH DISTRIBUTION

The primary reason for requiring multiple egress paths is the fact that in a fire or other emergency, it could be possible that at least one of the routes will be unavailable or blocked by fire. If the egress capacity was concentrated at one exit point, it could very easily be that path that affords the greatest portion of the egress capacity that would be lost. The resulting limitation on the occupants' ability to exit a building or portion thereof is simply unacceptable. In addition, the presence of two or more means of egress allows for a distribution of occupants which should provide for more efficient and orderly egress under emergency conditions. A third benefit of egress distribution is the potential reduction in the distance occupants must travel to reach an exit or exit access doorway.

Maintaining width. As stated earlier, it is the width of the most restrictive component that establishes the capacity of the overall exit system. To ensure that a design does not reduce the capacity at some point throughout the remainder of the egress system, this section stipulates that the capacity may not be reduced and that the design must accommodate any accumulation of occupants along the path. Therefore, once the required width is determined for any story (in fact, from any room or other space), that required width must be maintained until the occupants have reached the public way or ultimate safe place. It is important to remember that because different factors are used to determine the width requirements for stairways than for all other components, it is really the capacity, not the width, of the egress system that is not permitted to be reduced. An aisle, door, corridor, passageway or other horizontal egress component located at the bottom of any stairway may generally be reduced in width from what is required for the stairway. See Application Example 1005-4.

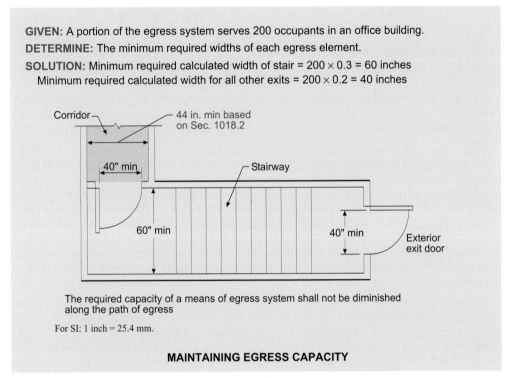

GIVEN: A portion of the egress system serves 200 occupants in an office building.

DETERMINE: The minimum required widths of each egress element.

SOLUTION: Minimum required calculated width of stair = 200 × 0.3 = 60 inches
Minimum required calculated width for all other exits = 200 × 0.2 = 40 inches

Corridor — 44 in. min based on Sec. 1018.2

40" min

Stairway

60" min

40" min

Exterior exit door

The required capacity of a means of egress system shall not be diminished along the path of egress

For SI: 1 inch = 25.4 mm.

MAINTAINING EGRESS CAPACITY

Application Example 1005-4

This is simply due to the width factor of the stairway being greater than the factor for other egress elements. It must again be mentioned that only the required width (capacity) needs to be maintained. The actual width of the means of egress may be reduced throughout the travel path as long as the required width is provided. A common application of this concept is shown in Figure 1005-1.

Width of exit stairways from a multistory building. In conjunction with Section 1004.4, these provisions also present the method for establishing the required widths of exit stairways in multilevel buildings. The basic provision is that as long as the occupants are traveling the same direction, there is no need to combine the loads from adjacent levels. The IBC assumes that in exiting multilevel buildings there will be occupants feeding into the exit stairs at various levels. This is accommodated by requiring stairways to be significantly wider than the paths of horizontal travel leading to them. As a result, flow rates should be equivalent. Also, the speed of exiting on a stair is substantially less than the speed of exiting on level or nearly level surfaces because of the forced reduction in normal stride, as the length of stride on a stair must coincide with the stair's run. A study of this difference shows that a stairway requires an increase in width of approximately 50 percent above that for horizontal travel. Therefore, in determining the required minimum width of a stair, the tributary occupant loads served by the stair are multiplied by the appropriate factor from

Section 1005.1 to yield the required minimum width measured in inches (mm). The end result should be checked to see if it complies with the minimum width requirements for stairways as governed by Section 1009.1.

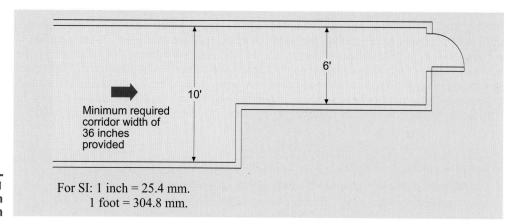

Minimum required corridor width of 36 inches provided

10'

6'

For SI: 1 inch = 25.4 mm.
1 foot = 304.8 mm.

Figure 1005-1
Reduction in actual width

At one time, building codes addressed a cascading effect when analyzing and determining the required width of stairs, but this concept is no longer applicable. The required width of a stair is calculated on a floor-level-by-floor-level application of the formula.

In the design of multistory buildings, it is quite common that different floor levels have different occupant loads. Thus, the occupant load calculated for each floor level must be considered. As a matter of fact, it is not uncommon for buildings to have assembly uses on the top floor. As a consequence of that configuration, it is entirely possible that the top floor of the building will have an occupant load greater than any other. Under such a condition, it will be necessary to calculate the required stairway width based only on the occupant load of the uppermost floor. This required width must be maintained through the successive levels until it serves a greater occupant load from a lower floor or until the occupants have reached the public way or ultimate safe place. Figure 1005-2 illustrates this requirement.

It should be noted that the same concept of determining stairway width that is discussed above applies where building occupants exit upward through the stairway. This is the situation in buildings with basements and sub-basements. Occupants of those below-grade floors must exit up the stairway, onto the landing on the ground floor, and then out through the exterior exit door from that landing. The largest capacity calculated from any floor level will be the controlling factor of the exit at ground level, as well as the exit doorway from the stairway. Of course, it is possible that occupants on the ground floor will exit through an exit enclosure. If so, the required width determined based on that condition may govern. However, the ground floor usually has adequate width of exits independent of any paths through the stairway enclosures. Thus, the occupant load of the ground floor is usually not an issue in the determination of the exterior exit doorway width from the stairway enclosure. In any case, the occupant load of the ground floor would not be added to any occupant load of floors above for determining egress width from the building. Where upper and lower floors converge at an intermediate level, see the discussion of Section 1004.5 on egress convergence.

Egress width in assembly spaces. As previously mentioned, where the provisions of Section 1028 are applicable for assembly occupancies, egress width shall be determined based upon such provisions. The requirements of Section 1028 regulate all assembly spaces containing seats, tables, displays, equipment or other material. Thus, it is typical that the egress width requirements of Section 1028.6 are to be followed rather than those of Section 1005.1. Section 1028.6 addresses egress width for assembly occupancies, based primarily on whether or not smoke-protected assembly seating is provided. A further discussion of this subject is provided in the analysis of Section 1028.

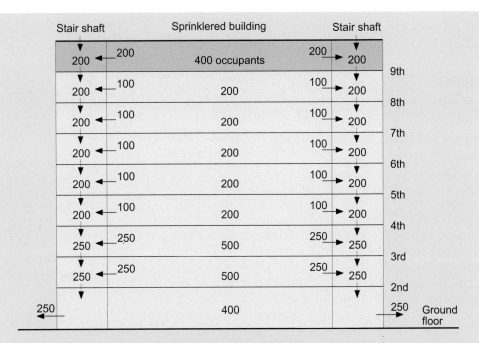

Note: First-story occupants may exit through stair enclosures or through independent exits. Width required at ground level is the same for each case.

Minimum stairway width required
4th through 9th stories: 200 × 0.3 = 60 inches
1st through 3rd stories: 250 × 0.3 = 75 inches

Minimum stair exit door to exterior clear width required
250 × 0.2 = 50 inches

For SI: 1 inch = 25.4 mm.

Figure 1005-2
Width of exits— multistory building

Door encroachment. Where doors open into the path of exit travel, they create obstructions that may slow or block egress. Therefore, the code limits the encroachment of doors into the required exit width. A door opening into a path of egress travel may not, during the course of its swing, reduce the width of the exit path by more than one-half of its required width. When fully open, the door may not project into the required width by more than 7 inches (178 mm). It is important to recognize that the provisions are based on the exitway's required width, not its actual width.

Again, as discussed in connection with Section 1008.1.6 for doors swinging over a stairway landing, it might be better to think of the permitted obstruction of a door during the course of its swing from a positive viewpoint. So stated, each door, when swinging into an egress path such as an aisle or a corridor, must leave unobstructed at least one-half of the required width of the path of travel during the entire course of its swing. At least one-half of the required width must always be available for use by the building occupants. When the door is in its fully open position, the required egress width, minus 7 inches (178 mm), must be available. See Figure 1005-3.

In applying the requirements for projections, the code imposes these limitations on a door-by-door basis. It is desirable that doors be arranged so as not to have two doors directly opposing each other on opposite sides of the exit path. Better design would avoid this arrangement. The intent of the code is that at least one-half of the required width of the exitway be available for use by the building occupant as illustrated in Figure 1005-4. The restrictions on door swing do not apply to doors within dwelling units and sleeping units of Groups R-2 and R-3.

1005.2

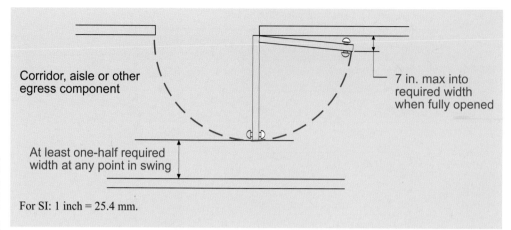

Corridor, aisle or other
egress component

At least one-half required
width at any point in swing

7 in. max into
required width
when fully opened

For SI: 1 inch = 25.4 mm.

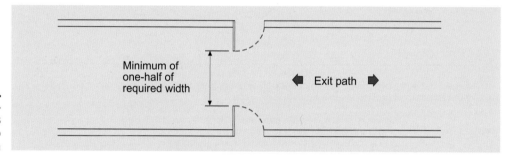

Minimum of
one-half of
required width

Exit path

Where nonstructural projections other than doors, such as trim and similar decorative features, extend into the required width of a means of egress component, the limit on their projection into the required width of egress components is $1^1/_2$ inches (38 mm) unless such projection is specifically prohibited by the code. In reviewing the provisions of aisles in Section 1017.1, corridors in Section 1018.3, and exit passageways in Section 1023.2, the required width must be unobstructed with only an exception for doors. Therefore, the allowance for a $1^1/_2$-inch (38 mm) nonstructural projection into the required egress width has limited, if any, application.

Section 1006 *Means of Egress*
Illumination

1006.1 **Illumination required.** In order for the exit system to afford a safe path of travel and for the building occupant to be able to negotiate the system, it is necessary that the entire egress system be provided with a certain minimum amount of lighting. Without such lighting, it would be impossible for building occupants to identify and follow the appropriate path of travel. The lack of adequate illumination would also be the cause of various other concerns, such as an increase in evacuation time, a greater potential for injuries during the egress process and most probably an increased level of panic to those individuals trying to exit the building. Therefore, the code requires that, except in a limited number of occupancies, the egress paths be illuminated throughout their entire length any time the building space served by the means of egress is occupied. The code intends that illumination be provided for those portions of the egress system that serve the parts of the building that are, in fact, occupied. Parts of the exiting system that would not be serving the occupants of the building need not, at that time, be illuminated. For obvious reasons, there are four exceptions that identify areas where continuous illumination during occupancy is not mandatory. Two exceptions address uses where sleeping is a common activity—dwelling units and sleeping units in Group R-1, R-2 and R-3 occupancies, and sleeping units in Group I occupancies. Another exception addresses utility structures designated as Group U, whereas a fourth exception exempts aisle accessways in Group A assembly uses.

The code emphasizes that the exit discharge—that portion of egress travel from the building to the public way—also be provided with adequate illumination. Although there are often numerous light sources at a building's exterior, such as lighting for landscaping, parking lots, city streets and adjacent buildings, it is important that the illumination be effective and reliable for use under this provision. It should also be noted that the requirements of this section are simply for general illumination of the entire egress system, and are not the higher level conditions for emergency lighting as mandated in Section 1006.3.

Illumination level. Such illumination must be capable of producing a light intensity of not **1006.2** less than 1 foot-candle (11 lux) at the walking surface throughout the entire path of travel through the system. An exception recognizes that such levels of illumination might interfere with presentations in such places as motion picture theaters and concert halls; therefore, the exception allows a reduction in such building uses to a level of not less than 0.2 foot-candle (2.15 lux). Such a reduced lighting level, however, is permitted only during a performance and would be brought up to the minimum 1 foot-candle (11 lux) level if a fire-alarm system was activated.

One foot-candle (11 lux) of light on a surface is not a great deal of light. It is probably not sufficient light to enable a person to read. However, it is sufficient light to allow a person passing through the exit system to distinguish objects and to identify obstructions in the actual path of travel. The light cast by a full moon on a clear night might approximate the 1 foot-candle (11 lux) light level. When the amount of light intensity is in doubt, it may be necessary to measure it with a light meter.

Illumination emergency power. Normally, the power for illumination of the egress path **1006.3** is provided by the premises' wiring system. However, where the potential life-safety hazard is sufficiently great, it is considered inadequate to solely provide the illumination of the exit system by such a system. In these cases, it is necessary that emergency power—a completely separate source of power—automatically provide illumination of the exit. In fundamental terms, separate sources of power are required in all occupancies in which two or more means of egress are required. Therefore, any space, area, room, corridor, exterior egress balcony or other portion of the egress system requiring access to at least two exits or exit access doors is to be provided with emergency lighting. Also included are exit stairways, both interior and exterior, and exterior landings at exit doors in buildings requiring a minimum of two means of egress. An example of this application is depicted in Figure 1006-1.

Where emergency power systems are required, they are to be supplied by storage batteries, unit equipment or an on-site generator. It is the intent that this power source be automatically available even in the event of the total failure of the public utility system. Therefore, a separate, independent source is generally required. Installation of the emergency power system is regulated by referring to the requirements of NFPA 70.

Performance of system. The initial illumination provided by emergency power along the **1006.4** path of egress at floor level shall average at least 1 foot-candle (11 lux), with a minimum level of illumination required to be 0.1 foot-candle (1 lux). Illumination levels are permitted to decline over the required 90-minute duration of the emergency power source to an average of 0.6 foot-candle (6 lux) with a minimum at any point of 0.06 foot-candle (0.6 lux). Recognizing the variation in light levels throughout the exit path, only the average illumination level needs be determined; however, an absolute minimum level of illumination must be attained. In no case shall the illumination uniformity ratio between the maximum light level and the minimum light level exceed 40 to 1.

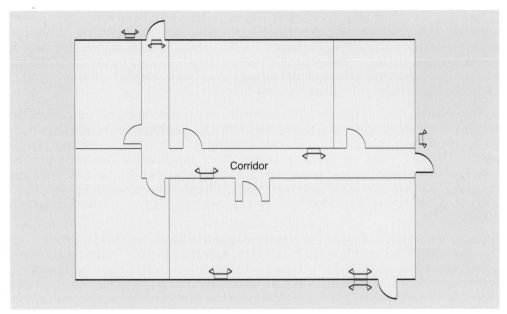

Figure 1006-1
Emergency
power for
egress
illustration

Section 1007 *Accessible Means of Egress*

1007.1   **Accessible means of egress required.** In addition to the access to buildings required by the provisions of Chapter 11, it is important that safe egress for physically disabled individuals is provided. Therefore, the code requires that accessible spaces be provided with accessible means of egress consisting of one or more of the following components:

1. Accessible routes complying with Section 1104.

2. Interior exit stairways complying with Sections 1007.3 and 1022.

3. Exterior exit stairways complying with Sections 1007.3 and 1026.

4. Elevators complying with Section 1007.4.

5. Platform lifts complying with Section 1007.5.

6. Horizontal exits complying with Section 1025.

7. Ramps complying with Section 1010.

8. Areas of refuge complying with Section 1007.6.

9. Where applicable by Exceptions 1 and 2, exterior areas for assisted rescue complying with Section 1007.7.

At least one accessible means of egress must be provided from all accessible spaces. Where more than one means of egress is required from any accessible space, at least two accessible means of egress are required. An example to illustrate this provision is a large department store requiring multiple exits. Although the number of exits from the store is addressed in Section 1021.1, only two accessible means of egress would be required from the accessible space. Therefore, the store might be required to provide three or more means of egress, but only two accessible means of egress need be provided.

Three exceptions reduce or eliminate the accessible means of egress requirements. Where an existing building is altered under the provisions of Section 3411, it is not necessary to provide any accessible means of egress. In addition, only one accessible means of egress is required from accessible mezzanines, as well as from sloped-floor or stepped assembly spaces with limited travel to all wheelchair spaces.

Continuity and components. As previously mentioned in the discussion of Section 1007.1, the code recognizes various accessible elements as components of an accessible means of egress. The accessible egress travel is required to extend beyond the building itself to the public way, unless an alternative means of protection is provided. If the egress route from the building to the public way is not accessible, it is acceptable to provide a complying exterior area of assisted rescue rather than create an accessible exit discharge path. Addressed further in the discussion of Section 1007.7, the exterior area of assisted rescue performs in much the same manner as an area of refuge inside the building. Exception 2 permits the use of an area of refuge on the interior of the building, as well as an exterior area of assisted rescue on the exterior side of an exit door served by an exterior exit stairway. The result of both exceptions is the same—the accessible means of egress may include an exterior area of assisted rescue or area of refuge as a portion of travel to the public way.

1007.2

Elevators required. Unlike the general provision found in Section 1003.7 that specifically prohibits considering an elevator as an approved means of egress, this section requires an elevator for rescue purposes under certain conditions. In buildings where a required accessible floor is four or more stories above or below a level of exit discharge, ramps and stairs cannot adequately serve as egress for individuals with a mobility impairment. Therefore, at least one complying elevator must be provided as an accessible means of egress. The elevator is not required to conform with Section 1007.4 in sprinklered buildings on those floors provided with a conforming ramp or horizontal exit.

1007.2.1

In the application of this provision, the second story of a typical building is considered the first story above the level of exit discharge. Accordingly, the building's fifth story is viewed as four stories above the level of exit discharge. See the discussion of Section 1002 for *level of exit discharge* and Figure 1007-1. Under such conditions, a minimum of one accessible means of egress must be a complying elevator.

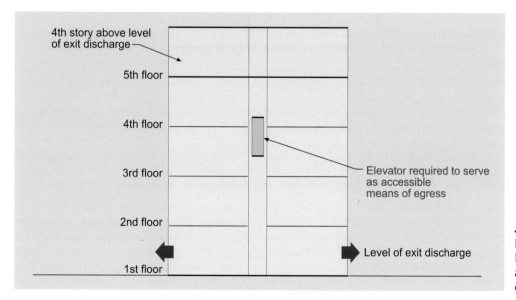

Figure 1007-1
Elevator as accessible means of egress

Exit stairways. Exit stairways are typically used as one or more of the required accessible means of egress in a multistory building. Increasing the width of stairs to 48 inches (1219 mm) between the handrails allows for the minimum amount of space needed to assist

1007.3

persons with disabilities in the event of a building evacuation. The provisions for an area of refuge or horizontal exit address the increased time needed for egress. A number of exceptions eliminate the general requirement for an area of refuge. In a fully sprinklered building, an area of refuge is not required at any complying unenclosed exit stairway that is considered part of the accessible means of egress. Areas of refuge are also not mandated at exit stairways serving open parking garages. In addition, areas of refuge are not mandated Group R-2 occupancies and in smoke-protected assembly seating areas as regulated by Section 1028.6.2. Another two exceptions remove the requirement for at least 48 inches (1219 mm) of clear width between handrails. The 48-inch (1219 mm) width is not mandated in sprinklered buildings, nor is it required for stairways accessed from a horizontal exit.

The most commonly-applied exception permits the omission of areas of refuge in buildings protected throughout by an automatic sprinkler system. The purpose of an area of refuge is to provide an area "where persons unable to use stairways can remain temporarily to await instructions or assistance during emergency evacuation." Much of the reasoning in exempting fully-sprinklered buildings from the requirement for areas of refuge comes from NISTIP 4770, a report issued by the National Institute of Standards and Technology (NIST) in 1992 titled *"Staging Areas for Persons with Mobility Impairments."* The primary conclusion of the report was that the operation of a properly designed sprinkler system eliminates the life threat to all occupants regardless of their individual abilities and can provide superior protection for persons with disabilities as compared to staging areas. The ability of a properly designed and operational automatic sprinkler system to control a fire at its point of origin and to limit production of toxic products to a level that is not life threatening to all occupants of the building, including persons with disabilities, eliminates the need for areas of refuge.

1007.4 Elevators. Although an elevator may be utilized as an accessible means of egress component in all multilevel facilities, it is only required as such in buildings regulated by Section 1007.2.1. Elevators utilized as accessible means of egress must comply with the operation and notification criteria of ASME A17.1, Section 2.27. In addition, standby power is required in order to maintain service during emergencies. The general requirement is that any elevator utilized as an accessible means of egress be accessed from an area of refuge or horizontal exit. However, the area of refuge and horizontal exit are not required in fully-sprinklered buildings, open parking garages, smoke-protected assembly seating areas, and where elevators are not required to be protected by shaft enclosures. Additional information is provided in the discussion of Section 1007.3. In such cases, the elevator is still considered an accessible means of egress when in compliance with the other criteria of this section.

1007.5 Platform lifts. Except in limited applications, a platform lift is specifically excluded as an acceptable element of a means of egress. The maintenance of the lift as well as the complexity and delay in utilizing a platform lift are considered substantial obstacles in providing acceptable means of egress for persons in wheelchairs. Section 1109.7 specifically sets forth the few instances where platform lifts are permitted for access and, with the exception of Item 10, egress purposes. Where a complying platform lift is utilized as an accessible means of egress component, it must be provided with standby power, much in the same manner as an elevator used for the same purpose.

1007.6 Areas of refuge. By definition, an area of refuge is an area "where persons unable to use stairways can remain temporarily to await instructions or assistance during emergency evacuation." Unfortunately, the term *temporary* is not defined, so a number of provisions are applied to an area of refuge to increase the level of protection for anyone using it. These provisions include a size large enough to accommodate wheelchairs without reducing exit width, smoke barriers designed to minimize the intrusion of smoke, two-way communications systems, and instructions on the use of the area under emergency conditions. The two-way communications system is intended to allow a user of the area of refuge to identify his or her location and needs to a central control point. Obviously, it is important that someone be available to answer the call for help when a two-way communications system is provided. The system shall have a timed automatic telephone dial-out capability to a monitoring location or 911 that can be used to notify the emergency services when the central control point is not constantly attended. Each area of refuge shall be identified by a sign with the international symbol of accessibility stating that it is an area of refuge.

The area of refuge must be located along the path of an accessible means of egress from any accessible spaces it may serve. The length of the travel path is limited to the maximum travel distance permitted for the occupancy in accordance with Section 1016.1. An area of refuge may be incorporated into an enlarged floor-level landing of an enclosed exit stairway. Access is also acceptable from an area of refuge directly to either an enclosed stairway complying with the provisions of Sections 1007.3 and 1022, or to an elevator complying with Section 1007.4.

Size. Each required area of refuge shall be sized to accommodate at least one wheelchair space not less than 30 inches by 48 inches (762 mm by 1219 mm). Where the occupant load of the refuge area and the areas served by the refuge area exceeds 200, additional wheelchair spaces must be provided. Because wheelchair spaces are not permitted to reduce the required exit width and should be located so as to not interfere with access to and use of the fire department hose connections and valves, the designer needs to consider access to fire protection equipment and exit width when placing wheelchair spaces in the area of refuge.

1007.6.1

Separation. The primary concern for individuals awaiting assistance in an area of refuge is the intrusion of smoke and toxic gases into the refuge area. Therefore, the code requires a physical separation between an area of refuge and the remainder of the building. The separation is to be a smoke barrier complying with the provisions of Section 710. The smoke barrier is not required where the area of refuge is located within a vertical exit enclosure. It is also permissible to create an area of refuge through the use of a horizontal exit.

1007.6.2

Two-way communication. Individuals awaiting assistance in an area of refuge must be provided with a communication means in order to contact a central control point. Where the central control point is not constantly attended, the area of refuge must be provided with a complying telephone with dial-out capability. Both audible and visible signals shall be provided.

1007.6.3

Exterior area for assisted rescue. An exterior space may be used as an exterior area for assisted rescue, provided it complies with this section. Sized in accordance with Section 1007.6.1, the exterior area for assisted rescue must be separated from the interior of the building in a manner similar to that addressed for egress courts and exterior exit stairways. Where the exterior area for assisted rescue is located within 10 feet (3048 mm) horizontally of the building's interior, a minimum 1-hour fire-resistance-rated wall with any openings protected for at least $^3/_4$ hour shall be provided from the ground level to a point at least 10 feet (3048 mm) above the floor level of the exterior area for assisted rescue. Such protection need only extend to the roof line if it is less than 10 feet (3048 mm) vertically above the floor level of the exterior area for assisted rescue. See Figure 1007-2. The required extent of an exterior area for assisted rescue is not described in the code, requiring an individual evaluation where the size of the exterior area for assisted rescue becomes quite large. For an outdoor space to be considered an exterior refuge area, it must be at least 50-percent open so that toxic gases and smoke do not accumulate. Where an exterior stairway serves the exterior area for assisted rescue, an adequate distance between the handrails is mandated. In addition, complying signage is required per Section 1007.9, Item 2 to identify the exterior area as an appropriate refuge location.

1007.7

Two-way communications. Unless provided in areas of refuge in multistory buildings, two-way communications systems must be located at the elevator landing of each accessible floor level other than the level of exit discharge. The system is intended to offer a means of communication to disabled individuals who need assistance during an emergency situation. Such a system can be useful not only in the event of a fire but also in the case of a natural or technological disaster by providing emergency responders with the location of individuals who will require assistance in being safely evacuated from floor levels above or below the discharge level. See Figure 1007-3.

1007.8

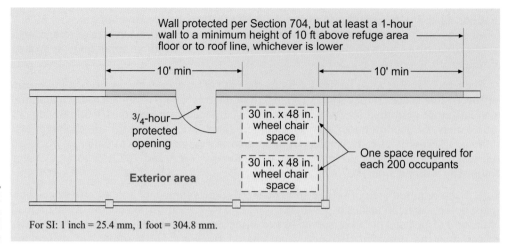

Figure 1007-2
Exterior area for assisted rescue

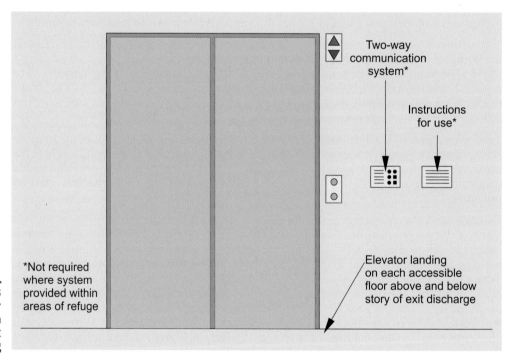

Figure 1007-3
Two-way communication system at elevator landing

The first exception exempts the requirement for locating the communication systems at the elevator landings where the building is provided with complying areas of refuge. Since areas of refuge are required by Section 1007.6.3 to be equipped with two-way communication systems, there is limited need to provide such additional systems at the elevator landings. However, where multistory buildings are not provided with areas of refuge, such as is the case with most sprinklered buildings, the installation of communications systems at the elevator landings is important to those individuals unable to negotiate egress stairways during an emergency. As a result, most sprinklered and nonsprinklered multistory buildings must be provided with the means for two-way communications at all accessible floor levels other than the level of exit discharge. A second exemption applies to floor levels that utilize exit ramps as vertical accessible means of egress elements. Where complying ramps are available for independent evacuation, such as occurs in a sports stadium, the two-way communications system is not required at the elevator landings. It should also be noted that multistory buildings without elevators, such as those identified in Section 1104.4, would not be regulated by this section.

The arrangement and design of the two-way communication system is specified in Section 1007.8.1. In addition to the required locations specified in Section 1007.6.3 for areas of refuge or Section 1007.8 for elevator landings, a communication device is also required to be located in a high-rise building's fire command center or at a central control point whose location is approved by the fire department. The term "central control point" is not a defined term. However, given the intent and function of the two-way communication system, a central control point is a location where an individual answers the call for assistance and either provides aid or requests aid for an impaired person. A central control point could be the lobby of a building constantly staffed by a security officer, a public safety answering point such as 9-1-1 center, a central supervising station, or possibly a nurses' station in a Group I occupancy. The key functions at the central control point are that an individual is always available to answer the call for assistance and can either provide assistance or is capable of requesting assistance. In addition, the communication system provides visual signals for the hearing impaired and audible signals to assist the vision impaired.

Guidance to the users of the two-way communication system is also specified. Operating instructions for the two-way communication system must be posted and the instructions are to include a means of identifying the physical location of the communication device. If a signal from a two-way communication system terminates to a public safety answering point, such as a fire department communication center, current 9-1-1 telephony technology only reports the address of the location of the emergency – it does not report a floor or area from the address reporting the emergency. The "identification of the location" posted adjacent to the communication system should ensure that most discrete location information can be provided to the central control point. This will aid emergency responders, especially in high-rise buildings or corporate campuses with multiple multistory structures.

Instructions. Instructions on the use of the area of refuge must be provided in those areas of refuge provided with a two-way communications system. The intent of the instructions is not only to provide directions on the use of the communications equipment, but also to alert the users as to other available means of egress. **1007.11**

Section 1008 *Doors, Gates and Turnstiles*

Doors. This section applies to doors or doorways that occur at any location in the means of egress system. The provision found in Section 1020.2 should also be noted insofar as it will require that at least one exterior door that meets the size requirements of Section 1008.1.1 be provided from every building used for human occupancy. As doors pose a potential obstruction to free and clear egress, they are highly regulated. **1008.1**

Additional doors. The IBC establishes criteria for all egress doors, including those that are not required by Chapter 10. Such additional egress doors must comply with all the provisions of Section 1008.1 for exit doors. Where the doors are installed for egress purposes, whether or not required, the building occupant would probably assume that they are a part of the means of egress system. Because the building occupant would then expect the door to provide a safe path from the space, it is imperative that such doors and doorways conform to all applicable code requirements.

Door identification. The primary gist of the provisions on door identification is that egress doors should be installed so that they are readily recognized as egress doors and are not confused with the surrounding construction or finish materials. It is important that they be easily discernible as doors provided for egress purposes. The corollary of this requirement is that exit doors should not be concealed. In other words, they should not be covered with drapes or decorations, nor should they be provided with mirrors or any other material or be arranged in a way that could confuse the building occupants seeking an exit.

1008.1.1 **Size of doors.** Every door used for egress purposes must comply with the width and height provisions of this section. It specifies that every required means of egress door opening be of such a size as to provide a clear width of at least 32 inches (813 mm) as illustrated in Figure 1008-1, with a minimum door height of 80 inches (2032 mm). Again, the code requires that the net dimension of clear width be provided by the exit component. Thus, when a swinging door is opened to an angle of 90 degrees (1.57 rad), it must provide a net unobstructed width of not less than 32 inches (813 mm) and permit the passage of a 32-inch-wide (813 mm) object, unless a projection into the required width is permitted. Where a pair of doors is installed without a mullion, only one of the two leaves is required to meet the 32-inch (813 mm) requirement. As a final requirement, a minimum $41^1/_2$-inch (1054 mm) means of egress doorway width to facilitate the movement of beds is mandated for Group I-2 occupancies.

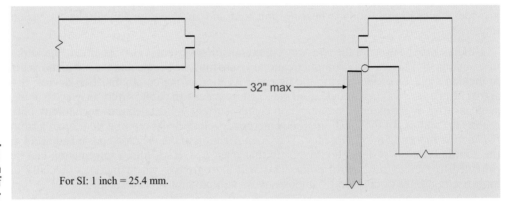

Figure 1008-1
Minimum clear width of egress door

For SI: 1 inch = 25.4 mm.

A number of reductions to the 32-inch (813 mm) door width requirement are found in the exceptions to this section. In Group I-3 occupancies, door openings to resident sleeping units need only have a clear width of at least 28 inches (711 mm). Accessible door openings within Type B dwelling units are permitted a minimum clear width of $31^3/_4$ inches (806 mm).

In addition, minimum door opening widths are totally unregulated in the following locations:

1. Door openings in Group R-2 and R-3 occupancies that are not part of the required means of egress.

2. Storage closet doors where the closet is less than 10 square feet (0.93 m²) in area.

3. Door leaves in revolving doors that comply with Section 1008.1.4.1.

4. In other than Group R-1, interior egress doors within a dwelling unit or sleeping unit not required to be an Accessible unit, Type A unit or Type B unit.

Throughout the rest of the code, the intent in specifying dimensions is to provide only minimum width. This particular section is at variance with that general approach, insofar as it limits the maximum width of any single swinging door leaf in a required egress doorway. As shown in Figure 1008-2, no such leaf may exceed 48 inches (1219 mm) in width. The reason for this is that doors often do not receive the maintenance necessary to ensure their continued proper operation. The issue being addressed is that door leaves should be reasonably limited in width because wide doors require substantially greater maintenance to ensure reasonable opening effort, and this maintenance is not often provided. The limitation on maximum door width does not apply to complying revolving doors, nor doors in Group R-2 or R-3 occupancies that are not a portion of the required means of egress.

The required clear width of door openings shall be maintained up to a height of at least 34 inches (864 mm) above the floor or ground. Projections may then encroach up to 4 inches

(102 mm) for a height between 34 inches (864 mm) and 80 inches (2032 mm). See Figure 1008-3. The maximum 4-inch (102 mm) limitation is based partially upon those accessibility provisions regarding protruding objects. Its application allows for the intrusion of panic hardware, or similar door opening devices, into the required clear width. At a height of 80 inches (2032 mm) or more above the walking surface, the projection is not regulated.

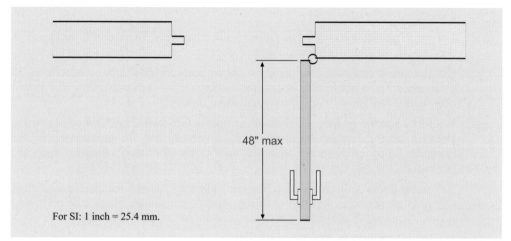

For SI: 1 inch = 25.4 mm.

Figure 1008-2
**Maximum
door leaf size**

Although the general requirement for door height is a minimum of 80 inches (2032 mm), the exceptions to Sections 1003.3.1 and 1008.1.1.1 permit door closers and stops to encroach into this clear height, provided a headroom clearance of at least 78 inches (1981 mm) is maintained. Door openings at least 78 inches (1981 mm) in height must be provided within a dwelling unit or sleeping unit. A minimum height of 76 inches (1930 mm) is required for all exterior door openings in dwelling units other than the required exit door. Only Exception 5 is applicable to the height reduction at a required exit door; therefore, required means of egress door openings must have a minimum height of 80 inches (2032 mm), 78 inches (1980 mm) at closers and stops, in other than a dwelling or sleeping unit.

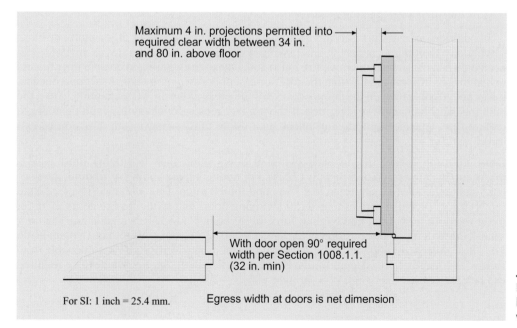

Maximum 4 in. projections permitted into required clear width between 34 in. and 80 in. above floor

With door open 90° required width per Section 1008.1.1. (32 in. min)

For SI: 1 inch = 25.4 mm.   Egress width at doors is net dimension

Figure 1008-3
**Egress door
width**

1008.1.2 **Door swing.** This section requires that every egress door, with exceptions, be of the pivoted or side-hinged swinging type. In most instances, it is necessary that the egress door encountered be of a type that is familiar to the user and easily operated. Therefore, swinging doors are required under all but the following conditions:

1. Private garages, office, factory and storage areas, and similar spaces where the occupant load of the area served by the doors does not exceed 10. Because of the limitation in occupant load and potential hazard, other types of egress doors are considered acceptable.

2. Detention facilities classified as Group I-3 occupancies. The security necessary in this type of use calls for special types of doors.

3. Critical care or intensive care patient room within suites of health care facilities. In these areas, it is preferable to utilize sliding glass doors to allow for visual observation and the efficient movement of equipment.

4. Within or serving an individual dwelling unit in Group R-2 and R-3 occupancies. Because of the limited occupant loads involved, and the familiarity of the occupants with the doors encountered, door types other than swinging doors are permitted.

5. Revolving doors conforming with Section 1008.1.4.1, where installed in other than Group H occupancies. In other than hazardous occupancies, the use of revolving doors is acceptable subject to the special conditions as set forth in the code.

6. Horizontal sliding doors complying with Section 1008.1.4.3, where installed in other than Group H occupancies. Conditions for the use of horizontal sliding doors make them equivalent to other doors used in egress situations.

7. Power-operated doors in compliance with Section 1008.1.4.2. Safeguards provided for power-operated doors create an acceptable level of safety.

8. Bathroom doors within individual sleeping units of Group R-1 occupancies. It is often beneficial to utilize sliding pocket doors to provide access to hotel bathrooms, particularly in those required to be accessible. Conflicts often occur between door swings and the required clearances for plumbing fixtures or clear floor space required at accessible doors.

9. The use of a typical horizontal sliding door that is operated manually, such as a "pocket" door or a sliding "patio" door, is deemed acceptable in those instances where the occupant load served by the door is very low.

In addition, an exit door serving an area with an occupant load of 50 or more, or those serving any Group H occupancy, shall swing with the flow of egress travel. In 1942, 492 people died in the Cocoanut Grove fire in Boston. One of the significant contributing factors to that loss of life was the fact that the exterior exit doors swung inward. As a consequence, it was not possible to open the doors because of the press of the crowd attempting to exit the building. This incident was identified as the primary reason for changing building codes to require that, under certain circumstances, exit doors must swing in the direction of exit travel.

1008.1.3 **Door opening force.** Interior side-swinging doors other than fire doors must have a maximum opening force of 5 pounds (22 N). For doors that are sliding or folding, the door latch shall release when subjected to a 15-pound (66 N) force. This limitation to a 15-pound (66 N) force level also applies to exterior swinging doors and swinging fire doors. In order to set the door in motion, a maximum force of 30 pounds (132 N) is mandated. The door shall swing to a fully open position when subjected to a force not greater than 15 pounds (66 N). These forces are applied to the latch side of the door. Most doors are openable with forces less than these maximum limits. However, when in doubt, the actual force required can be easily measured by use of a spring scale.

Special doors. Based on the provisions in Section 1008.1.2, the code generally requires that **1008.1.4**
doors in exiting systems be of the pivoted or side-hinged swinging type. In this section, five
different types of doors are identified that may be utilized under very specific conditions.

Revolving doors. Exception 5 of Section 1008.1.2 permits the installation and use of **1008.1.4.1**
revolving doors in all occupancies other than Group H when complying with this section.
Revolving doors are finding an ongoing use in buildings. Where once used primarily in cold
climates, they are now being installed in all regions, primarily as an energy-conservation
measure. However, it is not permissible to use revolving doors to supply more than 50
percent of the required egress capacity, nor be assigned a capacity greater than 50 persons.
Where used, the door must be an approved revolving door and comply with the specific
requirements listed.

When revolving doors are installed, they must be of a type where the door leaves will
collapse under opposing pressures to a book-fold position with the resulting parallel exit
paths providing at least 36 inches (914 mm) of aggregate width. Location of the door in
relationship to the foot or top of stairs or escalators is regulated, as is the maximum number
of revolutions per minute. At least one conforming exit door shall be located in close
proximity to the revolving door. In such an arrangement, the adjacent swinging door can be
used to satisfy exit capacity requirements. The maximum force levels required to collapse a
revolving door vary based upon whether or not the door is to be used as an egress
component.

Power-operated doors. Power-operated sliding or swinging doors are often used at the **1008.1.4.2**
main entry of a building, particularly in mercantile and business occupancies. The same
doors are also typically an important aspect in the overall exiting system for the building.
There are a number of different types of doors that utilize a power source to open a door or
assist in the manual operation of the door. This may include doors with a
photoelectric-actuated mechanism to open the door upon the approach of a person, or doors
with power-assisted manual operation. Where such doors are utilized as a portion of the
means of egress, they must be installed in accordance with this section. The main criterion
concerns the capability of the door being opened manually in the event of a power failure.
Essentially, doors shall have the capability of swinging, and they must be designed and
installed to break away from any position in the opening, and swing to the fully open
position when an opening force not exceeding 50 pounds (222 N) is applied at the normal
push-plate location. Doors that are fully power-operated must comply with BHMA
A156.10, whereas BHMA A156.19 applies to power-assisted and low-energy doors.

Horizontal sliding doors. Utilized as smoke or fire separation elements, these doors are **1008.1.4.3**
normally in a fully open position and hidden from view. Closing only under specific
conditions, they typically are part of an elevator lobby or similar protected area. Eight
provisions are identified in the IBC that regulate the use of horizontal sliding doors as a
component of means of egress. Fundamental to the use of horizontal sliding doors is that
manual operation of the normally power-operated doors must be possible in the event of a
power failure, and no special or complex effort or knowledge should be necessary to open
the doors from either side.

Access-controlled egress doors. Security concerns have prompted the need to provide **1008.1.4.4**
controlled access at entrance doors to buildings or tenant spaces of certain occupancies.
Therefore, this section permits the use of an approved entrance and egress access-control
system at the main entrance of a Group A, B, E, I-2, M, R-1 or R-2 occupancy.
Access-controlled egress doors will typically be locked from the exterior at all times.
Locking from the egress side is also permitted under certain conditions. In order to ensure
that the door is fully operable during a fire incident or other emergency, a number of criteria
have been developed. For the most part, when a problem situation is identified, the door
operates in a manner like any other egress door. Activation of the building fire alarm,
automatic sprinkler or fire detection system will automatically unlock the door. Unlocking
must also be possible from a location adjacent to the door, or occur when there is loss of
power to the access-control system. Where assembly, business, educational or mercantile
buildings utilize this type of door at the main entrance, it must remain unlocked from the

inside at any time the building is open to the general public. Very limited in application, this provision is potentially useful where there is a desire to have the entrance doors locked from the inside of the building during those periods of time where the public is not present.

1008.1.4.5 **Security grilles.** Because of the concern of exit doors being obstructed or even completely unusable, the use of security grilles is strictly regulated. By their nature, security grilles are difficult to operate under emergency conditions. Used frequently at the main entrances to retail sales tenants in a covered mall building, such grilles are also permitted in other Group M occupancies as well as Groups B, F and S. Security grilles, either horizontal sliding or vertical, are only permitted at the main entrance/exit. During periods of time when the space is occupied, including those times where occupied by employees only, the grilles must be openable from the inside without the use of a key or special knowledge. They must be secured in the fully open position during those times where the space is occupied by the general public. Where two or more means of egress are required from the space, a maximum of 50 percent of the exits or exit access doorways are to be equipped with security grilles.

1008.1.5 **Floor elevation.** The purpose of this section is to avoid any surprises to the person passing through a door opening, such as a change in floor level. Therefore, it is necessary that a floor or landing be provided on each side of a doorway. It is intended that such a floor or landing should be at the same elevation on both sides of the door. A permitted variation up to $^1/_2$ inch (12.7 mm) is permitted because of differences in finish materials. See Figure 1008-4. Landings are required to be level, except exterior landings may have a slope of not more than $^1/_4$ inch per foot (6.4 mm per m) for drainage purposes.

Exceptions for individual dwelling units of Group R-2 and R-3 occupancies. An allowance is provided for individual dwelling units where it is permissible to open a door at the top step of an interior flight of stairs, provided the door does not swing out over the top step. The reason for permitting this type of arrangement in dwelling units is that as a building occupant approaches such a door from the nonstairway side, he or she must back away from the door in order to open it. This creates the need for a minimum landing to be traversed before the occupant can proceed to step down onto the stairs. In this situation, with minimal occupant load and familiarity with the unusual condition, the opening may occur at the top of the stairs, but the door must swing toward the person descending the stairs. In an ascending situation, the stair user should have little difficulty in opening the door while standing on the stair treads, insofar as the door swings in the direction of travel. See Figure 1008-5. Also, in such occupancies it is permissible when screen doors or storm doors are installed, especially on the same jamb as the egress door, to swing them over stairs or landings.

In a Type B dwelling unit as provided in Chapter 11, a maximum drop-off of 4 inches (102 mm) is permitted between the floor level of the interior of the unit down to an exterior deck, patio or balcony. This limited elevation change is consistent with the level of accessibility provided throughout a Type B dwelling unit.

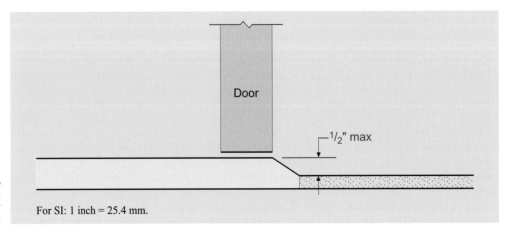

Figure 1008-4
Floor elevation

Door

$^1/_2$" max

For SI: 1 inch = 25.4 mm.

Exception for exterior doors. Reference is made to the first exception of Section 1003.5 regarding exterior doors in Group F, H, R-2, R-3, S and U occupancies. Where such exterior doors are not required to be accessible, a single step having a maximum riser height of 7 inches (178 mm) is permitted. The reference to Section 1020.2 is extraneous information and it is not applicable in regard to the exception.

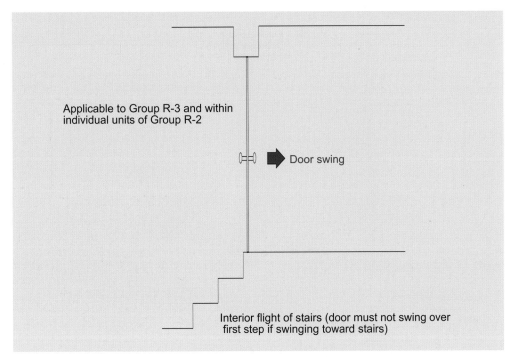

Applicable to Group R-3 and within individual units of Group R-2

Door swing

Interior flight of stairs (door must not swing over first step if swinging toward stairs)

Figure 1008-5
Floor level at doors

Landings at doors. This section contains the dimensional criteria for landings. It deals only with those landings where there is a door installed in conjunction with the landing. Landings at stairways and ramps are regulated by Sections 1009.4 and 1010.6. **1008.1.6**

Required width of landings. The minimum required width of a landing is determined by the width of the stairway or the width of the doorway it serves. Figure 1008-6 depicts these relationships. The requirement is that the minimum width of the landing be at least equal to the width of the stair or the width of the door, whichever is greater. The code is concerned that doors opening onto landings should not obstruct the path of travel on the landing. In this regard, the code establishes two limitations. The first states that when doors open onto landings, they shall not project into the required dimension of the landing by more than 7 inches (178 mm) when the door is in the fully open position. Second, whenever the landing serves an occupant load of 50 or more, doors may not reduce the dimension of the landing to less than one-half its required width during the course of their swing. Stated from the positive direction, it requires that doors swinging over landings must leave at least one-half of the required width of the landing unobstructed. Although the obstruction of one-half of the required width of the landing might seem excessive, it must be remembered that when the door is creating such an obstruction, it is in a position where it is free to swing and the obstruction is not fixed in place. These requirements are illustrated in Figure 1008-7.

Required length of landings. In addition to the width requirements, landings must generally have a length of at least 44 inches (1118 mm) measured in the direction of travel. Where the landing serves Group R-3 and U occupancies, as well as landings within individual units of Group R-2, the length need only be 36 inches (914 mm). These code requirements are illustrated in Figure 1008-8.

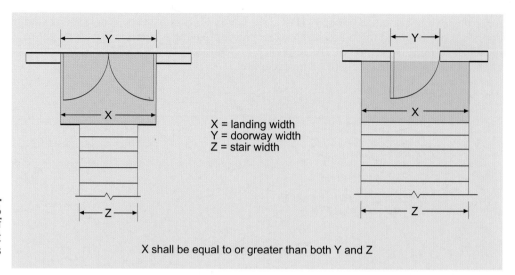

Figure 1008-6
Width of landings at doors

X = landing width
Y = doorway width
Z = stair width

X shall be equal to or greater than both Y and Z

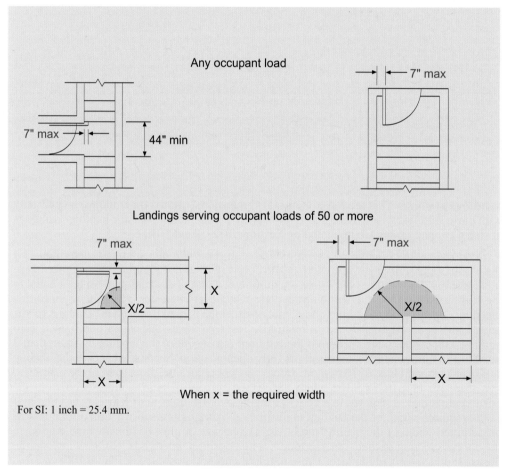

Any occupant load

7" max   44" min   7" max

Landings serving occupant loads of 50 or more

7" max   X/2   X

7" max   X/2   X

When x = the required width

For SI: 1 inch = 25.4 mm.

Figure 1008-7
Doors at landings

It should be noted that these minimum dimensions for landings in both width and length will be modified by the provisions in Chapter 11 where the door or doorway is a portion of the accessible route of travel.

1008.1.7   **Thresholds.** Raised thresholds make using doors more difficult for people with disabilities. In addition, thresholds with abrupt level changes present a tripping hazard. As a general rule, raised thresholds should be eliminated wherever possible. Where thresholds are

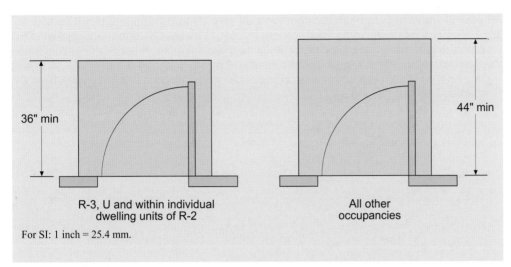

R-3, U and within individual
dwelling units of R-2

All other
occupancies

For SI: 1 inch = 25.4 mm.

Figure 1008-8
**Length of
landings at
doors**

provided at doorways, it is necessary to limit their height to provide easy access through the doorway. Changes in floor level and raised thresholds are limited to $^1/_2$ inch (12.7 mm) in height. Where raised thresholds or changes in floor level exceed $^1/_4$ inch (6.4 mm), the transition shall be achieved with a beveled slope of 1 unit vertical to 2 units horizontal (1:2) or flatter. See Figures 1008-9 and 1008-10. For a sliding door serving a dwelling unit, a maximum $^3/_4$-inch (19.1 mm) threshold is permitted. The threshold height at exterior doors may be increased to a maximum height of $7^3/_4$ inches (197 mm) in Group R-2 and R-3 occupancies, but only where such doors are not a required means of egress door, do not swing over the landing or step, are not on an accessible route and are not part of an Accessible unit, Type A unit or Type B unit.

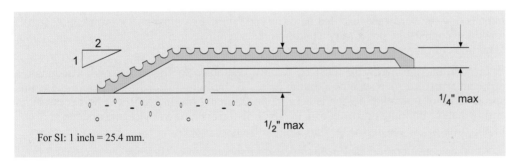

For SI: 1 inch = 25.4 mm.

Figure 1008-9
**Threshold
height**

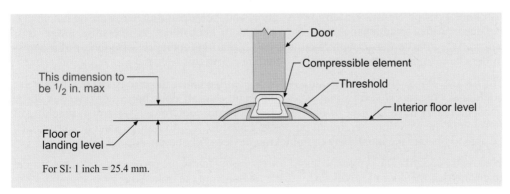

This dimension to
be $^1/_2$ in. max

Door

Compressible element

Threshold

Interior floor level

Floor or
landing level

For SI: 1 inch = 25.4 mm.

Figure 1008-10
**Threshold
height**

1008.1.8 **Door arrangement.** Adequate space must be provided between doors in a series to allow for ease of movement through the doorways. In other than dwelling units not considered Type A units, a minimum clear floor space of at least 48 inches (1219 mm) in length is sized for a wheelchair user to negotiate through the door arrangement. Where a door swings into the floor space, the clear length shall be increased by the width of the door. As shown in Figure 1008-11, doors in a series must swing in the same direction or swing away from the floor space between the doors.

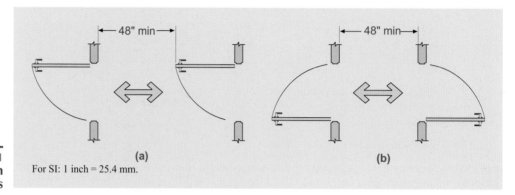

Figure 1008-11
Two doors in series

For SI: 1 inch = 25.4 mm.

(a)   (b)

1008.1.9 **Door operations.** This section, along with Section 1008.1.2, is particularly focused on the concept of ensuring that everything in the path of travel through the exit system, particularly doors, shall be under the control of and operable by the person seeking egress. Therefore, as a general statement, this section states that all doors in the egress system are required to be operable from the side from which egress is sought, without the need of a key or any special knowledge or special effort. If a key or special knowledge or effort is required, in all probability the door could not be readily openable by many building occupants. Such things as combination locks are also prohibited on doors in exiting systems. Essentially, the code intends that the hardware installed be of a type familiar to most users—something that is readily recognizable under any condition of visibility, including darkness, and under conditions of fire or any other emergency.

In addition, the hardware must be readily operable. At times, one will encounter a different type of device such as a thumb turn. The building official must determine if this special type of operating device is acceptable. In many instances, it will be necessary to ensure that the building occupant can, in fact, grip the operating device and operate it. Some thumb turns are so small that they are quite difficult to operate.

Another consideration that needs to be remembered in evaluating the acceptability of operating hardware is the fact that this hardware is going to be in place and in use over a substantial period of time. Unfortunately, doors and their operating hardware do not always get the constant maintenance that they should to keep them in operating order. It is imperative, in accordance with Section 3401.2, that the operation of doors in the egress system be maintained continuously in compliance with this section.

1008.1.9.1 **Hardware.** Where the door is required to be accessible under the provisions of Chapter 11, additional criteria come into play. It is important that all door hardware intended to be encountered by the door user be of a type that does not require tight grasping, pinching or twisting of the wrist for operational purposes. Individuals with limited hand dexterity must be able to operate the unlatching or unlocking device without any special effort.

1008.1.9.2 **Hardware height.** The proper height of the operating hardware on an egress door is critical to ensure that the door user can easily reach and operate the unlatching or opening device. Therefore, operating devices such as handles, pulls, latches and locks are to be installed at no less than 34 inches (864 mm) and no more than 48 inches (1219 mm) above the finished floor. These limitations are not imposed on locks installed strictly for security reasons.

Locks and latches. In regard to locking and latching door hardware, the IBC allows five **1008.1.9.3** significant exceptions to the general provision. In allowing these exceptions, it permits certain locking conditions that would appear to conflict with the basic requirement. However, in allowing these exceptions, the code often imposes certain compensating safeguards when the exit doors are to be locked or are provided with noncomplying hardware. It is the intent of the code that if the conditions are satisfied, the arrangement then essentially affords an equivalent level of safety as would be provided if the door were, in fact, readily openable at all times without the use of a key or any special knowledge or special effort.

1. This exception for locking hardware primarily concerns Group I-3 occupancies such as prisons, jails, correctional facilities, reformatories and similar uses, where individuals are restrained or secured. By their nature, it is necessary that the occupants in these facilities be limited in their movement. Therefore, the IBC allows alternate locks or safety devices when it is necessary to forcibly restrain the personal liberties of the inmates or patients.

   Where a portion of a building is used for the restraint or security of five or fewer individuals, it is not considered a Group I-3 occupancy. Rather, it is anticipated that the secured area would simply be classified the same as the occupancy to which it is accessory. For example, up to five individuals could be restricted in an area such as a merchandise viewing room for customers in a jewelry store. In such situations, the allowance provided by Item 1 would be applicable for the egress door from the viewing room. A lock or latch that would prevent the expected unlocking or unlatching operation of the egress door would be acceptable.

2. The provisions of this exception apply to the main exterior door or doors in Group A (having an occupant load of 300 or less), B, F, M and S occupancies, and in all places of religious worship. It permits the main exterior entrance/exit doors to be equipped with a key-operated locking device if several conditions are satisfied. In the occupancies listed, it is reasonable to assume that if the building is occupied, the main entrance/exit will, in all probability, be unlocked. The first condition states that the locking device be readily distinguishable as locked. The use of an indicator integral to the locking device may assist in determining when an unsafe condition is present. A second condition requires that there be a sign that is readily visible, permanently maintained and located on or adjacent to the door. This sign is required to read THIS DOOR TO REMAIN UNLOCKED WHEN BUILDING IS OCCUPIED. The letters must be at least 1 inch (25 mm) in height and placed on a contrasting background. Both of these requirements are for legibility. Although the language of the sign appears to apply without exception, there will be obvious situations where it should not be taken literally. For example, where an employee is working after hours, and may be the lone occupant in the building, it is not the intent that the main doors remain in an unlocked position.

   Obviously, the sign or the presence of the sign is not going to ensure that the door is unlocked. However, it does advise the occupant that whenever the space is occupied, the law does require in the interest of reasonable fire safety that the door be unlocked. In the event the door is not unlocked, the occupant is advised that his or her life may be at risk. The occupant should seek to alleviate that situation. The limitation imposed by this provision is typically applied only when the public is involved. For example, it is not reasonable to assume that the subject door be unlocked during a time period the occupancy is limited to a janitor or other personnel. Note that the use of this exception may be revoked by the building official for due cause.

3. Where egress doors are used in pairs, it is anticipated that each leaf in the pair of doors should be provided with its own operating hardware. This exception permits a special arrangement, however, when the pair of doors is equipped with automatic flush bolts that are designed so that the act of releasing one of the leaves of the pair releases both leaves. It is critical that the door leaf be provided with the automatic

flush bolts and have no door handle or other surface-mounted hardware. To ensure the immediate and reliable operation of the pair of doors, the unlatching of either leaf in the pair must be accomplished by not more than one operation.

4. The fourth exception refers to exit doors from individual dwelling or sleeping units of Group R occupancies. The general requirement of Section 1008.1.9.5 essentially prohibits the use of dead bolts or other security devices that would be installed in addition to the complying door hardware. This exception does permit, however, the use of a dead bolt, a security chain or a night latch when the occupant load is 10 or less, on the condition that the device be openable from the inside without the use of a key or any special tool. It follows from the basic requirement of this section, however, that the device must not require any undue effort in order to unlatch the door and gain egress.

5. The listed test procedures for a fire door include the disabling of the door operation mechanism. This exception clarifies that once the minimum elevated temperature has disabled the door's unlatching mechanism, the resulting prevention of the door's operation is acceptable.

1008.1.9.4   **Bolt locks.** This section specifically prohibits the use of manually operated flush bolts or surface bolts insofar as these clearly do not conform with the intent of Section 1008.1.9. The use of such latching and/or locking hardware on means of egress doors is typically prohibited due to the inability of users to quickly identify and operate such devices under emergency conditions. Exception 1 permits the use of these types of locking devices on doors in individual dwelling units and sleeping units, provided the doors are not required for egress purposes. The second exception recognizes that in certain instances, doorway widths are dictated by the need to pass equipment through the openings. As a consequence, doorways, such as those to a storage room or equipment room, are frequently larger than would be required for exiting purposes alone. Therefore, where that is the case for a normally unoccupied space, manually operated bolts may be used on the inactive leaf. As the space is not normally occupied, this exception presents no significant hazard to life safety. The other side of this coin, however, is that any door leaf that is part of the required egress width must comply with all the requirements that apply to exit doors.

Pairs of doors are often desired in commercial occupancies to allow for the movement of equipment and machinery. Automatic flush bolts and removable center posts can be easily damaged and difficult to maintain in areas of frequent door usage. Exceptions 3 and 4, applicable to Group B, F and S occupancies, address building functionality while maintaining high degree of occupant safety. In these moderate-hazard occupancies, the occupants are typically very familiar with the building and the means of egress system. It is expected that they are aware of the operational limits of the inactive door leaf and efficiently utilize the active leaf. In both exceptions, it is mandated that the inactive leaf not be provided with any hardware, such as levers or panic devices that might cause the user to assume the door is an active egress door. The presence of door hardware on the active leaf will provide the necessary expectation to the building occupants, as occupants will naturally approach the active leaf having the appropriate hardware. If the building is sprinklered throughout with an NFPA 13 system, there is no limit on occupant load assigned to the pair of egress doors other than that based upon the clear width of the active leaf. The inactive leaf cannot be assumed to provide for any required egress width. This ensures that the occupants have a fully complying door available for means of egress purposes. The mandate that the building be fully sprinklered further enhances occupant safety and provides recognition from a general perspective of the value of a fire suppression system. An allowance is also provided for pairs of doors that serve relatively small numbers of people in the occupancies identified. The limit of 49 occupants is consistent with various other means of egress requirements that allow for a reduced level of protection where the occupant load does not exceed 50. Under this exception, the building is not required to be sprinklered. An overview of the provisions is shown in Figure 1008-12.

A fifth exception addresses those patient room doors in Group I-2 occupancies where additional clear width is needed to allow for the efficient movement of patients and equipment. Where a pair of doors is utilized to provide the increased opening size, self-latching edge- or surface-mounted bolts may be installed on the inactive leaf. In order to distinguish that the inactive leaf is not an egress door, no lever device or other type of operating hardware is permitted. In addition, the inactive leaf cannot account for any of the minimum required egress width of 41 $\frac{1}{2}$ inches (1054 mm).

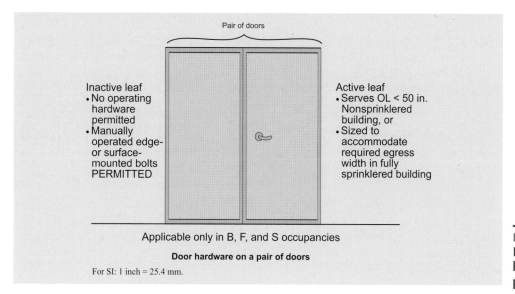

Applicable only in B, F, and S occupancies

Door hardware on a pair of doors

For SI: 1 inch = 25.4 mm.

Figure 1008-12
Door hardware on a pair of doors

Unlatching. The installation of multiple devices, or hardware requiring multiple operations, is inappropriate as well. Special effort and special knowledge is often necessary to open a door where more than one operation is required to unlock or unlatch the door. As a result, the multiple operations will typically result in an unacceptable delay in the egress efforts. Four exceptions set forth applications where multiple unlatching or unlocking operations are acceptable.

1008.1.9.5

Special locking arrangements in Group I-2. Many Group I-2 facilities house dementia and Alzheimer's patients. In order to balance the needs of the facility with the life safety of the occupants, the limitations on locking devices in these types of uses must allow for a safe and secure environment for these patients within the means of egress concepts of the code. Locks are permitted on means of egress doors that serve Group I-2 patients whose movement is restrained provided a number of conditions are met. Many of the conditions are similar to those set forth in Section 1008.1.9.7 for delayed egress locks. The building must be fully sprinklered or provided with an approved smoke or heat detection system. The doors must unlock upon actuation of the sprinkler or fire detection system, the loss of power to the lock or lock mechanism, or by a signal from the nursing station or other approved location. The staff must also have the means to unlock the doors when necessary.

1008.1.9.6

Where patients with mental disabilities are housed, it is often necessary that they be restrained or contained for their own safety. In such cases, the level of restraint must be maintained even if the fire protection systems are activated or the power to the lock fails. However, it is still important that the emergency preparedness plan be developed and the clinical staff has the ability to monitor and enable the evacuation. See Figure 1008-13.

Delayed egress locks. The building code provides for a degree of security to egress doors serving all occupancies other than Groups A, E and H. This section allows the use of a door that has an approved, listed egress-control device with a built-in time delay under specific conditions. These devices were introduced in the code to resolve the problem of an exit door being illegally blocked by building operators desperate to stop the theft of merchandise through unsupervised, secondary exits. Institutional and residential occupancies are included because it is perceived that they also have security problems that need to be

1008.1.9.7

addressed. The devices are sometimes needed in nursing homes or group-care facilities where facility operators must restrict patient egress while still maintaining viable exit systems.

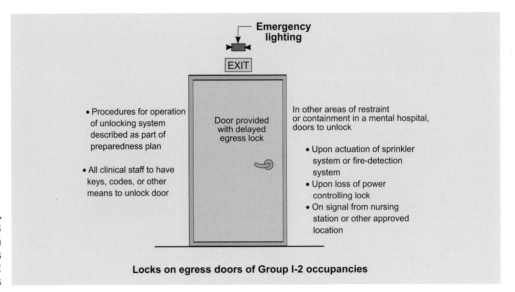

Figure 1008-13
Locks on
egress doors
of Group I-2
occupancies

It must be emphasized that under the conditions imposed by this section, and within the reliability of the automatic systems required, there will be no delay whatsoever at the exit in an actual fire emergency—the door will be immediately openable.

Delayed egress locking conditions. Several conditions should be emphasized at the outset:

1. Such devices may be used only in connection with the specifically listed occupancies.

2. The entire building in which the delayed egress locking system is installed must be completely protected throughout by either an approved automatic sprinkler system or an approved automatic smoke- or heat-detection system.

3. A building occupant shall not be required to pass through more than one door equipped with a delayed egress lock before entering an exit.

4. The door shall unlock in compliance with the following criteria: The device must immediately and automatically deactivate on activation of the sprinkler system or detection system, and on the loss of electrical power to the egress lock. There must also be a way of manually deactivating the device by the operation of a signal from the fire command center where provided. Where the operating device is activated, it must initiate an irreversible process that will cause the delayed egress lock to deactivate whenever a manual force of not more than 15 pounds (66 N) is applied for a minimum period of 1 second to the operating hardware. The irreversible process must achieve the deactivation of the device within a time period of not more than 15 seconds from the time the operating hardware is originally activated. Where approved by the building official, a delay of not more than 30 seconds is permitted. Upon activation of the operating hardware, an audible signal shall be initiated at the door so that the person attempting to exit the building will be aware that the irreversible process has been started.

A sign must be installed on the door above and within 12 inches (305 mm) of the operating hardware so that the person seeking egress can be informed as to the type and nature of the egress lock. The sign must read PUSH UNTIL ALARM SOUNDS. DOOR

CAN BE OPENED IN 15 (30) SECONDS. An additional requirement requires emergency lighting to be provided at the door where the delayed egress lock is used.

Delayed egress lock reactivation. The code emphasizes that, regardless of the means of deactivation, relocking of the device shall only be by manual means at the door. This requirement ensures that to relock the delayed egress lock, someone must go to the door itself, verify that the emergency no longer exists, and only then relock the door by manual means.

Electromagnetically locked egress doors. As a general rule, means of egress door hardware must be operable by manual operation to provide for occupant control of the egress system. Locking devices are typically prohibited as they can interfere or prevent efficient egress through the door during an emergency situation. However, owner concerns that must be considered sometimes require a greater degree of security. In specific occupancies, doors in the means of egress are permitted to be electromagnetically locked if equipped with listed hardware that incorporates a built-in switch that interrupts the power supply to the electromagnetic lock and unlocks the door. The use of this type of locking system provides for a greater degree of security than that offered by other methods addressed in the code, including delayed egress locking systems and access control egress systems.

1008.1.9.8

The allowance for electronically locked egress doors is limited to low and moderate hazard occupancies where security can be a major concern. The listed hardware that incorporates a built-in switch has been tested by UL under Special Locking Arrangements FWAX.SA6635. When the occupant prepares to use the door hardware, the method of operating the hardware must be obvious, even under poor lighting conditions. The operation shall be accomplished through the use of a single hand. This is consistent with the general requirement that the door be readily openable without the use of special knowledge or effort. The unlocking of the door must occur immediately on the operation of the hardware by interrupting the power supply to the electromagnetic lock. As an additional safeguard, the loss of power to the hardware shall automatically unlock the door.

These special provisions can only be utilized where the requirements of IBC Section 1008.1.10 regarding panic hardware are not applicable. Therefore, in Group A and E occupancies having occupant loads of 50 or more, electromagnetic locks are not permitted since the door hardware must comply with the requirements for panic hardware.

Stairway doors. The general requirement for interior stairway doors is that they be openable from both sides without the use of a key or special knowledge or effort. Such conditions allow for immediate access from the stairway enclosure to the adjacent floor area for emergency responders. In addition, in the unlikely event the stairwell becomes untenable during evacuation procedures, occupants may reenter a floor level as an alternative means of egress. However, three exceptions are provided to modify this requirement. Those doors that provide egress from the stair enclosure, discharging to the exterior or an egress component leading to the exterior, are permitted to be locked only on the side opposite the direction of egress travel. In high-rise buildings, stairway doors, other than exit discharge doors, may be locked from the stairway side. Under these conditions, such doors must be capable of being unlocked simultaneously without unlatching upon a signal from the fire command center. Although this exception is not limited to use in high-rise buildings, it is most commonly applied in such situations. When used in buildings that are not considered high-rise, the criteria of Section 403.5.3 may also be applied. Doors are also permitted to be locked from the side opposite the egress side for stairways serving four or fewer stories, where emergency personnel have the ability to simultaneously unlock the door. This action must be accomplished upon a signal from a single interior location at the building's main entrance, with the specific location for the actuating device likely approved by the fire code official. Where the building is provided with a fire command center, the signal must be actuated from within the center. It is important that the unlocking signal not deactivate the latching devices of the stairway doors. The doors must remain latched in order to maintain their integrity as fire door assemblies.

1008.1.9.10

1008.1.10 **Panic and fire exit hardware.** Basically, panic hardware is an unlatching device that will operate even during panic situations, so that the weight of the crowd against the door will cause the device to unlatch. This provision of the code, in harmony with the need to swing the door in the direction of egress travel, is intended to prevent the type of disaster experienced in the Cocoanut Grove Fire in Boston in 1942. When a panic hardware device is installed on a door leaf, the press of the crowd, which prevented the opening of the door in the Cocoanut Grove Fire, will ensure the automatic opening of the door.

Where panic hardware is provided, it is necessary that the activating member of the device extend for at least one-half of the width of the door leaf. This minimum length ensures that the unlatching operation will take place when one or more individuals impact the door. In addition, the device must be arranged so that a horizontal force not exceeding 15 pounds (66 N), when applied in the direction of exit travel, will unlatch the door.

Where required. Because of the large concentration of people in an assembly occupancy that may reach an exit door at about the same time during an emergency, the Group A occupancy is one of those occupancy groups that requires the installation of panic hardware listed in accordance with UL 305. Such hardware is also required in Group E occupancies for essentially the same reason. Therefore, in these two occupancies, any egress door provided with a latching or locking device that serves an area having an occupant load of 50 or more must be provided with panic hardware. In addition, in all Group H occupancies, panic hardware is required on every egress door regardless of the occupant load served. The potential life-safety hazard in these Group H occupancies is such that when it is necessary to evacuate this type of use, exiting from such spaces must be almost immediate. To facilitate the rapid escape from these occupancies, egress doors shall not be provided with a latch or lock unless it is panic hardware. Again, it should be noted that if a door is used in a situation where panic hardware might otherwise be required, it is not necessary to install panic hardware if the door has no means for locking or latching. If the door is free to swing at all times, there is no need to install panic hardware to overcome a lock or latch. Panic hardware is also not mandated on those doors considered as the main exit in a Group A occupancy having an occupant load of 300 or less where the provisions of Section 1008.1.9.3, Item 2, are applied permitting the use of key-operated locking devices.

Panic hardware on balanced doors. Special care must be taken when installing panic hardware on balanced doors. In this instance, push-pad-type panic hardware is to be used, and it must be installed on the one-half of the door width nearest the latch side to avoid locating the pad too close to the pivot point.

Section 1008.1.2 requires that exit doors should be side-hinged or pivoted swinging doors. The typical pivoted door has its top and bottom pivot points located near the edge of the door frame opposite the latch side. Balanced doors are nothing more than a specialized type of pivoted door in which the pivot point is located some distance inboard from the door edge, creating a counter-balancing effect. The length of the panic bar is limited for balanced doors because the door cannot be opened if the opening force is applied too close to the pivot point or beyond. Limiting the panic bar length to one-half the door width ensures that those who use the door will apply opening pressure at a distance sufficiently removed from the pivot point to allow the door to open. See Figure 1008-14.

1008.2 **Gates.** This section serves as a reminder that gates within the means of egress system must comply with all of the requirements for doors. The single exception, applicable only to fences and walls surrounding stadiums, overrides the general door requirement that the doors must swing, and that the width of any leaf cannot exceed 4 feet (1219 mm).

1008.2.1 **Stadiums.** Facilities such as stadiums may be enclosed by fencing or similar enclosures. The requirement for panic hardware does not apply to gates through such enclosures, provided that the gates are under constant and immediate supervision while the stadium is occupied. However, there must be a safe dispersal area of a size sufficient to accommodate the occupant load of the stadium based on 3 square feet (0.28 m^2) per person, and located between the stadium and the fence or other enclosure. Such a dispersal area must not be less than 50 feet (15 240 mm) from the stadium it serves.

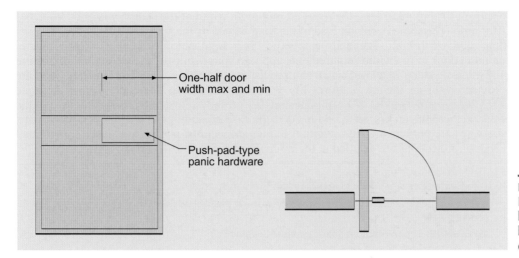

One-half door width max and min

Push-pad-type panic hardware

Figure 1008-14
Panic hardware on balanced doors

Turnstiles. In order to address safety concerns that are due to the use of turnstiles or similar devices that may be placed along the path of exit travel, this section regulates their use when located in a manner to restrict travel to a single direction. The general rule requires that each turnstile be credited with a maximum capacity of 50 occupants, and then only when specific conditions are met. When primary power is lost, the device shall turn freely in the direction of egress travel. Release shall also occur upon manual operation by an employee in the area. When determining the overall egress capacity, turnstiles may only be considered for up to 50 percent of the required capacity. Each device is limited to 39 inches (991 mm) in height and must have a minimum clear width of $16^{1}/_{2}$ inches (419 mm) at and below the height of 39 inches (991 mm). Above the 39-inch (991 mm) height, a clear minimum width of 22 inches (559 mm) is necessary. Obviously, variations in these requirements are necessary where the turnstile is located along an accessible route of travel. Where turnstiles exceed 39 inches (991 mm) in height, they are regulated in a manner consistent with revolving doors. **1008.3**

To address the concern for use of such devices in large occupancies, Section 1008.3.2 requires a side-hinged swinging door for devices other than portable turnstiles. Required at a point where the occupant load served exceeds 300, a swinging door must be located within 50 feet (15 240 mm) of each turnstile. Portable turnstiles are designed to be moved out of the way for large occupancies such as sporting events.

Section 1009 *Stairways*

In order to apply the requirements of this section on stairways in an appropriate manner, the scope of the provisions must be determined. The definition of a stairway is critical for this determination. Found in Section 1002.1, the definition consists of two parts. First, a stair is considered a change of elevation accomplished by one or more risers. Second, one or more flights of such stairs make up a stairway, along with any landings and platforms that connect to them. Based on these two definitions, a single step would also be considered a stairway under the IBC. Both interior and exterior stairways are regulated by the provisions of Section 1009.

Stairway width. The provisions concerning width of stairways are analogous to the provisions relating to corridors discussed in Section 1018.2. If the stair is subject to use by a sufficiently large occupant load, the minimum required width of the stair is determined by using the formula stated in Section 1005.1. Otherwise, the minimum stairway width cannot be less than the width established by this section. In general terms, the minimum required width of any stair must be at least 44 inches (1118 mm). In the event the stairway serves an **1009.1**

occupant load of 49 or less, the required minimum width of the stairway is only 36 inches (914 mm). The entire occupant load of the floor served by the stairway is considered, rather than divided by all available stairways. See Application Example 1009-1. Other modifications to the width requirements apply to spiral stairways as addressed in Section 1009.9, aisle stairs regulated under the provisions of Section 1028 and stairways that are provided with an incline platform lift or stairway chairlift. Generally, when the code specifies a required width of a component in the egress system, it intends that width to be the clear, net, usable unobstructed width. As is the case of landings in conjunction with doors, stairways are permitted to have certain limited projections per Section 1009.5. Permitted handrail projections are detailed in the discussion on Section 1012.8.

Application Example 1009-1

GIVEN: A two-story building with two stairways serving the second floor. The occupant load of the second floor is 68.

DETERMINE: The minimum required width of the stairway.

SOLUTION: Because both of the stairways are required to serve the floor, the minimum required width of each stairway is 44 inches. The occupied load is not divided (as in the case of distributing calculated width) between the two stairways.

For SI: 1 inch = 25.4 mm.

1009.2 **Headroom.** A minimum headroom clearance of 6 feet, 8 inches (2032 mm) is required in connection with every stairway. Such required clearances shall be measured vertically from the leading edge of the treads to the lowest projection of any construction, piping, fixture or other object above the stairs, and shall be maintained for the full width of the stairway and landing. See Figure 1009-1. This specific height requirement overrides the general means of egress ceiling height requirement found in Section 1003.2 and is modified for spiral stairways by Section 1009.9.

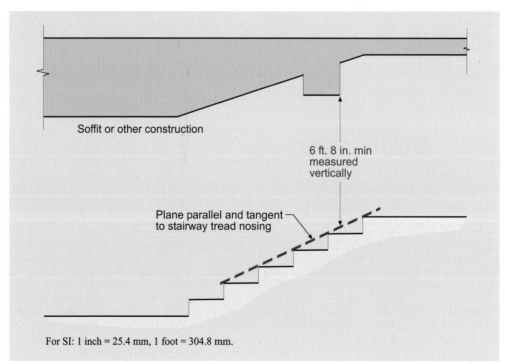

Soffit or other construction

6 ft. 8 in. min measured vertically

Plane parallel and tangent to stairway tread nosing

Figure 1009-1
Stairway headroom clearance

For SI: 1 inch = 25.4 mm, 1 foot = 304.8 mm.

Riser height and tread depth. This section provides for a maximum riser height of 7 **1009.4.2** inches (178 mm), a minimum riser of 4 inches (102 mm) and a minimum tread run of 11 inches (279 mm) for each step on any stairway. These limiting dimensions are identified in Figure 1009-2. Variations in the requirements for treads and risers apply to alternating tread devices, spiral stairways and stairs serving as aisles in assembly seating areas. These variations are discussed elsewhere in this chapter. Another exception allows $7^3/_4$ inches (197 mm) maximum and 10 inches (254 mm) minimum for rise and run, respectively, for stairways in Group R-3 occupancies, within dwelling units in Group R-2 occupancies and in Group U occupancies accessory to a Group R-2 or R-3 dwelling unit.

The 7-inch (178 mm) rise and 11-inch (279 mm) run figures for the steps are based primarily on safety in descending the stairs and are the result of much research. Probably at no prior time in the history of codes has the proportionate of stairs enjoyed a better foundation in research.

As one descends a stairway, balance is essential for safety. Therefore, the tread run must be of such a dimension as to permit the user to balance comfortably on the ball of the foot. The appropriate combination of riser height and tread run provides the proper geometry to enable the user to accomplish the necessary balance to descend the stairway with

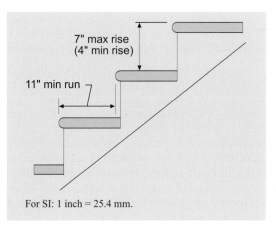

For SI: 1 inch = 25.4 mm.

Figure 1009-2
Rise and run

reasonable safety. Consistent with the importance of the tread dimension, the method of measurement of the tread is expressly stated. Specifically, tread depth (or run) is that distance measured horizontally between vertical planes passing through the foremost projections of adjacent treads. As such, the tread dimension is the net gain in the run of the stair. Tread dimension is measured in this manner because any tread surface underneath the overhang of a sloping riser or nosing on the tread above is not available to the person descending the stair. Because descending is the more critical direction, proper dimension of the tread is of paramount importance.

Studies of people traveling on stairways have shown that probably the greatest hazard on a stair is the user. Inattention has been identified as the single factor producing the greatest number of missteps, accidents and injuries. Inattention frequently results from the user being overly familiar with the stair and its surroundings. It often results from a variety of distractions. It is critical to stair safety that the stair user be attentive to the stair, although attentiveness cannot be codified or dictated. However, stair design and geometry, which usually trigger human error, can be controlled.

Curved stairways, along with spiral stairways, represent somewhat of an exception to what is normally considered a traditional stairway. Where the typical stairway is required to have treads of a consistent and uniform size and shape, these two stairs may have different dimensional characteristics from adjacent treads and vary from one end of the tread to the other. Alternating tread devices are another type of stair whose design is inconsistent with that of a typical stairway. The use of these stairways as a portion of the means of egress system varies; however, the only one of these stairs that may be used as a part of the means of egress in all occupancies and locations is the curved stair. These stairs are addressed in Sections 1009.8 through 1009.10.

Winder treads. Winders are defined as treads that have nonparallel edges. Spiral stairways **1009.4.3** and circular stairways are both examples of stair systems that utilize winders throughout the stair run and are regulated independent of the general requirements. However, another use of winders is in combination with a straight run of stairs. In such a condition, permitted by

Exception 2 to Section 1009.3.2, the winders may provide for a change in direction without the need for a landing.

The use of winders is particularly advantageous in those buildings where floor area is at a premium. Not surprisingly, the most common usage of winder treads is within dwelling units. In fact, dwelling units are the only locations where winders can be utilized in the means of egress. In all cases, winders must be a minimum of 10 inches (254 mm) in tread depth at the walkline with a minimum winder tread depth of 6 inches (152 mm).

1009.4.4 **Dimensional uniformity.** A significant safety factor relative to stairways is the uniformity of risers and treads in any flight of stairs. The section of a stairway leading from one landing to the next is defined as a flight of stairs. It is very important that any variation that would interfere with the rhythm of the stair user be avoided. Although it is true that adequate attention to the use of the stair can compensate for substantial variations in risers and treads, it is all too frequent that the necessary attention is not given by the stair user.

To obtain the best uniformity possible in a flight of stairs, the maximum variation between the highest and lowest risers and between the widest and narrowest treads is limited to $^3/_8$ inch (9.5 mm). This tolerance is not intended to be used as a design variation, but it does recognize that construction practices make it difficult to get exactly identical riser heights and tread dimensions in constructing a stairway facility in the field. Therefore, the code allows the variation indicated in Figure 1009-3. Although the code allows for a tolerance in both the tread depth and riser height, this tolerance is not intended to permit a reduction in the minimum tread depth or increase in the maximum riser height established by the IBC. For example, a tread depth of $10^5/_8$ inches (270 mm) is not permitted if a minimum of 11 inches (279 mm) is required, nor is a riser height of $7^3/_8$ inches (188 mm) acceptable where the code limits riser height to 7 inches (178 mm).

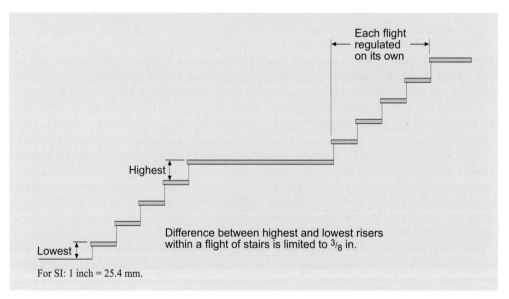

Figure 1009-3
Stair
tolerances

For SI: 1 inch = 25.4 mm.

Under the provisions of Section 1028.11.2, riser height nonuniformity is permitted for aisle stairs where changes in the gradient of the adjoining assembly seating area are necessitated in order to maintain adequate lines of sight. Another exception permits the transition between a typical straight run stairway and consistently shaped winders under the conditions of Section 1009.4.2.

With respect to variation, it is recognized that stairs occasionally descend or rise to areas where the ground or the finished surface is sloping. Where this occurs on private property, the code anticipates that the landing of the stairs be level so that there will not be any variation in the riser height across the width of the stair at that point. However, from time to time, stairs will land on spaces that are not under the control of the property owner, such as a public sidewalk. Therefore, a certain degree of slope across the width of the stair is permitted, resulting in a variation of the height of the riser from one side of the stair to the other. Where this occurs, the height of such a riser may be reduced along the slope to less than 4 inches (102 mm), and the maximum permitted slope shall not exceed 1 unit vertical in 12 units horizontal (8.3 percent slope). Figure 1009-4 shows this condition. It should be clarified that the sloping surface is intended to be an established grade, such as a walkway, public way or driveway.

Profile. Tread nosings are limited to a maximum radius or beveling of $^9/_{16}$ inch (14.3 mm) and shall not extend more than $1^1/_4$ inches (32 mm) beyond the tread below. Nosing projections are to be consistent throughout the stair flight, including where the nosing occurs at the floor at the top of a flight. Risers are to be solid and, if sloped, slope no more than 30 degrees (0.52 rad) from the vertical. See Figure 1009-5 for an overview of these provisions.

1009.4.5

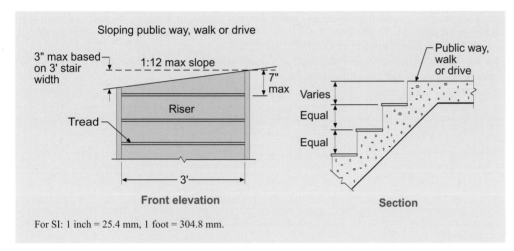

Figure 1009-4
Sloping landings

For SI: 1 inch = 25.4 mm, 1 foot = 304.8 mm.

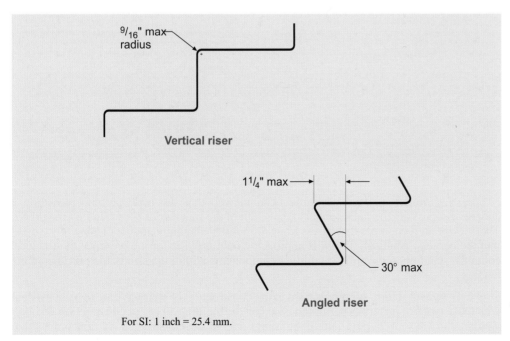

Figure 1009-5
Stair nosings

For SI: 1 inch = 25.4 mm.

There are four exceptions that exempt the requirement for solid risers. First, the risers need not be solid where the stairway does not need to comply with the provisions of Section 1007.3 for a stairway utilized as an accessible means of egress. However, some method of construction must be utilized to limit the openings between treads to less than 4 inches (102 mm). Second, solid risers are not required in Group F, H, I-3 and S occupancies. The third and fourth exceptions indicate that solid risers are not required for complying spiral stairways and alternating tread devices.

1009.5   **Stairway landings.** Landings are discussed to some extent under Section 1008.1.6, which covers landings that are used with adjoining doors. This section covers landings associated with stairs in creating a stairway. The basis for determining the required dimensions of landings is simple; every landing must be at least as wide as the stair it serves. It must also have a dimension measured in the direction of travel not less than the width of the stairway. However, in those instances where the stair has a straight run, the landing length need not be more than 48 inches (1219 mm) measured in the direction of travel. As a stairway changes direction at a landing, it is important that the actual width of the stairway be maintained throughout the travel even if the actual width is greater than the required width. Where the stairway reaches capacity across its width during egress, a landing of reduced size will create an obstruction to the flow pattern that has been established. Because this condition does not occur in a straight run of stairs, a limitation on length is permitted. The dimensional criteria for stair landings are illustrated in Figure 1009-6. An exception provides that aisle stairs need only comply with Section 1028.

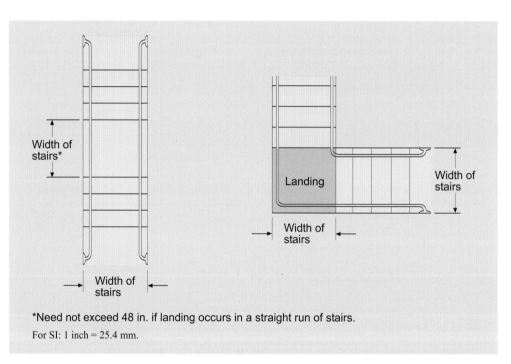

Figure 1009-6
**Landing
dimensions**

*Need not exceed 48 in. if landing occurs in a straight run of stairs.

For SI: 1 inch = 25.4 mm.

Shape of landing. It has generally been viewed that the code permits providing a complete curve with the radius equal to the width of the stairway as shown in Figure 1009-7. This viewpoint recognizes that egress travel through the landing would not generally extend into the corners. Application of this approach would provide for the presence of structural columns, standpipes and similar building components within the nonusable portions of the landing.

1009.6   **Stairway construction.** Materials used in the construction of stairways are to be consistent with those types of materials permitted based on the building's type of construction. In other words, stairways in Type I and II buildings are to be constructed of noncombustible materials. In buildings classified as Type III, IV or V construction, combustible or noncombustible materials may be used for stairway construction. In all types of construction, wood stairway handrails are permitted. These requirements are applicable to the structural elements of the stairway, including stringers and treads. Finish materials on stairways are regulated under the provisions of Section 804.

It is important that both ascending and descending portions of stairway flights end at relatively level landings, having maximum slopes of 1:48. In addition, the treads themselves are limited to a slope no steeper than 1:48 in any direction. For increased safety and ease of use, the treads and landings are to have a substantially solid surface. However, the use of gratings or similar open-type walking surfaces are permitted for use in treads and landing platforms in manufacturing buildings, warehouses and hazardous occupancies. Stairways in such uses are not typically accessible to the public. The extent of the openings in the walking surfaces of such stairs and landings are limited to where a sphere with a diameter of $1^1/_8$ inches (29 mm) cannot pass through. In all occupancies, small openings are also permitted provided they do not permit the passage of a $^1/_2$-inch (12.7 mm) sphere. Where the opening is elongated, the long dimension must be placed perpendicular to the direction of stairway travel.

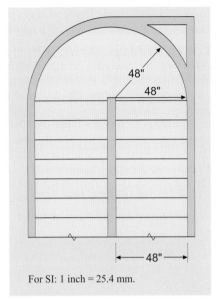

For SI: 1 inch = 25.4 mm.

Figure 1009-7
Alternate shape of landing

Because of the hazards associated with accumulated water on exterior stairways, outdoor stairs and outdoor approaches to such stairs are to be designed in a manner so that water will not stand on the walking surfaces.

To protect the integrity of this very important element of the exit system, any enclosed usable space under either an unenclosed or enclosed stairway shall be protected by fire-resistance-rated construction. For this section to be applicable, both conditions must exist; the space under the stairway must be enclosed, and it also must be accessible, such as by a door or access panel. The level of protection of the walls and soffits within the enclosed space shall be equivalent to the fire-resistance rating of the stairway enclosure, but in no case less than 1 hour. In order to better ensure the integrity of the enclosure, access to the usable space cannot occur from within the stair enclosure. Within an individual dwelling unit in Group R-2 or R-3, the mandated fire-resistance-rated construction is not required; however, a minimum of $^1/_2$-inch (12.7 mm) gypsum board protection must be provided

Much like interior stairways, enclosed usable space under exterior egress stairways must be protected by fire-resistance-rated construction. The minimum level of fire resistance is 1 hour, regardless of the number of stories the exterior stairway serves. Where there is open space below exterior stairways, such space is not to be used for any purpose.

Vertical rise. Negotiating stairs can sometimes become difficult, particularly for persons **1009.7** not accustomed to using stairs, or for the elderly or disabled. So that this difficulty does not become excessive, the code limits the maximum vertical rise between landings serving stairs to 12 feet (3658 mm). Aside from the physical exertion necessary, stairs of exceptional height can be intimidating. It is necessary at vertical intervals not exceeding 12 feet (3658 mm) to provide for places in the stairway where the user can rest. The limitation of 12 feet (3658 mm) does not apply to aisle stairs in compliance with Section 1028 and up to 20 feet (6096 mm) of vertical rise between landings is permitted for alternating tread devices when such devices are used as a means of egress.

The 12-foot (3658 mm) dimension is not unreasonable. In most instances, it will more than accommodate a single-story height so that in most buildings a single flight of stairs could be utilized, if desired, to negotiate travel from one floor to the next adjacent floor.

Even though a single flight of stairs is permitted by the code, stairs having an intermediate landing between floors are utilized in the majority of buildings.

1009.8 **Curved stairways.** Curved stairs are one of the special types of stairways that the code allows as an alternate to the typical straight stair. Figure 1009-8 depicts this type of alternative stair. It is essentially circular in configuration. The basic requirement is that the inside, or least, radius should be at least twice the required width of the stair. This rule ensures a certain limited degree of curvature deemed acceptable in circular stairs. The only other criterion specified for circular stairs is that treads comply with the winder tread provisions of Section 1009.4. In designing the curved stair and relating the inside radius to the width of the stair, it is important that only the required width of the stair need be used. Because of the stair's geometry, this is a fairly comfortable stairway to use. The geometry of the curved stairway for Group R-3 occupancies and within dwelling units of Group R-2 would only be regulated by the winder tread provisions of Section 1009.4.

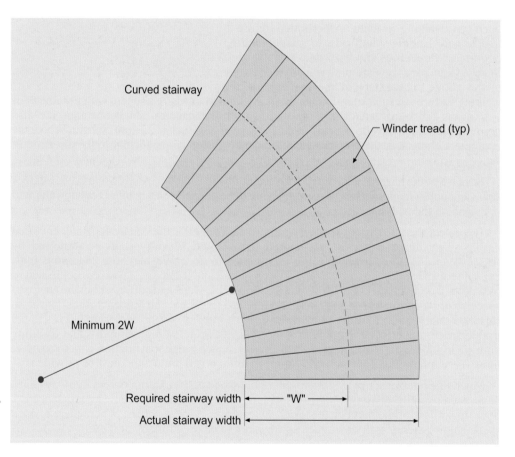

Figure 1009-8
Curved
stairways

1009.9 **Spiral stairways.** Spirals stairways are a special type of stair that the code permits in limited applications. As a means of egress component, spiral stairways may be used within dwelling units, or from spaces having an occupant load of five or less and not exceeding 250 square feet (23 m²) in area. Spiral stairways are also permitted as egress elements from galleries, catwalks and gridirons associated with a stage.

A spiral stairway is one where the treads radiate from a central pole. Such a stair must provide a clear width of at least 26 inches (660 mm) at and below the handrail. Each tread must have a minimum dimension of 7¹/₂ inches (191 mm) at a point 12 inches (305 mm) from its narrow end. The stair must have at least 6 feet, 6 inches (1981 mm) of headroom measured vertically from the leading edge of the tread. The rise between treads can be as much as, but not more than, 9½ inches (241 mm). The required dimensions of a spiral stairway are depicted in Figure 1009-9.

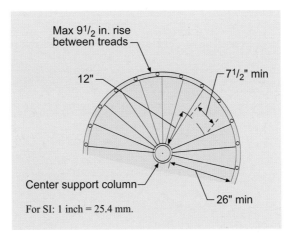

Max 9½ in. rise between treads

12"

7½" min

Center support column

For SI: 1 inch = 25.4 mm.

26" min

Figure 1009-9
Spiral stairway

Alternating tread devices. An alternating tread device is a unique type of stairway that also has some characteristics of a ladder. Because it is considered difficult to utilize for egress purposes, an alternating tread device may only be used as a means of egress in a limited number of occupancies. In factories, warehouses and high-hazard occupancies, this device can only be used for egress from a small mezzanine serving a limited number of occupants. In Group I-3 occupancies, an alternating tread device may be the egress path from a small guard tower or observation area. See Figure 1009-10.

1009.10

Alternating tread devices must have complying handrails on both sides. In addition, treads are regulated based upon whether or not the alternating tread device is used as a means of egress component. Where it is not used for egress purposes, but rather is for convenience only, the minimum tread depth may be reduced and the maximum riser height may be increased.

Handrails. Probably the most important safety device that can be provided in connection with stairways is the handrail. It will never be known how many missteps, accidents, injuries or even fatalities have been prevented by a properly installed, sturdy handrail. Basically, a handrail should be within relatively easy reach of every stair user. In general, all stairways are required to have handrails on each side. However, the code has the following specific conditions allowing the use of only one handrail:

1009.12

1. At aisle stairs where a center handrail is provided

2. At aisle stairs serving seating on only one side

3. On stairways within dwelling units

4. On spiral stairways

Handrails may be omitted under four conditions. Where a single change in elevation occurs at a deck, patio or walkway, handrails are not required, provided a complying landing area is present. In Group R-3 occupancies, handrails are not required at a single riser serving an entrance door or egress door. In individual dwelling units and sleeping units of Group R-2 and R-3 occupancies, it is also permissible to provide a change in room elevation of three or fewer risers and not install a handrail. Handrails may also be omitted in assembly seating areas as established in Section 1028.13.

Stairway to roof. To provide for easy access to roof surfaces and to facilitate fire fighting in buildings four or more stories in height, at least one stairway is required to extend to the roof. This stairway to the roof must comply with all code requirements for stairways, unless the roof is considered unoccupied. Access to an unoccupied roof may be accomplished by an alternating tread device as described in Section 1009.10. However, the stairway or

1009.13

alternating tread device to the roof is not a requirement on steeper roofs where the slope exceeds 4 units vertical in 12 units horizontal (33 percent slope).

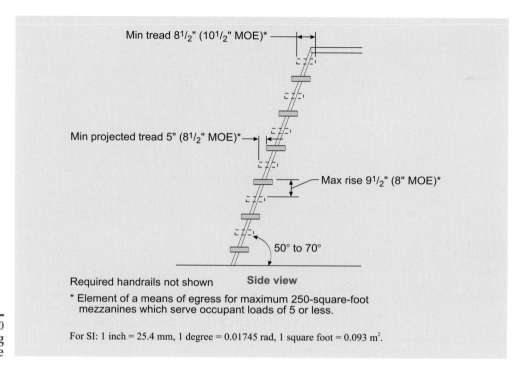

Figure 1009-10
Alternating tread device

1009.13.1 **Roof access.** In buildings having an occupied roof, a penthouse complying with the provisions of Section 1509.2 must be provided for stairway access to the roof. Where the roof is considered unoccupied, access need only be provided through a roof hatch of the minimum size prescribed.

1009.13.2 **Protection at roof hatch openings.** Roof hatches are permitted as a means of access to unoccupied roofs where such access is required in buildings four or more stories in height. In addition, roof-hatch openings are often provided in low-rise buildings for varied purposes, including access to rooftop equipment. This provision addresses the hazard created where the roof hatch is located very close to the roof edge. See Figure 1009-11. It is necessary that persons accessing the roof by a roof hatch be protected from falling off a roof as a result of a trip or misstep. Occasionally, these roof accesses are used during inclement weather, emergency situations, or times of darkness. It is during these conditions when the hazard level is even higher.

1009.14 **Stairway to elevator equipment.** Where elevator equipment is located on the roof or in a rooftop penthouse and such equipment must be accessed for maintenance purposes, a stairway must be provided to the roof or penthouse. Consistent with the provisions of ASME A17.1 *Safety Code for Elevators and Escalators*, the requirement mandates a code-complying stairway. As a result, the use of alternating tread devices or ladders as the only means to access the elevator equipment is prohibited.

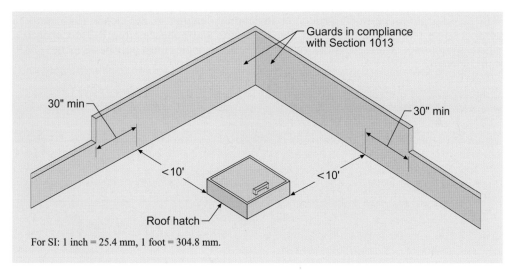

Guards in compliance
with Section 1013

30" min

30" min

<10'

<10'

Roof hatch

For SI: 1 inch = 25.4 mm, 1 foot = 304.8 mm.

Figure 1009-11
**Protection at
roof-hatch
openings**

Section 1010 *Ramps*

Scope. With the exception of some ramped aisles in assembly rooms, curb ramps and vehicle ramps in parking garages used for pedestrian exit access travel, whenever a ramp is used as a component anywhere in a means of egress it is necessary that the ramp comply with the provisions of this section. Note that where ramps are used as part of the egress system, every ramp must meet these standards, regardless of the occupant load served. Ramped aisles in assembly occupancies, other than those to and from accessible wheelchair spaces, need only conform to the provisions of Section 1028.

1010.1

In addition to those egress provisions, all ramps located within an accessible route of travel shall be made to conform to the requirements found in ICC A117.1.

Slope. Whenever any ramp is used in an exit system, the slope of such a ramp shall not be steeper that 1 unit vertical in 12 units horizontal (8.3 percent slope). Those pedestrian ramps not considered a portion of the means of egress may have a greater slope, but may not be steeper than a slope of 1 unit vertical in 8 units horizontal (12.5 percent slope). For areas of assembly seating, Section 1028.11 allows a 1:8 slope for means of egress ramps not on an accessible route.

1010.2

Minimum dimensions. Because requirements for the widths of ramps are identical to those requirements for widths of corridors, see discussion of Section 1018.2. The net clear width between the handrails is to be no less than 36 inches (914 mm), differing from the stairway provisions that allow for handrail projections into the required width. In fact, all projections are prohibited into the minimum width requirements of the ramp and its landings. Where the width of an egress ramp is established, it shall not be diminished at any point in the direction of travel. Where doors enter onto or swing over ramp landings, the doors may not, during the course of their swing, reduce the minimum dimension of the landing to less than 42 inches (1067 mm). As depicted in Figure 1010-1, in all cases there must be at least 42 inches (1067 mm) of width available on the ramp even while the door is swinging.

1010.5

Landings. Any ramp, defined as having a slope steeper than 1 unit vertical in 20 units horizontal (5 percent slope), must have landings at the top and bottom, at any changes of direction, at points of entrance or exiting from the ramp, and at doors. All landings are required to have a dimension measured in the direction of the ramp run not less than 5 feet (1524 mm). These larger-than-normal dimensions are required to reasonably ensure that disabled persons in wheelchairs will have sufficient space to maneuver on any intermediate landings as well as in the area at the top and bottom approaching the ramp. See Figure

1010.6

1010-2. Exceptions permit a reduction in landing lengths to 36 inches (914 mm) in nonaccessible Group R-2 individual dwelling units and nonaccessible R-3 occupancies, whereas ramps not on an accessible route in other occupancies must only be 48 inches (1220 mm) in length.

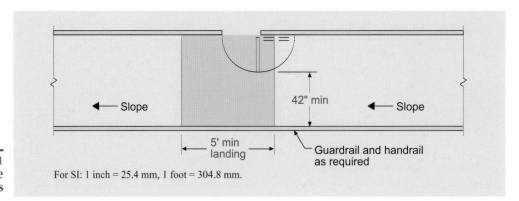

Figure 1010-1
Intermediate ramp landings

For SI: 1 inch = 25.4 mm, 1 foot = 304.8 mm.

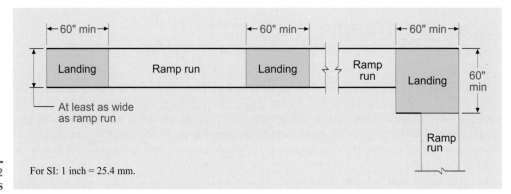

Figure 1010-2
Ramp landings

For SI: 1 inch = 25.4 mm.

1010.7   **Ramp construction.** Consistent with the provisions for stairway construction found in Section 1009.6, ramps shall be built of materials consistent with the building's type of construction. In addition, exterior ramps and their approaches should be designed in a manner to avoid water accumulation.

Because a ramp has a sloping surface, the potential for accidents resulting from slips and falls is greatly increased. It is critically important that the surface of the ramp be made slip-resistant to help prevent such accidents. It is imperative that very careful attention be given to the selection of materials for ramp surfaces and the methods of finishing such surfaces. Certainly, slick-finished materials should be avoided on ramps, unless it can be ensured that they have been made slip-resistant by some process.

There are a number of products on the market available for treating ramp surfaces that are not sufficiently slip-resistant. In many instances, the use of these materials has not proven to be completely satisfactory. One method is to install carborundum strips across the width of the ramp at appropriate intervals. Other available products can be installed on the ramp surface with an adhesive. It is essential that slip-resistant treatments and materials be of reasonably permanent nature and securely attached so they can perform the function for which they were installed, while at the same time not becoming a potential tripping hazard. Therefore, the slip resistance should be proven by tests and be an integral part of the ramp surface.

1010.8   **Handrails.** Whenever any ramp in the means of egress has a rise greater than 6 inches (152 mm), the ramp shall be provided with handrails on both sides. The detailed requirements for the handrails are found in Section 1012. There are no general provisions that allow a handrail on only one side of a ramp, nor is there any requirement for intermediate handrails.

It should be remembered that handrails for ramped aisles in assembly occupancies are regulated by Section 1028.13.

Edge protection. In order to further protect the user of a ramp, safeguards are required along ramp edges as well as along each side of ramp landings. Two different methods as shown in Figure 1010-3 are available for providing compliant edge protection at ramp runs and ramp landings. Either approach may be taken to provide the necessary level of protection. It is acceptable to utilize a curb with a minimum height of 4 inches (102 mm) to satisfy the first method. A rail located just above the ramp or landing surface will also suffice, as will a wall or similar barrier. As an alternative methodology, the surface of the ramp or landing may extend a minimum distance of 12 inches (305 mm) beyond vertical plane established from the inside face of a complying handrail. Both approaches are deemed adequate for preventing a ramp user from an unacceptable drop-off at the landing or ramp edge. **1010.9**

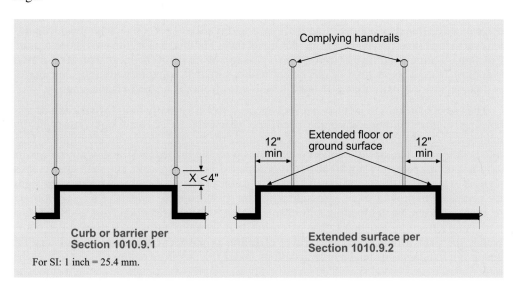

For SI: 1 inch = 25.4 mm.

Figure 1010-3
Ramp edge protection

Section 1011 *Exit Signs*

Where required. To properly identify the egress path through a building, it is necessary to provide exit signs. Although somewhat vague, these provisions are probably more performance-oriented than most requirements. The basic provision is that signs are required so that the exiting path is clearly indicated. By combining this requirement with Exception 1, it will mean that, in general, when two or more means of egress are required from any portion of a building, such as from a room, area, floor or other space, exit signs must be installed at all required exit and exit access doorways, and at any other location throughout the exiting system where deemed necessary by the building official to clearly identify the path of travel to the exit. **1011.1**

Care must be taken in the placement of the exit signs to ensure they are properly located and properly oriented to the direction of travel in the egress system. They need to be easily read by persons seeking the exit. Particular care should be taken to see that nothing occurs in the means of egress that might tend to obscure or screen the exit sign, or that might cause confusion in identifying the exit sign. It is not uncommon in many building plans to find that exit signs are shown at what appear to be the appropriate locations, only to later find they are actually installed behind ducts, equipment or other building elements in such a manner that they are not really visible to building occupants, or near or over doors that are not the exit. Additionally, the presence of banners, signs and other movable elements may obstruct the view of required exit signs. In the placement of exit signs, the use of the space and its impact on clear sight lines to exit signs must be considered.

In certain instances, the exit signs may be present, but they are not oriented to the path of travel and to the approaching building occupant. Therefore, they are not visible and are certainly not legible to the persons seeking the exit. It is strongly advised that, as one of the last points of inspection, and before approving the occupancy of any building, the building inspector or building official carefully walk the egress path to ensure the proper installation and effective location and orientation of exit signs. It is also important that ceiling-suspended exit signs or exit signage extending from a wall do not project below the minimum permitted headroom height of 80 inches (2032 mm) for projecting elements as required by Section 1003.3.1.

As occupants proceed along the path of travel through a corridor or exit passageway, they typically assume that their direction of travel is taking them to an exit. However, where the corridor or exit passageway length is quite extensive, the users may at some point make a determination that they are traveling in the wrong direction. This could cause the occupants to reverse their direction and seek another travel path. For this reason, the provisions require exit sign placement within a corridor or exit passageway at points no more than 200 feet (60 960 mm) apart. This method of placement will ensure that no point within the corridor or exit passageway is more than 100 feet (30 480 mm) from the nearest visible sign.

The use of exit signs to identify paths of egress travel are usually limited to the exit access portions of the building. Once the occupants reach the exits, such signs are typically unnecessary as the paths are often direct and single directional. However, in buildings with more complicated means of egress systems, it is possible that egress travel within the exits may not be immediately apparent to the occupants. For this reason, the mandate for exit signs is extended to those portions within exits, such as exit passageways, where such signs are necessary to provide clear egress direction for the occupants. Evacuees may be hesitant or even confused when traveling within an exit that involves transition from a vertical to a horizontal direction and horizontal extension that includes turns and intervening doors within the path of egress. Where travel direction is not clear within an exit, it creates uncertainty and causes delays in evacuations under threatening conditions. Therefore, the direction of egress travel should be identified by exit signs where such direction may not be clearly understood, and all intervening means of egress doors within the exits must also be provided with complying exit signs. An example of the appropriate use of exit signs within exits is shown in Figure 1011-1.

Omission of exit signs. Although the installation of exit signs is of paramount importance in most instances, there will be cases where the means of egress is abundantly obvious to any building occupant. Where the exit point is clearly identifiable, such as at main exterior entrance/exit doors, the exit is essentially its own identifying sign. Where approved by the building official, it may not be necessary to install an exit sign at such an obvious exit. In addition, exit signs may be omitted in Group U occupancies, and within individual sleeping and dwelling units in Group R-1, R-2 and R-3 occupancies. Because of the residential nature of such uses, it is felt that exit signs are unnecessary. Exit signs may also be omitted in day rooms, sleeping rooms and dormitories of a Group I-3 occupancy, as requiring exit signs for such an occupancy is considered unnecessary. The first exception, which eliminates the need for exit signs in areas that need access to only one exit or exit access element, has application to a great number of areas within most buildings. It is assumed that the exit path is obvious because of its use as the typical entrance to the area or room. It should be noted that the exception applies even where multiple means of egress are provided, as long as only one of them is required. A final exception applies to assembly seating arrangements that occur in Groups A-4 and A-5. Exit signs may be omitted in the seating area, provided that the direction to the concourse area is easily identified and adequately illuminated. Once reaching the concourse, exit signs identifying the exit path and egress doors must be provided.

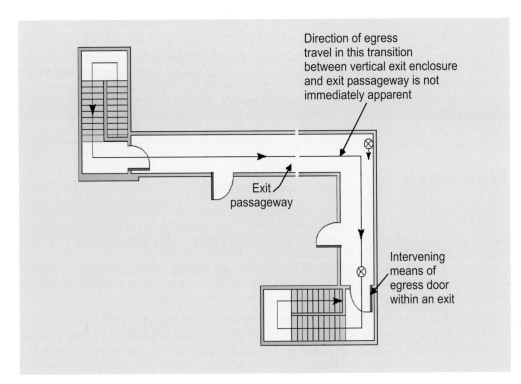

Direction of egress
travel in this transition
between vertical exit enclosure
and exit passageway is not
immediately apparent

Exit
passageway

Intervening
means of
egress door
within an exit

Figure 1011-1
**Exit sign
locations**

Illumination. Where exit signs are required, they must be illuminated. The source of the illumination is not material, provided the level of illumination at the sign meets the minimum requirements of the code. Tactile exit signs, required by Section 1011.3, are exempted from the illumination requirement for obvious reasons.

1011.2

Internally illuminated exit signs. Internally illuminated signs include all exit signs that generate their own luminosity. Electrically powered exit signs, including LED, incandescent, fluorescent and electroluminescent signs, in combination with those signs considered as self-luminous and photoluminescent, represent the full range of product types currently in the market. All such exit signs must be listed and labeled in accordance with the provisions of UL 924, *Standard for Safety Emergency Lighting and Power Equipment.* Consistent with most other installation requirements found in the code, the manufacturer's instructions must be followed. The value of visible exit signs is demonstrated by the requirement that exit signs must be illuminated at all times.

1011.4

Externally illuminated exit signs. Although it is uncommon to provide illumination for an exit sign from an external source, there are occasions where such methods are utilized. This section regulates the graphics, illumination and emergency power supply for such signs.

1011.5

Although no particular color is specified for exit signs, it is required that the color and design of the signs, the lettering, the arrows and other symbols on the sign provide good contrast so as to increase legibility. The letters of the word *exit* are required to be at least 6 inches (152 mm) in height and have a width of stroke of not less than $^{3}/_{4}$ inch (19.1 mm). By specifying the letter spacing, the code ensures that the letters are not placed too close together and that the signs will be legible and more effective as a result. Additionally, when a larger sign is provided, it is important the lettering be increased proportionally.

When the illumination of the sign is from an external source, that source must be capable of producing a light intensity of at least 5 foot-candles (54 lux) at all points over the face of the sign. Measurement of the 5 foot-candles (54 lux) from an external source can be made by a light meter. The measurement of the equivalent luminance, however, from an internal source is more difficult. That is why the code relies on a testing laboratory that has examined, certified and labeled the exit sign. In labeling such a sign, the testing agency is certifying, among other things, that the light level is at least that required by the code.

Externally illuminated exit signs are required to be illuminated at all times and must provide continued operation for a minimum of one- and one-half hours after the loss of the normal power supply. This backup power supply shall be from a complying storage battery system, unit equipment, or another approved on-site, independent source. The exception addresses illumination that is provided independent of external power supplies and, therefore, does not require compliance with the emergency power provisions.

Section 1012 *Handrails*

A handrail is defined in Section 1002 as "a horizontal or sloping rail intended for grasping by the hand for guidance or support." The IBC mandates, with limited exceptions, that handrails be provided to assist the users of stairways and ramps during normal travel conditions. In addition, a handrail must be available for support in case of a misstep or other occurrence that might cause the user to stumble and fall. Numerous criteria for the design and installation of effective handrails are set forth in this section.

1012.2 **Height.** In past building codes, the height of handrails was traditionally established within a range of at least 30 inches (762 mm) and not more than 34 inches (864 mm) vertically above the leading edge of treads of the stairway that the handrails serve. However, research has shown that handrails better serve stair users if located in a range higher than this traditional location. Higher handrails can be more readily reached by adult stair users and, interestingly enough, handrails at the higher elevation are also more usable by very small persons, including toddlers. Where handrails are required, they must be located at least 34 inches (864 mm) and not more than 38 inches (965 mm) vertically, measured from the nosing of the stair treads to the top of the rail. See Figure 1012-1 for an illustration of IBC requirements for stairs, including handrail height. A reduction in the handrail height range is provided for alternating tread devices and ship ladders due to the steepness of travel.

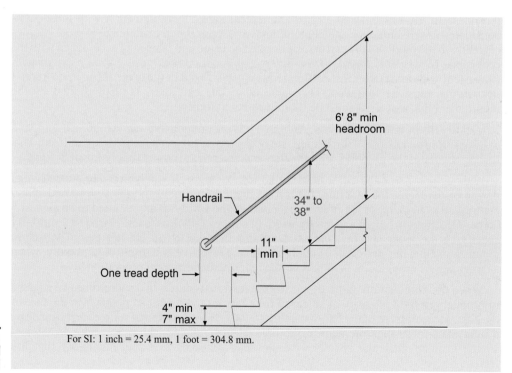

For SI: 1 inch = 25.4 mm, 1 foot = 304.8 mm.

Figure 1012-1
Handrail and stair detail

In many instances, stairways and ramps are constructed in conjunction with landings, balconies, porches and other building components. A common condition occurs with the typical switchback stairway, where a substantial elevation change can occur from one flight to a lower adjacent flight. When those components or conditions are more than 30 inches (762 mm) above the adjacent ground level or surface below, they must be protected by guards meeting the requirements of Section 1013, which requires guards to be a minimum of 42 inches (1067 mm) in height. Therefore, compliance with requirements for handrail heights in Section 1012.2 is not considered sufficient by the code for guard protection. The handrail must be supplemented with an additional element at a greater height in order to meet the provisions for guards. See Figure 1012-2. As a note, one item often missed when looking at the provisions is that the glazed side of a stairway must be protected by a guard unless the glazing meets the strength and attachment requirements of Section 1607.7.

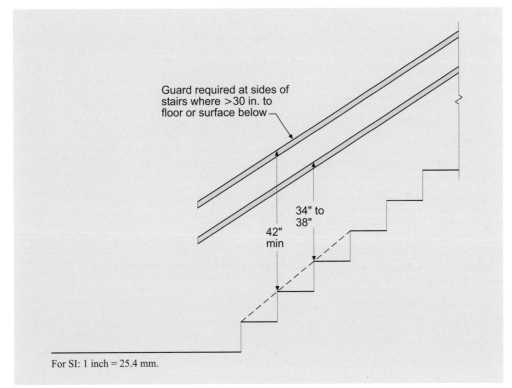

Guard required at sides of stairs where >30 in. to floor or surface below

34" to 38"

42" min

For SI: 1 inch = 25.4 mm.

Figure 1012-2
Guard and handrail

Handrail graspability. To be truly effective, handrails must be graspable. Therefore, the handgrip portion of the handrail must not be less than $1^1/_4$ inches (32 mm) or more than 2 inches (51 mm) in a circular cross-sectional dimension as shown in Figure 1012-3, or the configuration of the handrail must be such that it provides an equivalent, graspable shape. Where the cross section of the handrail is not circular, an alternate shape is described by the code and illustrated in Figure 1012-4. Such handrails are considered as Type I handrails and are permitted for use in all occupancies. Handrail profiles having a perimeter dimension greater than $6^1/_4$ inches (160 mm), identified as Type II handrails, are also acceptable where installed in the specified residential applications.

1012.3

Research has shown that Type II handrails have graspability that is essentially equal to or greater than the graspability of handrails meeting the long-accepted and codified shape and size defined as Type I. The key features of the graspability of Type II handrails are graspable finger recesses on both sides of the handrail. These recesses allow users to firmly grip a properly proportioned grasping surface on the top of the handrail, ensuring that the user can tightly retain a grip on the handrail for all forces that are associated with attempts to arrest a fall. Examples of complying Type II handrails are illustrated in Figure 1012-5.

Many persons are incapable of exerting sufficient finger pressure on a plane surface, such as that provided by a rectangular handrail. To get adequate support or to adequately grasp the handrail, it is necessary to provide those persons with one of the shapes that the code specifies. Where a handrail with a complying profile is installed, it is possible for those persons to actually wrap their fingers around that portion of the handrail and, thereby, obtain better support.

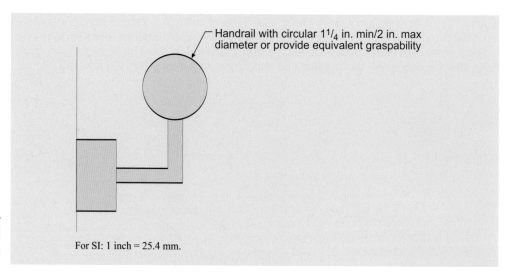

Handrail with circular 1¹/₄ in. min/2 in. max diameter or provide equivalent graspability

For SI: 1 inch = 25.4 mm.

Figure 1012-3
Circular handrail

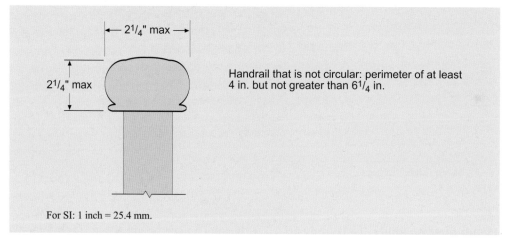

2¹/₄" max

2¹/₄" max

Handrail that is not circular: perimeter of at least 4 in. but not greater than 6¹/₄ in.

For SI: 1 inch = 25.4 mm.

Figure 1012-4
Noncircular handrail

1012.4 Continuity. The handrails must be continuous for the full length of the ramp or the flight of stairs, thereby affording the user support throughout the entire flight. Within dwelling units, handrails are permitted to terminate at a starting newel or ramp or stair volute that is located on the first tread. In addition, a newel post may interrupt the handrail continuity at a stair landing. These types of terminations have been found in residences for years without a record of accidents or lawsuits for an unsafe condition.

Handrail brackets or balusters are not considered obstructions, provided they are attached to the bottom surface of the handrail and do not project horizontally beyond the sides of the handrail within 1¹/₂ inches (38 mm) of the bottom of the rail as shown in Figure 1012-6. This provision is used to regulate the method of support so that the handrail is graspable at any point along its length. A lesser vertical clearance at the handrail's bottom surface is permitted where the perimeter of the handrail exceeds 4 inches (102 mm). As the perimeter of the handrail increases, the vertical clearance may be reduced.

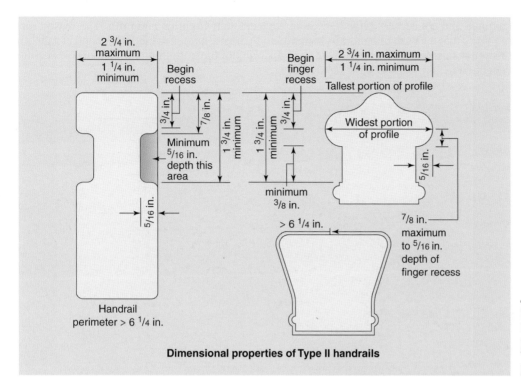

Dimensional properties of Type II handrails

Figure 1012-5
**Dimensional
properties of
Type II
handrails**

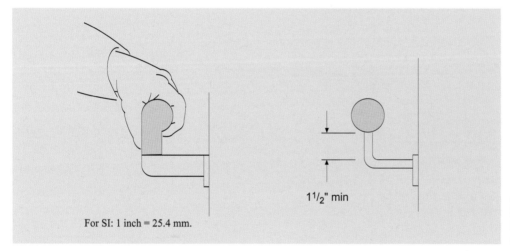

For SI: 1 inch = 25.4 mm.

1½" min

Figure 1012-6
**Handrail
continuity**

Handrail extensions. Other than locations where handrails are continuous from one flight of stairs to the next, the handrails must extend horizontally not less than 12 inches (305 mm) beyond the top riser. In addition, under such conditions they must also continue to slope beyond the bottom riser for a distance equal to the depth of one tread. See Figure 1012-7. The purpose of the extension of the handrail at the top and bottom of the stairs is to provide minimal additional facility in assisting the stair user. As such, the handrail must extend in the same direction as the travel along the stairway or ramp. This extension is not required for aisle handrails in assembly areas of Group A and E occupancies, for handrails within a dwelling unit that is not required to be accessible, as well as handrails for alternating tread devices and ship ladders. Somewhat similar conditions are applicable where ramp handrails do not continue from one ramp run to the next. For such noncontinuous handrails, a minimum extension of 12 inches (305 mm) is required as shown in Figure 1012-8. Handrails for both stairs and ramps must return to a wall, guard or the walking surface unless they

1012.6

continue and connect to a handrail of an adjacent stair flight or ramp run. The reason for requiring that the ends of handrails be returned to the wall, floor or landing, or end at a guard or similar safety terminal, is to avoid the possibility of loose clothing or other articles being caught on the projection of the handrail.

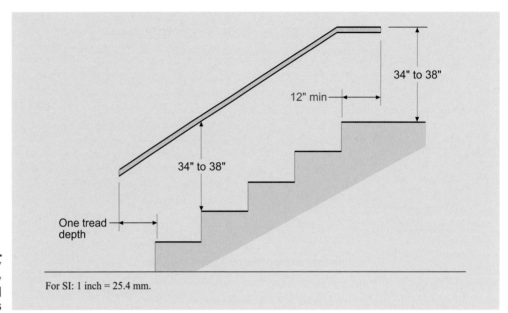

Figure 1012-7
Stairway handrail extensions

For SI: 1 inch = 25.4 mm.

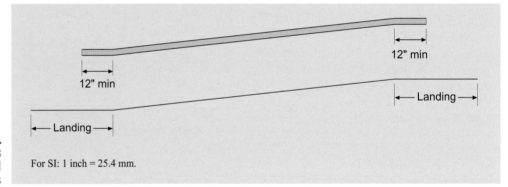

Figure 1012-8
Ramp handrail extensions

For SI: 1 inch = 25.4 mm.

1012.7 **Clearance.** In view of the desire to make the handrail graspable, and considering the requirement that the handrail be continuous, it is necessary to provide a clear space of at least $1^1/_2$ inches (38 mm) between the handrail and any abutting construction to avoid injury to fingers. See Figure 1012-9.

1012.8 **Projections.** As stated earlier, when the code specifies a required width of a component in the egress system, it generally intends that width to be clear and unobstructed. One permitted projection is for handrails, which may project a maximum distance of $4^1/_2$ inches (114 mm) from each side of the ramp or stair as shown in Figure 1012-10. In addition, stringers, trim and other features are permitted to project into the required width up to $4^1/_2$ inches (114 mm) at or below the handrail height. Again, unless a projection into the minimum required width of the stairway is expressly allowed, none is permitted.

Unlike the permissible projections of handrails into the required stairway width, such projections into the required ramp width are limited. In no case may the distance between ramp handrails be less than 36 inches (914 mm).

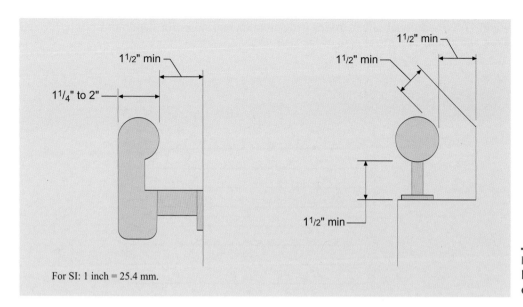

For SI: 1 inch = 25.4 mm.

Figure 1012-9
Handrail clearance

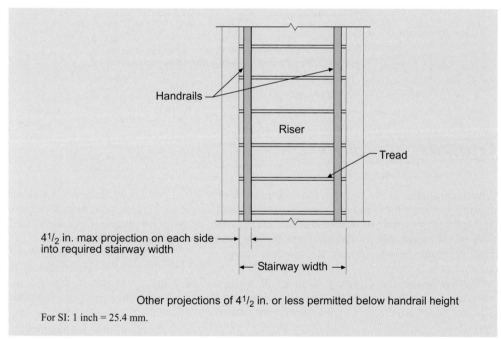

Handrails

Riser

Tread

$4^1/_2$ in. max projection on each side into required stairway width

← Stairway width →

Other projections of $4^1/_2$ in. or less permitted below handrail height

For SI: 1 inch = 25.4 mm.

Figure 1012-10
Projections into stairway width

Intermediate handrails. Where the occupant load served by a stairway becomes significant, additional handrails may be necessary to assist stair users. The requirement is based upon the required width of the stair established by Section 1005.1, not the actual width, and mandates that at no point shall the required stairway width be more than 30 inches (762 mm) from a handrail. See Figure 1012-11. It is difficult to determine the exact point at which intermediate handrails are required, as the handrail projection into the required width can vary from one design to the next. It should be noted that the measurement is to be taken in regards to the handrail location, which is permitted to extend a maximum of $4^1/_2$ inches (114 mm) into the required width. Where the maximum encroachment occurs on each side of the stairway, an intermediate handrail must be provided where the required width exceeds 69 inches (1752 mm). A lesser required width would apply where the handrails do not extend the full $4^1/_2$ inches (114 mm) into the minimum required stairway

1012.9

width. As an additional safeguard for wide monumental stairs, the handrails must be located along the anticipated travel path of the stair users.

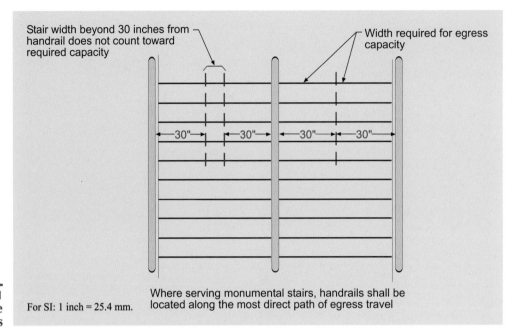

Figure 1012-11
Intermediate
handrails

For SI: 1 inch = 25.4 mm.

Stair width beyond 30 inches from handrail does not count toward required capacity

Width required for egress capacity

Where serving monumental stairs, handrails shall be located along the most direct path of egress travel

Section 1013 *Guards*

1013.1   **Where required.** In this section, the code provides for guard protection at open sides along walking surfaces, mezzanines, stairways, ramps and landings that are more than 30 inches (762 mm) above grade or a floor surface below. Also, protection is required for the glazed sides of stairways, ramps and landings located more than 30 inches (762 mm) above the floor or grade below, unless the glazing complies with the strength and attachment provisions in Section 1607.7 for live loads. The need for guards in these circumstances is evident, although the arbitrary limit of 30 inches (762 mm) is subject to conjecture. Nevertheless, in the case of the IBC, it is assumed that the maximum height of 30 inches (762 mm) does not create a significant safety hazard.

In the determination of the difference in elevation, an objective method is established for measuring the height of the walking surface above the grade below. Rather than taking this measurement to the ground level or floor directly below the edge of the walking surface, the code requires the height of the walking surface above grade to be based on the lowest point within 36 inches (914 mm) horizontally from the edge of the deck, porch or other element. This approach recognizes that a sloped site sometimes occurs adjacent to a deck, porch or similar walking surface, increasing the level of hazard. This method of measurement is illustrated in Figure 1013-1.

There are seven exceptions that identify locations or situations where guards complying with Section 1013 are not required. These exceptions fall into essentially three categories where, for obvious reasons, guards would be inappropriate:

1. Commercial and industrial applications. Guards are not required on the loading side of loading docks or piers, nor along vehicle-service pits inaccessible to the public. Such guards would severely restrict the work that takes place in these areas.

2. Stage and platform areas. Because of the nature of activities involving performance stages or platforms, it is impractical to provide guards in various locations. Guards may be omitted on the audience side of stages and raised platforms, at runways or side stages used for presentations, at any vertical openings in the performance area, and at elevated walking surfaces used to access or utilize lighting or equipment. Guards also may be omitted along any steps that may lead from the auditorium area to a stage or platform; however, one or more handrails must still be provided under the provisions of Section 1012.

3. Assembly seating areas. In order to achieve adequate lines of sight for the assembly audience seated in a tiered configuration, it is often necessary to provide for guards of a height lower than required by the provisions of Section 1013. Therefore, reduced guard heights are permitted under the provisions of Section 1028.14.

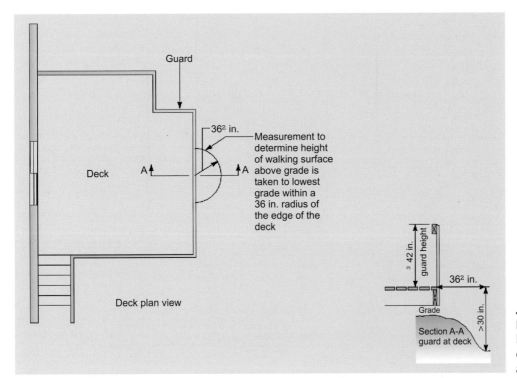

Figure 1013-1
Determination of guard applicability

Height. Where required, the guard must be of adequate height to prevent someone from falling over the edge of the protected area. Therefore, the code establishes 42 inches (1067 mm) as a minimum height that is acceptable for guard protection. In the case of a guard adjacent to a stairway, the minimum height of 42 inches (1067 mm) is measured vertically above the leading edge of the tread. An exception allows the top rail of a stairway handrail in Group R-3 occupancies and within individual dwelling units in Group R-2 occupancies to be considered an adequate guard height, provided the handrail is located between 34 inches (864 mm) and 38 inches (965 mm) measured vertically from the leading edge of the stair tread nosing. Additional modifications are also provided in Section 1028.14 for stadiums, arenas, theaters and other assembly areas to address the impact guard heights have on the lines of sight in some spectator locations. Exception 4 allows alternating tread devices and ship ladders are permitted to have guards with top rails serving as handrails located at a reduced height above the tread nosings.

1013.2

Although the minimum height is typically measured from a walking surface to the floor or ground below, fixed seating adjacent to a guard is also considered to have the same safety concerns as a walking surface in the regulation of such guards. The minimum 42-inch (1067 mm) guard height must be measured from the surface of the seating where such seating is considered as fixed-in-place. See Figure 1013-2. This method of measurement addresses concerns regarding children climbing or playing on the seating area with the potential to fall over an adjacent guard whose minimum height was measured from the floor surface. Although it is recognized that movable furniture and planters may present the same hazard, there is no realistic way to regulate such furniture.

It is important to note that the minimum height criteria for guards only apply to those guards that are required by Section 1013.1. The minimum required guard height does not apply to optional barriers that are installed.

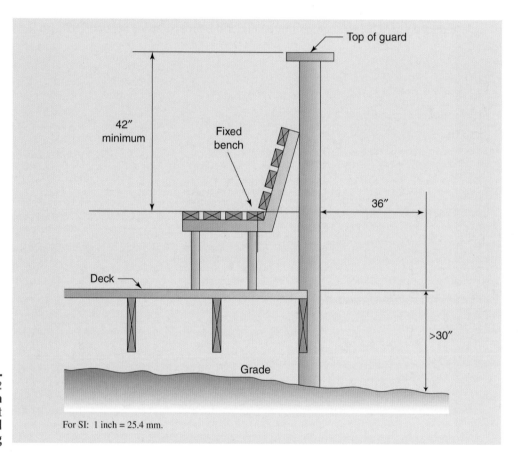

Figure 1013-2
**Minimum
guard height
at fixed
seating**

For SI: 1 inch = 25.4 mm.

1013.3   **Opening limitations.** Along with a minimum height requirement for guards, the code also requires that for open-type rails, intermediate members or ornamentation be provided so that a sphere 4 inches (102 mm) in diameter cannot pass through any opening; a requirement that prevents individuals, particularly small children, from falling or climbing through the guard assembly. This limitation applies to the lower 36 inches (914 mm) of the guard. At a height above 36 inches (914 mm) to a height of 42 inches (1067 mm), the opening size may be increased, provided a sphere $4^3/_8$ inches (111 mm) in diameter cannot pass through the opening. See Figure 1013-3.

Several exceptions increase the maximum opening size permitted in an open guard. The triangular area formed by the tread, riser and bottom rail of the guard is limited in size to where a sphere of 6 inches (152 mm) in diameter cannot pass through the opening. Because of the unusual configuration of a stairway and its required guard, as well as the location of the triangular opening, an increased size is deemed reasonable. This configuration is shown in Figure 1013-4.

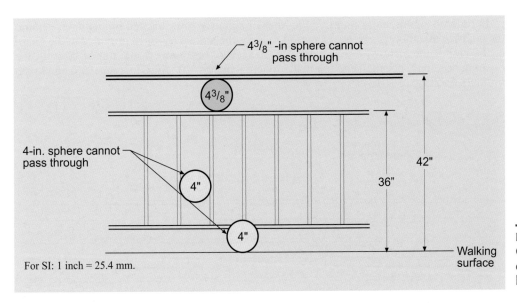

For SI: 1 inch = 25.4 mm.

Figure 1013-3
Guard opening limitations

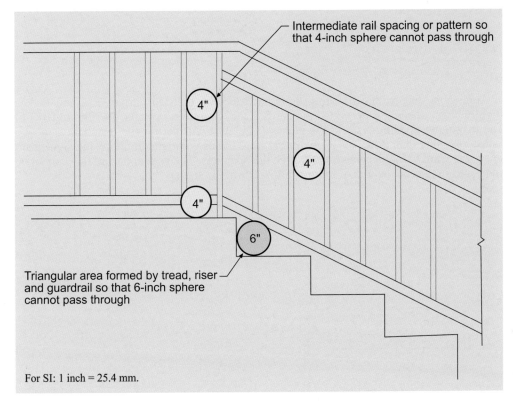

For SI: 1 inch = 25.4 mm.

Figure 1013-4
Guardrail openings at stairs

In commercial, industrial and security uses where the public is not invited (therefore, the guard is typically not subject to children crawling or falling through), open guards may have intermediate members or ornamentation spaced so that a 21-inch (533 mm) diameter sphere cannot pass through. This exception applies to those elevated walking surfaces utilized for access to areas containing electrical, mechanical, or plumbing systems or equipment, as well as platforms provided for the use of such systems or equipment. It also applies to elevated areas in occupancies of Group I-3, F, H or S where access to the public is not available.

In order to significantly improve the lines of sight for rows located immediately behind the guard in assembly areas, an exception reduces the amount of infill provided in the guard at the end of the aisles. From a height of 26 inches (660 mm) to the top of the rail, the opening need only be small enough that a sphere 8 inches (203 mm) in diameter cannot pass through. See Figure 1013-5.

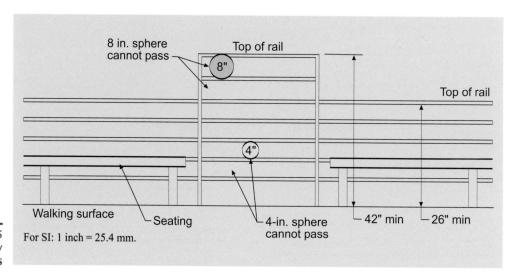

Figure 1013-5
Assembly seating guards

For SI: 1 inch = 25.4 mm.

Applicable only within dwelling units and sleeping units in Groups R-2 and R-3, a slightly larger opening is permitted in guards at the open sides of stairs. See Figure 1013-6. The maximum opening size of $4^3/_8$ inches (111 mm) will permit the installation of two balusters, rather than three, on 10-inch (254 mm) treads without compromising significant safety in regard to infants crawling through the openings. The greater hazard that cannot be addressed in the code is that of infants falling down the stairs.

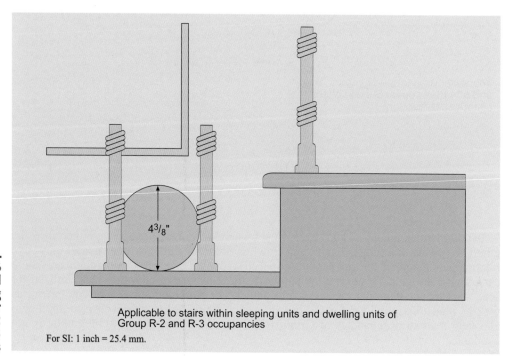

Figure 1013-6
Guard opening limitations for Group R-2 and R-3 occupancies

Applicable to stairs within sleeping units and dwelling units of Group R-2 and R-3 occupancies

For SI: 1 inch = 25.4 mm.

Section 1014 *Exit Access*

The exit access is identified as the initial component of the means of egress system—that portion between any occupied point in a building and an exit. Leading to an exterior exit door at the level of exit discharge; an exterior egress stairway or ramp; or the door of an exit passageway, vertical exit enclosure or horizontal exit, the exit access makes up the vast majority of any building's floor area.

General. The key design provisions related to means of egress are typically found in this section. Issues relating to the number and arrangement of exit paths are addressed, as is travel through intervening spaces and travel distance. The concept of a common path of egress travel is also presented. **1014.1**

Egress through intervening spaces. Basically, the code intends that access to exits should be direct from the room or area under consideration. This section, however, makes some modifications where, under certain circumstances, exit paths may be arranged through adjoining rooms or spaces rather than directly into corridors or exit elements, such as exit enclosures or exit passageways, or through exterior doors. **1014.2**

It is permissible to provide egress through an adjoining room or space, provided the adjacent rooms or spaces are accessory to each other. In this context, the term *accessory* describes an interrelationship between the adjoining spaces based on their use, and not necessarily their size. Where egress must occur through an intervening space, it is important that such a space be under the same control as the initial space. Egress through such an intervening area to an exit must be direct and obvious so that the occupant is well aware of the exit path. It is assumed that a discernible egress path through an area that essentially is an extension of the area served poses no significant hazard to exiting, provided travel does not enter a hazardous area. Egress travel is not permitted to pass through a room or space classified as a Group H occupancy unless permitted under the conditions of the exception. See Figure 1014-1.

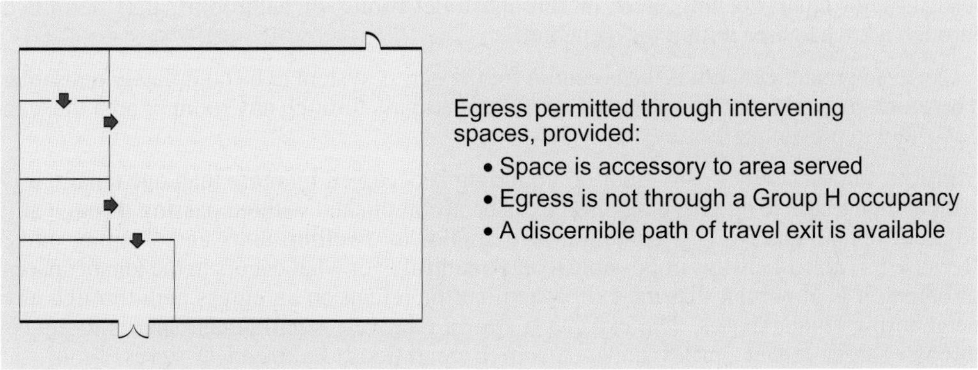

Egress permitted through intervening spaces, provided:
- Space is accessory to area served
- Egress is not through a Group H occupancy
- A discernible path of travel exit is available

Figure 1014-1
Egress through intervening spaces

Where the room of origin is a Group H, S or F occupancy, an exception permits the means of egress to pass through other accessory rooms or spaces designated as equal or lesser hazards. Because the hazard level is not increased along the egress path through the intervening room, the general prohibition does not apply.

This section is also concerned with the arrangement of the exit path in that it puts restrictions on certain spaces that are considered to present an undue probability of obstruction to free egress travel. Therefore, the code prohibits the exit path from passing through kitchens, store rooms, closets and spaces used for similar purposes where the probability of things obstructing the path of travel is substantially greater. An exception permits travel through a stockroom, provided four conditions are met as illustrated in Figure 1014-2.

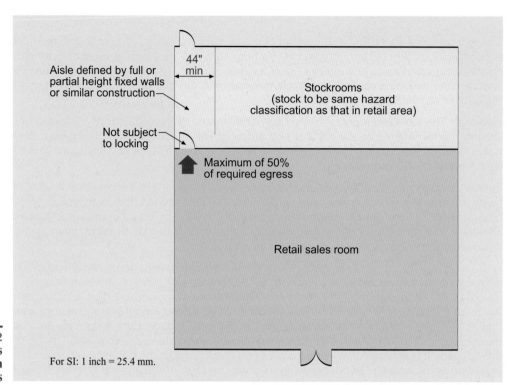

Figure 1014-2
**Egress
through
stockrooms**

As is so frequently the case throughout the code, exceptions are made for rooms within dwelling and sleeping units. Although egress from dwelling units or sleeping units shall not lead through other sleeping areas or through toilet rooms or bathrooms, it is permitted through a kitchen area within the same unit.

An even greater concern is the potential for the egress system to be completely unusable. Therefore, the code specifies that exit access cannot pass through any room or area that can be locked to prevent egress.

1014.2.1 **Multiple tenants.** Where any floor of a building is occupied by more than one tenant, it is critical that access to all of the required exits be accomplished without passing through any adjacent tenant spaces. This condition also applies to dwelling units and sleeping units. Because it is almost always impossible to have control over what occurs in the tenant spaces of others, it is important that the exit system not be reliant on an egress path through any neighboring tenant spaces. The exception allows relatively small tenant spaces to egress through larger tenant spaces under specified conditions. Independent egress from the smaller tenant spaces is not required where such spaces are limited in size, provided they are classified to an occupancy group similar to that of the main occupancy and their means of egress cannot be locked. An example might be a branch bank or fast-food restaurant tenant that is located within a large retail store.

1014.2.2 **Group I-2.** Special means of egress provisions are provided for health care suites in Group I-2 occupancies. The definition of "Suite" in Section 202 identifies the scope of such special provisions. The concept of suites recognizes those arrangements where staff must have more supervision of patients in specific treatment and sleeping rooms. Therefore, the general corridor width and fire-rating requirements are not appropriate under such conditions. The special allowances for suites are not intended to apply to day rooms or business functions of the health care facility.

As a general rule, exiting from habitable rooms or suites in Group I-2 occupancies must be directly to a corridor. The potential for obstructions is high where egress in such occupancies must pass through other use areas. There is an exception where the door to the room or suite opens directly to the exterior at ground level.

Suites in patient sleeping areas. Travel through one intervening room is permitted in **1014.2.3** suites of patient sleeping room areas provided one of two specified conditions is met:

1. Patient sleeping rooms are permitted to egress through one intervening space, provided no more than eight patient beds are served.

2. For special nursing suites or for patient sleeping room suites with more than eight beds, the code allows for egress travel through one intervening room where direct and continuous supervision is provided. This need for such supervision is not required where Condition 1 is met.

Limitations on floor area, single means of egress and travel distance are shown in Figure 1014-3.

Suites in areas other than patient sleeping areas. Where the area of the suite does not **1014.2.4** contain patient sleeping areas, travel through intervening rooms is limited to one of the following conditions:

1. Occupants of rooms, other than those used for patient sleeping purposes, that are located within a suite may travel through a single intervening room if the travel distance does not exceed 100 feet (30 480 mm).

2. Occupants of rooms within suites designed for other than sleeping purposes may travel through up to two intervening rooms, provided the travel distance is restricted to 50 feet (15 240 mm).

Additional requirements are illustrated in Figure 1014-3.

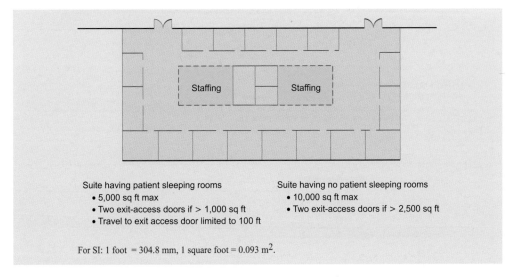

Suite having patient sleeping rooms
• 5,000 sq ft max
• Two exit-access doors if > 1,000 sq ft
• Travel to exit access door limited to 100 ft

Suite having no patient sleeping rooms
• 10,000 sq ft max
• Two exit-access doors if > 2,500 sq ft

For SI: 1 foot = 304.8 mm, 1 square foot = 0.093 m^2.

**Figure 1014-3
Health care
suites**

Health care suites are often utilized to eliminate the requirement for smoke partitions that define a corridor in Group I-2 occupancies. This is necessary because privacy curtains or sliding glass doors are often desired for more efficient operation of the facility. By simply considering the connected space an intervening room rather than a corridor, the limitations imposed by the corridor provisions do not apply. It should be noted that for Group I-2 occupancies, the threshold for the minimum number of means of egress is based on floor area as opposed to occupant load. Typically, surgery recovery areas (post-op) and intensive care units (ICU) are not considered patient sleeping rooms as they are under constant monitoring and supervision.

1014.3 **Common path of egress travel.** The definition of a common path of egress travel is found in Section 1002.1. Described as "that portion of exit access which the occupants are required to traverse before two separate and distinct paths of egress travel to two exits are available," a common path of travel is that portion of the exit system where no egress options are available to the occupant. The length of the common path is measured from the most remote point of a room or area to the nearest location where multiple exit paths to separate exits are available. See Figure 1014-4. The definition of a common path of egress travel also indicates that exit paths that merge are also considered common paths of egress travel. An example of such a situation is shown in Figure 1014-5.

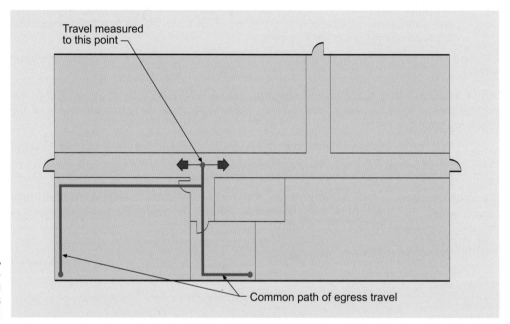

Figure 1014-4
Common path of egress travel

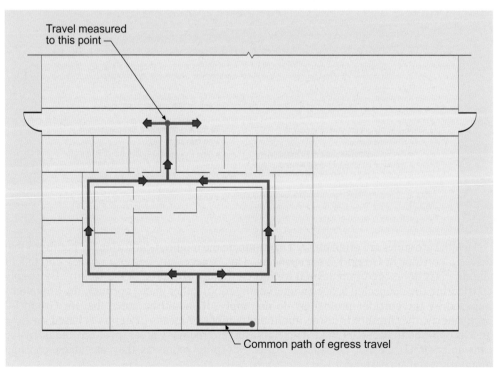

Figure 1014-5
Merging paths of travel

	Nonsprinklered Buillding	Sprinklered Building
H-1, H-2, H-3	25 feet	25 feet
B[1], F, S[1]	75 feet	100 feet
I-3	not permitted	100 feet
A, E, H-4, H-5, I-1, I-2, I-4, R[2], M, U[1]	75 feet	75 feet

[1] Tenant space having an occupant load of 30 or less may have a common path of egress travel of 100 feet.
[2] In a fully sprinklered (NFPA 13 or 13R) Group R-2 occupancy, common path limited to 125 feet.

For SI: 1 foot = 304.8 mm.

Figure 1014-6
Common path of egress travel

As a general rule, the maximum length of a common path of egress travel is 75 feet (22 860 mm). Because of unique risks potentially encountered in a high-hazard occupancy, the common path of travel is limited to 25 feet (7620 mm) in Groups H-1, H-2 and H-3. Four exceptions are provided that increase the common path of travel to a distance of 100 feet (30 480 mm) or more. These exceptions are included with the general requirements in Figure 1014-6.

The most obvious example of a "common path" condition is a room with a single exit or exit access doorway. Although there are numerous paths that may lead to the doorway, eventually they all end up at the same point. Where two or more complying exits or exit access doorways are provided, common path conditions do not exist.

Section 1015 *Exit and Exit Access Doorways*

Exit or exit access doorways required. It would seem obvious that every occupied portion of a building must be provided with access to at least one exit or exit access doorway. It is assumed that if buildings are occupied, then the occupants obviously have a method of entering the various building spaces. Therefore, that same entrance is available to serve as the means of egress. Under many conditions, however, the use of the entrance as the only egress point is insufficient. The basic reason for requiring multiple means of egress is that in a fire or other emergency, it is very possible that the entry door will be obstructed by the fire and, therefore, not be usable for egress purposes. A second exit or exit access doorway can provide an alternative route of travel for occupants of the room or area. However, it is often unreasonable to require multiple egress paths from small spaces or areas with limited occupant loads. It is also seldom beneficial because of the relatively close proximity in which such exits or exit access doorways must be located. Therefore, the code does not require a secondary egress location from all rooms, areas or spaces. **1015.1**

In this section, the code deals with the number of exits or exit access doorways that are going to be required to accommodate the occupant load to be served. In addition, where the distance of travel to a single exit or exit access doorway is excessive, an alternative egress route is also required. The IBC establishes two basic criteria for providing adequate egress for occupants from any space within the building. First, at least two exits or exit access doorways must be provided when the occupant load of the space exceeds the values set forth in Table 1015.1. Second, two or more egress doorways are required when the common path of egress travel exceeds the limitations found in Section 1014.3. See Application Examples 1015-1 and 1015-2. Third, the provisions of Sections 1015.3 through 1015.6 may mandate a minimum of two egress doorways for specific mechanical equipment areas. The only exception to the general provisions involves rooms in Group I-2 occupancies, which are regulated by Sections 1014.2.2 through 1014.2.7.

As seen in Table 1015.1, the threshold for requiring a second exit or exit access doorway from a space varies based upon the occupancy designation of the space. The variations are caused by conditions associated with the specific uses that occur within the room or area. Factors that contribute to the differences in occupant load include the concentration of occupants, occupant mobility and the presence of hazardous materials. There is one specific exception to Table 1015.1 which allows a single means of egress from Group R-2 and R-3 occupancies in fully-sprinklered buildings for occupant loads of 20 or less.

Application Example 1015-1

GIVEN: A sprinklered office tenant space as shown with a total occupant load of 73.

DETERMINE: The minimum number of exit access doorways required.

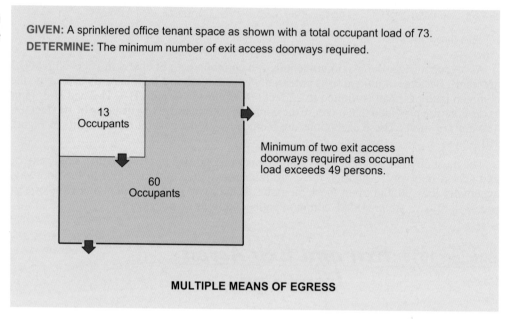

MULTIPLE MEANS OF EGRESS

Application Example 1015-2

GIVEN: A nonsprinklered office tenant space as shown with a total occupant load of 46.

DETERMINE: The minimum number of exit access doorways required.

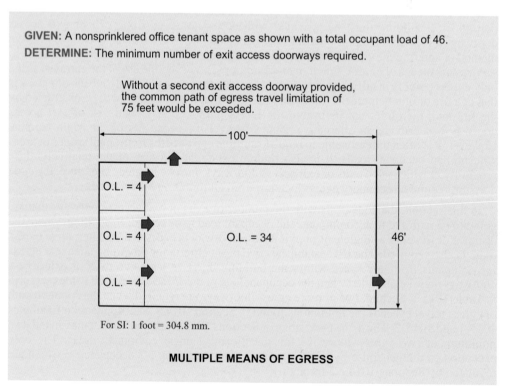

MULTIPLE MEANS OF EGRESS

Three or more exits or exit access doorways. Where any room, space or other floor area **1015.1.1**
has an occupant load exceeding 500, at least three exits or exit access doorways are
mandated. A minimum of four exits or exit access doorways are required where the
occupant load exceeds 1,000.

Exit or exit access doorway arrangement. In addition to providing multiple means of **1015.2**
egress, it is imperative that egress paths remain available and usable. To ensure that the
required egress is sufficiently remote, the code imposes rather strict requirements relative to
the location or arrangement of the different required exits or exit access doorways with
respect to each other. The purpose here is to do all that is reasonably possible to ensure that
if one means of egress should become obstructed, the others will remain available and will
be usable by the building occupants. As a corollary, this approach assumes that because the
remaining means of egress are still available, there will be sufficient time for the building
occupants to use them to evacuate the building or the building space.

Required separation of two exits. This remoteness rule in the IBC is sometimes referred
to as the one-half diagonal rule. The one-half diagonal rule states that if two exits or exit
access doorways are required, they shall be arranged and placed a distance apart equal to not
less than one-half of the maximum overall diagonal of the space, room, story or building
served. Such a minimum distance between the two means of egress, measured in a straight
line, shall not be less than one-half of that maximum overall diagonal dimension. See Figure
1015-1 for examples of the application of this rule.

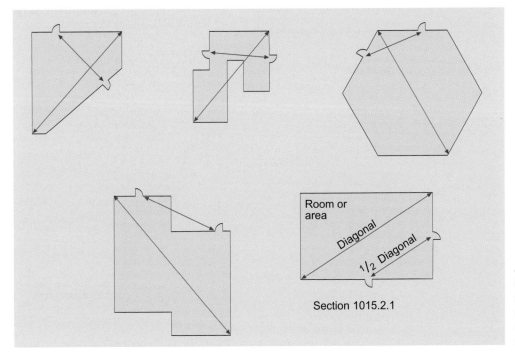

Section 1015.2.1

Figure 1015-1
**Separation of
exits or
exit-access
doorways**

The code does not specifically state the manner in which the straight-line measurement
should be taken. In practice, different building officials determine the length of the
straight-line measurement for egress separation in different ways. In some instances,
building officials measure that distance from the near edge of one egress door to the near
edge of the other. Other building officials apply the rule as measuring the distance between
the center lines or far edges of the two required egress doors. Based upon the lack of
specifics in the code, it would seem logical to consider the distance to be measured as
between the center lines of exits or exit access doorways. It should be noted that, by
definition, the term *exit access doorway* includes any point of egress where the occupant has

a single point that must be reached prior to continued travel to the egress door. See Figure 1015-2.

The use of the one-half diagonal rule has been beneficial to code users for many years. It quantifies the code's intent when the code requires that separate means of egress be remote. It does not leave the building official with a vague performance-type statement that can, in many instances, result in a situation where egress separation would be dictated more by the design or desired layout of the building rather than by a consideration for adequate and safe separation of the means of egress.

In applying the one-half diagonal rule to a building constructed around a central court with an egress system consisting of an open balcony that extends around the perimeter of the court, it is important to take the measurement of the diagonal from which the one-half diagonal dimension is derived at the proper locations. Refer to Figure 1015-3 for examples.

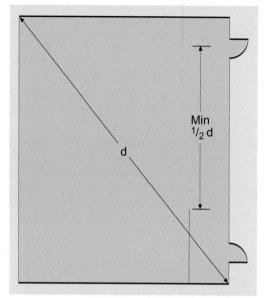

Figure 1015-2
Egress separation

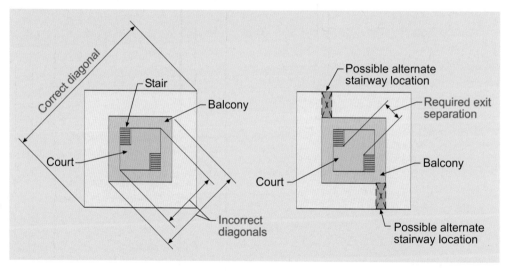

Figure 1015-3
Required egress separation

Figure 1015-4 illustrates Exception 1 to the one-half diagonal rule for those buildings, such as core buildings, where the means of egress are sometimes arranged in rather close proximity. The code recognizes the benefits of such a floor arrangement and makes a specific exception in the event there is such a design. If the exits or exit access doorways are connected by a fire-resistance-rated corridor, the distance determined by one-half of the maximum overall diagonal of the space served may be measured along the path of travel inside the corridor between the two exits. It is specific that the connecting corridor is to be of 1-hour fire-resistant construction.

A second exception reduces the minimum length of the overall diagonal dimension between remote exits or exit access doorways in those buildings equipped throughout with automatic sprinkler systems. Because of the presence of sprinkler protection, the separation distance need only be one-third of the length of the overall diagonal dimension. The use of this exception results in a reduction of the required distance between exits or exit access doorways by $33^{1}/_{3}$ percent. See Figure 1015-5.

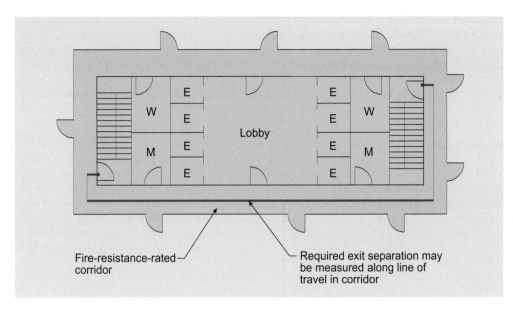

Fire-resistance-rated corridor

Required exit separation may be measured along line of travel in corridor

Figure 1015-4
Core arrangement of exit enclosures

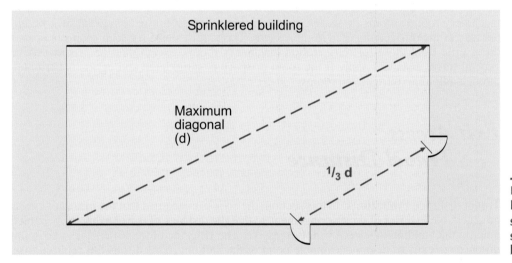

Sprinklered building

Maximum diagonal (d)

$^{1}/_{3}$ **d**

Figure 1015-5
Exit separation—sprinklered building

Three or more exits or exit access doorways. When more than two means of egress are required, the remoteness rule takes on more of a performance character. In such an instance, at least two of the required exits or exit access doorways shall be arranged to comply with the one-half diagonal rule (one-third in fully sprinklered buildings). Although not specifically stated, the other means of egress should be arranged at a reasonable distance from the other egress points so that if any one of the required exits or exit access doorways becomes blocked by a fire or any other emergency, the others will be available. Obviously, this decision will require some very careful evaluation and judgment on the part of the building official. There may be a sufficient basis for applying that same rule when considering each possible pair of egress components in a multi-exit situation. The code is silent in this particular aspect, and proper code administration does require substantial, careful evaluation and judgment on the part of the building official in ensuring that the number of means of egress required is sufficiently remote so that it is not likely that the use of more than one access to exit will be lost in any fire incident.

1015.2.2

1015.3 Boiler, incinerator and furnace rooms. Depending upon the capacity of the fuel-fired equipment located in a large boiler room or similar area, it may be necessary to provide a secondary egress doorway because of the potential hazards. At least two exit access doorways are required where the room exceeds 500 square feet (46 m²) and any single piece of fuel-fired equipment such as a furnace or boiler exceeds 400,000 Btu (422 000 KJ) input capacity. The requirement is based on the size of a single piece of equipment, not the aggregate total of all fuel-fired equipment in the room. Access to one of the two exit access doorways may be accomplished through the use of a ladder or an alternating stair device described in Section 1009.10. As with other multiple-exit situations, the two doorways must be adequately separated, in this case by a horizontal distance no less than one-half the maximum overall diagonal.

1015.4 Refrigeration machinery rooms. The *International Mechanical Code®* (IMC®) mandates when refrigeration systems must be contained in a refrigeration machinery room. It can be based on the type of refrigerant, the amount of refrigerant, the type of equipment or other factors. Once a refrigeration machinery room is required by the IMC, the IBC provides the exiting criteria. Where larger than 1,000 square feet (92.9 m²), the room must have at least two exit access doorways, accessed and separated in the same manner as discribed for boiler and furnace rooms. Egress doors from the room must be tight-fitting, self-closing and swing in the direction of travel. Travel distance is also restricted to 150 feet (45 720 mm). It is evident that the presence of multiple exitways and a more limiting travel distance is necessary in order to address the hazards associated with areas containing refrigerants. In addition, provisions are made to provide moderate enclosure of the room through the use of self-closing, tight-fitting doors.

1015.5 Refrigerated rooms or spaces. Considered a bit less of a concern than refrigerated machinery rooms, rooms or spaces that are refrigerated still pose somewhat of a hazard because of the refrigerants used in the system. The requirements for such rooms or spaces differ little from those for refrigeration machinery rooms, except that the less restrictive general travel distances apply if the room or space is sprinklered and egress is permitted through adjoining refrigerated rooms.

Section 1016 *Exit Access Travel Distance*

1016.1 Travel distance limitations. In this section, the code is concerned that the means of egress be accessible in terms of their arrangement so that the distance of travel from any occupied point in the building to an exit is not excessive. The IBC, therefore, establishes maximum distances to exits from any occupiable point of the building. This distance is referred to as the travel distance. Travel distance is the distance that a building occupant must travel from the most remote, occupiable portion of the building to either the door of a vertical exit enclosure, an exit passageway or a horizontal exit; to an exterior egress stairway or exterior egress ramp; or to an exterior exit door located at the level of exit discharge. A travel limit is only imposed to the nearest exit component, not to all required exits from the room, floor or building.

Each of these six elements in a means of egress system is considered to represent a sufficient level of safety such that the code is no longer concerned about the distance that the building occupant must travel to reach the eventual safe place. On the top floor of the world's tallest buildings, the distance to exits, referred to as the travel distance, is measured on that floor from the most remote occupiable point to the point where the building occupant enters the enclosed exit stairway. The fact that the building occupant then has dozens of floors of stairway to traverse before exiting the building is not a consideration in dealing with distance to exits. Indirectly, in establishing a maximum distance of travel to a point of reasonable safety, the code is imposing a time factor on the ability of the building occupant to travel from the point of occupancy to a relatively safe place either outside or within the building.

Measurement of travel distance. Travel distance is one of the most difficult features of the egress system to determine in either the design or the plan review stage. Travel distance is intended to be measured along the natural, unobstructed path available to the building occupant. See Figure 1016-1. That path is often determined by the location of partitions, doors, furniture, equipment and similar objects. Many of these objects are reasonably portable and, as a consequence, the actual path available is frequently and easily altered. Although it is obvious that travel is to be measured around permanent construction and building elements, how to measure travel in areas with tables, chairs, furnishings, cabinets and similar temporary or movable fixtures or equipment is debatable.

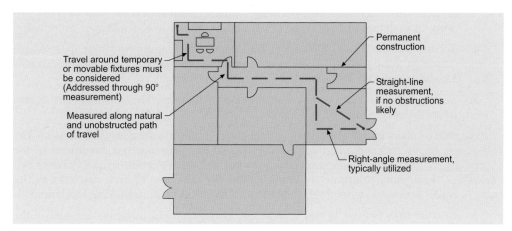

Figure 1016-1
Measurement of travel distance

The preferred approach, conservative in nature, would dictate using the right-angle method for measuring travel distance. This method recognizes that obstacles such as desks, shelving, modular furnishings, etc., would cause travel to negotiate around such objects. The increased distance determined by right-angle measurement would account for such travel. On the other hand, the straight-line method would assume there are no obstructions along the measured travel path. Although such conditions at times exist, it is seldom that a straight-line path of travel will always be available. Although within private offices and similar small areas the straight-line approach would seem to cause no great concern, the method of measurement could be critical in larger spaces such as home improvement centers or department stores. Care should be taken to measure the travel distance in a manner that best represents the actual means of egress through the space.

Maximum travel distance. Basically, the code states that travel distance to an exit may not exceed the distances found in Table 1016.1. For most occupancies, the travel limitation is 200 feet (60 960 mm) in nonsprinklered buildings and 250 feet (76 200 mm) in buildings provided throughout with an automatic sprinkler system. However, there are numerous modifications to the general requirements. Fully sprinklered Group B occupancies are permitted travel distances of 300 feet (91 440 mm). In low-hazard factories and warehouses, as well as utility buildings, the maximum travel distance is also 300 feet (91 440 mm) and 400 feet (121 920 mm) where the building is fully sprinklered. Travel distance varies for high-hazard occupancies from a low of 75 feet (22 860 mm) in a Group H-1 to a high of 200 feet (60 960 mm) in a Group H-5. Those institutional occupancies classified as Groups I-2, I-3 and I-4 are permitted a maximum travel distance of 200 feet (60 960 mm) in a sprinklered building.

To alert code users to the fact that there are also other code sections that affect travel distance requirements, the footnotes to Table 1016.1 identify several locations where special requirements may be found, including:

1. Section 402.4.4. Travel distance within mall tenant spaces or within the mall itself. The section limits travel distance within the mall to 200 feet (60 960 mm). The travel distance within individual tenant spaces is also limited to 200 feet (60 960 mm).

2. Section 404.9. Travel distance within the atrium is limited to 200 feet (60 960 mm) on balconies or other egress paths not located on the lowest level of an atrium.

3. Section 1028.7. Travel distance within an assembly building having smoke-protected assembly seating or open-air seating. Where smoke-protective assembly seating is provided, the travel distance from every seat to the nearest concourse entrance is limited to 200 feet (60 960 mm). Travel from the concourse entrance to a stair ramp or walk on the exterior of the building is also limited to 200 feet (60 960 mm). Where the seating is open to the exterior, the maximum travel distance from each seat to the building's exterior shall not exceed 400 feet (121 920 mm). In such buildings of Type I or II construction, travel distance is unlimited.

4. Section 1021.2. Travel distance in buildings requiring only a single exit. Where buildings are permitted to have only one exit, the maximum travel distance is regulated by Table 1021.2.

5. Section 3103.4. Travel distance in temporary structures. The maximum exit access travel distance permitted in a temporary structure is 100 feet (30 480 mm).

Exceptions 3 and 4 intend to clarify that unenclosed stairways are permitted as part of the exit access travel distance when connecting two stories provided the conditions of the exceptions are met. Although the exceptions are not located in Section 1022 regulating exit enclosures, their primary purpose is to indicate that the enclosure of such stairways by fire barriers and/or horizontal assemblies is not required subject to the following limitations:

3. In other than Groups H and I, enclosure is not required for up to one-half of the number of egress stairways that serve only one adjacent floor. Any two such atmospherically interconnected floors shall not communicate with other floors.

4. In other than Groups H and I, all interior egress stairways that serve only the first and second stories are not required to be enclosed in a fully sprinklered building. A minimum of two means of egress must be provided from both floors served by the unenclosed stairways.

1016.3   **Exterior egress balcony increase.** As depicted in Figure 1016-2, the travel limitations specified in Section 1016.1 may be increased by an additional 100 feet (30 480 mm) if the increased travel distance is the last portion of the travel distance on an exterior egress balcony. As an example, for a Group B occupancy, either the 200 feet (60 960 mm) in a nonsprinklered building or the 300 feet (91 440 mm) in a sprinklered building may be increased to 300 feet and 400 feet (91 440 mm and 121 920 mm), respectively, if in each instance the last 100 feet (30 480 mm) of travel distance is on an exterior egress balcony. Simply stated, all travel, up to a maximum of 100 feet (30 480 mm), that occurs beyond that permitted by Section 1016.1 must occur on an exterior egress balcony.

Section 1017 *Aisles*

As a portion of the exit access, aisles are primarily regulated for minimum width purposes. Aisles typically serve occupant travel from adjoining aisle accessways and often merge into a main aisle. Aisles are commonly found in those buildings that contain furniture, fixtures or equipment

1017.1   **General.** This section requires that aisles be provided in all occupied portions of the exit access that contain seats, tables, furnishings, displays and similar fixtures or equipment. Primarily, this would have application to occupied-use areas or rooms where it is necessary

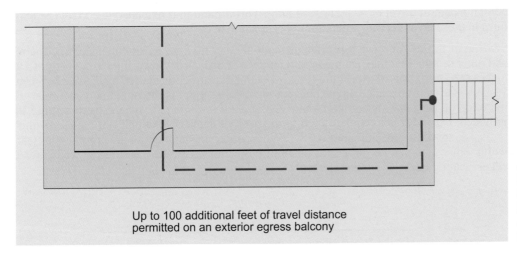

Up to 100 additional feet of travel distance
permitted on an exterior egress balcony

**Figure 1016-2
Egress
balcony travel
distance
increase**

to provide a circulation system so that building occupants will have reasonable means for moving around in the occupied spaces, as well as have access to corridors and other components of the egress system. It is customary to think of aisles in such facilities as theaters where an aisle system is installed to serve the fixed-seating areas. However, this section does not apply to such uses. Aisles in assembly areas, other than those with seating at tables, such as in restaurants, are regulated solely under the provisions of Section 1028. In addition to dining areas and similar uses with seating at tables, the provisions of this section apply to circulation systems through open office areas, retail sales rooms, manufacturing areas and other spaces with similar features. As mentioned, aisles serving assembly areas, including grandstands and bleachers, are to comply with Section 1028. Also, all aisles located within an accessible route of travel must comply with Chapter 11.

The minimum required width of aisles may vary according to the occupancy in which the aisles are located; the nature of the use area that the aisles serve; the occupant load served by the aisles; and even, in some instances, the type of occupant served. In public areas of Group B and M occupancies, such as open offices, retail sales areas and similar spaces, the minimum required clear width of any aisle is 36 inches (914 mm). In nonpublic areas, where the number of employees or other people served by the aisle is less than 50 and the aisle is not required to be accessible, a minimum required width of only 28 inches (711 mm) is mandated. All aisles are to be unobstructed, except for those permitted projections such as nonstructural trim and doors.

It is important to realize that the first element of any means of egress system is typically an aisle accessway, which then leads to an aisle. The code regulates aisle accessways to some degree in Group A and M occupancies, as well as where seating occurs at tables, but it is silent for all other occupancy classifications. The building official must determine at what point an aisle accessway becomes an aisle and must be regulated for minimum width in accordance with this section.

Aisle accessways in Group M. Aisle accessways are regulated within merchandise pads of **1017.3** Group M mercantile occupancies. A merchandise pad is defined as the merchandise display area that contains multiple counters, shelves, racks and other movable fixtures. Every element within a merchandise pad must adjoin a minimum 30-inch-wide (762 mm) aisle accessway on each side. Travel within a merchandise pad is also limited, with a maximum common path of travel of 30 feet (9144 mm). This limitation is extended to 75 feet (22 880 mm) in those areas serving a maximum occupant load of 50.

1017.4   **Seating at tables.** Because it is often difficult to determine the exact dimensions of chairs and other seating that may encroach into an aisle or aisle accessway, a measurement of 19 inches (483 mm) is used to account for any potential obstruction caused by the seating. The 19-inch (483 mm) distance, applicable for movable seating, is measured perpendicular to the edge of the table or counter where the seating is provided. Where fixed seating is provided, the width is to be measured to the back of the seat. Examples of the use of this provision are illustrated in Figure 1017-1. Where other side boundaries are present, the clear width, other than that permitted for handrail projections, is to be measured to walls, tread edges, seating edges or similar elements.

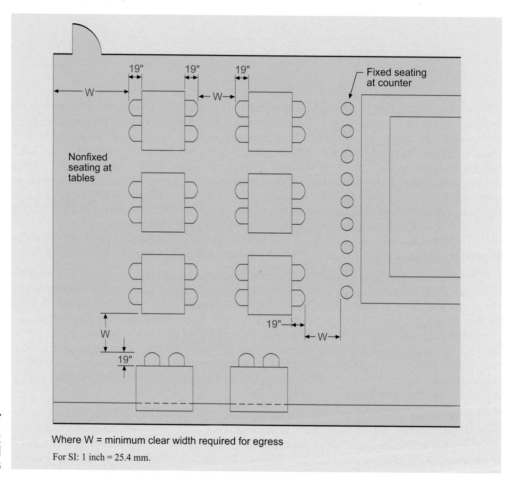

Figure 1017-1
**Seating at
tables and
counters**

Where W = minimum clear width required for egress

For SI: 1 inch = 25.4 mm.

The clear width of aisle accessways serving various arrangements of seating at tables or counters is based upon 12 inches (305 mm) plus 0.5 inch (12.7 mm) of additional width for each 1 foot (305 mm) or fraction thereof, beyond 12 feet (3658 mm) of aisle accessway length. This length is measured from the center line of the seat farthest from the aisle.

Minimum clear width = 12 inches + 0.5 inch (x – 12 feet),

WHERE:

    x = the distance in feet between the aisle and the center line of the most remote seat.

It should be remembered that in no case should the width be less than that based upon the capacity requirements of Section 1005.1, or, if a Group A occupancy or an assembly area of a Group E occupancy, Section 1028.6. On the other hand, there is no minimum width required for those portions of aisle accessways serving four or fewer persons and 6 feet (1829 mm) or less in length.

Consistent with other provisions for means of egress, single-directional travel along aisle accessways is also limited in length. The distance from any seat to the point where two or more paths of egress travel are available to separate exits is limited to 30 feet (9144 mm). In application, the 30-foot (9144 mm) limitation would be the maximum length of a dead-end aisle accessway. Within that distance, the aisle accessway must reach an aisle or other means of egress element where travel would be limited based upon the common path provisions of Section 1014.3. Various examples of the width provisions for aisle accessways serving seating and tables are shown in Figure 1017-2.

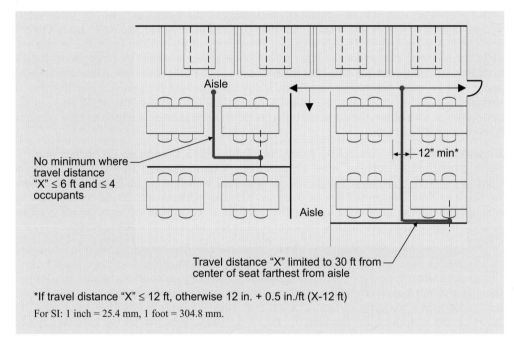

No minimum where travel distance "X" ≤ 6 ft and ≤ 4 occupants

12" min*

Aisle

Aisle

Travel distance "X" limited to 30 ft from center of seat farthest from aisle

*If travel distance "X" ≤ 12 ft, otherwise 12 in. + 0.5 in./ft (X-12 ft)

For SI: 1 inch = 25.4 mm, 1 foot = 304.8 mm.

Figure 1017-2
Egress width at tables

Section 1018 *Corridors*

The IBC contains a definition of the term *corridor*. Defined as "an exit access component that defines and provides a path of egress travel to an exit," the determination as to when a corridor exists is essentially left to the building official.

For the purpose of the code, a corridor is typically a space where the building occupant has very limited choices as to paths or directions of travel. The available path is restricted and is usually bordered by other occupied-use spaces. As a consequence, it is potentially exposed to fires that might occur in those enclosed spaces unknown to anyone in the corridor. Generally speaking, in a building space of this type, the occupant has only two choices as far as direction of travel through the exiting system is concerned. For that reason, it is sometimes necessary for the building official to evaluate the planned layout of an area and determine whether the space presents a potential fire-hazard exposure to building occupants as any regular, well-defined corridor might. If the determination is that the fire-exposure potential is the same, the building space should be made to comply with requirements for corridors.

To provide for a greater degree of consistency in the identification or determination of corridors, some jurisdictions have established a set of guidelines that expand on the definition in the IBC. An excellent example addresses four common characteristics of corridors as regulated in the code.

1. It is a space formed by enclosing walls or construction over 6 feet (1828 mm) in height,

2. It has a length to width ratio greater than 3 to 1,

3. Its primary function is for the movement of occupants in the means of egress system, and

4. It has a length greater than that permitted for a dead-end condition.

All four conditions must be present for the element to be considered a corridor. There are many rooms that meet the first, second and fourth conditions, but their primary use is for something other than the movement of occupants in the egress system. An open office system may have spaces that meet Conditions 2, 3 and 4; however, the walls are not over 6 feet (1828 mm) in height. In such a space, the egress paths would be regulated as aisles or aisle accessways. It also makes little sense to designate a space as a corridor where it conforms to Conditions 1, 2 and 3 but is very limited in length, as the travel time through the space will be quite short. This approach is just one method for providing uniformity in defining a corridor; there are undoubtedly others that also help in the application of the code's intent. Of course, the definition of a corridor is not as critical where it is not required to be fire-resistance rated.

1018.1 **Construction.** The thresholds found in Table 1018.1 indicate at what point a fire-resistance-rated corridor must be provided. Where required by the table, walls of corridors shall be considered fire partitions, and as such shall be regulated by the provisions of Section 709. The provisions in Section 709 also address the lid of the corridor in relationship to the continuity of the fire partitions. Examples of appropriate corridor fire-resistance-rated construction are illustrated in the discussion of Section 709.

In a fully sprinklered building, a corridor must only be fire-resistance rated in Groups H, R, I-1 and I-3. In Groups H-1, H-2, H-3, I-1 and I-3, the protection is required regardless of the occupant load served by the corridor. Where a corridor serves an occupant load of more than 10 in a sprinklered Group R occupancy, it must have a minimum $1/_2$-hour rating. Where a corridor serves an occupant load of more than 30 in a Group H-4 or H-5 occupancy, the walls of the corridors are required to be of not less than 1-hour fire-resistance-rated construction.

In buildings not equipped with an automatic sprinkler system throughout, corridors are required to be fire-resistance rated in all occupancies, based upon the occupant load served. When a corridor serves an occupant load greater than 10 in a Group R occupancy or more than 30 in a Group A, B, E, F, M, S or U occupancy, 1-hour fire-resistance-rated corridor walls are required. Where the occupant load reaches the levels specified, it is appropriate to afford those persons in the exit corridor some additional protection from potential fire occurring in the enclosed spaces bordering the corridor. Therefore, a minimum separation for the corridor of 1-hour fire resistance is deemed necessary.

Fire-resistance-rated corridor exceptions. A series of exceptions to the 1-hour corridor requirement addresses a number of special circumstances where it is felt that the fire-resistance-rated separation is not necessary. The first applies to corridors in Group E occupancies. Such corridors need not comply with the 1-hour fire-resistance requirement when:

1. Every classroom that the corridor serves has at least one egress door leading directly to the outside at ground level, and

2. Rooms that are served by the corridor and are used for assembly purposes have at least one-half of their required egress doors leading directly to the outside at ground level.

Exception 2 exempts corridors within dwelling units or sleeping units of Group R occupancies from the fire-resistance rating. Exceptions 3 and 4 indicate that fire-resistance-rated corridors are not required in open parking garages, nor are they required in spaces of Group B occupancies requiring only a single means of egress.

Corridor width. The minimum required width of a corridor is regulated like any other **1018.2** component of the means of egress. The calculated width, determined by Section 1005.1, must be compared with the minimum component width (in this case, determined by this section). The greater of the two required widths must be available for egress purposes.

For component width purposes, a corridor is required to be at least 44 inches (1118 mm) in width. A number of exceptions reduce or increase this minimum width to the following dimensions:

1. Twenty-four inches (610 mm) is permitted in those nonpublic areas where access is necessary to service or utilize electrical, mechanical or plumbing systems or equipment. These areas are seldom occupied, and then typically only by a single occupant.

2. Thirty-six inches (914 mm) is adequate where the corridor serves as a means of egress for 49 occupants or fewer, or where the corridor is located in a dwelling unit. Small occupant loads can easily egress through corridors of this minimum width. It also allows adequate width for circulation purposes and the movement of furnishings and equipment.

3. Seventy-two inches (1829 mm) is the minimum width required for school corridors serving at least 100 students and for corridors in surgical health-care areas where the movement of gurneys is anticipated. The increased width for Group E occupancies provides for better circulation during peak usage, whereas the increase in surgical areas allows the limited movement of nonambulatory patients and equipment.

4. Ninety-six inches (2438 mm) is required in Group I-2 occupancies for the movement of patients in wheelchairs, gurneys or in standard hospital beds. This width is necessary to safely accommodate this kind of use.

As stated, this section provides for the minimum required widths of corridors. The code is not totally clear on how the required corridor width is determined when the occupant load is large enough to require corridor widths in excess of these minimums. A common approach to determining the required minimum widths of corridors is that the width should be related to the required width of the exit to which the corridor leads. The most logical procedure for determining corridor widths is to determine the required width of the exits at the end of the corridor and to size the corridor to provide that same minimum width. See Figure 1018-1. The controlling concept is to mutually equate the required width of the corridors and the required width of the exits that the corridor serves. However, the minimum required width of 44 inches (1118 mm) must be maintained for the corridor unless an exception is applicable.

Dead ends. This section establishes the limitations on dead-end corridors. Basically, **1018.4** wherever more than one exit or exit access doorway is required, the exit access shall be arranged so that occupants can travel in either direction from any point to a separate exit. However, dead ends in corridors are permitted up to a maximum length of 20 feet (6096 mm) as illustrated in Figure 1018-2. In the event that the conditions of building occupancy permit access to only a single exit, the dead-end limit does not apply. When the basic requirement is for travel in only a single direction to one exit, travel in the opposite direction is not considered to be in violation of the dead-end requirements. However, the maximum distance of travel will be regulated by the common path provisions of Section 1014.3.

Two exceptions permit dead ends of up to a length of 50 feet (15 240 mm). In Group I-3 occupancies where free movement is allowed to some degree (Conditions 2, 3 or 4), and in low-hazard and moderate-hazard occupancies housed in fully sprinklered buildings, the increased length of the dead-end condition has been found to be acceptable. An additional exception permits a dead end of unlimited length, provided that such length of the dead-end corridor does not exceed two- and one-half times the least width of the dead-end portion of the corridor. See Figure 1018-3.

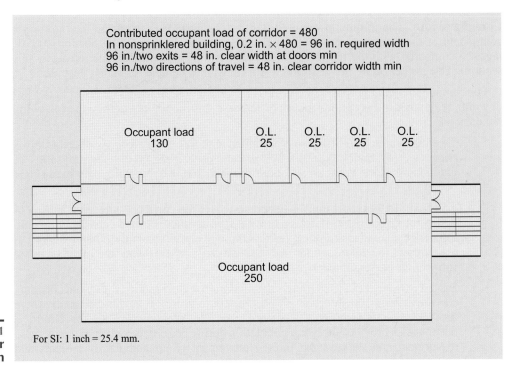

Contributed occupant load of corridor = 480
In nonsprinklered building, 0.2 in. × 480 = 96 in. required width
96 in./two exits = 48 in. clear width at doors min
96 in./two directions of travel = 48 in. clear corridor width min

Occupant load
130

O.L.
25

O.L.
25

O.L.
25

O.L.
25

Occupant load
250

For SI: 1 inch = 25.4 mm.

Figure 1018-1
Corridor
width

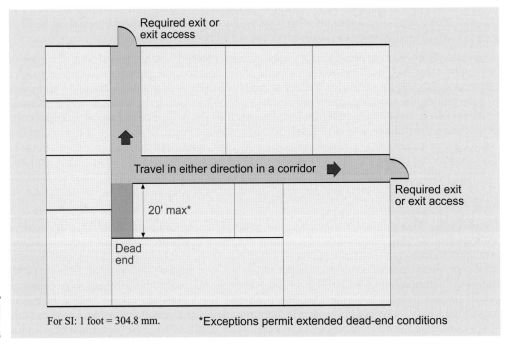

Required exit or
exit access

Travel in either direction in a corridor

Required exit
or exit access

20' max*

Dead
end

For SI: 1 foot = 304.8 mm.   *Exceptions permit extended dead-end conditions

Figure 1018-2
Dead-end
corridors

The limitation on dead ends is directed toward avoiding portions of the corridor system that could result in entrapment of the building occupant by creating a situation where the building occupant, following a proper path of travel through the means of egress system might, under fire conditions, take a wrong turn into a portion of the system from which there is no outlet. In such a situation, it is possible that the building occupant will have to proceed all the way to the end of the dead end before learning that there is no way out and, thereafter, will have to retrace steps to arrive at an exit.

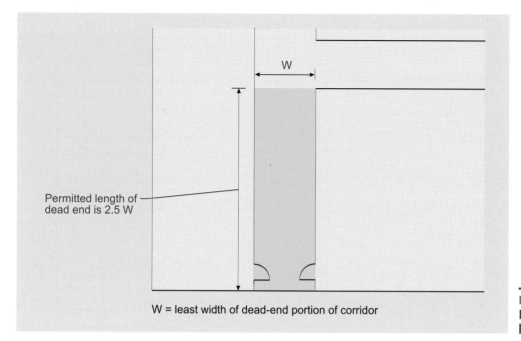

Permitted length of dead end is 2.5 W

W = least width of dead-end portion of corridor

Figure 1018-3
Dead-end limits

Air movement in corridors. Corridors are not to be utilized as a portion of the heating, ventilating and air-conditioning system by serving as plenums or ducts. They also may not be used for relief or exhaust venting purposes. As corridors are relied upon to be an important component of the egress system, it is not appropriate to provide potential avenues for the spread of smoke and gases. Exceptions to the general restriction are provided where the corridor is a source of makeup air for exhaust systems in adjacent spaces, a corridor in a dwelling unit, and a corridor within a tenant space not exceeding 1,000 square feet (93 m²) in area. It should be noted that this provision is applicable to all corridors, not just those that are required to be fire-resistance rated. However, there appears to be inconsistency in the code insofar as adjoining spaces are permitted to be open to a nonrated corridor. Where a lack of physical separation occurs between the corridor and surrounding spaces, how can air be prevented from moving between them? In application, the provision merely is intended to require that those spaces open to the corridor be mechanically designed independent of the corridor (supply, return, exhaust and make-up air) and not rely on the corridor to move air to or from the space.

1018.5

Corridor continuity. It is required that fire-resistance-rated corridors should not be interrupted by intervening rooms, and that they be continuous to an exit. This provision carries out the basic concept, which states that once a building occupant progresses to a certain level of safety—in this case, the safety afforded by a fire-resistance-rated corridor—that level of safety is not thereafter reduced as the building occupant proceeds through the remainder of the means of egress. Therefore, the building occupant, having once reached such a corridor, is not thereafter brought out of the corridor and introduced into another occupied use space of the building; the corridor must be continuous. However, the code emphasizes that it is possible to permit fire-resistance-rated corridors to be conducted through foyers, lobbies and reception rooms. These are not to be considered intervening spaces as long as they are constructed in accordance with the requirements for the corridor they serve. See Figure 1018-4. The code is silent regarding the maximum size or permitted uses for a lobby or reception room. An extreme example would be the lobby of a large hotel. It is not uncommon to see seating areas, piano bars, breakfast buffets and similar

1018.6

occupiable areas open to the fire-resistance-rated corridor. The building official must evaluate the appropriateness of all areas open to the extended corridor. It would be desirable for exiting purposes to provide all direct egress from fire-resistance-rated corridors; however, lobbies, foyers and reception areas are an accepted entrance element that must be considered.

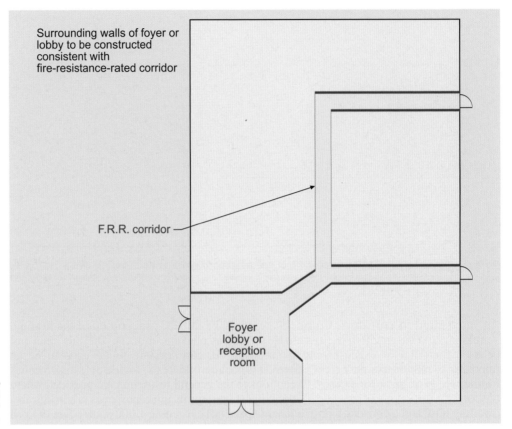

Surrounding walls of foyer or lobby to be constructed consistent with fire-resistance-rated corridor

F.R.R. corridor —

Foyer lobby or reception room

Figure 1018-4
Corridor
continuity

Section 1019 *Egress Balconies*

Balconies used for egress purposes are similar in many ways to corridors and, as such, are required to meet the corridor provisions addressing width, headroom, dead-ends and projections.

1019.2 **Wall separation.** Because there is a potential exposure to a fire condition for occupants utilizing an exterior egress balcony during evacuation of a building, a minimum level of fire protection is mandated. This section requires that the balcony be separated from the interior of the building in a manner consistent with the separation requirements for corridors. However, where two stairs serve the egress balcony, and travel from a dead-end condition to a stair does not occur past an unprotected opening, the separation is not required.

1019.3 **Openness.** One of the features of any egress balcony is its openness to the atmosphere, limiting the amount of smoke and toxic-gas accumulation. To qualify as part of the exit system, the balcony must be at least 50 percent open on the long side. It is also necessary that the open areas above the guardrail be distributed to allow for adequate natural ventilation.

Section 1020 *Exits*

Exits constitute those portions of the means of egress where the occupant first achieves a significant level of fire protection. An exit is expected to provide a mandated level of protection, and that level cannot be reduced until arrival at the exit discharge. Travel within exits is not limited and is typically single directional. Therefore, care is taken to ensure that the exit component is available for use by the building occupant during a fire or other emergency. This section makes it clear that the primary function of an exit component is to provide egress, and any other use cannot interfere with that function. Many of the design requirements for exits are addressed under the general provisions of Sections 1003 through 1013; however, provisions specific to the exit portion of the means of egress are found in Sections 1020 through 1026.

Exterior exit doors. Once the provisions of Section 1008 for doors have been reviewed, **1020.2** there are relatively few requirements remaining to address. It should be noted that at least one exterior door that meets the size requirements of Section 1008.1.1 must be provided for every building used for human occupancy.

Section 1021 *Number of Exits and Continuity*

Exits from stories. As a general requirement, every floor area shall be provided with access **1021.1** to at least two exits. The discussion of Section 1021.2 will address those buildings permitted to have only a single exit. Table 1021.1 also indicates that, based solely on occupant load, three, or even four, exits may be required. Any story of a building that has an occupant load in excess of 500, up to and including 1,000, shall be provided with access to at least three exits. Any story having an occupant load in excess of 1,000 must be provided with access to not less than four exits. Under no circumstances does the IBC require more than four exits for any building or portion thereof based on the number of persons present. It must be noted, however, that additional exits will sometimes be required to satisfy the other egress requirements of Chapter 10.

Buildings with one exit. The code recognizes that there are instances where the life-safety **1021.2** risk is so minimal that it is reasonable to permit a single means of egress. Under limited conditions, this allowance extends to three-story conditions. Table 1021.2 identifies those stories where a single exit is permitted. The table is based on four criteria, including occupancy, number of stories above the grade plane, occupant load and travel distance. Examples illustrating the use of this table are shown in Figure 1021-1. It should be noted that special consideration is given in the footnotes to certain Group B, F, S and R-2 occupancies. All Group R-3 buildings may have a single exit, as well as single-level buildings complying with Section 1015.1 as a space with one means of egress.

Also illustrated in Figure 1021-1 is a permitted condition for a basement where a single exit is permitted. Conditions of this section limit the building to a single level below the first story. Travel distance and occupant load are also both limited.

It is important to understand that the code first looks at exiting from each individual room, space or area within the building. The means of egress is then regulated for each story as evidenced by the provisions of Section 1021.1 and this section. As an added note, once two or more exits are required from a story, all occupants of the story must have access to all of the required exits.

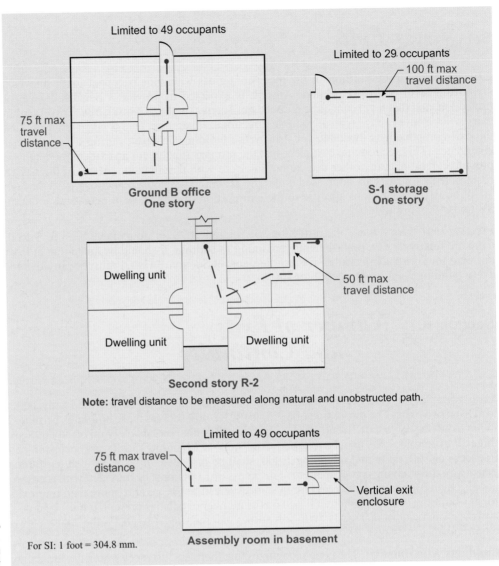

Limited to 49 occupants

75 ft max travel distance

**Ground B office
One story**

Limited to 29 occupants

100 ft max travel distance

**S-1 storage
One story**

Dwelling unit

Dwelling unit

Dwelling unit

50 ft max travel distance

Second story R-2

Note: travel distance to be measured along natural and unobstructed path.

Limited to 49 occupants

75 ft max travel distance

Vertical exit enclosure

Assembly room in basement

For SI: 1 foot = 304.8 mm.

Figure 1021-1
**Stories with
one exit**

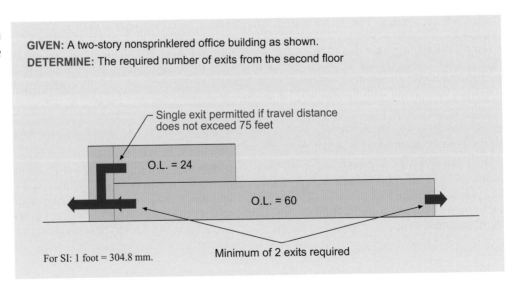

Application
Example
1021-1

GIVEN: A two-story nonsprinklered office building as shown.
DETERMINE: The required number of exits from the second floor

Single exit permitted if travel distance does not exceed 75 feet

O.L. = 24

O.L. = 60

Minimum of 2 exits required

For SI: 1 foot = 304.8 mm.

Only one exit is required from a story where permitted by Table 1021.2 regardless of the number of exits required from other stories in the building. For example, a Group B occupancy on the second floor of a multi-story building is only required to have one exit from the story provided its occupant load does not exceed 29 and the maximum travel distance to an exit does not exceed 75 feet. The number of occupants and travel distances on the other stories do not affect the determination of the second story as a single-exit story. See Application Example 1021-1. Where applicable, other stories are also regulated independently as to the number of exits.

Where multiple tenants or occupancies are located on a specific story, they are to be regulated independently for single-exit determination. The provisions can be applied to specific portions of the story, rather than the story as a whole. As an example, the second story of a building houses two office tenants, each with their own independent means of egress. Each tenant would be permitted a single, but separate, exit provided each had an occupant load of less than 30 and a travel distance not exceeding 75 feet. This portion-by-portion philosophy also applies to a mixed-occupancy condition provided each of the individual occupancies does not exceed the limitations of Table 1021.2. See Figure 1021-2.

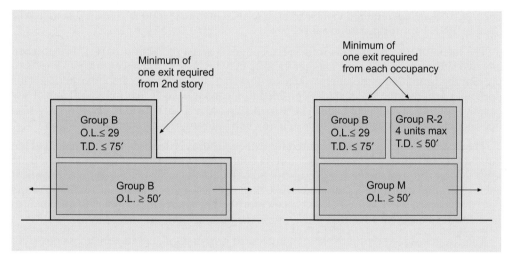

Figure 1021-2
Stories with one exit

Section 1022 *Exit Enclosures*

Enclosures required. Consistent with the general requirements for the enclosure of vertical openings as set forth in Section 708, all vertical openings for every interior stairway or ramp must be similarly enclosed within fire-resistance-rated construction. Because vertical openings provide probably the most readily available paths for fire spreading upward from floor to floor through buildings, it is extremely important that such vertical openings be adequately enclosed. This enclosure is required to protect and separate the vertical exitway from potential fire and products of combustion in other spaces of the building.

1022.1

However, several exceptions from the enclosure requirements are provided:

1. In occupancies other than Groups H and I, an open stairway from a story directly above or below the level of exit discharge is permitted when serving an occupant load of less than 10.

2. No enclosure is required for exits from bleachers, grandstands and other Group A-5 occupancies, provided that all portions of the means of egress are essentially open to the outside.

3. Stairways within an individual dwelling unit or sleeping unit of Group R-1, R-2 or R-3 need not be enclosed.

4. Stairs in open parking garages that serve only the parking structure need not be enclosed.

5. In a Group I-3 institutional building where persons are restrained or secured, a reduction in the enclosure requirements is permitted for one of the required exit stairways per Section 408.3.8.

6. Open stairways may be used for egress from backstage areas as specified in Sections 410.5.3 and 1015.6.1.

7. Unenclosed stairways are permitted for egress purposes from balconies, galleries and press boxes as established in Section 1028.5.1.

Additional allowances for unenclosed stairways are set forth Exceptions 3 and 4 of Section 1016.1. Where either of these exceptions is utilized, exit enclosures complying with Section 1022 are not required to the extent described in the exceptions. See the discussion of Section 1016.1 for further information.

The intent of Exception 1, as well as Exceptions 3 and 4 of Section 1016.1, is that when any two floors are open to one another, neither may be open to yet another floor. This is to prevent the formation of an unprotected vertical shaft through more than two stories. This does not mean that the floors under consideration cannot have access to other floors. They can, but complying enclosures must be provided in order to do so. However, this provision does not allow for the enclosure of a stairway to be interrupted by unenclosed flights.

The degree of fire resistance required for exit enclosures is dependent on the height of the building. In buildings four stories or more in height, the enclosing construction must be at least 2-hour fire-resistance-rated construction. In all other instances, required vertical enclosures of exits must be a minimum of 1-hour fire-resistance-rated construction. For the determination of the required fire resistance of the enclosure, the number of stories also includes any basements that may exist within the building, but not mezzanines.

The fire-resistance rating of the exit enclosure is also regulated in a manner consistent with shaft enclosures in regard to the rating of the floor(s) being penetrated by the enclosure. Where the floor construction penetrated by the exit enclosure has a fire-resistance rating, the exit enclosure must have the same minimum rating. For example, an exit enclosure that penetrates a 2-hour floor assembly must have a minimum fire-resistance rating of 2 hours, regardless of the number of stories the enclosure connects. The fire-resistance rating of an exit enclosure need never exceed 2 hours. If the floor assembly penetrated requires a minimum 3-hour fire-resistance rating, the exit enclosure rating is only required to be 2-hour fire-resistance rated.

1022.3 **Openings and penetrations.** Because exit enclosures are so fundamental to the safety of building occupants and their ability to safely exit a multistory building during a fire emergency, the code is careful to protect the integrity of these vertical enclosures in every way possible. In addition to the fire-resistance-rated construction required of the vertical exit enclosure, the openings that are permitted to penetrate the fire-resistant enclosure are narrowly limited. This section very clearly establishes that only those openings necessary to provide exit facilities for occupants of the building spaces and allowed openings in the exterior walls are permitted.

Because the exterior walls of an enclosed exit enclosure are not protecting the exit from other building spaces, openings through the exterior wall into the atmosphere are permitted. In fact, in buildings that are located on the lot so that there would be no requirement for the fire-resistance-rated construction of the exterior wall or for the protection of the openings in such wall, the exterior wall of an exit enclosure could be eliminated entirely. However, such openings must comply with the proper requirements of the code relating to the location of those openings with respect to lot lines or to other potential exterior fire exposures.

This provision also makes it clear that it is the intent of the code to prohibit openings from typically unoccupied spaces directly into the exit enclosure. Therefore, it is not permitted to provide openings from such spaces as store rooms, toilet rooms, equipment rooms, machinery rooms, electrical rooms and similar rooms directly into an exit enclosure. In addition, elevators are specifically prohibited from opening into an exit enclosure.

Where openings for exit doorways are provided in vertical exit enclosures, it is necessary that they be protected with a fire-rated assembly. Fire-door assemblies and other opening protectives permitted under this section are addressed in Section 715. Where an exit passageway is utilized as an extension to the exterior of the interior exit enclosure, Section 1022.2.1 indicates that a fire door is required to separate the exit enclosure from the exit passageway. See Figure 1022-1. Such a fire door shall have a fire-protective rating in accordance with Section 715.4.

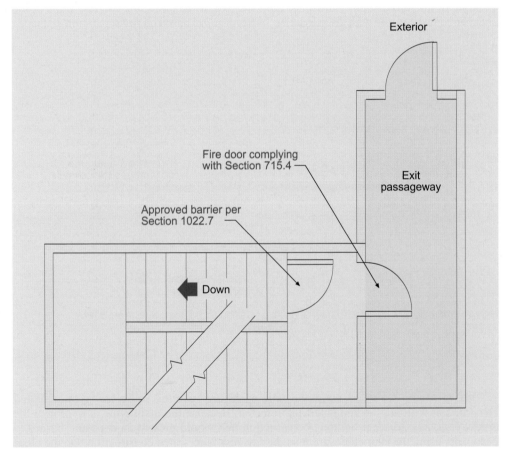

Figure 1022-1
Extent of enclosure

Penetrations. Penetrations into a vertical exit enclosure are prohibited unless necessary to service or protect the exit component. Acceptable penetrations include sprinkler piping, standpipes and electrical conduits serving the exit enclosure. All such penetrations shall be made in a manner that will maintain the structural and fire-resistance integrity of the enclosure or passageway. Under no circumstances shall there be communicating openings or penetrations between adjacent exit enclosures. **1022.4**

Ventilation. Equipment and ductwork necessary for independent pressurization of a vertical exit enclosure is permitted under specific conditions. Where ventilation of the vertical exit enclosure is desired, the ventilation systems shall be independent and isolated **1022.5**

from the other building ventilation systems. There are three methods set forth to regulate the installation of the ventilation equipment and ductwork. The equipment and ductwork may be located at the exterior of the building, within the enclosure or within the building. In all cases, the provisions of Section 708 for shaft enclosures will regulate the separation of the equipment and ductwork from the remainder of the building.

1022.6 **Exit enclosure exterior walls.** Whenever a stairway is installed as a component in an exiting system, it is important to protect the stair user from potential exposure to any fire that might occur in the building. Therefore, where exterior walls of exit enclosures are nonrated and openings in such walls are unprotected, the location and protection of adjacent portions of the building must be considered. Only those walls and openings within 10 feet (3048 mm) horizontally and located at an angle less than 180 degrees (3.14 rad) from the vertical enclosure walls need be protected. Such walls shall have a minimum fire-resistance rating of 1 hour and any openings shall be protected at least $^3/_4$ hour. The extent of the protected construction shall be from the ground to a point at least 10 feet (3048 mm) above the topmost landing of the stairway, or to the roof line if it is lower than 10 feet (3048 mm). An example of this provision is shown in Figure 1022-2. In all cases, the exterior walls of the enclosure must comply with the provisions of Section 705 based on fire separation distance.

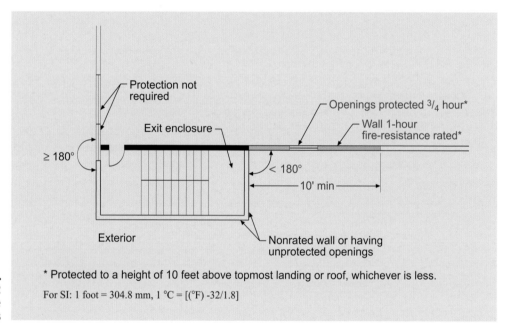

Figure 1022-2
**Exit enclosure
exterior walls**

This provision is based upon the exit enclosure exterior walls being of nonrated construction with any openings left unprotected. Because the main issue is to protect the egress path, any adjacent building construction that would present an exposure hazard to occupants using the stair enclosure must be fire-resistance-rated construction. On the other hand, the enclosure walls and openings of the vertical exit enclosure could be fully protected on all sides, including those on the exterior of the building, under the provisions of Section 1020.1 and, therefore, the requirements of this section would not be applicable. Whether the protection is at the stair enclosure or at the location of the hazard is ultimately immaterial.

1022.7 **Discharge identification.** The barrier required by this section is intended to prevent persons from accidentally continuing into the basement. The design and location details of the barrier must be approved by the building official. Directional exit signs as specified in Section 1011 are also required, in addition to the physical barrier.

1022.8 **Floor identification signs.** This section specifies a system whereby any persons, particularly fire fighters, inside a stairway enclosure in a building will be provided with information telling them where they are in the building and where the stairway leads to both above and below that point. This required sign can be critically important for fire-fighting purposes and is frequently useful to other building occupants. Exit enclosures connecting three stories or fewer are exempt from the identification requirements.

As set forth in this provision, this sign is to be positioned 5 feet (1524 mm) above the floor landing in such a manner that it is readily visible whether the door is open or shut. Information to be provided on the sign includes:

1. The floor level.
2. The top terminus of the enclosure.
3. The bottom terminus of the enclosure.
4. The identification or location of the stairway.
5. The story of exit discharge.
6. The direction of exit discharge.
7. The availability of roof access from the stairway.

A sample sign complying with these requirements is shown in Figure 1022-3. In addition, tactile signage in compliance with ICC A117.1 must be provided to identify the floor level.

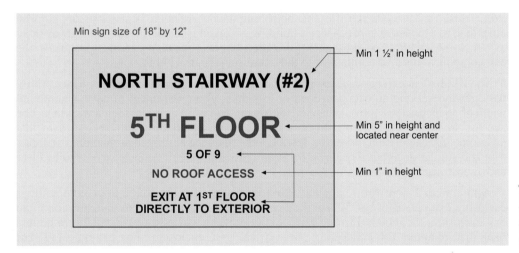

Figure 1022-3
Typical stairway floor number sign

Smokeproof enclosures. Under certain conditions of building occupancy, protection of the vertical exitways over and above that required in the more usual situations is warranted. In such situations, it is necessary that the vertical exitway be constructed not only with a vertical enclosure, but also with either an outside balcony, a ventilated vestibule, or by pressurization meeting the requirements of Section 909.20. A smokeproof enclosure is a special arrangement of the vertical exitway to minimize, if not prevent, the infiltration of smoke and other products of combustion into the actual stairway. Therefore, a smokeproof enclosure is one where the building occupant does not enter the stair enclosure directly from the occupied space of the building, unless the stair pressurization alternative addressed in Section 909.20.5 is utilized.

1022.9

Only those buildings regulated by Section 403 (high-rise buildings), Section 405 (underground buildings) or Section 412.1 (airport traffic control towers) are required to utilize smokeproof enclosures in their means of egress. It is required that all exits in such buildings be smokeproof enclosures or pressurized stairways for each of the exits that serve stories where the floor surface is located more than 75 feet (22 860 mm) above the level of fire department vehicle access or more than 30 feet (9144 mm) below the level of exit discharge serving such floor levels. Details of the various features of smokeproof enclosures are found in the discussion of Section 909.20.

Where a smokeproof enclosure or pressurized stairway reaches the level of exit discharge, it is to exit directly to a public way, or to a yard or open space providing access to a public way. As an option, egress is permitted to travel through areas on the level of exit discharge if the conditions of Section 1027 are met. Where the enclosure is not located at the exterior wall, an exit passageway may be utilized to maintain the protection of the exit way. The exit passageway must be of 2-hour fire-resistance-rated construction and have no openings that could adversely affect the integrity of the enclosure. An exception permits openings where they are adequately fire protected and the exit passageway is pressurized in the same manner as the smokeproof enclosure. Pressurized stairways are offered the same exception. As previously mentioned, the access to a stairway within a smokeproof enclosure must be made through a vestibule or an open exterior balcony. The one exception is the use of a pressurized system complying with the provisions of Section 909.20.

Section 1023   *Exit Passageways*

In many ways, the role of an exit passageway is identical to that of an exit enclosure. Therefore, the provisions regulating exit passageways are very similar to those governing exit enclosures. The code specifies that both components are not to be used for any purpose other than as a means of egress. Once a building occupant is inside an exit passageway or an exit enclosure, there is no subsequent limitation on travel distance. In most instances, the exit component must be continuous to the exit discharge or a public way.

The width of an exit passageway is regulated in the same manner as for a corridor. Those passageways serving an occupant load of less than 50 are permitted to have a minimum width of 36 inches (914 mm). Where the occupant load is 50 or more, such a width must be at least 44 inches (1118 mm). In no case, however, shall the width be less than that determined by calculation in Section 1005.1. Other than the permitted projections for doors, trim and similar decorative features, the required width of exit passageways is to be clear and unobstructed.

A minimum 1-hour fire-resistance rating is required for the walls, floors and ceilings enclosing exit passageways. Where the exit passageway is provided as an extension of an exit enclosure, the required fire-resistance rating of the exit passageway shall not be less than that required for connecting the vertical exit enclosure. For example, an exit passageway serving as a horizontal extension of stairway travel in a 2-hour fire-resistance-rated exit enclosure must also be 2-hour enclosed.

Openings and penetrations that occur in the enclosure elements of an exit passageway are strictly controlled. In fact, the limitations are consistent with provisions regulating openings into and penetrations through vertical exit enclosures. See the discussion of Sections 1022.3 and 1022.4 for an analysis of the provisions that are applicable to both exit passageways and vertical exit enclosures.

Uses of exit passageways. Exit passageways are commonly used in several different exiting situations. In many cases, it is required that interior stairways be enclosed and that the enclosure extend completely to the exterior of the building, including, if necessary, an exit passageway on the floor of the level of exit discharge. When used in this configuration, the exit passageway assumes the same fire-protection requirements as for the stairway enclosure it serves.

A continuing use of exit passageways is in connection with covered mall buildings. An example of the use of exit passageways in covered mall buildings results from the fact that travel distance in the mall is also limited to a maximum of 200 feet (60 960 mm). It will occasionally be necessary between major exits from the mall to introduce an exit passageway or provide an additional exit to satisfy the requirements limiting travel distance. By the use of such exit passageways, it is possible to locate the main entrance/exit points to the mall building at substantially greater intervals. In addition, the use of an exit passageway can potentially eliminate dead-end conditions that occur where back-of-tenant egress is provided along the same path as mall egress.

A historic use of exit passageways is in buildings that have very large floor areas. In such buildings, it is sometimes not possible to get the building occupants to the exits within the limitations of the permitted travel distance. Therefore, an exit passageway is used to literally bring the exit to the interior of the spaces and to the building occupant so that it is possible for any building occupant to reach and enter the exit passageway within the permitted travel distance. This type of exit passageway is frequently accomplished by constructing, in effect, a special type of fire-resistive corridor. It can also be accomplished by constructing either an overhead, fire-resistant, enclosed passageway or a tunnel. By these latter means, it is possible to avoid manufacturing processes and other functions at the floor level within the building.

Another use of exit passageways occurs when a separation of multiple exit doors is insufficient. By extending the points of egress through the use of one or more exit passageways it is possible to relocate the exit doors to the point where they comply with Section 1015.2. The exit separation distance would then be measured in a straight line between the exit doors, which could each enter into an exit passageway.

Again, it should be noted that once in an exit passageway, the building occupant is considered to be in a relatively safe location and travel distances within the exit passageway are not limited, just as travel distances within enclosed exit stairways are not limited.

In most cases, the primary difference between an exit passageway and an exit corridor lie in their respective requirements for opening protection and the fact that the passageway requires a complete enclosure, including ceiling and floor, of at least 1-hour fire-resistance-rated construction. Permitted openings and penetrations in exit passageways are also much more limited than what is allowed into a corridor.

Section 1024 *Luminous Egress Path Markings*

Improving the visibility of stair treads and handrails under normal and emergency conditions is a significant factor in raising the level of occupant safety for individuals negotiating stairs during egress of a high-rise building. A second source of emergency power for exit illumination, exit signs and stair shaft pressurization systems in smokeproof enclosures is mandated for high-rise buildings. In the event of an emergency that disconnects utility power, the emergency power source should engage, causing the stair shaft to be illuminated and maintained smoke-free by the pressurization system. Unfortunately, such systems can fail under demand conditions. The mandate for luminous egress path markings adds an additional level of safety to the egress activity. The installation of photoluminescent or self-illuminating marking systems which do not require electrical power and its associated wiring and circuits provide an additional means for ensuring that occupants can safely egress a building via exit stairs even if the emergency power supply and system fails to operate.

The use of photoluminescent or self-illuminating materials to delineate the exit path is required in Group A, B, E, I, M and R-1 occupancies having occupied floors more than 75 feet (22 860 mm) above the lowest level of fire department vehicle access. In such high-rise buildings, the required use of these markings is limited to exit enclosures and exit passageways. The selected materials must meet the requirements of UL 1994, *Luminous*

Egress Path - Marking Systems or ASTM E 2072, *Standard Specification for Photoluminescent (Phosphorescent) Safety Markings*.

All markings are required to be solid and continuous stripes. A key requirement for marking systems is that their design must be uniform. The placement and dimensions of markings shall be consistent throughout the same exit enclosure. By specifying standard marking dimensions, the requirements ensure that the marking is visible during dark conditions and provides consistent and standard application in the design and enforcement of exit path markings. Markings installed on stair steps, perimeter demarcation lines and handrails must have a minimum width of 1 inch (25 mm). For stair steps and perimeter demarcation lines, their maximum width cannot exceed 2 inches (51 mm). The provisions for stair steps, perimeter demarcation lines and handrails allow the width of the marking to be reduced to less than 1 inch (25 mm) when marking stripes are listed in accordance with UL 1994.

Markings are required along the entire length of the leading edge of each stair step and along the leading edge of stair landings. Markings are also required along the perimeter of stair landings and other floor areas within the exit enclosure. These demarcation lines serve to identify the transition from the stair steps to the landing, which is important to minimize the risk of a fall inside of a stairway enclosure that is not illuminated. In order to discern the transition from the stair to the floor, the demarcation line is located either across the bottom of the door or on the floor in front of the door.

Selected materials used in the construction of the luminous egress path markings must comply with either UL 1994 or ASTM E 2072. ASTM E 2072 allows the use of paints and coatings, which can be useful because it avoids a potential tripping hazard, especially in locations where the surface substrate may not be even. The luminescence of the selected marking system must provide an illumination of at least 1 foot-candle (11 lux), which is consistent with the requirement in IBC Section 1006.2 for the general illumination of walking surfaces. This degree of illumination must be provided for at least 60 minutes.

Analogous to rechargeable batteries, many photoluminescent and self-illuminating egress path markings require exposure to light to perform properly. Thus, luminous egress path markings must be exposed to a minimum 1-foot candle (11 lux), of light energy at the walking surface for at least 60 minutes prior to the building being occupied. The charging rate for luminous egress path markings is based on the wattage of lamps used to provide egress path illumination. Therefore it is important to verify that the specified lamps have sufficient wattage to meet the specified time period.

Section 1025 *Horizontal Exits*

The horizontal exit may well be the least understood and most under-utilized component in an exit system. It can be a very effective method for providing adequate required exiting capacity while at the same time realizing some very substantial construction cost and space savings.

A horizontal exit consists essentially of separating a floor into parts by dividing it with construction having a fire-resistance rating. The construction of one or more horizontal exit walls divides the floor into fire compartments. A horizontal exit may also be located between two buildings where travel occurs from one building to an area in another building at approximately the same level. This would include travel through a fire wall. The concept of the horizontal exit is to permit each of these fire compartments to serve as an area of refuge for occupants in one or more of the fire compartments in the event of a fire emergency. Building occupants in the compartment of fire origin may then pass through the fire-resistance-rated horizontal exit into the compartment of refuge, thereby gaining sufficient protection and sufficient time for either the extinguishment of the fire and elimination of the fire threat, or the orderly use of the remaining exits from the compartment serving as the area of refuge.

The horizontal exit is utilized effectively in hospitals, as well as in detention and correctional facilities, where total evacuation from the building may present numerous physical and other problems. If in a health-care facility it is necessary to move patients from their rooms in a fire emergency, it is desirable to avoid the need for moving them vertically by stairs. Therefore, if an arrangement can be provided whereby patients would only be subject to horizontal movement, the safety of the building occupant can be far more easily achieved. Although horizontal exits are most frequently used only where the more traditional forms of egress design are not satisfactory, they can often be used quite effectively in many situations. In fact, in those instances where fire walls having a fire-resistance rating of not less than 2 hours are provided, the resulting arrangement is often tailor-made for use as a horizontal exit.

Horizontal exits. When properly constructed and installed, a horizontal exit may serve as a required exit. It may, in fact, be substituted on a one-for-one basis for other types of exits. However, in only very limited situations may horizontal exits be used as the only exit from a portion of a building, or to provide more than one-half the total required number of exits from any building space. See Figure 1025-1 for the general rule. In a Group I-2 occupancy, horizontal exits may comprise up to two-thirds of the required exits from any building or floor area, and Group I-3 occupancies may utilize horizontal exits for 100 percent of the required egress.

1025.1

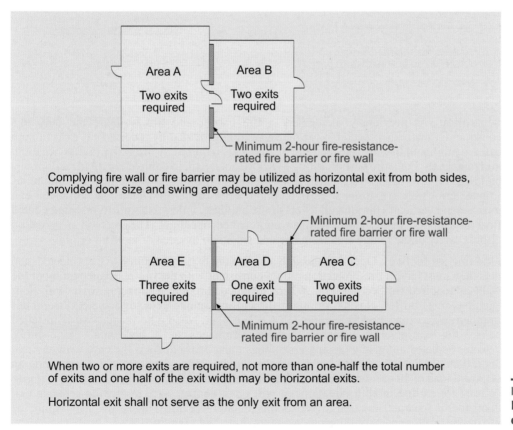

Complying fire wall or fire barrier may be utilized as horizontal exit from both sides, provided door size and swing are adequately addressed.

When two or more exits are required, not more than one-half the total number of exits and one half of the exit width may be horizontal exits.

Horizontal exit shall not serve as the only exit from an area.

Figure 1025-1
**Horizontal
exits**

In the design of a horizontal exit system, it must be emphasized that it is not necessary to accumulate the total occupant load of the two compartments and then provide exit capacity for that total occupant load from each of the compartments. Were that the case, the horizontal exit would often provide little or no benefit. In designing a horizontal exit arrangement, one simply treats the separate compartments as if they were, in fact, separate buildings. Each compartment is provided with an exit system that is in compliance with all of the various criteria for a total exit system from a building. The main difference is that in this configuration, one of the exits from the separate compartments is a horizontal exit into the compartment of refuge. Figure 1025-2 depicts possible arrangements for horizontal exits.

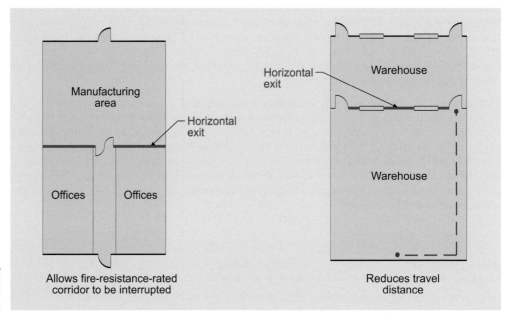

Figure 1025-2
**Horizontal
exit uses**

Allows fire-resistance-rated
corridor to be interrupted

Reduces travel
distance

1025.2 **Separation.** As previously mentioned, a fire barrier and fire wall are the only two construction methods acceptable for the separation provided by a horizontal exit. In both cases, a minimum fire-resistance rating of 2 hours is required. A fire wall provides a natural horizontal exit. For a fire barrier, acting as a horizontal exit, to completely divide a floor of a building into two or more separate refuge areas, the fire barrier walls must be continuous from exterior wall to exterior wall. In addition, a fire barrier utilized as the horizontal exit must extend vertically through all levels of the building, unless 2-hour fire-resistance-rated floor assemblies are provided with no unprotected openings. This method of isolation affords safety in the refuge area from fire and smoke in the area of incident origin.

1025.3 **Opening protectives.** Openings through a horizontal exit are required to be protected, and thus they must have a fire-protection rating, consistent with the fire-resistance rating of the wall, as required by Section 715. Where installed through a horizontal exit wall, door openings must be self-closing, or automatic closing smoke detector-actuated assemblies installed in accordance with Section 715.4.8.3. In fact, when a horizontal exit is installed across the corridor, or when a corridor terminates at a horizontal exit, smoke detector-actuated automatic closing assemblies must be used. As is the case with any exit door, doors in horizontal exits must swing in the direction of exit travel when serving an occupant load of 50 or more as provided in Section 1008.1.2. When the horizontal exit is to be used for exiting in both directions, it will often be necessary to provide separate exit doors for the separate directions to satisfy this requirement. Most likely, this will require the use of one or more pairs of opposite-swinging doors. See Figure 1025-3.

1025.4 **Capacity of refuge area.** The area of refuge in a horizontal exit configuration must be sized to provide sufficient space for the original occupant load of the compartment of refuge plus a partial occupant load from the fire compartment for a limited period of time. The occupant load assigned from the fire compartment is determined by calculating the capacity of the horizontal exit doors that enter the area of refuge. It is necessary for building occupants to remain in the area of refuge only long enough to permit the extinguishment of the fire and elimination of the fire threat, or to allow the combined occupant load of the two compartments to utilize the remaining exit facilities from the compartment of refuge.

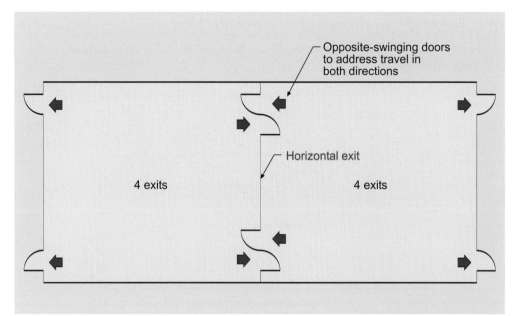

Figure 1025-3
Opposite-swing doors

To reasonably accommodate the combined occupant loads in the compartment of refuge, the code requires that at least 3 square feet (0.28 m²) of net clear floor area be provided for each occupant. In Group I-2 occupancies, it is required that there be provided at least 15 square feet (1.4 m²) be provided per occupant for each ambulatory person and at least 30 square feet (2.8 m²) for each nonambulatory occupant. As in hospitals, such nonambulatory occupants will frequently be brought into the compartment of refuge either in a hospital bed or on a gurney. In Group I-3 occupancies, it is required that there be at least 6 square feet (0.56 m²) of net floor area provided per occupant. An example of calculating refuge area capacity is illustrated in Application Example 1025-1.

Although these area figures will permit a rather dense occupancy in the area of refuge, it must be remembered that the occupancy of that space is only temporary and that the occupants of the area of refuge will continue to evacuate the area of refuge by use of the remaining exit or exits. In a fire emergency, such space per person is considered adequate for a short period of time.

It is also important to provide such a required refuge area for occupants in spaces that will, in fact, be available to the occupants of the building as they enter the compartment of refuge. Such spaces can be provided in corridors, lobbies and other public areas, as long as they are sufficient to accommodate the total occupant load at the appropriate rate of area per person. Spaces in the refuge area occupied by the same tenant as in the area of fire origin are also permitted.

GIVEN: A fully sprinklered building contains an office area and a large assembly room.

A horizontal exit is provided between the two areas.

DETERMINE: The clear office area floor space required when used as the refuge area.

The office area is to be the refuge from the assembly room. Therefore, it must provide clear floor space of 555 sq ft to adequately house 185 occupants,* based on 3 sq ft per person.

*(20 original + 165 capacity through single exit door entering area of refuge based on 33 inches of clear width).

Sprinklered building

Office
20 occupants

Assembly room
320 occupants

Horizontal exit

33" of clear width provides an egress capacity of 165 (33 ÷ 0.20*)

*Sec. 1025.6.1, Item 4

For SI: 1 inch = 25.4 mm, 1 square foot = 0.093 m².

Section 1026 *Exterior Exit Ramps and Stairways*

1026.2 **Use in a means of egress.** An exterior exit stairway or ramp may serve as an exit component in the means of egress system in all occupancies other than Group I-2. The use of an exterior exit stairway or ramp as a required means of egress element is limited to buildings with a maximum of six stories, with no occupied floors more than 75 feet (22 860 mm) above the lowest level of fire department vehicle access.

1026.3 **Open side.** To be classified as an exterior stair, it must be open on at least one side. The open side must then adjoin open areas such as yards, courts or public ways. In order to qualify as an open side, there must be at least 35 square feet (3.3 m²) of aggregate open area adjacent to each floor level and at the level of each intermediate landing. In addition, the required open area must be at least 42 inches (1067 mm) above the adjacent floor or landing level. See Figure 1026-1. By limiting the amount of enclosure by the exterior walls of the building, an exterior stair will be sufficiently open to the exterior to prevent accumulation of smoke and toxic gases. Any stairway that does not comply with these criteria is considered an interior stairway.

1026.6 **Exterior ramps and stairway protection.** In order to adequately protect the building occupants as they travel an exterior stairway or ramp during egress, the exterior exit path must be adequately separated from the interior of the building. With exceptions, this section requires that an exterior stairway or ramp be provided with protection in the same manner and to the same degree as is provided by an exit enclosure. See the discussion of Section

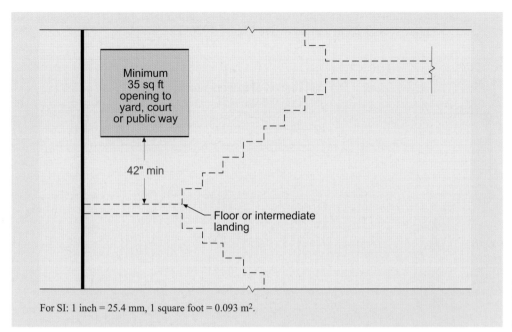

Minimum 35 sq ft opening to yard, court or public way

42" min

Floor or intermediate landing

For SI: 1 inch = 25.4 mm, 1 square foot = 0.093 m².

Figure 1026-1
Exterior exit stairways

1022.1. Consistent with these provisions, openings are not permitted unless necessary for egress from normally occupied spaces to the exterior stair. There are four situations where the separation between the exterior exit ramp or stairs and the building's interior is not required:

1. In buildings no more than two stories above grade plane housing other than Group R-1 and R-2 occupancies, separation is not required where the level of exit discharge is at the first story. In such a scenario, only one story of vertical exit travel would be required. This would limit the exposure to a degree that protection is deemed unnecessary.

2. If an open exterior balcony provides access to at least two remote exterior stairways or other exits, the separation is not required. See Figure 1026-2. To be considered open, the balcony must be open to the exterior for at least 50 percent of its perimeter. This length must then be open vertically at least 50 percent of the wall height, with openings extending at least 7 feet (2134 mm) above the balcony. Where the building occupant has an alternate choice of exit travel, the occupant is not forced to utilize the stairway or ramp adjacent to the hazard. Therefore, wall and opening protection is not required.

3. The provisions of Sections 1022.1 and 1016.1 permit unenclosed interior stairways under certain conditions. If the conditions warrant the elimination of any fire protection for a stairway inside the building, there should be no reason to require fire protection for a similar exterior situation.

4. An open-ended corridor is simply a corridor that is open to the outside at the exterior of the building, leading directly to an exterior stairway or ramp at each end with no intervening doors or enclosures at the exterior wall. Where open-ended corridors are utilized, such as in a breezeway design, the code identifies under what conditions the fire separation is not necessary between the building interior and the exterior stairway or ramp. First, the building, including the corridor and stairs, must be sprinklered throughout. Second, the corridor shall meet all of the corridor provisions of Section 1018. Third, the exterior stairways at the ends of the open-ended corridor are to comply with the general exterior exit stairway and ramp provisions of Section 1026. Fourth, where a change in direction of more than 45 degrees (0.79 rad) occurs in the corridor, an exterior stairway, exterior ramp or openings to the exterior shall be provided. The openings shall be such that the

accumulation of smoke and toxic gases will be minimized, but in no case less than 35 square feet (3.25 m²) in area. See Figure 1026-3.

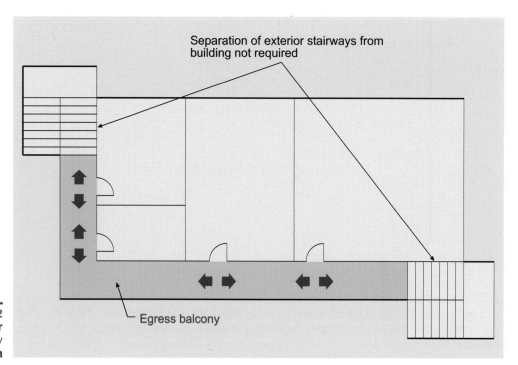

Separation of exterior stairways from building not required

Egress balcony

Figure 1026-2
Exterior stairway protection

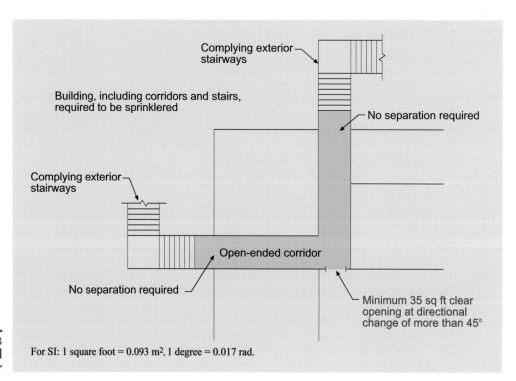

Complying exterior stairways

Building, including corridors and stairs, required to be sprinklered

No separation required

Complying exterior stairways

Open-ended corridor

No separation required

Minimum 35 sq ft clear opening at directional change of more than 45°

For SI: 1 square foot = 0.093 m², 1 degree = 0.017 rad.

Figure 1026-3
Open-ended corridor

Section 1027 *Exit Discharge*

This section regulates the exit discharge portion of the egress system. This is the last portion of the three-part means of egress system and is the portion between the point where an occupant leaves an exit and continues until a public way is reached. As stated earlier, essentially all exterior travel at grade level is considered a part of the exit discharge.

General. Exits are intended to discharge directly to the exterior of the building. Four exceptions permit the exit path to include a portion of the building beyond the exit component. An exception to the requirements for the continuity of exit enclosures is permitted where a maximum of 50 percent of the exits pass through areas on the level of exit discharge. The path of travel to the exterior must be unobstructed and easily recognized. Sprinkler protection is required for the egress path between the termination of the exit enclosure to the building's exterior, as is fire-resistance-rated construction isolating any areas below the discharge level. See Figure 1027-1. Effectively, all portions of the discharge level that provide access to the egress path must be sprinklered as well. A second exception permits egress from an exit enclosure to enter a vestibule where limited to 50 percent of the number and capacity of the total enclosures provided. The vestibule must be separated from the areas below by fire-resistance-rated construction, be limited in size and shape, cannot be used for purposes other than egress, and must lead directly to the exterior. A minimum degree of construction providing fire and smoke protection, at least the equivalent of approved wired glass in steel frames, shall separate the vestibule from other portions of the level of exit discharge. See Figure 1027-2. Although it is acceptable to utilize both of these exceptions in the same building, their combined use cannot exceed 50 percent of the number and capacity of the required exits. Therefore, this limitation will typically permit only one of the exceptions to be applied.

1027.1

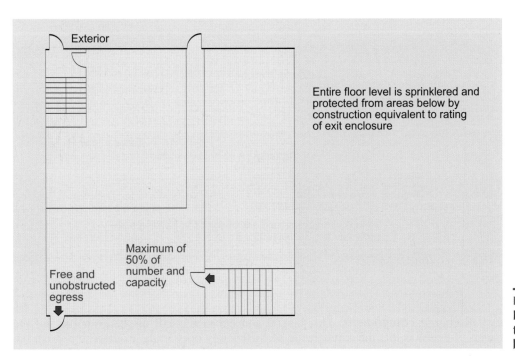

Exterior

Entire floor level is sprinklered and protected from areas below by construction equivalent to rating of exit enclosure

Maximum of 50% of number and capacity

Free and unobstructed egress

Figure 1027-1
Exit discharge through building

Exit discharge is prohibited from reentering a building to ensure that the user does not go back into any portion of the building once they have reached the exit discharge. This prohibition can be viewed as excluding the reentry into exit enclosures, exit passageways or any portion that would be considered a component of the exit.

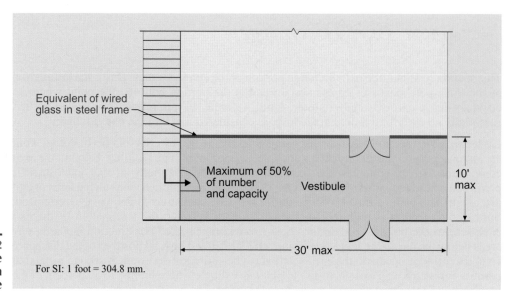

Figure 1027-2
Exit discharge through vestibule

For SI: 1 foot = 304.8 mm.

1027.3 Exit discharge location. Because of exposure potential from adjacent property, exterior balconies, stairways and ramps are prohibited within 10 feet (3048 mm) of an adjacent lot line. Such components must also maintain a minimum 10-foot (3048 mm) separation from other buildings on the same lot unless the exterior walls and openings of the adjacent building are protected in accordance with Section 705 based on fire separation distance. See Figure 1027-3.

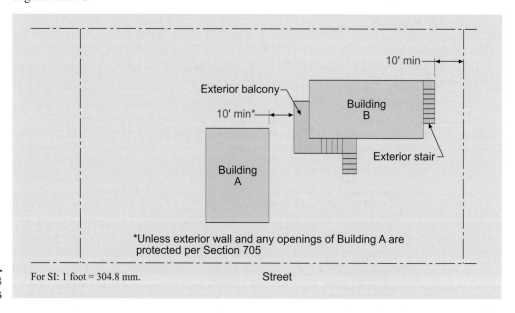

Figure 1027-3
Egress courts

For SI: 1 foot = 304.8 mm.   Street

1027.4 Exit discharge components. The general concept of the exit discharge portion of the means of egress is that the components be sufficiently open to the exterior to prevent the accumulation of smoke and toxic gases. As occupants reach the exterior of the building at grade level, they expect to have arrived at a point of relative safety. Where adequate natural cross ventilation is available to disperse any smoke or gases that may be present, one of the major fire- and life-safety concerns is assumed to have been addressed.

Egress courts. An egress court is defined as any court or yard that provides access to a **1027.5** public way for one or more required exits. As such, the code requires that every egress court discharge into a public way. See Figure 1027-4.

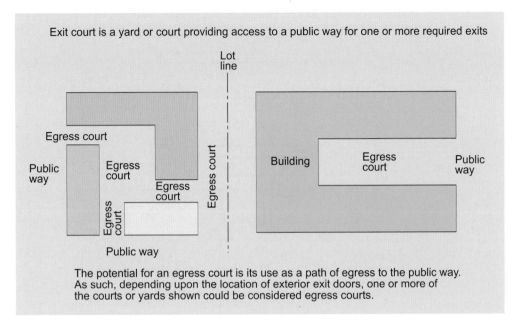

Exit court is a yard or court providing access to a public way for one or more required exits

Lot line

Egress court

Public way

Egress court

Egress court

Egress court

Egress court

Building

Egress court

Public way

Public way

The potential for an egress court is its use as a path of egress to the public way. As such, depending upon the location of exterior exit doors, one or more of the courts or yards shown could be considered egress courts.

Figure 1027-4
Egress courts

Width. The minimum required width of an egress court is determined in a similar manner to **1027.5.1** that of a corridor. An egress court must provide at least 44 inches (1118 mm) of clear width in all occupancies except Groups R-3 and U, where the width may be reduced to 36 inches (914 mm). When egress courts are subject to use by a sufficiently large occupant load, the required width may be wider than 44 inches (1118 mm). Such a greater width would be determined in accordance with the applicable provisions of Section 1005.1 based on the occupant load served. As is the case for most other egress components, limited encroachments into the required width are permitted for doors. Whatever the required width of the egress court, there must be at least 7 feet (2134 mm) of unobstructed headroom.

In no event may the minimum required width be less than that required by the above paragraph. However, should the actual width of an exit court be greater than the minimum width required, and it becomes necessary to reduce the width of the court as the building occupants proceed toward the public way, such a reduction cannot be an abrupt change. The changes must be made by making an angle of not more than 30 degrees (0.52 rad) with the axis of the egress court so that the reduction in width is a gradual one.

Construction and openings. Because an egress court is a component in a means of egress **1027.5.2** system, building occupants utilizing that component must be afforded sufficient protection to reasonably ensure that they will reach the safety of the public way. Therefore, in other than a Group R-3 occupancy, any time an egress court serves an occupant load of 10 or more and is less than 10 feet (3048 mm) in width, the walls of the exit court must be of 1-hour fire-resistance-rated construction for a minimum height of 10 feet (3048 mm) above the floor of the court. By this means, a fire-resistant separation is maintained between the persons in the court and the occupied use spaces of the building. Should any openings occur in the portion of the egress court wall required to be fire-resistance-rated construction, those openings must be protected by assemblies having a fire-protection rating of not less than $^3/_4$ hour. See Figure 1027-5.

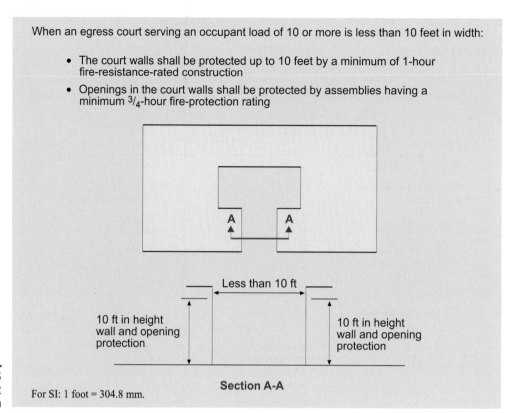

When an egress court serving an occupant load of 10 or more is less than 10 feet in width:

- The court walls shall be protected up to 10 feet by a minimum of 1-hour fire-resistance-rated construction
- Openings in the court walls shall be protected by assemblies having a minimum ³/₄-hour fire-protection rating

Less than 10 ft

10 ft in height wall and opening protection

10 ft in height wall and opening protection

Section A-A

For SI: 1 foot = 304.8 mm.

Figure 1027-5
Egress court construction

Section 1028 *Assembly*

Because of the potential for large occupant loads in concentrated areas, assembly uses are regulated for egress a bit differently than for other occupancies. This section addresses a variety of issues that are specific to Group A occupancies, including exit capacity, aisle widths and smoke-protected assembly seating. Although the provisions of this section do not apply to all assembly-type uses, they are applicable to most. The means of egress for any Group A containing seats, tables, displays, shelving, equipment, fixtures or similar elements must comply with this section.

Section 1028 is also applicable to assembly uses accessory to Group E occupancies. The classification of assembly spaces as Group E occupancies where accessory to educational facilities classified as Group E is based on the assembly areas being subsidiary to the school function. Typical examples include libraries and media centers that are used almost exclusively by students of the school. School gymnasiums and auditoriums may also be included where their primary function is an extension of the educational activities. Although the classification of such areas may be based upon the educational function of the building, it is important to recognize that from a means of egress perspective the large assembly uses still function as assembly spaces. All of the hazards involved with assembly-type functions are present in school assembly rooms, regardless of whether the room is classified as Group A or E. Therefore, the specific means of egress provisions for Group A occupancies are also applicable to those assembly spaces classified as Group E.

Bleachers, grandstands, folding seating and telescoping seating are not regulated by Section 1025, but rather by ICC 300, *Bleachers, Folding and Telescoping Seating, and Grandstands*. This standard, developed by the ICC Consensus Committee on Bleacher

Safety, addresses the means of egress for such special types of seating arrangements. Simply an extension of the IBC, ICC 300 also addresses structural design and construction features.

Assembly main exit. This section provides special exiting requirements for assembly **1028.2**
occupancies. The first specification is that every assembly occupancy having an occupant load greater than 300 must be provided with a main exit. The minimum required width of this main exit is determined by two criteria—calculated width and component width. The rule that would produce the larger required width for the main exit is the one that would govern the required minimum size.

First, in any assembly occupancy having such a sizable occupant load, the designated main exit must have sufficient width to accommodate at least one-half of the total occupant load. Second, its width shall not be less than the total required width of all means of egress such as aisles, corridors, exit passageways and stairways that lead to the main exit. The main exit must always be connected by appropriate exit components to ensure that the occupants have continuous and unobstructed access to a public way.

Basically, the requirement that the main exit be adequate to accommodate 50 percent of the total occupant load takes into account a characteristic of human nature. The majority of the occupants are, in all probability, not completely familiar with the facility and its exit system. As a consequence, in the event of an emergency it is typical that people will attempt to reach the exit through which they entered. This natural tendency could put an unduly large load on the main exit if it were not sized according to this requirement.

An exception permits distribution of exits where there is no well-defined main exit or where multiple main exits are provided. In no case, however, may the total width of egress be less than 100 percent of the required width. Should there be multiple points of entry into the building, it would be unreasonable to size each point as if it were a main exit. However, where a main exit is specifically identified, the 50 percent rule is applicable.

Assembly other exits. In addition to requiring that the main exit accommodate 50 percent **1028.3**
of the total occupant load, it is also necessary that every assembly occupancy with an occupant load of more than 300 be provided with additional means of egress. They are required to be of sufficient capacity to accommodate at least 50 percent of the total occupant load served by that level. As expected, these additional means of egress must also comply with Section 1015.2 addressing doorway arrangement.

Foyers and lobbies. In specific types of Group A occupancies, typically theaters and **1028.4**
similar uses, persons often occupy a lobby to await a presentation in the major-use area. It is important that while the lobby is occupied, the required clear egress width from the major-use area is maintained through the lobby to the exit from the building.

Interior balcony and gallery means of egress. Consistent with the basic requirement for **1028.5**
the number of exits in a Group A occupancy contained in Table 1015.1, every balcony, gallery and press box that has an occupant load of 50 or more must be provided with a minimum of two remote, separate exits. In some Group A occupancies, more than one such feature is provided. These requirements for exits apply to any and all balconies, galleries and press boxes in the building. In all cases, at least one of the required means of egress must lead directly to an exit. Although the general provisions mandate exit enclosures, the exception indicates the means of egress may include open stairways between the balcony, gallery or press box and main assembly floor in uses such as theaters, churches and auditoriums.

Width of means of egress for assembly. The method for calculating egress widths has two **1028.6**
distinct categories. These are buildings with, and buildings without, smoke-protected assembly seating. Thus, the first thing that must be established is the category in which the egress must be analyzed. Section 1002.1 defines smoke-protected assembly seating as "seating served by means of egress that is not subject to smoke accumulation within or under a structure." Section 1028.6.2 regulates smoke-protected assembly seating in regard to smoke control, roof height and automatic sprinklers.

If it is determined that a building has smoke-protected assembly seating, a Life Safety Evaluation complying with NFPA 101 must be conducted. An acceptable evaluation would include such criteria as pedestrian flow rates, movement characteristics of persons using exit systems, the nature of the means of egress system beyond the aisles, location of potential hazards, staff-response capabilities, facility preplanning and training, and other items relating to occupants' safety. Should it be determined that a building is not smoke protected, or that no life-safety evaluation has been performed, the aisle width will need to be calculated using the more stringent nonsmoke-protected conditions.

1028.6.1 **Without smoke protection.** For egress on stairs, two modifications to the basic width factor of 0.3 inch per occupant (7.6 mm) are to be considered. The general requirement of 0.3 inch (7.6 mm) is based on stairs having riser heights of no more than 7 inches (178 mm). Where the riser height exceeds 7 inches (178 mm), at least 0.005 inch (0.127 mm) of additional stairway width for each occupant shall be provided for each 0.10 inch (2.5 mm) of riser height above 7 inches (178 mm). The following formula represents this method of width increase:

$$W = 0.3 + 10(R - 7.0)(0.005)$$

where:

W = Required width in inches per occupant

R = Riser height in inches

Formula 10-1

The second potential modification applies only where egress is accomplished on the stair in a descending fashion. Under such a condition, an increase of at least 0.075 inch (1.9 mm) of additional width per occupant must be provided where no handrail is provided within a horizontal distance of 30 inches (762 mm). This egress-width increase is illustrated by the following formula:

$$W = 0.3 + (0.75) = 0.375$$

where:

W = Required width in inches per occupant

Formula 10-2

It is also very possible that both increases will be applied to the same aisle stairs. Where a stair has a riser height exceeding 7 inches (178 mm), and a portion of the required stair width is more than 30 inches (762 mm) from a handrail, both modifications to the basic requirement of 0.3 inch (7.6 mm) per occupant are applicable. The increase in required egress width can be shown as:

$$W = 0.3 + 10(R - 7.0)(0.005) + (0.075)$$

where:

W = Required width in inches per occupant

R = Riser height in inches

Formula 10-3

Where a ramp is used as a portion of the means of egress, the difference in calculated width is based upon the slope of the ramp. Ramps having a slope steeper than 1 in 12 (1:12) shall have a minimum clear width based upon 0.22 inch (5.6 mm) per occupant served. Note that this condition would only apply to ramps not regulated by Chapter 11 for accessibility. The minimum clear width of level or ramped egress paths having a slope of 1 in 12 (1:12) or less is based on 0.20 inch (5.1 mm) for each occupant.

The required increases in egress width are all based on one of the primary IBC concepts regarding means of egress: where it is anticipated that users of the egress system may be slowed in their egress travel, the path is required to be widened to compensate for the reduced travel speed. By increasing the width of the exit path, the occupants are expected to continue to travel at a relatively consistent rate through the exit system. For stair travel, a riser height over the customary maximum of 7 inches (178 mm) creates an uncomfortable condition for most stair users. In such cases, it is natural for persons to slow their travel in order to more safely use the stairway. Occupants also tend to travel more slowly when descending a stairway where a handrail is not within easy reach. Without the ability to easily grasp a handrail in case of a misstep, the stair user tends to be a bit more cautious. On the other hand, the presence of an adjacent handrail provides a degree of security that encourages faster stair travel. In ramp travel, a slope steeper than that normally encountered also creates a small degree of hesitancy, requiring a greater width to compensate for the reduction in travel speed.

The increases in required egress width prescribed by this section are only intended to apply where the minimum width is calculated based on the number of occupants served by the egress path. The width increases are not applicable to the component widths established in Section 1028.9.1. For example, a minimum width of 36 inches (914 mm) is mandated for aisle stairs with seating on only one side. This condition would typically allow for a single handrail, which would be located more than 30 inches (762 mm) horizontally from a portion of the required aisle width. It is not appropriate to increase the minimum 36-inch (914 mm) required width based upon the factors established in Item 3 of this section. Additionally, if the aisle stair included risers that exceeded a height of 7 inches (178 mm), the adjustment found in Item 2 would also not be applied to the minimum component width as established in Section 1028.9.1.

Smoke-protected seating. Calculation of egress width in an assembly space exposed to a smoke-protected environment is based upon Table 1028.6.2. As the total number of seats in the smoke-protected assembly occupancy increases, egress-width requirements continue to decrease until reaching an end point of 25,000 or more seats. As addressed earlier, a Life Safety Evaluation must be done for a facility utilizing the reduced width requirements of Table 1028.6.2. Application Example 1028-1 shows how the calculated width of exit provisions would be applied for smoke-protected seating and seating without smoke protection. The permitted reduction in egress width applies to all elements in the means of egress system (vomitories, concourses, stairways, etc.), but only to the extent that they too are smoke-protected. Any egress elements that are not provided with complying smoke protection are subject to the greater widths established for areas that are not smoke-protected. **1028.6.2**

Smoke control. To maintain an essentially smoke-free means of egress system, a smoke-control system complying with Section 909 must be provided. As stated in Section 909.1, a smoke-control system should be designed to provide a tenable environment for the evacuation or relocation of occupants. When it can be satisfactorily demonstrated to the building official, a design incorporating a natural venting system is permitted. The natural ventilation must be designed to maintain the smoke level at a point 6 feet (1829 mm) or more above the floor level of any portion of the means of egress within the smoke-protected assembly seating area. **1028.6.2.1**

Roof height. Whenever smoke-protected assembly seating is covered by a roof, a minimum clearance of 15 feet (4572 mm) is required between the highest aisle or aisle accessway to the lowest portion of the roof. In an outdoor stadium, the roof canopy need only be 80 inches (2032 mm) or more above the highest aisle or aisle accessway, provided there are no projections or obstructions below the 80-inch (2032 mm) level. By providing an adequate roof height above the occupiable portion of the building or structure, a smoke-containment **1028.6.2.2**

area is created. Smoke control or removal would then limit smoke migration into the egress environment.

Application Example 1028-1

GIVEN: A 10,000-seat arena with seating sections as shown.

Case I—Smoke-protected assembly seating for which a life-safety evaluation has been provided.

Case II—Nonsmoke-protected assembly seating.

DETERMINE: Required aisle width for both cases.

SOLUTION:

Case I—Use Table 1028.6.2. For a 10,000-seat arena, the width equation for stairs is 0.130 inch per seat served, and the required width of aisle becomes:

$$400 \times 0.130 = 52 \text{ inches or } 4.33 \text{ feet (1321 mm)}$$

Since 52 inches (1321 mm) exceeds the minimum required width of 48 inches (1219 mm), it becomes the governing width.

Case II—For nonsmoke-protected assembly seating, the width is based on 0.3 inch per seat served. However, since a center handrail will exceed the 30-inch rule, an increase of 0.075 inch is required. Thus, the factor is 0.375 inch per seat and the required width of aisle becomes:

$$400 \times 0.375 = 150 \text{ inches}$$

Thus, the aisle required for nonsmoke-protected assembly seating is almost three times as wide as the aisle required for smoke-protected assembly seating.

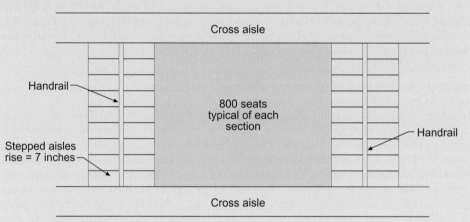

For SI: 1 inch = 25.4 mm, 1 foot = 304.8 mm.

1028.6.2.3 **Automatic sprinklers.** Another condition for the use of the liberal egress width provisions for smoke-protected assembly seating areas is the installation of an approved automatic sprinkler system. In general, the sprinkler system is required in all areas enclosed by walls and ceilings in buildings or structures containing smoke-protected assembly seating. However, three exceptions identify locations where sprinklers may be omitted. Where the area on the assembly room floor is utilized for low fire-hazard uses such as performances, contests or entertainment, sprinklers may be omitted, provided the roof construction is more than 50 feet (15 240 mm) above the floor level. Small storage facilities and press boxes under 1,000 square feet (92.9 m²) in area are also exempt. A third exception clarifies that sprinkler protection is not required in the seating area of outdoor facilities where the means of egress for the seating area is essentially open to the outside.

1028.6.3 **Width of means of egress for outdoor smoke-protected assembly.** Where the facilities are outdoors, such as in a stadium, and the egress system for the assembly seating is considered smoke-protected owing to the natural ventilation available, the clear width is based on one of two factors. Where the egress is by aisles and stairs, the width factor is 0.08 inch per occupant, whereas it is 0.06 inch for ramps, corridors, tunnels or vomitories. An example of this calculation is shown in Application Example 1028-2. If, however, the width calculated through the use of Table 1028.6.2 for all types of smoke-protected seating is determined to be a lesser width, such a lesser width is acceptable. As indicated in the exception to Section 1028.6.2, the point where Table 1028.6.2 applies for outdoor seating is 18,000 occupants.

GIVEN: An outdoor smoke-protected assembly seating area having stairs serving an occupant load of 200. The total number of seats in the facility is 2,600.

DETERMINE: The required aisle width.

CASE 1: Using Table 1028.6.2, the width per person is 0.280* inch. 200 persons (0.28 inch) = 56 inches.

CASE 2: Using Section 1028.6.3, the width per person is 0.08 inch. 200 persons (0.08 inch) = 16 inches < 48 inches minimum required for seating on both sides.

* Interpolated between 0.3 inch and 0.2 inch.

∴ Minimum required width is 48 inches

Application Example 1028-2

Travel distance. Travel distance in assembly occupancies is regulated under one of three conditions; seating without smoke protection, smoke-protected seating and open-air seating. The measurement of this distance shall be along the line of travel, including along the aisles and aisle accessways, from each seat to the nearest exit door. It is improper to measure this travel distance over or on the seats. Inside a building without the benefit of smoke protection, travel distance is limited to 200 feet (60 960 mm) in nonsprinklered buildings and 250 feet (72 200 mm) in sprinklered buildings. In smoke-protected assembly seating, a maximum travel distance of 200 feet (60 960 mm) is permitted from each seat to the nearest entrance to an egress concourse. Up to another 200 feet (60 960 mm) of travel is permitted from the egress concourse entrance to an egress stair, ramp or walk at the building exterior. Where the assembly seating is located in an outdoor facility and all portions of the means of egress are open to the outside, the maximum travel distance is 400 feet (121 920 mm). This distance is measured from each seat to the exterior of the building. When the seating facilities are of Type I or II noncombustible construction, the travel distance may be unlimited.

1028.7

Common path of travel. By providing persons in an assembly occupancy a choice of travel paths, egress opportunities will be enhanced. The distance an occupant must travel prior to reaching a point where two paths are available is more limited than permitted by the general provisions of Section 1014.3. In most assembly seating areas, the common path of travel is limited to 30 feet (9144 mm). In smoke-protected areas, up to 50 feet (15 240 mm) of common path travel is permitted. Where the seating area has a limited occupant load, additional travel is also permitted.

1028.8

Path through adjacent row. Because the common path of travel is most often limited to 30 feet (9144 mm), single-access seating areas are often required to egress through rows of the adjoining seating area to reach another aisle in order to comply. Such seating areas are served by an aisle on only one side, and the aisle is single directional (top-loading or bottom-loading only). Where this condition occurs, the code (1) limits the maximum number of seats in the adjoining row to 24, and (2) increases the minimum required width of the aisle accessway serving the row. A similar concept is applied in Section 1028.9.5. for increasing the maximum permitted length of dead-end aisles.

1028.8.1

Assembly aisles are required. This section requires that aisles be provided in all occupied portions of any assembly occupancy that contains seats, tables, displays and similar fixtures or equipment. Although the intent of the aisle provisions is to provide safe access to

1028.9

components of the egress system such as exits or exit access doorways, the provisions would also have application to occupied-use areas where it is necessary to provide a circulation system so that building occupants will have reasonable means for moving around in the occupied spaces. Egress travel within restaurants, classrooms and similar uses with seating at tables is regulated by Section 1017.4.

1028.9.1 **Minimum aisle width.** Where seating is arranged in rows, the clear width shall not be less than the following, while still conforming to the calculated aisle-width provisions:

1. Forty-eight inches (1219 mm) for aisle stairs where seating is provided on both sides of an aisle. A 36-inch (914 mm) aisle is permitted where the aisle serves less than 50 seats.

2. Thirty-six inches (914 mm) for aisle stairs where seating is provided only on one side of an aisle.

3. Twenty-three inches (584 mm) between an aisle stair handrail and the nearest seat where the rail is within the aisle.

4. Forty-two inches (1067 mm) for level or ramped aisles where seating is provided on both sides of an aisle. A 36-inch (914 mm) aisle is permitted where the aisle serves less than 50 seats, and a minimum of 30 inches (762 mm) is allowed where serving 14 seats or less.

5. Thirty-six inches (914 mm) for level or ramped aisles where seating is provided on only one side of an aisle. Only 30 inches (762 mm) of aisle width is required where the aisle serves no more than 14 seats.

6. Twenty-three inches (584 mm) between an aisle stair handrail and the nearest seat where an aisle serves no more than five rows on only one side.

This section must be used in concert with the other provisions of Section 1025 to completely define and arrange an assembly seating area.

1028.9.2 **Aisle width.** The occupant load served by an aisle is the determining factor in establishing its minimum required width. In this determination, it is assumed that the egress travel is distributed evenly among the adjacent travel paths. The tributary occupant load would be assigned to each aisle proportionally, based upon the arrangement of the means of egress.

1028.9.3 **Converging aisles.** Where aisles converge to form a single aisle, the capacity of that single aisle shall not be less than the combined required capacity of the converging aisles. There is no penalty for providing aisles that are wider than the minimum code requirement.

1028.9.4 **Uniform width.** Where egress is possible in two directions, the shape of an aisle cannot be of an hourglass configuration. The clear width shall be uniform throughout. A tapered aisle is allowed only for dead-end aisle conditions.

1028.9.5 **Assembly aisle termination.** In the arrangement of a seating area, all aisles serving the seating area must end in a cross aisle, foyer, doorway, vomitory or concourse. In large facilities, it is not uncommon to find a number of aisles leading to a number of cross aisles. The required egress capacity of a cross aisle is the same as that for converging aisles, the combined required capacities of the aisles leading to the cross aisle. Although dead-end aisles are allowed, their length is limited to 20 feet (6096 mm) except where the seats served by the dead-end aisle are not more than 24 seats from another aisle measured along a row having a minimum clear width of 12 inches plus 0.6 inch (305 mm plus 15 mm) for each additional seat over seven in a row. For smoke-protected assembly seating, up to 21 rows are permitted in a dead-end vertical aisle except where the seats served by the dead-end aisle are not more than 40 seats from another aisle measured along a row having a minimum clear width of 12 inches plus 0.3 inch (305 mm plus 7.6 mm) for each additional seat over seven in a row.

1028.10 **Clear width of aisle accessways serving seating.** The minimum clear width between rows of seats is 12 inches (305 mm) measured between the rearmost projection of the seat in the forward row and the foremost projection of any portion of the seat in the row behind, where the rows have 14 or fewer seats. If automatic or self-rising seats are used (such as in movie theaters) the minimum clear width shall be measured with the seat up. If any chair in the row does not have an automatic- or self-rising seat, the measurement must be made with the chair in the down position. See Figure 1028-1. Seats with folding tablet arms are regulated in a special manner. Unless the tablet arm is of a type that automatically returns to the stored position by gravity when raised manually to the vertical position, the row spacing measurement must be taken with the arm in the position in which it is used.

Such clear width must be increased whenever the number of seats in a row exceeds 14, but in no case shall the number of seats in any row exceed 100. The increased width when seating rows are served by aisles or doorways at both ends of a row is the minimum 12 inches plus 0.3 inch (305 mm plus 7.6 mm) for each additional seat over 14; however, the width need not exceed 22 inches (559 mm). The establishment of the minimum required width is illustrated in the following formula:

Clear width, in inches = 12 inches (305 mm) + 0.3 inch (7.6 mm) (x − 14 seats)

WHERE:

x = the number of seats in a row.

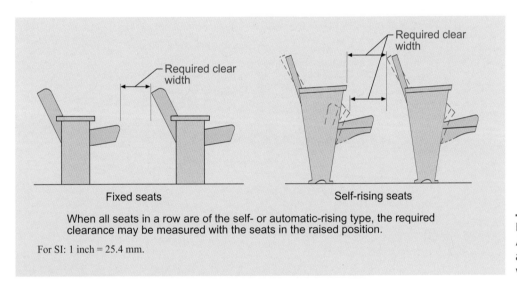

When all seats in a row are of the self- or automatic-rising type, the required clearance may be measured with the seats in the raised position.

For SI: 1 inch = 25.4 mm.

Figure 1028-1
Aisle accessway width

By interpreting the formula, we find that the maximum required clear width occurs when the number of seats in a row reaches 48. See Figure 1028-2.

In smoke-protected facilities, the maximum number of seats permitted for a 12-inch-wide (305 mm) aisle accessway with dual access is permitted to exceed the 14-seat limit for seating areas without smoke protection. With a maximum that varies up to 21 seats, the limitations indicated in Table 1028.10.1 are based on the total occupant load devoted to assembly seating. Increases on single-access aisle accessways are also available. A similar version of the previous formula can be used with Table 1028.10.1 to determine the minimum required aisle accessway width when the number of seats permitted for a 12-inch-wide (305 mm) aisle accessway is exceeded in a building provided with smoke-protected assembly seating. As an example, assume a 16,000-seat arena provided with smoke-protected assembly seating. A dual-access row contains 34 seats. The minimum required aisle accessway (row) width is determined by the following formula:

12 inches + [(0.3 inch)(x − y)]

where:

x   = the number of seats in the row (34 in this example), and

y   = the maximum number of seats permitted with a 12-inch aisle accessway (19 in this example)

12 + [(0.3)(34 − 19)] = 12 + 4.5 + 16.5 inches minimum width

For rows of seating served by an aisle or doorway at only one end of a row, the formula for the clear width is the minimum 12 inches plus 0.6 inch (305 mm plus 15 mm) for each seat over seven, but the clear width need not exceed 22 inches (559 mm). This is similar to the previous provision with one major difference; a maximum 30-foot (9144 mm) path of travel is permitted from the occupant's seat to a point where there is a choice of two directions of travel to an exit. This can be to the adjacent two-way aisle or along a dead-end aisle to a cross-aisle or doorway. See Figure 1028-3. Where one of the two paths of travel is across the aisle through a row of seats to another aisle, the maximum of 24 seats rule described previously is in effect. For smoke-protected assembly seating, single direction travel distance can be increased per Table 1028.10.1.

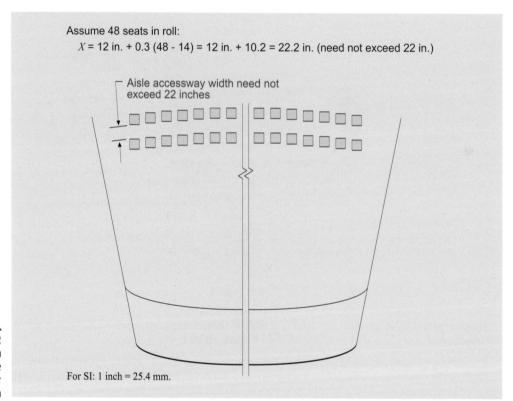

Assume 48 seats in roll:

X = 12 in. + 0.3 (48 - 14) = 12 in. + 10.2 = 22.2 in. (need not exceed 22 in.)

Aisle accessway width need not exceed 22 inches

For SI: 1 inch = 25.4 mm.

Figure 1028-2
Maximum required aisle accessway width

1028.11   Assembly aisle walking surfaces. Where aisles have a slope of 1 unit vertical in 8 units horizontal (12.5 percent slope) or less, steps in the aisle are prohibited because occupants in low-slope aisles have a tendency to not notice steps as readily as they would in the steeper aisles. Continuous surfaces are safer surfaces.

Where an aisle has a slope steeper than 1 unit vertical in 8 units horizontal (12.5 percent slope), a series of risers and treads must be used. These risers and treads shall extend the entire width of the aisle, shall have a rise of no more than 8 inches (203 mm) nor less than 4 inches (102 mm), and shall be uniform for the entire flight. The tread shall not be less than

11 inches (279 mm) and shall be uniform throughout the flight. Variations in run or height between adjacent treads or risers shall not exceed ³/₁₆ inch (4.8 mm).

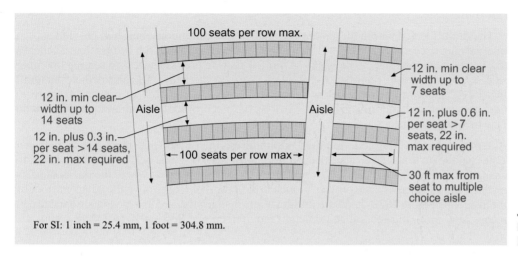

For SI: 1 inch = 25.4 mm, 1 foot = 304.8 mm.

Figure 1028-3
Row spacing

One provision that helps the user of an aisle notice a step is the provision requiring a contrasting strip or other approved marking on the leading edge of each tread. Designed to identify the edge of each tread when viewed in descent, this marking strip may be omitted where it can be shown that the location of each tread is readily apparent.

An exception permits variations in rise or height to exceed ³/₁₆ inch (4.8 mm) between risers, provided the exact location of such a variation is clearly identified with a marking strip at the nosing or leading edge adjacent to the nonuniform risers. This edge marking strip, having a width between 1 inch and 2 inches (25 mm and 51 mm) wide, shall be distinctively different from the contrasting marking strip required on each tread. In another exception to the riser height provision, riser heights may be increased to 9 inches (229 mm) where it can be demonstrated that lines of sight would otherwise be impaired.

Seat stability. Because of the potential obstructions to the paths of egress travel caused by loose seating, this section requires seats in assembly occupancies to be securely fastened to the floor. The following six exceptions identify conditions under which the securing of seating is either impractical or unnecessary: **1028.12**

1. Where 200 or fewer seats are provided on a flat floor surface

2. Where seating is at tables and the floor surface is flat

3. Where more than 200 seats are fastened together in groups of three or more on a flat floor surface

4. Where seating flexibility is critical to the function of the space, and 200 or fewer seats are provided on tiered levels

5. Where level seating is separated by railings or similar barriers into groupings of 14 seats or fewer

6. Where seating is separated by railings or similar barriers and limited to use by musicians or other performers

Handrails. All aisles having a slope steeper than 1 unit vertical in 15 units horizontal (6.7 percent slope), and all aisles stairs, shall have handrails complying with Section 1012. The handrails can be placed on either side of, or down the center of, the aisle served and can project into the required width no more than 4¹/₂ inches (114 mm). **1028.13**

Handrails may be omitted where the slope of the aisles is not greater than 1 unit vertical in 8 units horizontal (12.5 percent slope) with seating on both sides, or where a guard that conforms to the size and shape requirements of a handrail is located at one side. The first exception intends the seating to be a substitute for handrails. The second exception permits a graspable top rail of a guard to be utilized as the handrail where a drop-off occurs on the one side of the aisle.

Handrails located within the aisle width shall not be continuous, but shall provide gaps at intervals not exceeding five rows. The width of these gaps should not be less than 22 inches (559 mm), nor more than 36 inches (914 mm). This is to provide access to seating on either side of the rail and to facilitate the flow of users on the aisle. An intermediate handrail located 12 inches (305 mm) below the main handrail is required to prevent users from ducking under the handrail and hindering flow. Also, it provides a handrail for toddlers who may be using the aisle.

1028.14 **Assembly guards.** The code requires minimum 26-inch-high (660 mm) guards between aisles parallel to seats (cross aisles) and the adjacent floor or grade below where an elevation change of 30 inches (762 mm) or less occurs. An exception exists where the backs of seats on the front of the cross aisle project 24 inches (610 mm) or more above the adjacent floor of the aisle. See Figure 1028-4. Where the elevation change adjacent to a cross aisle exceeds 30 inches (762 mm), the general guard height requirements of Section 1013.2 shall apply.

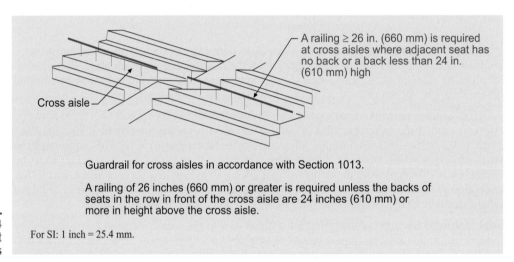

A railing ≥ 26 in. (660 mm) is required at cross aisles where adjacent seat has no back or a back less than 24 in. (610 mm) high

Cross aisle

Guardrail for cross aisles in accordance with Section 1013.

A railing of 26 inches (660 mm) or greater is required unless the backs of seats in the row in front of the cross aisle are 24 inches (610 mm) or more in height above the cross aisle.

For SI: 1 inch = 25.4 mm.

Figure 1028-4
Guards at cross aisles

The intent of this provision is to provide a certain degree of protection from falls that may occur while occupants are using a cross aisle adjacent to a drop-off. Even if the drop is minimal, the conditions of egress from an assembly use, particularly in low light, dictate the need for an increased level of safety. In addition, where the top of the seat backs are less than 24 inches (610 mm) above the aisle floor, an unintentional impact of the seat back could cause a fall over the seats.

In order to provide for proper viewing in auditoriums, theaters and similar assembly uses where the floor or footboard elevation is more than 30 inches (762 mm) above the floor or grade below, a guard in front of the first row of fixed seats, and which is not at the end of an aisle, may be 26 inches (660 mm) in height. Under such conditions, a guard height of at least 36 inches (914 mm) high shall extend the full width of the aisle at the foot of the aisle. In addition, the top of the guard shall be located at least 42 inches (1067 mm), measured diagonally, from the nosing of the nearest tread as depicted in Figure 1028-5.

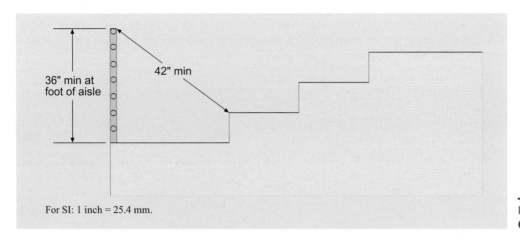

For SI: 1 inch = 25.4 mm.

Figure 1028-5
Guard heights

Section 1029 *Emergency Escape and Rescue*

General. Because so many fire deaths occur as the result of occupants of residential and institutional buildings being asleep at the time of a fire, the IBC requires that basements and all sleeping rooms below the fourth story have windows or doors that may be used for emergency escape or rescue. Applicable in Group R and I-1 occupancies, the requirement for emergency escape and egress openings in sleeping rooms is because a fire will usually have spread before the occupants are aware of the problem, and the normal exit channels will most likely be blocked. The reason for the requirement in basements is that they are so often used as sleeping rooms. Two exceptions eliminate the requirement for emergency escape and rescue openings in other than Group R-3 occupancies—buildings equipped throughout with an approved automatic sprinkler system, and sleeping rooms having direct access to a fire-resistance-rated corridor, provided the corridor has access to at least two remote exits in opposite directions. By providing sprinkler protection or a protected exit way leading to multiple exits, the opportunity for safe egress from the interior of the building is greatly enhanced, as is the chance for rescue.

1029.1

The code intends that the openings required for emergency escape or rescue be located on the exterior of the building so that rescue can be affected from the exterior or, alternatively, so that the occupants may escape from that opening to the exterior of the building without having to travel through the building itself. Therefore, where openings are required, they shall open directly onto a public street, public alley, yard or court. This provision ensures that continued egress can be accomplished after passing through the emergency escape and rescue opening. An exception permits such openings to lead directly to a balcony within an atrium complying with the requirements of Section 404. The atrium must provide access to an exit, and the dwelling unit or sleeping room must be provided with an additional means of egress that is not open to the atrium. These conditions create a situation equal to or better than the general requirement.

Minimum size. The dimensions prescribed in the code, and as illustrated in Figure 1029-1 for exterior wall openings used for emergency egress and rescue, are based in part on extensive testing by the San Diego Building and Fire Departments to determine the proper relationships of the height and width of window openings to adequately serve for both rescue and escape. The minimum of 20 inches (508 mm) for the width was based on two criteria—the width necessary to place a ladder within the window opening and second, the width necessary to admit a fire fighter with full rescue equipment. The minimum 24-inch (610 mm) height dimension was based on the minimum necessary to admit a fire fighter with full rescue equipment. By requiring a minimum net clear opening size of at least 5.7

1029.2

square feet (0.53 m²), the code ensures that an opening of adequate dimensions is provided. Where the opening occurs at grade level, the opening need only be 5 square feet (0.46 m²) because of the increased ease of access from the exterior.

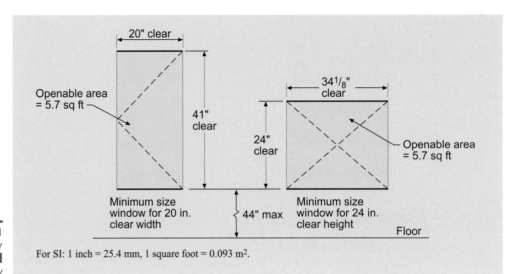

Figure 1029-1
Emergency escape and rescue window

1029.3 **Maximum height from floor.** In order to be relatively accessible from the interior of the sleeping room or basement, the emergency escape and rescue opening cannot be located more than 44 inches (1118 mm) above the floor. The measurement is to be taken from the floor to the bottom of the clear opening.

1029.4 **Operational constraints.** As stated in the code, these openings used for emergency escape or rescue must be operational from the inside of the room. Where windows are utilized, the intent is that they be of the usual double-hung, horizontal sliding or casement windows operated by the turn of a crank. The building official should evaluate special types of windows other than those just described based on the difficulty of operating or removing the windows. If no more effort is required than that required for the three types of windows just enumerated, they could be approved as meeting the intent of the code as long as no keys or tools are required.

The ever-increasing concern for security, particularly in residential buildings, has created a fairly large demand for security devices such as grilles, bars and steel shutters. Unless properly designed and constructed, the security devices over bedroom windows can completely defeat the purpose of the emergency escape and rescue opening. Therefore, the IBC makes provisions for security devices, provided the release mechanism has been approved and is operable from the inside without the use of a key, tool or force greater than that which is required for normal operation of the escape and rescue opening. Furthermore, in this case, the code requires that the building be equipped with smoke detectors in accordance with Section 907.2.11. Fire deaths have been attributed to the inability of the individual to escape from the building because the security bars prevented emergency escape.

The very essence of the requirement for emergency escape openings is that a person must be able to effect escape or be rescued in a short period of time because the fire will have spread to the point where all other exit routes are blocked. Thus, time cannot be wasted in figuring out means of opening rescue windows or obtaining egress through them. Therefore, any impediment to escape or rescue caused by security devices, inadequate window size, difficult operating mechanisms, etc., is not permitted by the code.

1029.5 **Window wells.** Window wells in front of emergency escape and rescue openings also have minimum size requirements. These provisions address those emergency escape windows that occur below grade. Obviously, just providing the standard emergency escape window criteria to these windows will get occupants through the window, but the window well may actually trap them against the building without providing for their escape from the window well or providing for fire-fighter ingress.

The minimum size requirements in cross section are similar in intent to the emergency escape and opening criteria; that is, to provide a nominal size to allow for the escape of occupants or ingress of fire fighters. See Figure 1029-2. The ladder or steps requirement is the main difference.

Emergency escape openings below the fourth story are not required to have an escape route down to grade; however, those openings below adjacent grade are so required. When the depth of a window well exceeds 44 inches (1118 mm), a ladder or steps from the window well are required. The details for construction of steps are not identified in the provisions; however, the design of the ladder is specifically addressed. Rungs are to have a minimum interior width of 12 inches (305 mm), shall project at least 3 inches (76 mm) from the wall and be spaced no more than 18 inches (457 mm) on center vertically for the full height of the window well. Because ladders and steps in window wells are provided for emergency use only, they are not required to comply with the provisions for stairways found in Section 1009.

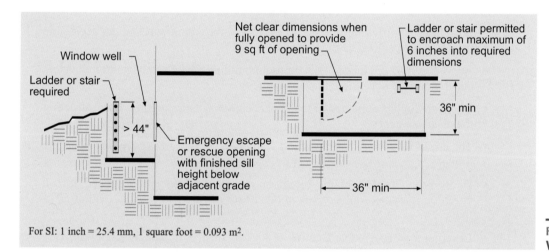

For SI: 1 inch = 25.4 mm, 1 square foot = 0.093 m².

Figure 1029-2
Window wells

KEY POINTS

- The means of egress is an exiting system that begins at any occupied point in a building and continues until the safety of the public way is reached.

- Three distinct elements compose the means of egress—the exit access, the exit and the exit discharge.

- Occupant load, the driving force behind the design of an exiting system, must be determined for the expected use or uses of a building.

- Components along the path of egress travel must be sized to accommodate the expected occupant load served by the components.

- Specific minimum component widths, such as those provided for doors, aisles, corridors and stairways, often dictate the capacity of the means of egress.

- Where multiple complying exit ways are provided, the calculated width may be dispersed among the various exits or exit-access doorways.

- With limited exceptions, the means of egress must have a minimum clear height of 7 feet 6 inches (2286 mm) throughout the travel path.

- The code regulates the means of egress so that there is no change in elevation along the path of exit travel that is not readily apparent to persons seeking to exit under emergency conditions.

- As a general rule, exit signs are required from rooms or areas requiring access to two or more paths of exit travel.

- Requiring continuous illumination, exit signs must be provided with a secondary source of power.

- In those rooms or areas requiring access to at least two exitways, a second source of power is required for maintaining illumination to the exit path.

- Guards must be designed to reduce the probability of falls from one level to a lower level that exceeds 30 inches (762 mm) in elevation difference.

- Guards are to be at least 42 inches (1067 mm) in height.

- In all public areas, guards must have limited openings to prevent individuals from falling or climbing through the guard assembly.

- In addition to providing proper access to and through a building, an accessible means of egress must be developed.

- Areas of refuge are mandated for certain buildings where stairs and/or elevators occur along the accessible means of egress.

- Doors are highly regulated in the IBC because of their potential for obstructing the means of egress.

- The use of revolving, overhead and sliding doors for egress purposes is strictly limited.

- Doors swinging toward the direction of egress travel are mandated for all hazardous uses, as well as areas in other uses having an occupant load of 50 or more.

- Criteria for an acceptable latching or locking device on an egress door are very basic in that no key, special effort or special knowledge is necessary to open the door.

- Where security issues are as important as those addressing fire- and life-safety, the IBC permits the installation of delayed egress locks.

- Panic hardware is mandated in Group A and E occupancies having an occupant load of 50 or more, as well as in all Group H occupancies.

- Gates located in the means of egress are regulated in a manner similar to doors.

- A stair is considered a change of elevation accomplished by one or more risers, whereas one or more flights of such stairs make up a stairway, along with any landings that connect to them.

- Treads and risers must be appropriately sized and uniform throughout the stair flight.

- Spiral stairways, curved stairways and alternating tread devices are limited in their use because of the uniqueness of their configurations.

- Handrail design is regulated for height, size, shape and continuity.

- Ramps must be designed for egress purposes as well as for accessibility.

- Exit access describes the vast majority of a building's floor area that provides the access necessary to reach a protected area (an exit).

- Access to at least two exits is typically required from floor levels above the first story, and rooms or areas having sizable occupant loads or excessive travel distance.

- Multiple exit paths must be arranged in order to minimize the risk of a single fire blocking all of the exit ways.

- Egress from a room through a nonhazardous accessory area is permitted, provided there is a discernible egress path that is direct and obvious.

- Travel distance is limited within the exit access portion of the means of egress; however, once an exit is reached, travel distance is no longer regulated.

- In rooms where seating is at tables, additional limitations are placed upon the aisles and aisle accessways utilized for egress purposes.

- Corridors are intended to be used for circulation and egress purposes, and at times must be constructed as a protected element for use as a path for egress travel.

- The exit is the portion of the means of egress that provides a degree of occupant protection from fire, smoke and gases.

- Horizontal exits, exit passageways, exit enclosures, exterior exit doors at grade level, exterior exit ramps and exterior exit stairways are the exit components addressed in the IBC.

- In some buildings, two floors may be interconnected without the need for vertical exit enclosures.

- Exit enclosures are to be constructed of either 1-hour or 2-hour fire-resistance-rated construction with protected openings.

- Openings and penetrations into an exit enclosure are strictly limited because of the hazards involved with vertical egress.

- An exit passageway is similar to a corridor, but built to a higher level and limited in much the same manner as a vertical exit enclosure.

- The concept of a horizontal exit is the creation of a refuge area to be used by occupants fleeing the area of fire origin.

- The use of an exterior exit stairway as a required means of egress element is limited to buildings not exceeding six stories or 75 feet (22 860 mm).

- In high-rise buildings, luminous egress path markings are required in exit enclosures and exit passageways of Group A, B, E, I, M and R-1 occupancies.

- Egress travel outside of the building at grade level is considered exit discharge, continuing until the public way is reached.

- An egress court, open so that smoke and toxic gases will not accumulate, is an exit discharge component.

- Egress courts of limited width must be provided with a minimum level of fire protection in order to protect occupants as they pass through the egress court.

- Larger auditoriums, theaters and similar assembly spaces are uniquely regulated as to the design of the egress system.

- In assembly occupancies, the method for calculating aisle widths is modified where smoke-protected assembly seating is provided.

- Grandstands and bleachers, although similar in many aspects to typical assembly seating, are regulated by unique provisions found in ICC 300.

- Emergency escape and rescue openings are not required in fully sprinklered Group R and I-1 occupancies.

11

ACCESSIBILITY

This chapter addresses accessibility and usability of buildings and their elements for persons having physical disabilities. Where a facility is designed and constructed in accordance with this chapter and other related provisions throughout the IBC, it is considered accessible.

A historical perspective. In 1961, the American National Standards Institute (ANSI) published ANSI Standard A117.1. The President's Committee on Employment of the Handicapped and the National Easter Seal Society were designated as the secretariat for the standard. Since that time, a number of historic events have occurred that have brought accessibility and usability issues to the forefront of not only building code enforcement, but of society as a whole.

In the early 1970s, all three United States legacy model code groups approved code changes to make buildings more accessible and usable for people with disabilities. These independent developments resulted in confusion in the regulatory design and construction community. As a result, the Council of American Building Officials (CABO) requested that the Board for the Coordination of Model Codes (BCMC) review the regulations and suggest provisions to all of the model codes that would result in uniformity. In addition, ANSI requested that CABO become the secretariat for ANSI A117.1.

In October 1987, BCMC began its assignment to provide regulations that set forth when, where and to what degree access must be provided (commonly referred to as scoping) for persons with disabilities. The ANSI A117.1-1986 standard contained design specifications intended to provide buildings and facilities accessible to and usable by people with disabilities but did not specify scoping provisions. Authorities who chose to employ ANSI A117.1 found it necessary to adopt amendments to establish when, where and to what degree its provisions applied. During the BCMC work on accessibility, it became apparent that safe egress for people with disabilities was essential if access to buildings was to be increased. Therefore, the final BCMC report addressed both accessibility and egress for people with disabilities. While BCMC was working on scoping provisions for ANSI A117.1-1986, the ANSI A117.1 committee continued to study revisions to their standard for public review.

In 1988, the United States Congress passed the Fair Housing Amendments Act to cover multifamily housing of four units or more on a site. On July 26, 1990, President George Bush signed the Americans with Disabilities Act (ADA), which set forth comprehensive civil rights protection to individuals with disabilities in the areas of employment, public accommodations, state and local government services, and telecommunications. One of the reasons legislators supported the ADA was the recognized inadequacy, limited application and nonuniformity of existing protection for individuals with disabilities. One year later, on July 26, 1991, the United States Department of Justice (DOJ) issued its final rules, the Americans with Disabilities Act Accessibility Guidelines (ADAAG), that provided for access and usability for disabled persons in public accommodations and commercial facilities. Both acts were born from the Civil Rights Act of 1964. The ADA set forth statutory deadlines for when certain requirements became effective. One of these requirements was that new facilities designed and constructed for first occupancy after January 26, 1993, must be accessible.

The public review draft of revisions to the ANSI A117.1 standard dated January 24, 1992, was submitted to the DOJ with a request for technical assistance. A staff comparison of ANSI A117.1 and ADAAG yielded only a few areas in which ADAAG was deemed to provide greater accessibility. Generally, the differences found between ADAAG and ANSI A117.1 indicated that the ANSI standard provided for greater overall accessibility. At the BCMC meetings in May and June of 1992, the committee reviewed suggestions that would incorporate ADA guidelines into the ANSI A117.1 standard and other regulations. At the BCMC meeting on June 8, 1992, the committee finalized its report. From June 9 through 11, 1992, the ANSI A117.1 committee finalized its standard.

The final BCMC report of June 8, 1992, and the final draft of the ANSI A117.1 standard were soon adopted by all of the model code groups as their accessibility requirements. The final BCMC report of June 8, 1992, and the CABO/ANSI A117.1-1992 standard were submitted to the DOJ for a technical review. The resulting letter received in November 1995 described nine general problems, many of which were differences in philosophy. Noncode items such as laboratory equipment, automated teller machines and telephones were also a concern of the DOJ. During the DOJ review, a joint task force was established to make recommendations for changes to both the CABO/ANSI A117.1, *Accessible and Usable Buildings and Facilities* and ADAAG. This harmonization effort over many months resulted in suggested revisions to both documents. In an effort to reduce conflict, confusion and frustration among all users, the results of the harmonization report were accepted in July 1996 by the ADAAG Review Advisory Board. That board presented its final report to the Architectural and Transportation Barriers Compliance Board, which in turn resulted in additional rulemaking by the Access Board.

The CABO/ANSI revisions were finalized and subsequently approved by ANSI's Board of Standards Review on February 13, 1998. Because CABO was incorporated into the International Code Council (ICC) in November 1997, the resulting standard was retitled ICC A117.1-1998. The 2003 edition of the standard is now referenced in the *International Building Code* (IBC), Chapter 11, for the design and construction of accessible buildings and facilities. The 2003 edition of ICC A117.1 was published in mid-2004 and its technical provisions even more closely parallel those technical requirements in the current ADAAG. The provisions have also moved towards coordination with the proposed "new" ADA/ABA AG (Americans with Disabilities Act/Architectural Barriers Act Accessibility Guidelines).

The first major rewrite of ADAAG since 1990 was published in the Federal Register on July 23, 2004, resulting in the reconciliation of most differences that occurred between the A117.1 standard and ADAAG. Although the coordination effort is not fully completed, it is expected that the efforts of all parties will result in a system that greatly enhances compliance with the recently released ADA Accessibility Guidelines. It should be noted that although the new ADAAG was released in 2004, at the time of publication of this handbook it has not received the final approval from the Department of Justice. It has, however, been used by several of the federal agencies for their own projects since its release in 2004.

On another front, efforts were initiated addressing the differences between those provisions of the IBC, ICC A117.1 and the federal Fair Housing Accessibility Guidelines (FHAG). In 2000, the Department of Housing and Urban Development (HUD) reviewed the IBC and ICC A117.1-1998 for compliance with FHAG. Based on HUD's report, a series of modifications were proposed as part of the 2000 code change cycle. The proposed modifications were accepted by the voting members and were incorporated into the 2001 Supplement to the IBC. As a result of these changes, HUD issued a press release that stated the 2000 IBC with the 2001 Supplement and ICC A117.1-1998 could be considered "safe harbor" for anyone wanting to comply with the FHAG requirements. The 2003 and 2006 editions of the IBC incorporated those provisions, along with other appropriate modifications, and they have both been recognized as a safe harbor in compliance with the Fair Housing requirements. At the time of this publication, HUD is reviewing the 2009 IBC to determine if it will also be granted "safe harbor" status. Given the previous approvals and the way HUD and ICC have worked together to coordinate the IBC, it is anticipated that the 2009 edition of the IBC will also be determined to be equivalent when addressing the housing accessibility requirements.

Section 1101 *General*

1101.2 **Design.** This section adopts ICC A117.1, more specifically the 2003 edition, as the adopted design standard to be used to ensure that buildings and facilities are accessible to and usable by persons with disabilities. With this section providing the accessible design and construction standards for buildings, the remaining sections of the chapter provide the scoping provisions that set forth when, where and to what degree access must be provided.

The importance of this section and the requirements it imposes should not be overlooked. As previously stated, buildings must be designed and constructed to the minimum provisions of Chapter 11, along with the other applicable provisions of the IBC and ICC A117.1, to be considered accessible. Therefore, prior to applying the code provisions for accessibility, it is important for the code user to review the technical requirements found in ICC A117.1 Although these items are not completely addressed in the IBC, they are an important part of making a building accessible. Such elements include, but are not limited to, space allowance and reach ranges, accessible route, protruding objects and ramps.

Space requirements can vary greatly depending on the nature of the disability, the physical functions of the individual, and the skill or ability of the individual in using an assistive device. However, it is generally accepted that spaces designed to accommodate persons using wheelchairs will be functional for most people.

It is important to note that not all portions of ICC A117.1 are referenced by the IBC. For example, the criteria set forth in Section 504 for stairways as well as the stair handrail provisions of Section 505 have no application insofar as accessible stairways are not specifically scoped by IBC Chapter 11. As a result, the provisions of IBC Sections 1009 and 1012 solely regulate all stairways and associated handrails for accessibility purposes. As another example, accessible telephones and automatic teller machines are addressed in ICC A117.1, Sections 704 and 707, respectively. However, these elements are only regulated in the IBC by Appendix E, which must be specifically adopted to be in effect.

Section 1102 *Definitions*

It is important to note that the definitions of this section are specifically for this code and may have different meanings from definitions in other accessibility provisions or regulations such as ADAAG. Additional definitions that apply to these provisions can be found in Section 106 of ICC A117.1.

The term *accessible* takes on a very broad meaning by requiring compliance with Chapter 11. Space requirements must be addressed for all portions of the building to be considered accessible or to provide an accessible route within a building. Space requirements apply to adequate maneuvering space, clearance width for doors and corridors, and height clearances.

Location of controls, switches and other forms of hardware becomes a function of forward and parallel reach ranges if they are to be considered accessible. To assist persons with limited dexterity, controls and other forms of hardware need to be operable without tight gripping, grasping or twisting of the wrist.

People with visual impairments are provided with accessible routes, by the inclusion of provisions for clear and unobstructed routes that are free of protrusions created by benches, overhanging stairways, poles, posts and low-hanging signs. Many hazards can be eliminated by using different textural surfaces in the accessible route to alert sight-impaired persons. Visually impaired or partially sighted persons can be assisted with proper signage of the correct size, surface and contrast. Signage is also provided for the hearing impaired. Directional signage should always be clear, concise and appropriately placed.

An accessible route is defined in both the IBC and ICC A117.1, and is described in Section 402 of the latter document. Accessible routes have a number of components, such as walking surfaces with a slope not steeper than 1:20, marked crossings at vehicular ways, clear floor spaces at accessible elements, access aisles, ramps, curb ramps and elevators. Each component must comply with specific applicable standards and requirements.

Several key elements to review when addressing accessible routes are the requirements related to protruding objects and ramps. An accessible route may contain a number of protruding objects that can affect its use. These objects include ordinary building elements such as telephones, water fountains, signs, directories and automatic teller machines. Protruding objects and other such obstructions are regulated by Section 307 of ICC A117.1. Similar provisions governing the means of egress system are located in IBC Section 1003.3. Ramps with proper slopes may serve as acceptable means of egress, as well as part of an accessible route. A sloped surface that is steeper than 1:20 is considered to be a ramp and is required to comply with both the requirements of Section 1010 for egress paths and Section 405 in ICC A117.1. To be considered acceptable, ramps shall have a slope not steeper than 1:12, with a maximum rise for any ramp not to exceed 30 inches (762 mm) and a minimum width of at least 36 inches (914 mm).

Three types of dwelling units and sleeping units are defined: Accessible, Type A and Type B. The Accessible units are deemed to be fully accessible and must be in compliance with the IBC and Section 1002 of ICC A117.1. Type A units are considered dwelling units and sleeping units that are designed and constructed for accessibility in accordance with Section 1003 of ICC A117.1. Designed and constructed in accordance with Section 1004 of ICC A117.1, a Type B dwelling unit or sleeping unit is intended to be consistent with the technical requirements of HUD's Fair Housing Guidelines. A Type A unit is considered to provide a significant degree of accessibility, whereas a Type B unit provides for only a minimum level.

Several other definitions assist in clarifying the intent of the provisions. An employee work area is limited to those spaces used directly for work activities and does not include those common use areas such as toilet rooms and break rooms. Public use areas are identified as those spaces utilized by the general public and may be interior or exterior.

Section 1103 *Scoping Requirements*

In general, access to persons with physical disabilities is required for all buildings and structures, whether temporary or permanent. There are, however, certain conditions under which sites, buildings, facilities and elements are exempt from the provisions where specifically addressed. Various modifications to the general requirements for accessibility are also found throughout other areas of Chapter 11. For example, an exception to Section 1104.4 eliminates the requirement for an accessible route of travel to levels having relatively small floor areas, applicable in all but a few types of uses. There are also a number of exceptions that apply only within dwelling units.

Where the building under consideration is existing, only the requirements found in Section 3411 apply. Intended to include historic buildings, the provisions apply to the maintenance, alteration or change in use of an existing structure. An exception relating to Type B dwelling units and sleeping units states that they need not be provided in existing buildings.

In commercial applications, individual employee workstations are not required to be fully accessible. They must, however, comply with the appropriate provisions for visible alarms, accessible means of egress and common use circulation paths. In addition, such work areas shall be located on an accessible route in order to provide access to, into and out of the work area. Modifications to each individual work station can then be made in order to address the specific needs of the employee. None of these limited provisions are applicable for small, elevated work areas where the area must be elevated because of the nature of the work performed. A number of other employee-use areas are also considered spaces that need not be provided with accessibility. Raised areas used for security or safety purposes, limited access spaces, equipment spaces and single-occupant structures have been identified as those types of areas where it seems unreasonable, if not impossible, to provide full access.

Observation galleries, prison-guard towers, fire towers, lifeguard stands and similar elevated observation areas need not be accessible or served by an accessible route. Ladders, catwalks, crawl spaces, freight elevators, very narrow passageways or tunnels, and any other space deemed to be nonoccupiable require no access. Spaces such as elevator pits and penthouses; mechanical, electrical or communications equipment rooms; equipment catwalks; water or sewer treatment pump rooms and stations; electric substations in transformer vaults; and any other areas accessed solely by personnel for maintenance, repair or monitoring of equipment are not required to be accessible. In addition, single-occupant structures where access occurs only by passageways elevated above grade or buried below grade are not required to be accessible. An example would be a toll booth accessed only by an underground tunnel, or a bank teller booth reached from an overhead enclosure. Accessibility is also not required to walk-in coolers and freezers, provided they are strictly employee-access.

Section 1103.2.6 recognizes that some activities directly associated with construction projects will not be safe for persons with certain physical disabilities. Therefore, structures, sites and equipment used or associated with construction are not required to be accessible. The limited scope of this provision is important insofar as the accessibility provisions generally do apply to the temporary buildings. Although accessibility would not be required to a construction trailer, it must be provided to sales trailers that are common in new subdivisions or multifamily projects. It is also necessary to know that where pedestrian protection is required by Chapter 33 and Table 3306.1, such walkways are required to be accessible.

For the most part, occupancies designated as Group U are exempt from the provisions of Chapter 11. One exception requires agricultural buildings that are associated with the general public to be provided with access to those public areas. An example might be a produce stand set up just inside the entry of the agricultural greenhouse. In addition, all paved work areas for agricultural buildings must be provided with access to such areas. A second exception mandates that where accessible parking is provided in private garages or carports, such parking structures shall be accessible.

A live/work unit, primarily regulated by Section 419, is considered to be a dwelling unit or sleeping unit in which a significant portion of the space includes a nonresidential use operated by the tenant. Although the entire unit is classified as a Group R-2 occupancy in the IBC, for accessibility purposes it is viewed more as a mixed-use condition. The residential portion of the unit is regulated differently for accessibility purposes than the nonresidential portion. The floor area of the dwelling unit or sleeping unit that is intended for residential use is regulated under the provisions of Section 1107.6.2 for Group R-2 occupancies. The requirements for an Accessible Type A or B unit would be applied based upon the specific residential use of the unit and the number of units in the structure. The exceptions for Type A and B units set forth in Section 1107.7 would also exempt such units where applicable. In the nonresidential portion of the unit, full accessibility would be required based upon the intended use. For example, if the nonresidential area of the unit is utilized for hair care services, all elements related to the service activity must be accessible. This would include site parking where provided, site and building accessible routes, the public entrance, and applicable service facilities. In essence, this portion of the live/work unit would be regulated in the same manner as a stand-alone commercial occupancy.

Section 1104 *Accessible Route*

An accessible route is defined as a continuous unobstructed path that complies with the provisions in Chapter 11. This route connects all accessible elements or spaces of the building or facility, including corridors, aisles, doorways, ramps, elevators, lifts and clear floor space at fixtures. Code users should review the accessible route provisions in Chapter 4 of ICC A117.1. In addition, exterior portions of accessible routes must be evaluated and may include parking access aisles, curb ramps, crosswalks at vehicular ways, walks, ramps and lifts.

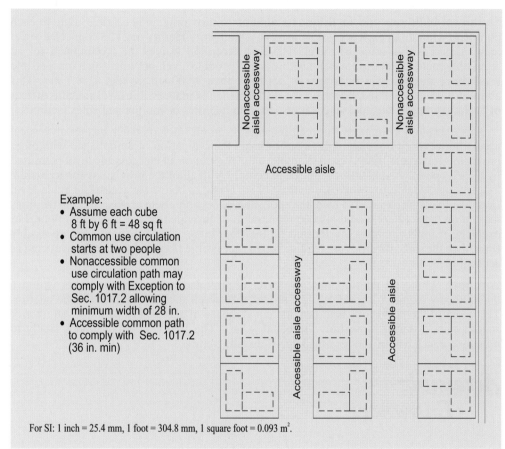

Example:
• Assume each cube 8 ft by 6 ft = 48 sq ft
• Common use circulation starts at two people
• Nonaccessible common use circulation path may comply with Exception to Sec. 1017.2 allowing minimum width of 28 in.
• Accessible common path to comply with Sec. 1017.2 (36 in. min)

For SI: 1 inch = 25.4 mm, 1 foot = 304.8 mm, 1 square foot = 0.093 m².

Figure 1104-1
Employee work areas

Within a site. The clear intent of these provisions is to provide, on sites with single or multiple buildings or facilities, access to each accessible element from all parking areas, as well as from one accessible element to another on the same site. Where the only means of access between accessible facilities on a site is a vehicular way, an accessible route is not required between such facilities. Should a sidewalk, walking path or similar circulation route be provided connecting the site elements, the route is to be designed and constructed as an accessible route. **1104.2**

Connected spaces. Those portions of a building that are required to be accessible must be connected by at least one accessible route of travel. Such route shall connect to all accessible entrances, and lead to accessible walkways connecting other accessible site elements and potentially the public way. One exception clarifies that in assembly areas with fixed seating an accessible route must only be provided to the accessible seating areas. The other **1104.3**

exception modifies the maneuvering clearance requirements of ICC A117.1 for doors to Group I-2 sleeping units.

1104.3.1 **Employee work areas.** Although employee work areas are typically exempt from the accessibility provisions of the code, the circulation paths within such areas that are used by multiple employees are regulated. The paths must be designed and constructed as a complying accessible route unless exempted by one of the three exceptions. In those cases where an exception is applicable, it is still necessary to connect the work area to an accessible route such that physically-disabled persons can approach, enter and exit the area. For the purpose of applying any of the three exceptions, it is important to note that *common use* is defined as nonpublic areas shared by two or more individuals. An example is shown in Figure 1104-1.

1104.4 **Multilevel buildings and facilities.** In addition to providing an accessible route to each portion of a building, each accessible level in a multistory facility must be connected via at least one accessible route of travel. The first exception waives the requirement for an accessible route to floors above and below any accessible level, provided such inaccessible levels have an aggregate floor area of not more than 3,000 square feet (278.7 m²). This allowance is not permitted, however, where the vertically inaccessible level contains offices of health-care providers, passenger transportation facilities or multitenant sales facilities.

This exception eliminates the requirement for an elevator or other means of vertical access; however, it does not reduce or eliminate the obligation of compliance with other provisions for accessibility. As an example, a toilet room on an inaccessible level permitted by the exception would still be required to comply with all of the provisions for accessible toilet rooms. In addition, whereas the floor either above or below the accessible grade-level floor would not be required to meet the applicable accessible route provisions, any facilities located on these floors would also be required to be provided on the accessible floor. For example, if toilet facilities are located on the floor either above or below the accessible floor, then the same facilities are required on the accessible floor and shall be constructed as accessible facilities.

1104.5 **Location.** Where an interior route of travel is provided between floor levels within a building, any required accessible route provided between such levels shall also be interior. The intent of the provisions is to provide equal means of access. For example, an interior stairway and exterior ramp fails the equality test. The first exception applies solely to parking garages within and serving Type B dwelling units.

Section 1105 *Accessible Entrances*

In general, at least one public entrance, but not less than 60 percent of all such entrances to a building or individual tenant space, must be accessible.

In Figure 1105-1, two entrances (Doors A and B) are considered to be public entrances to the entire building. As such, both entrances must be accessible. If Doors C and D are public tenant entrances they must be accessible in addition to the accessible entrances provided into each tenant space from the common lobby. One easy requirement to remember is that if there are only one or two entrances into a building or tenant, those entrances must be accessible. In addition to the general provisions for public entrances, the code mandates accessible entrances under other conditions, such as between a parking garage and the building served by the garage. Even though the minimum number of accessible entrances has been provided at other locations, the direct access between the parking garage and the building must be accessible. Exceptions are provided for entrances used exclusively for loading and service, as well as entrances to spaces not required to be accessible.

An important aspect of accessibility and accessible entrances is obviously the door assembly and its related components, including the threshold, hardware, closers and opening force. Important requirements that affect doors and their accessibility are in IBC Section 1008 and Section 404 of ICC A117.1.

Landings on both sides of doorways are also important accessibility features. When access for persons with disabilities is required, a floor or landing shall not be more than $\frac{1}{2}$ inch (12.7 mm) [$\frac{3}{4}$ inch (19.1 mm) for sliding doors] lower than the threshold of the doorway according to Section 1008.1.7 of the IBC. Where the level change or threshold exceeds $\frac{1}{4}$ inch (6.4 mm), it is required to be beveled at a slope of one unit vertical in two units horizontal (1:2) or less. Similar provisions are found in Sections 303 and 404.2.4 of the A117.1 standard.

Door hardware, including handles, pulls, latches, locks or any other operating device on accessible doors, is required to be of the shape that is easy to grasp with one hand and does not require a tight pinching, tight grasping or twisting of the wrist to operate. Many individuals have great difficulty operating door hardware that does not include push-type, U-shaped handles or lever-operated mechanisms.

Door closers with delayed action capability are also important and allow a person more time to maneuver through a door. These closers are required to be adjusted so that from an open position of 90 degrees (1.57 rad), the time required to move the door to an open position of 12 degrees (0.21 rad) will be a minimum of 5 seconds. Door closers are required to have minimum closing forces in order to close and latch the door; however, for other than fire doors and exterior doors, maximum force levels are set to limit the force levels for pushing or pulling open doors. Opening forces and the methods used to measure them are specified in Section 404.2.8 of ICC A117.1.

Minimum maneuvering clearances at doors, other than those that are for automatic or power-assisted doors, are based on a combination of forward- and side-reach limitations, the direction of approach, and minimum clear width required for wheelchairs. They also permit enough space that a slight angle of approach can be gained, which then provides additional leverage or opening force by the user. Without these required clearances, there is a possibility of interference between the edge of the door and the footrest on the wheelchair. This could render the door inaccessible to someone using a wheelchair.

In addition to the traditional provisions for doors, ICC A117.1 contains many detailed provisions in illustrations that are found in Section 404.

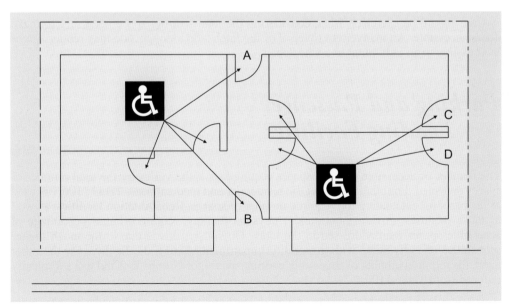

Figure 1105-1
Accessible entrances

Application Example 1106-1

GIVEN: A 440-space parking garage and 160-space surface parking area.

DETERMINE: The minimum required number of accessible parking spaces in each facility.

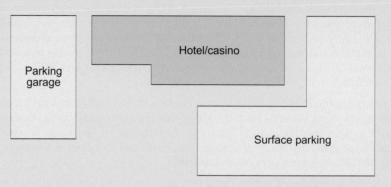

SOLUTION: Per Table 1106.1, the minimum required number of accessible parking spaces is:

	Total spaces	Van spaces
Parking garage	9	2
Surface parking	6	1
	15	3

Of the minimum 15 required accessible parking spaces, at least 3 must be van accessible. They need not be located in the individual facilities as calculated, provided substantially equivalent or greater access is provided in terms of distance from an accessible entrance or entrances, parking fee and user convenience.

CALCULATION OF ACCESSIBLE PARKING SPACES

Care should be exercised when considering the maneuvering space necessary to make doors accessible. An inadvertent reversal of the latch side to hinge side may render a door inaccessible to an individual in a wheelchair.

Section 1106 *Parking and Passenger Loading Facilities*

The number of accessible parking spaces required on a site varies by the total number of spaces provided. For other than specific residential and medical uses, Table 1106.1 is used as the basis for calculating the required number of spaces. Rehabilitation facilities, as well as those facilities providing out-patient physical therapy, require a larger percentage of accessible spaces than addressed in the table. Obviously, this is due to the much higher probability of individuals with a mobility impairment visiting the facility. On the other hand, the required number of accessible parking spaces for Groups R-2 and R-3 is typically reduced from Table 1106.1.

For every six accessible parking spaces, at least one of the spaces must be an accessible van space. As an example, consider a parking lot with 23 total parking spaces. According to Table 1106.1, at least one accessible parking space is to be provided, and it must be designed and constructed to be van accessible. As another example, a parking lot with 202 parking spaces must be provided with a minimum of seven accessible spaces. At least two of those seven spaces shall be van accessible.

An indicator of the critical nature of site development is the requirement for the shortest accessible route of travel. On a site with multiple buildings and a number of requirements for each, accessibility may pose complex site-design problems related to direct routes from parking areas. Where the parking facilities do not necessarily serve a particular building, the provisions require the accessible parking spaces to be located as closely as possible to the accessible pedestrian entrance to the parking facility. Early involvement or early attention to the location of multiple accessible elements during site development may tend to eliminate most, if not all, related problems.

Where multiple distinct parking facilities are provided on a site, such as a parking garage and a surface parking lot, the provisions of Section 1106.1 require that the total minimum required number of accessible parking spaces be determined individually. See Application Example 1106-1. However, Section 1106.6 allows accessible spaces to be relocated from remote lots to locations near accessible building entrances. This addresses a concern that in a large facility, the dispersion of accessible parking spaces into remote lots may result in decreased access for persons with disabilities.

If a passenger loading zone is provided, it shall have an adjacent access aisle that is part of the accessible route to the building. In accordance with ICC A117.1, the space shall have a vertical clearance of at least 98 inches (2490 mm) at the zone and along the vehicle access route on the site. There are only two conditions under which the code mandates the installation of a passenger loading zone, where valet parking services are provided, and at accessible entrances of specified medical facilities.

Section 1107 *Dwelling Units and Sleeping Units*

This section is limited to the accessibility provisions related to dwelling units and sleeping units as defined in Chapter 2. It provides guidance as to the conditions under which Accessible units, Type A units and Type B units are mandated. These scoping provisions generally indicate where some degree of accessibility is required, and to what extent.

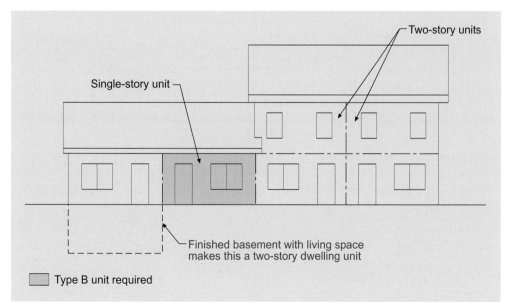

Figure 1107-1
One-story and multistory units

1107.2 **Design.** As addressed under the discussion of Section 1102, there are three types of dwelling and sleeping units that provide varying degrees of accessibility. Accessible units are provided with the most comprehensive accessibility requirement and required to comply with those applicable provisions of Section 1002 of ICC A117.1. Type A and B units, regulated by ICC A117.1 Sections 1003 and 1004, respectively, provide not only a reasonable degree of accessibility, but also allow for the use of adaptive features. This section reflects how it is always acceptable to design and construct to a higher degree of accessibility than that required by the code.

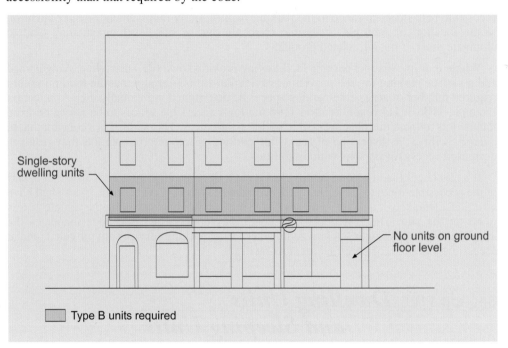

Figure 1107-2
One story of Type B units required

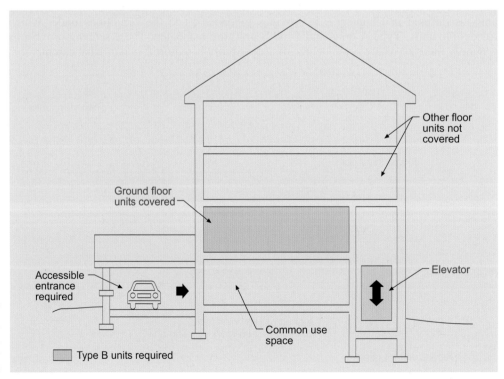

Figure 1107-3
Elevator service to lowest story with units

Group I. Dwelling units and sleeping units in a variety of Group I occupancies, including **1107.5** nursing homes, hospitals, nurseries, assisted-living facilities, group homes and care facilities, are regulated for accessibility. In Group I-1 occupancies, at least 4 percent of the dwelling and sleeping units shall be Accessible units. In the Group I-2 category, however, the percentage of Accessible patient rooms varies based upon the type of institutional facility. Hospitals and rehabilitation facilities that specialize in the treatment of conditions affecting mobility are required to have all patient rooms, including the toilet rooms and bathrooms, designed and constructed as Accessible units. General-purpose hospitals, psychiatric facilities and detoxification facilities are required to have at least 10 percent of their patient rooms be Accessible units for their patient population. It is assumed that, in most cases, not more than 10 percent of the facility's patients would need accessible rooms. In nursing homes and long-term care facilities, at least 50 percent of the dwelling and sleeping units are required to be Accessible units. This increase in the percentage of accessible rooms recognizes that the patients or residents of these facilities may be ambulatory on admission, but may become nonambulatory or have further mobility limitations during their stay. In Group I-3 facilities, at least two percent of the resident dwelling units and sleeping units must be Accessible units.

Group R-1. Unless intended to be occupied as a residence, a Group R-1 occupancy is **1107.6.1** typically regulated for Accessible units only. The minimum required number of Accessible units is easily determined from Table 1107.6.1.1. For example, a hotel with 185 guestrooms must provide a minimum of eight guestrooms that comply as Accessible units. The number of buildings in which the guestrooms, referred to in the code as dwelling units or sleeping units, are located does not impact the result. Where more than one building on the site contains guestrooms, the aggregate number of rooms is utilized to determine the minimum requirements. Where multiple types of dwelling or sleeping units are provided, the Accessible units must be represented in each room type. However, it is not necessary to provide for additional Accessible units above and beyond the number required by Table 1107.6.1.1. As an example, a motel requiring two Accessible units and providing three room types (such as a double, king and king suite) would only require two of the three room types to be Accessible units.

Although all of the Accessible units must be provided with accessible bathing facilities, only a portion are required to have roll-in showers. The table indicates the minimum number of Accessible units that must contain roll-in showers as described in Section 608 of ICC A117.1. The code mandates that such shower facilities be provided with a permanently mounted folding shower seat to address various disabilities. The minimum required number of units provided with roll-in showers is complemented by requiring a minimum required number of Accessible units without roll-in showers. The intent is to provide persons with physical disabilities a range of options equivalent to those available to other persons served by the facility. If the standard rooms have bathtubs, then some of the Accessible units should also be provided with bathtubs and a small percentage of rooms incorporating roll-in showers. Likewise, if all of the standard rooms have shower compartments, then most of the Accessible units should have transfer showers, with again a small percentage provided with roll-in showers. It has been shown that accessible bathtubs are preferred by many people with mobility impairments for both security when sitting and the therapeutic relief from a warm bath.

Group R-2. The provisions for Group R-2 occupancies are divided into two general **1107.6.2** categories, those applicable to typical apartment buildings and those applicable to dormitories, fraternity houses and sorority houses. Apartment houses, along with monasteries and convents, are regulated for Type A and B units. On the other hand, dormitories and similar congregate living facilities do not need to contain Type A units, but do require one or more Accessible units.

The mandate for Type A units in apartment buildings applies where 21 or more dwelling units or sleeping units are contained within the building. If two or more buildings are located on the same site, the aggregate number of units is used to determine the minimum required number of Type A units. Assuming a site contains four apartment buildings, each containing

30 units, the provisions are based on the sum total of 120 units. Using the 2 percent rule, at least three of the dwelling units are required to be designed and constructed as Type A units. Assuming that throughout the four buildings three types of units (studio, one-bedroom and two-bedroom) are represented, a minimum of one unit of each type shall be a Type A unit. All three Type A units are permitted in the same building, and all may be located on the same floor level, which would typically be at grade. Those units that are not required to be Type A must be designed and constructed as Type B units unless exempted by the provisions of Section 1107.7.

In dormitories, sorority houses, fraternity houses and boarding houses, Accessible units are required in the same manner as Group R-1 occupancies. At least one of the dwelling or sleeping units within the building must be an Accessible unit, with additional Accessible units as required by Table 1107.6.1.1. As with other Group R-2 occupancies, the remaining units must be designed and constructed as Type B units unless allowed to be reduced by Section 1107.7.

1107.7 **General exceptions.** The required number of Type A and B dwelling units and sleeping units required in Section 1107 may be reduced under the provisions of this section. However, there is no provision allowing the reduction or elimination of Accessible units mandated by Sections 1107.6.1.1, 1107.6.2.2.1 or 1107.6.4.1. Five exceptions to these requirements are included in this section. The first four exceptions relate to buildings with limited or no elevator service and are relatively self-explanatory as worded in the code. The fifth exception relates to buildings that are required to have raised floors at the primary entrances as measured from the elevations of the vehicular and pedestrian arrival points in order to accommodate the base-flood elevation. Where no such arrival points are within 50 feet (15 240 mm) of the primary entrance, the closest arrival points shall be used.

As noted, most of the exceptions are only applicable to buildings with no elevator service. It is expected that once an elevator is installed to access floors above the grade level, the additional measures to make the units more accessible are appropriate. However, the provisions do not mandate the installation of an elevator simply to make the upper floors accessible. Several of the applications of this section are shown in Figures 1107-1 through 1107-3.

Section 1108 *Special Occupancies*

In addition to those requirements that apply to all occupancies in a general fashion, a number of provisions are specific to certain occupancy groups. Assembly, institutional and storage occupancies are uniquely regulated because of the special characteristics of their use.

1108.2 **Assembly area seating.** In stadiums, theaters, auditoriums and similar occupancies, the number of accessible wheelchair spaces to be provided is addressed in Table 1108.2.2.1. Unlike parking requirements where each lot is to be considered separately in the calculation of the minimum required number of accessible parking spaces, the minimum required number of wheelchair spaces for grandstands and bleachers serving a single function should be based on the aggregate number of seats. For example, the total number of seats provided in a high school football facility would be the basis for determining the minimum required number of wheelchair spaces even though the bleachers may be located on opposite sides of the playing field. However, the accessible spaces must be dispersed in an appropriate manner to accommodate individuals on both sides of the field.

The requirements for wheelchair locations can be found in Section 802 of ICC A117.1, including a provision mandating that at least one seat be provided beside each wheelchair space to allow for companion seating. The required dispersion of wheelchair spaces is comprehensively addressed in the 2003 edition of ICC A117.1. Wheelchair spaces should be an integral feature of any seating plan to provide individuals with physical disabilities with a choice of admission prices and a line of sight comparable to that provided to the general public. There are two exceptions to the general rule for the location of wheelchair spaces in multilevel facilities that permit all wheelchair spaces to be located on the main level where the second floor or mezzanine level is limited in capacity.

Designated aisle seats. The intent of providing designated aisle seats is to permit individuals who might find it difficult to negotiate an aisle accessway the opportunity to use a seat directly adjacent to the aisle. The requirements of ICC A117.1 call for signs to identify the designated aisle seats. In addition, where armrests are installed, they must be retractable or folding on the aisle side of the seat. **1108.2.5**

Assistive listening systems. Assistive listening systems are required to be installed where audible communications are a necessary part of the assembly room's use. If the type of assembly use does not necessitate the installation of an audio-amplification system, an assistive listening system is not required. The number of assistive listening receivers is based on the total seating capacity of the assembly area and determined by Table 1108.2.7.1. As an example, an 800-seat theater would require at least 30 receivers, of which at least eight must be hearing-aid compatible, as shown in the following calculation: **1108.2.7**

20 receivers + 1 receiver [(800 seats – 500 seats) / 33 seats per receiver] = 29.1 = 30 receivers.

25% of 30 receivers = 7.5 = 8 hearing aid compatible receivers.

These systems are intended to augment a standard public address or other audio system by providing signals that are free of background noise to individuals who use special receivers or their own hearing aids. Further provisions found in Section 706 of ICC A117.1 require that individual fixed seating served by an assistive listening system must be located to provide a complete view of the stage, playing area or cinema screen. AM and FM radio-frequency systems, infrared systems, induction loops, hard-wired earphones and other equivalent devices are all permitted.

Dining areas. The general rule for accessibility in dining rooms is that the total floor area be accessible. The first exception to this section eliminates the requirement for an elevator or ramp system to a mezzanine seating area in a dining room under strict conditions. Caution should be exercised when determining the same services mentioned in this exception. It is generally accepted that the same services not only address actual food and drink service but also include decor, views and ambiance, etc., and it is equally important that a specific accessible space or area not be set aside and restricted for use only by people with disabilities. **1108.2.9**

For spaces at accessible fixed tables or counters, at least one table in the facility shall be provided for wheelchairs. When more than one is required by this section, then they shall be equally distributed around the facility so as not to isolate an accessible area from the rest of the establishment.

Self-service storage facilities. The number of individual storage spaces that must be accessible at self-service storage facilities is determined from Table 1108.3. Assuming a total of 280 spaces in the facility, at least 12 accessible storage spaces must be provided as shown: **1108.3**

10 spaces + [2% (280 spaces – 200 spaces)] = 11.6 =12 storage spaces.

Where different sizes or classes of storage space are available, the individual accessible spaces shall be appropriately dispersed among the sizes or classes available, but only up to the total number of accessible spaces that are required. All accessible storage spaces complying with this section are permitted to be located in a single building.

Judicial facilities. Courtrooms, holding cells and visitation areas of judicial facilities are required to be accessible to the degree mandated by this section. Courtrooms typically have elements unique to their use, such as witness stands, jury boxes, judge's benches and similar areas. Provisions are in place to provide for a limited degree of accessibility to such areas. The spectator area of a courtroom is regulated under the provisions for assembly seating. Work stations are regulated differently based upon their use. Employee work stations, such **1108.4**

as the judge's bench, bailiff's station and court reporter's station, need to be on an accessible route but the portion of the route leading to such elevated work areas is not required at the time of construction. It is only necessary to provide adequate space and support such that a complying ramp, platform lift or elevator can be installed in the future without extensive reconstruction. Work stations for other than employees, such as the litigant's station, counsel station and lectern, must provide full accessibility.

Section 1109 *Other Features and Facilities*

This section provides scoping provisions to ensure that certain elements or areas within buildings, where specific services are provided or activities are performed, are made accessible. Certain components in ICC A117.1 or other accessibility regulations have not been included in this section. Items such as telephones, automatic-teller machines and fare machines are referenced in Appendix E, insofar these are not typically considered items regulated by the building code.

1109.2 **Toilet and bathing facilities.** As a general rule all toilet rooms and bathing facilities shall be accessible. A number of exceptions revise, reduce or eliminate the requirements for toilet rooms and bathing facilities under specific conditions. Within the facility, at least one of each type of fixture, element, controls or dispenser must be accessible. There is an allowance for a nonaccessible urinal where only a single urinal is located within a toilet room. The term *accessible* applies to the doors, fixtures, clear floor space, operable parts, towel dispensers and mirrors, among other things.

Typically, the main components of toilet facilities are the water closet, the toilet stall and the lavatory; and the main accessibility issues are the clear space, door swing, transfer capability, height of fixtures, grab bars and controls. In bathing facilities, a number of items related to the shower or bathtub and its location, controls and grab bars need to be considered. The ICC A117.1 standard contains details and many illustrations that depict these requirements.

1109.2.1 **Family or assisted-use toilet and bathing rooms.** Specific provisions are provided in this section on accessibility requirements for family or assisted-use bathing and toilet rooms. Such bathing and toilet rooms are required to comply with this section and ICC A117.1. The primary issue relative to family or assisted-use toilet/bathing facilities is that some people with disabilities may require the assistance of persons of the opposite sex and, therefore, require a toilet or bathing facility that accommodates both persons.

Bathing facility requirements for recreational facilities require that an accessible family or assisted-use bathing room be provided where separate-sex bathing facilities are provided. If each separate-sex bathing facility has only one shower fixture, family or assisted-use bathing facilities will not be required. Accessible family or assisted-use toilet room requirements for Group A and M occupancies are mandated by this section where an aggregate of six or more male and female water closets are required. These occupancies generally have high occupant loads with a minimum stay of approximately 1 hour for the occupants; thus, a high probability exists that there will be occupants who will need the use of such facilities. In determining the total number of fixtures required in a building as mandated by Chapter 29, those fixtures located in family or assisted-use facilities may be included in the total fixture count.

Family or assisted-use bathing rooms are to be provided with a water closet and lavatory in addition to the single shower or bathtub fixture. Family or assisted-use toilet rooms are limited to a single water closet, lavatory and optional urinal. A complying family or assisted-use bathing room may be considered the family or assisted-use toilet room. In order to provide the appropriate privacy for a family or assisted-use facility, doors to a family or assisted-use toilet room or bathing room must be capable of being secured from the inside.

Family or assisted-use toilet and bathing rooms shall be located on an accessible route. Family or assisted-use toilet rooms are to be located not more than one story above or below separate-sex toilet facilities. The accessible route from any separate-sex toilet room to a family or assisted-use toilet room must not exceed 500 feet (152 400 mm). Additionally, in passenger transportation facilities and airports, the accessible route from separate-sex toilet facilities to a family or assisted-use toilet room shall not pass through security checkpoints. The restriction regarding crossing through security checkpoints at airports and similar facilities is intended to eliminate any potential delays that may cause missed flights and/or connections.

Water closet compartment. At least one wheelchair-accessible compartment shall be provided in all toilet rooms or bathing facilities where compartments are installed. Also, an ambulatory-accessible water closet compartment must be provided in addition to the wheelchair-accessible compartment in those toilet rooms having an aggregate total of six or more water closet compartments and urinals. The ambulatory-accessible compartment benefits those individuals who, although not wheelchair users, have physical limitations or impairments that make it difficult to utilize other types of toilet compartments. Both wheelchair-accessible and ambulatory-accessible compartments are to be in compliance with ICC A117.1. It should be noted that the threshold for this provision is based on a room-by-room evaluation, not the aggregate number of fixtures throughout the facility.

1109.2.2

Lavatories. The provisions specific to lavatories are a subsection to the other provisions for toilet and bathing facilities. The requirement for a minimum of 5 percent of the lavatories to be accessible results in a single mandated accessible lavatory in almost every toilet room and bathing room. Only where more than 20 lavatories are provided must additional accessible sinks be provided.

1109.2.3

Additional accessibility features are required where the total number of lavatories in a toilet room or bathing room exceeds five. Where six or more lavatories are provided, a minimum of one lavatory must be provided with enhanced reach ranges. It is permissible for the lavatory with the enhanced reach range to serve as the required accessible lavatory. The technical requirements for such lavatories with enhanced reach ranges are found in ICC A117.1 Section 606.5. These types of lavatories are usable by individuals who have a limited obstructed reach depth. Such individuals can often only reach faucets and soap dispensers up to a reach depth of 11 inches (279 mm) in lavatories with a height of 34 inches (864 mm). The maximum 11-inch (279 mm) depth is possible by locating the faucet controls to the side of the bowl while leaving the spout towards the back, mounting the faucet on a sidewall, installing the faucet on the side of the bowl, or other potential locations that will provide the necessary access.

Drinking fountains. This section ensures that all of the drinking fountains provided are made accessible. This is another of the many provisions where it is important to focus on the word *provided* and note that the provision does not require drinking fountains to be installed.

1109.5

Drinking fountains must be accessible based on use from a wheelchair while still providing access to people with a limited ability to bend or stoop. The technical provisions set forth in Section 602 of ICC A117.1 address drinking fountains designed for wheelchair access, as well as those intended for use by standing persons. Other than the requirement for clear floor space, the provisions of Section 602 are applicable to both heights of fountains. Where a single water fountain (combination *hi-lo* unit) can comply with the requirement for both user groups, it is permitted to be utilized in lieu of two individual fountains.

The minimum required number of drinking fountains is established by the *International Plumbing Code®* (IPC®). At times, drinking fountains are installed even though they are not required by the code. It is also not uncommon for the actual number of fountains provided to exceed the minimum required. Under such conditions, one-half of the number provided must comply with the provisions of ICC A117.1 for wheelchair users, and the remainder shall accommodate standing persons. If an odd number of fountains is installed, an additional drinking fountain is not required in order to meet the 50 percent criteria. The remaining water fountain after the 50/50 split is permitted to be of either type. For example, if three drinking fountains are provided on the first floor of a building, at least one, but not more than two, must be wheelchair accessible.

1109.7   **Lifts.** The provisions of this section address the limited acceptance of platform (wheelchair) lifts as a portion of an accessible route. To use platform lifts in lieu of an elevator or ramp, they must comply with Section 408 of ICC A117.1 and be installed in one of the 10 listed locations. A companion provision, Section 1007.5, allows platform lifts to be utilized as an accessible means of egress for the first nine locations established in this section. However, the platform lift cannot be used as a portion of an accessible means of egress where the lift is used along the accessible route because of existing exterior site constraints as described in Item 10. Because of the slow operation of a platform lift, it is considered inappropriate for egress purposes where there is the potential for a large number of lift users.

1109.9   **Detectable warnings.** Detectable warnings must be provided on edges of passenger transit platforms where a drop-off occurs, unless platform screens or guards are present. Not applicable to bus stops, detectable warnings should be standardized to assist in the universal recognition and reaction of persons using these platforms. Detectable and tactile warning devices also serve guide dogs and should contrast visually with the adjoining surfaces in a light-on-dark or dark-on-light application.

1109.11   **Service facilities.** This section provides nominal accessibility provisions that are intended to allow persons with disabilities to use these facilities without much assistance. Other features commonly found in these uses would have to comply with the other provisions in this chapter.

1109.12   **Controls, operating mechanisms and hardware.** To be able to operate controls, operating mechanisms and hardware, a clear floor space and reach ranges for either forward or parallel approaches must be provided. This section also requires that, in other than kitchens or bathrooms, an accessible operable window be provided in those rooms in Accessible units and Type A units where operable windows are located.

1109.14   **Recreational facilities.** Where recreational facilities are provided serving Type A or B units in Group R-2 or R-3 occupancies, at least 25 percent, but not less than one, of each type of such facilities shall be accessible. All recreational facilities of each type on a site shall be considered to determine the total number of each type that is required to be accessible. This requirement recognizes that not all recreational facilities need to be accessible, nor would such a requirement be feasible. However, at least one of every four recreational facilities shall be available to someone with a disability. Where multiple residential buildings are on a site, each type of recreational facility, such as a basketball court, a handball court, a weight room, an exercise area, a game room or a television room serving each building must be fully accessible. Each type of recreational facility on a site, such as a racquetball court in an apartment complex, is considered to determine the total number required to be accessible. Thus, if there were 12 racquetball courts on the site, at least three of the courts would need to be accessible. If they were evenly distributed in two separate buildings, one building would need to contain at least one accessible court, and another would need at least two accessible courts to meet the overall 25-percent rule. If the courts were located in four separate buildings, each building would require at least one accessible facility, even if the total exceeded the overall 25-percent requirement.

Section 1110   *Signage*

This section lists those required accessible elements and locations that must be identified with appropriate signage. Elements such as accessible parking spaces and loading zones, accessible areas of refuge, accessible dressing rooms and family or assisted-use toilet rooms shall be provided with the International Symbol of Accessibility as shown in Figure 1110-1. Directional signage, including the International Symbol of Accessibility, is to be used to indicate the nearest route to a like accessible element. Directional signage must be provided at inaccessible building entrances, inaccessible public toilets and bathing facilities, elevators not serving an accessible route, separate-sex toilet and bathing facilities (to indicate the location of the nearest family or assisted-use facility), and exits and exit stairways serving a space required to be accessible but not providing an accessible means of egress. Additional signage will be needed where assistive-listening systems are available for assembly occupancies: at doors to egress stairways, exit passageways and exit discharge; at areas of refuge; at areas of rescue assistance; and at two-way communication systems.

Figure 1110-1
International Symbol of Accessibility

KEY POINTS

- Chapter 11 of the IBC provides the scoping provisions for accessible spaces and elements within buildings.

- ICC A117.1-2003 is the referenced standard for regulating the facilities and elements of buildings that are required to be accessible by IBC Chapter 11.

- Virtually all buildings, both temporary and permanent, must be designed and constructed for accessibility and usability.

- Accessible routes must be provided to connect all accessible elements within a building, as well as all accessible elements on the site.

- At least 60 percent of all public building entrances must be accessible.

- The number of accessible parking spaces required on a site varies by the total number of spaces provided.

- All occupancies require some degree of accessibility, with special provisions for assembly, institutional, residential and utility uses.

- Assembly areas are further regulated for wheelchair spaces and assistive-listening systems.

- Both Type A and B dwelling units may be required in a large apartment building.

- A variety of building elements are regulated as facilities required to be accessible, including toilet rooms, bathing facilities, drinking fountains, elevators, stairs, platform lifts, fixed or built-in seating, storage, customer-service facilities, controls and operating mechanisms, and alarms.

- Under specific conditions, accessible family or assisted-use bathing rooms and toilet rooms must be provided in addition to the accessible separate-sex facilities.

- Appropriate signage is mandated in order to properly identify the various accessible elements.

INTERIOR ENVIRONMENT

This chapter is designed to address those issues related to the interior environment aspects of a building's use, such as ventilation, lighting, temperature control, yards and courts, sound transmission and room dimensions.

Section 1203 *Ventilation*

Ventilation in buildings is regulated based upon the ventilating method utilized. This section addresses the use of natural ventilation, whereas the use of mechanical ventilation is regulated by the *International Mechanical Code®* (IMC®).

1203.2 **Attic spaces.** During cold weather, condensation is deposited on cold surfaces when, for example, warm, moist air rising from the interior of the building and through the attic comes in contact with the roof deck. This alternate wetting and drying that is due to condensation creates dry rot in the wood, and preventive measures are required. In attic areas of noncombustible construction, it is also important to ventilate the area, particularly in light-gage steel construction. Therefore, enclosed attics and enclosed rafter spaces formed where ceilings are applied directly to the underside of roof-framing members, such as in cathedral ceiling applications, are to have cross ventilation for each separate space. Ventilation of the attic prevents moisture condensation on the cold surfaces and, therefore, will prevent dry rot on the bottom surfaces of shingles or wood roof decks. Figure 1203-1 provides three examples of attic ventilation. In the areas where the moisture condensation is a particular problem or where the normal requirement for attic ventilation cannot be provided, ventilation of the attic by mechanical exhaust fans may be required. Exhaust fans are particularly beneficial in all cases because of the extra movement of the air provided.

The method and arrangement of providing ventilated openings is an important aspect in the proper ventilation of attic spaces. It is critical that any such openings be protected against the entrance of rain and snow. In addition, blocking and bridging that is installed must be located so as to not interfere with the movement of air. At least 1 inch (25 mm) of airspace must be provided between the insulation and the roof sheathing as shown in Figure 1203-2. As illustrated in Figure 1203-3, the net free ventilating area must be at least $^1/_{300}$ of the area of the space ventilated. In addition, at least 50 percent of the required ventilating area must be provided by ventilators located in the upper portion of the space.

Something often overlooked when sizing attic vents is that the code requires that the area provided be the net free area. The net free area can be as much as 50 percent less than the gross area. For example, one manufacturer's 24-inch (610 mm) square gable vent [gross area equals 576 square inches (0.37 m²)] is listed in their catalog as having a net free area of 308 square inches (0.20 m²), which is about 53 percent of the gross area. The manufacturer's literature for the specific vents being utilized needs to be consulted in order to obtain accurate free area information.

1203.2.1 **Openings into attic.** Exterior ventilation openings are required by the code to be screened in order to prevent entry of birds, squirrels, rodents and other similar creatures. A mesh size between $^1/_{16}$-inch (1.6 mm) and $^1/_4$-inch (6.4 mm) is required to address both the problem of smaller openings being blocked by debris and spider webs, and of larger openings permitting access to small rodents. In addition to the use of corrosion-resistant-wire cloth screening, it is also permissible to utilize hardware cloth, perforated vinyl or any other similar material that will prevent unwanted entry. A cross reference is also provided to IMC Chapter 7 to remind users that there are special requirements where combustion air is obtained from the attic area.

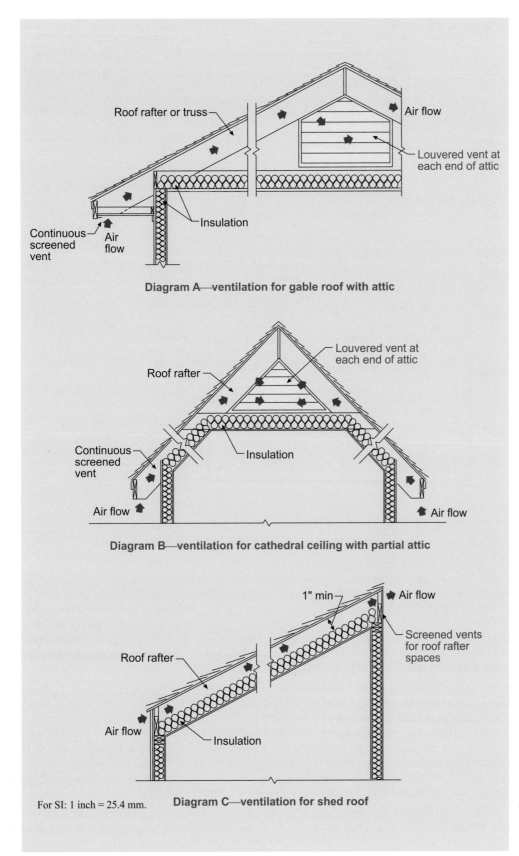

Diagram A—ventilation for gable roof with attic

Diagram B—ventilation for cathedral ceiling with partial attic

Diagram C—ventilation for shed roof

For SI: 1 inch = 25.4 mm.

Figure 1203-1
Attic ventilation

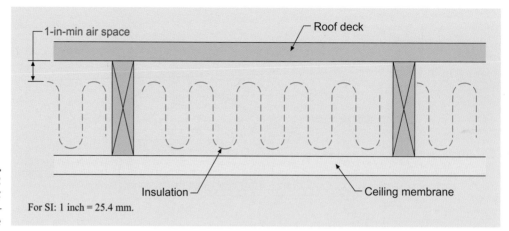

Figure 1203-2
Attic ventilation— air space

For SI: 1 inch = 25.4 mm.

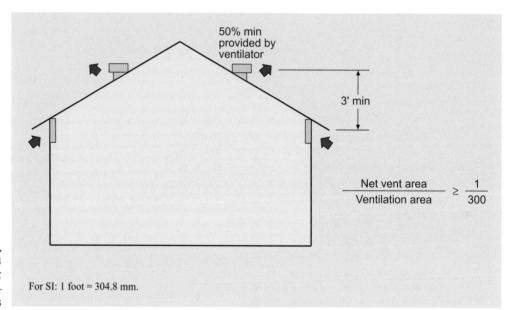

Figure 1203-3
Attic ventilation— calculations

For SI: 1 foot = 304.8 mm.

$$\frac{\text{Net vent area}}{\text{Ventilation area}} \geq \frac{1}{300}$$

1203.3 **Under-floor ventilation.** In order to ventilate the space below the building between the bottom of the floor joists and the ground, ventilation openings shall be provided through foundation walls or exterior walls. The provisions apply to areas such as crawl spaces, rather than those occupiable areas such as basements. Under certain climatic conditions, it is possible to ventilate the under-floor space into the interior of the building, or continuously operated mechanical ventilation may be provided in lieu of ventilation openings where the ground surface is covered with an approved vapor retarder.

To properly determine the minimum net area of ventilation openings, at least 1 square foot (0.0929 m²) shall be provided for each 150 square feet (13.9 m²) of crawl space area. The openings shall be located so as to provide cross ventilation in the under-floor area. See Figure 1203-4. An exception permits a dramatic reduction in the amount of ventilation opening area, provided the ground surface is treated with Class I vapor-barrier material. In this case, the total area of ventilation openings need not exceed $^{1}/_{1500}$ of the under-floor area, provided such openings are located to provide for adequate cross ventilation of the under-floor space. In this case, the vents may have operable louvers.

It is critical that ventilation openings be completely covered with a substantial material to prevent the entrance of insects and animals. Corrosion-resistant wire mesh, with the least dimension not exceeding $^1/_8$ inch (3.2 mm), is one of six materials identified by the code to address this concern.

Where the under-floor space is conditioned, it is unnecessary to provide ventilation openings. It has been shown that by insulating the perimeter walls, covering the ground surface with a Class I vapor barrier and conditioning the space in accordance with the *International Energy Conservation Code®* (IECC®), unvented crawl spaces outperform vented under-floor areas.

Natural ventilation. Where buildings are not provided with adequate mechanical **1203.4** ventilation as specified in the IMC, natural ventilation through openings directly to the exterior must be provided. In order to determine the amount of ventilation air required, the minimum openable area to the outdoors is based upon 4 percent of the floor area being ventilated.

In those cases where rooms or spaces do not have direct openings to the exterior, it is still necessary to ventilate the interior space, often through an adjoining room. In this case, the opening to the adjoining room should be unobstructed and have an area not less than 8 percent of the floor area of the interior room or space. In no case should the opening between the rooms be less than 25 square feet (2.3 m²). Where the intervening room is a thermally isolated sunroom or patio cover, the minimum openable area of 8 percent is still applicable; however, the opening need only be 20 square feet (1.86 m²).

As previously mentioned, in calculating the total openable area to the outdoors, such opening area shall not be less than 4 percent of the total floor area being ventilated. In those conditions where openings that provide the natural ventilation are located below grade, the outside horizontal clear space measured perpendicular to the opening is required to be at least one- and one-half times the depth of the opening.

Where rooms contain bathtubs, showers, spas and similar bathing fixtures, natural ventilation is not an acceptable method. Because of the common reluctance to open exterior windows, particularly in cold-weather conditions, a mechanical system provides for more consistent ventilation. Therefore, bathrooms and similar spaces must be mechanically ventilated in accordance with the IMC. Where flammable and combustible hazards or other contaminant sources are present within an interior space, ventilation-exhaust systems shall be provided as required by the IMC and the *International Fire Code®* (IFC®).

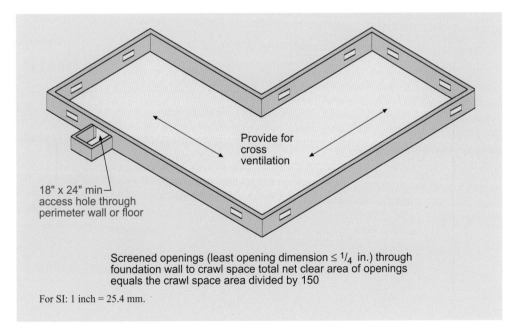

Provide for cross ventilation

18" x 24" min access hole through perimeter wall or floor

Screened openings (least opening dimension ≤ $^1/_4$ in.) through foundation wall to crawl space total net clear area of openings equals the crawl space area divided by 150

For SI: 1 inch = 25.4 mm.

Figure 1203-4
Underfloor ventilation

Section 1204 *Temperature Control*

For those interior spaces where the primary purpose is associated with human comfort, it is important that a minimum indoor temperature of 68°F (20°C) can be maintained, measured at 3 feet (914 mm) above the floor on the design heating day. Although the code does not require that this temperature be constantly provided, it does mandate that such interior spaces be provided with equipment or systems having the capability of maintaining the desired temperature.

Section 1205 *Lighting*

Almost every occupancy requires some level of lighting that is due to its use. Means of egress illumination is also required by Section 1006. In spite of the obvious need for interior light, the code mandates that some degree of lighting, whether artificial or natural, be provided to every occupiable space.

1205.2 Natural light. Where glazing to the exterior is utilized as the method for providing natural light, the exterior openings shall open directly onto a public way, yard or court in compliance with Section 1205. Exterior wall openings used to provide natural light must have an area that is computed based on the net glazed area for windows and doors, and not the nominal size of the opening. Where a room is not located on an exterior wall, the provisions of this section permit the borrowing of light from an adjoining room. Figure 1205-1 illustrates the requirements for this condition.

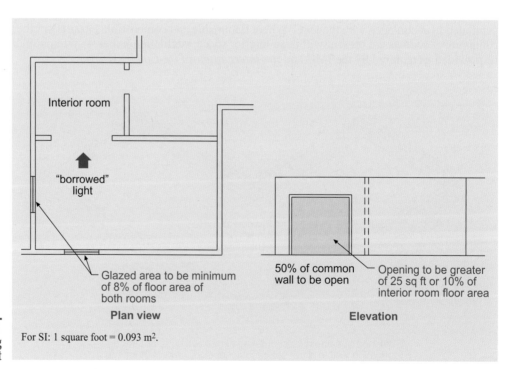

Figure 1205-1
Borrowing natural light

For SI: 1 square foot = 0.093 m².

Section 1206 *Yards or Courts*

The *International Building Code* (IBC®) contains provisions for yards and courts where they are used to provide required light and ventilation to exterior openings in the building. Most modern-day zoning ordinances also have requirements for yards and courts, and quite often these are more than adequate to gain the lighting and ventilation required by the code. In addition, where the alternatives of artificial light and mechanical ventilation as provided by Sections 1205.3 and 1203.1, respectively, are utilized in lieu of natural light and ventilation, the provisions of Section 1206 are not applicable.

To be considered providing adequate natural light and ventilation, each yard or court must have a minimum width of 3 feet (914 mm). In addition, these yards and courts must be increased in width, depending on the height of the building. See Figure 1206-1. The intent for the tall building is to have an increased court width so that light coming into the court will be able to reach the lower stories of the building, and this is only possible where the width of the court is in proper relationship to the height of the court. The requirements in the code are an obvious compromise between optimum light at the bottom of a court and the need to build as much building area on the lot as possible for economic purposes.

Inner courts that are enclosed by the walls of the buildings, sometimes referred to as light wells, obviously need some means to remove accumulated trash at the bottom, provide for adequate drainage and provide for circulation of air for ventilation purposes. In keeping with this intent, the code requires that an air intake be provided at the bottom of courts for buildings more than two stories in height, that grading and drainage be addressed, and that all courts be provided with access for cleaning.

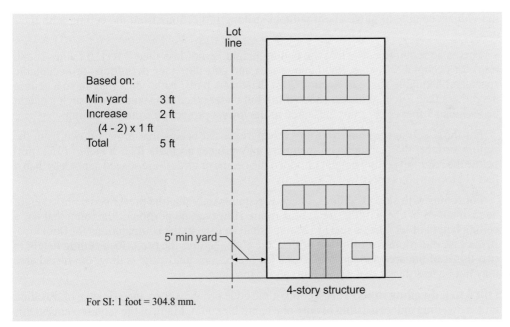

Based on:

Min yard	3 ft
Increase	2 ft
(4 - 2) x 1 ft	
Total	5 ft

Lot line

5' min yard

4-story structure

For SI: 1 foot = 304.8 mm.

Figure 1206-1
Yards and courts

Section 1207 *Sound Transmission*

Applicable only to buildings containing dwelling units, this section is intended to provide regulations covering sound transmission control in residential buildings. It pertains to wall and floor/ceiling assemblies separating dwelling units from each other and from public space, such as interior corridors, stairs or service areas. These must be provided with airborne-sound insulation for the walls and both airborne and impact-sound insulation for the floor/ceiling assemblies.

For airborne-sound insulation, the separating walls and floor/ceiling assemblies must be provided with insulation equal to that required for a Sound Transmission Class (STC) of 50 (45 when field tested) as defined by ASTM E 90. As an alternative for concrete masonry and clay masonry assemblies, the sound transmission class may be calculated per TMS 0302. Penetrations or openings through the assemblies must be sealed, lined, insulated or otherwise treated to maintain the required ratings. Dwelling unit entrance doors from corridors only need to be tight fitting to the frame and sill. Floor/ceiling assemblies between separate dwelling units must also provide impact-sound insulation equal to that required to meet an impact insulation class of 50 (45 when field tested) as defined in ASTM E 492.

Section 1208   *Interior Space Dimensions*

Room size, tightness of construction, minimum ceiling height, number of occupants and ventilation all interact with each other to establish the interior living environment insofar as odors, moisture and transmission of disease are concerned. Therefore, the IBC regulates room sizes to assist in maintaining a comfortable and safe interior environment, and the minimum room sizes become increasingly important as buildings become even tighter in their construction because of energy-conservation requirements.

1208.2   **Minimum ceiling heights.** Section 1208.2 regulates ceiling height, not only to assist in maintaining a comfortable indoor environment, but also to provide safety for the occupants of the building. As our population becomes increasingly taller, it is important that tall individuals be able to move about without striking projections from the ceiling with their heads.

The basic requirement is that the ceiling height be not less than 7 feet, 6 inches (2286 mm) for occupiable spaces, habitable spaces and corridors (see definitions of occupiable space and habitable space in Section 202). Kitchens, halls, baths, etc., may have a ceiling height less than 7 feet, 6 inches (2286 mm), but under no circumstances may such a height be less than 7 feet (2134 mm) measured to the lowest projection from the ceiling.

For those ceilings within dwellings having exposed beams that project down from the ceiling surface, the ceiling beam members may project no more than 6 inches (152 mm) below the required ceiling height, provided the beams or girders are spaced at not less than 4 feet (1219 mm) on center.

For rooms with sloped ceilings, the code requires only that the prescribed ceiling height be maintained in one-half the area of the room. However, no portion of the room that has a ceiling height of less than 5 feet (1524 mm) shall be used in the computations for floor area. In the case of a room with a furred ceiling, the code requires the prescribed ceiling height in two-thirds of the area and, as in all cases for projections below the ceiling, the furred area may not be less than 7 feet (2134 mm) above the floor.

1208.4   **Efficiency dwelling units.** This section of the code provides for a specific type of dwelling unit, a dwelling unit consisting of only one habitable room. Many of the requirements in this section are redundant, as this chapter already requires many of these provisions. However, there are some requirements that are unique to the efficiency dwelling unit:

    1.   The living room (which also serves as a bedroom and kitchen) is to have not less than 220 square feet (20.4 m^2) of floor area. It is the intent of the code that this floor area be the total gross floor area, less the area occupied by built-in cabinets and other built-in appliances that are not readily removed and that preclude any other use of the floor space occupied by the built-in cabinets and fixtures.

2. The minimum room size shall be increased by 100 square feet (9.29 m^2) of floor area for each intended occupant in the unit in excess of two.

3. A closet is required.

4. A kitchen sink, cooking appliance and refrigeration facilities are required, each providing a clear working space of not less than 30 inches (762 mm) in front.

5. A separate bathroom containing a water closet, lavatory, and bathtub or shower is required.

Section 1209 *Access to Unoccupied Spaces*

Access to crawl spaces and attic spaces is regulated by this section. Though typically unoccupied, it is sometimes necessary that these normally concealed areas be accessed for various reasons.

Crawl spaces. This section of the code mandates that under-floor areas be accessible by a minimum 18-inch by 24-inch (457 mm by 610 mm) access opening. Where the access opening opens to the exterior of the building, the code intends it to be screened or covered to prevent the entrance of insects and animals. Also, it is the intent of the code that all portions of the under-floor area be accessible and access be provided beneath or around obstructions created by pipes, ducts, etc. **1209.1**

Attic spaces. Because enclosed attics provide an avenue for the undetected spread of fire in a concealed space, the code requires that access openings be provided into the attic so that fire-fighting forces may gain entry to fight the fire. To be of any value, the access openings must be of sufficient size to admit a fire fighter with fire-suppression gear and must also have enough headroom so that entry into the attic may be secured. Although not specified, the access should be located in a readily accessible location. A public hallway is the best location for attic-access openings. Fire department personnel will not then have to open private offices, apartments or hotel rooms in order to enter the attic. Attic access may be provided through a wall as well as through a ceiling. In split-level buildings with multiple attics, an attic-access opening must be provided to each attic space. **1209.2**

Attic access is only required by the IBC for those attic areas having a clear height greater than 30 inches (762 mm). Where such conditions occur, an attic-access opening of not less than 20 inches by 30 inches (559 mm by 762 mm) shall be provided. However, if the attic contains mechanical equipment, the opening may need to be enlarged to gain compliance with the IMC.

Section 1210 *Surrounding Materials*

The primary thrust of this section is to provide easily cleanable, sanitary and water-resistant surfaces in toilet rooms and showers.

Floors. Except for dwelling units, the code requires toilet room and bathing room floors to have a smooth, hard, nonabsorbent surface. Finishes such as concrete and ceramic tile are certainly acceptable, as are other approved materials that may also be used. It is the intent of the code that the building official determine the suitability of the proposed floor surface insofar as cleanability and water resistance are concerned. **1210.1**

Whatever floor finish is used, the code requires that it extend upward onto the walls at least 6 inches (155 mm). The intent here is that the flooring form an integral cove so that there will be no sharp joint at the floor/wall intersection. If sheet materials are approved for use in toilet or bathing rooms, the top set base that is so often used would not satisfy the intent of the code.

Toilet and bathing room floor-finish requirements apply to all uses and occupancies except for dwelling units. Because motel and hotel rooms are typically not dwelling units, toilet room flooring in these uses must often comply with this section.

1210.2 **Walls and partitions.** Walls and partitions within 2 feet (610 mm) of the front and sides of urinals and water closets are required to have a smooth, hard, nonabsorbent finish. Required finishes shall extend to a height of at least 4 feet (1219 mm) above the floor and shall be of a type not adversely affected by moisture. Water-resistant gypsum backing board may be used as a base for tile or wall panels, provided it satisfies the limitations set forth in Section 2509. Although the code does not specifically state that concrete walls must be sealed, concrete is not, strictly speaking, a nonabsorbent material and needs some type of surface treatment. Sealing is particularly important for concrete block walls.

Special wall finishes at water closets and urinals are not required for dwelling units, sleeping units and toilet rooms not accessible to the public that contain only one water closet. Note that the exceptions for floor finishes and wall finishes are not the same. A private toilet room containing one water closet would be required to have flooring that complies with Section 1210.1, but there would be no special requirements for wall finishes.

In all occupancies, including dwelling units, penetrations of the water-resistant surfacing of the walls for the installation of accessories such as grab bars, towel bars, etc., are required to be sealed to protect the structural elements from moisture. The intent of the code is that because the structural elements are not required to be moisture resistant, the penetrations should be sealed to protect the structural elements. Although sealing is required for all walls in a toilet room, sealing is obviously most critical in areas of water splash such as in showers or behind or adjacent to lavatories.

1210.3 **Showers.** All showers must have floor and wall finishes that are smooth, nonabsorbent and not affected by moisture. Wall finishes must extend not less than 70 inches (1778 mm) above the drain inlet.

KEY POINTS

- The method and arrangement of providing ventilated openings is an important aspect in the proper ventilation of attic spaces.

- Under-floor ventilation is to be provided through foundation walls or exterior walls in order to adequately ventilate the space below the building.

- Ventilation of interior spaces may be accommodated through exterior openings or by a mechanical system.

- The IBC mandates that some degree of lighting, whether artificial or natural, is to be provided to every occupiable space.

- Wall and floor/ceiling assemblies separating dwelling units in residential buildings must be provided with sound insulation.

- In specific areas of a building, floors and walls are required to have a smooth, hard, nonabsorbent finish.

ENERGY EFFICIENCY

This chapter of the *International Building Code®* (IBC®) references the *International Energy Conservation Code®* (IECC®) for provisions regulating the design and construction of buildings for energy efficiency. The IECC sets forth minimum requirements for new and existing buildings by regulating their exterior envelopes. In addition, it addresses the selection of heating, ventilating and air-conditioning; service water-heating; electrical distribution; and illumination systems and equipment in order to provide for efficient and effective energy usage.

The provisions of this chapter apply to both residential and commercial buildings; however, specific energy requirements are provided in the *International Residential Code®* (IRC®) to address structures regulated by the IRC.

EXTERIOR WALLS

This chapter establishes the basic requirement for exterior wall coverings, namely that they shall provide weather protection for the building at its exterior. The other important item necessary for complete weather protection, the roof system, is addressed in the next chapter. Chapter 14 presents general weather-protection criteria and special requirements for veneer. In addition, exterior wall coverings, balconies and similar architectural appendages constructed of combustible materials are regulated in Section 1406.

Section 1402 *Definitions*

The code intends by the definition of veneer that it be a nonstructural facing of masonry, concrete, metal, plastic or similar approved material. It is intended to provide ornamentation, protection or insulation. On this basis, face brick that are laid with common brick so as to provide a composite structural assembly are not considered veneer, as they act structurally along with the common brick. To be considered veneer, a material must not act structurally with the backing insofar as the consideration of structural strength of the assembly is concerned. However, in many cases the veneer does act structurally with the backing, and problems can result if not constructed properly.

Veneer can be either adhered or anchored and either exterior or interior, depending on its method of attachment to the backing and whether or not it is applied to a weather-exposed surface.

Section 1403 *Performance Requirements*

This section provides specifications for the protection of the interior of the building from weather; therefore, there are requirements for:

1. Protection of the interior wall covering from moisture that penetrates the exterior wall covering.

2. Flashing of openings in, or projections through, exterior walls.

3. Vapor retarders used to resist the transmission of water vapor through the exterior envelope.

It is also necessary to provide for the removal of any water that may accumulate behind the exterior veneer. The code requires a means for draining water to the exterior that enters the assembly. The intent of the provision is to merely utilize the felt building paper and flashing to drain the water to the outside, rather than provide an air space between the siding and the water-resistant barrier. Thus, the practice of placing siding directly over the building paper is acceptable, provided flashing is correctly installed to direct water within the wall assembly to the exterior.

The code exempts the need for a weather-resistant wall envelope over concrete or masonry walls designed in accordance with Chapters 19 and 21, respectively. The *International Building Code*® (IBC®) requirements for weather coverings, flashings and drainage are also not applicable where the exterior wall envelope has been tested in accordance with ASTM E 331 and shown to resist wind-driven rain. In addition, where an exterior insulation and finish system (EIFS) is installed in accordance with Section 1408.4.1, the requirements of Section 1403.2 need not be followed.

The provisions of Section 1405.3 are referenced in regards to protection against condensation in exterior walls. Where framed walls, floors and ceilings are not ventilated in a manner that will allow moisture to escape, it is mandated that a complying vapor barrier be installed on the warm-in-winter side of the insulation. A vapor barrier is not required for masonry or concrete exterior walls insofar as the provision is only applicable to framed walls. In addition, vapor barriers are not required in warmer climatic areas, specifically Climate Zones 1, 2, 3, and 4 (other than Marine 4) and where the value of such barriers is negligible because of the lack of damage any moisture will create. Vapor barriers may also be omitted in basement walls and any other portions of walls located below grade.

Section 1404 *Materials*

As most exterior wall coverings are permeable by moisture, the code requires that there be a water-resistive barrier placed under the exterior wall covering and over the sheathing. The purpose of the barrier is to prevent moisture infiltration to the sheathing and subsequently the interior wall surfaces. The required water-resistive barrier is identified as at least one layer of No. 15 asphalt felt complying with ASTM D 226 for Type I felt. All necessary flashing must be installed as described in Section 1405.3 in order to complete the water-resistant envelope.

Section 1405 *Installation of Wall Coverings*

Exterior wall coverings are to be installed in accordance with the provisions of this section. Of primary concern are the types and thicknesses of the weather protection provided by the wall coverings. In addition, it is important that flashing be installed in those areas prone to leakage. Various types of veneer are addressed, including wood, masonry, stone, metal, glass, vinyl, fiber and cement. The specifics for each veneer type are described in detail.

1405.2
Weather protection. A broad range of exterior wall-covering materials is listed in Table 1405.2 as being satisfactory when utilized as protection against the elements of nature. Covering types addressed include wood and particleboard siding, wood shingles, stucco, anchored and adhered masonry veneer, aluminum and vinyl siding, stone and structural glass. For each of the 35 types listed, the minimum required thickness for weather protection is mandated. In the case of particleboard and plywood without sheathing, reference is made to Chapter 23.

1405.3
Vapor retarders. In order to protect against condensation in framed exterior wall assemblies of buildings constructed in the specified Climate Zones, complying vapor retarders are mandated. Table 301.1 in the *International Energy Conservation Code®* (IECC®) establishes the appropriate Climate Zones in the United States based upon counties. Those counties designated as Zone 5, 6, 7 or 8, as well as those classified as Zone 4 Marine, must be provided, with exceptions, with either a Class I or II vapor retarder on the interior side of any exterior framed walls. The use of Class III vapor retarders is also acceptable provided such retarders and Climate Zones are set forth in Table 1405.3.1. In addition to the use of assemblies tested and identified as Class I, II or III vapor retarders, Section 1405.3.2 also provides a prescriptive listing of materials considered as compliant.

1405.4
Flashing. The code requires that all points subject to the entry of moisture be appropriately flashed. Roof and wall intersections, as well as parapets, are especially troublesome, as are exterior wall openings exposed to the weather, and particularly, those exposed to wind-driven rain. Even though the code may not cover every potential situation that might occur, it intends that the exterior envelope of the building be weatherproofed so as to protect

the interior from the weather. Furthermore, for buildings of human occupancy, the interior must be sanitary and livable. Therefore, whether prescribed in the code or not, any place on the envelope of the building that provides a route for admission of water or moisture into the building is required to be properly protected.

1405.5 Wood veneers. The use of wood veneer in buildings is limited when installed on buildings of other than Type V construction. The height of the veneer cannot exceed three stories, four stories where fire-retardant-treated wood is used. The backing must be noncombustible, with any open or spaced veneer projecting no more than 24 inches (610 mm) from the building wall.

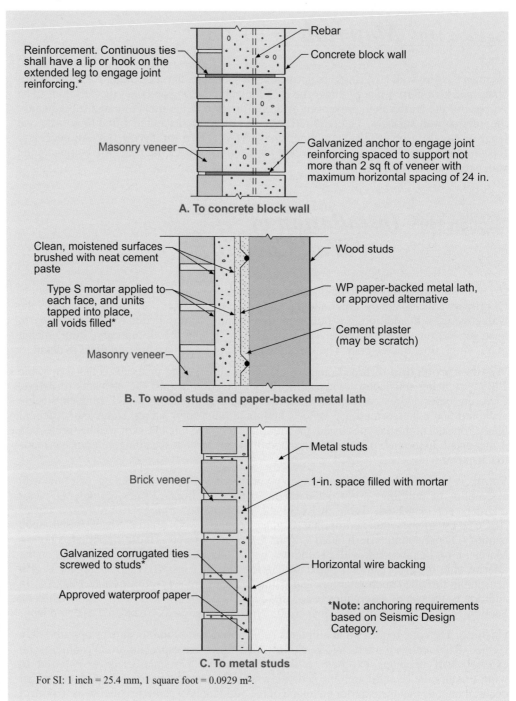

Figure 1405-1
Generic application of anchored masonry veneer (5-inch maximum thickness)

For SI: 1 inch = 25.4 mm, 1 square foot = 0.0929 m².

Anchored masonry veneer. In addition to complying with the applicable provisions of   **1405.6**
Chapter 14, anchored masonry veneer must also comply with Sections 6.1 and 6.2 of TMS
402/ACI 530/ASCE 5. Anchored masonry veneer must typically be supported by
noncombustible corrosion-resistant structural framing with the exception that this section
does permit the use of wood construction supports when in conformance with four criteria.
In addition to the structural considerations, the deflection limitation and joint requirements
are intended by the code to limit the differential interaction of the veneer with its backing so
as to prevent cracking, spalling and buckling of the veneer surfaces. Figure 1405-1 shows
examples of anchored masonry veneer.

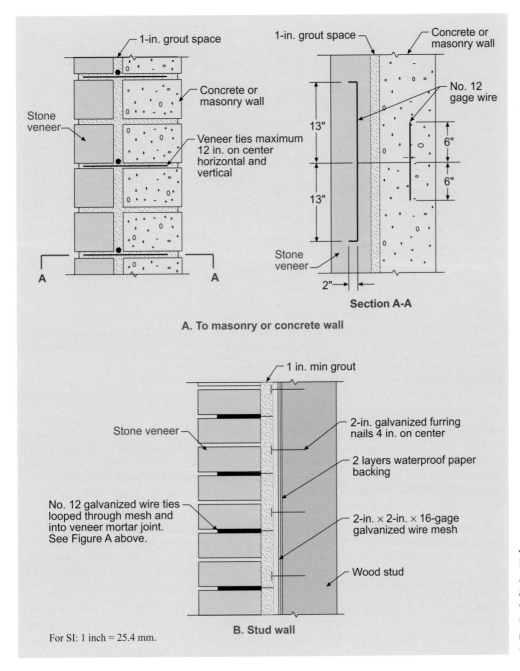

A. To masonry or concrete wall

Section A-A

B. Stud wall

For SI: 1 inch = 25.4 mm.

Figure 1405-2
**Application of
anchored stone
veneer units up
to 10-inch
maximum
thickness**

1405.7   Stone veneer. Two methods are detailed for the application of stone veneer having a maximum thickness of 10 inches (254 mm). One method is applicable where the stone is anchored directly to masonry or concrete construction, whereas the second is to be utilized for stud construction. Figure 1405-2 details several of the requirements for both methods of attachment.

1405.8   Slab-type veneer. When limited in thickness to 2 inches (51 mm), slab-type veneer units are to be anchored directly to concrete, masonry or stud construction. The maximum size of each unit is limited to 20 square feet (1.9 m²). Attachment shall be made with dowels and ties that are corrosion-resistant in a manner prescribed by this section. An illustration of the application of slab-type veneer units is shown in Figure 1405-3.

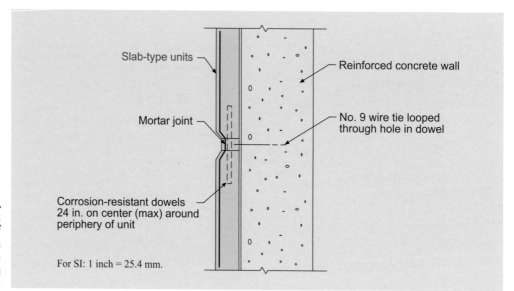

Figure 1405-3
**Application of
slab-type units
(2-inch
maximum
thickness)**

1405.9   Terra cotta. Where anchored terra cotta or ceramic units are utilized as exterior veneer, they must be at least $1^5/_8$ inches (41 mm) in thickness. They shall be anchored directly to masonry, concrete or stud construction in a manner consistent with the requirements of this section as shown in Figure 1405-4.

1405.10   Adhered masonry veneer. Adhered masonry veneer must comply with the requirements of TMS 402/ACI 530/ASCE 5 as well as the provisions of Section 1405.10.1. Figure 1405-5 provides an example of the application of adhered veneer.

1405.11   Metal veneers. To help ensure the integrity and durability of the exterior wall covering material, metal veneer must be manufactured of corrosion-resistant materials. As an alternative, the metal panels may be completely encased in porcelain enamel or treated in some other manner so that the metal is resistant to corrosion. The nominal thickness of sheet steel-metal veneer is to be at least 0.0149 inch (0.378 mm), with the veneer to be mounted on furring strips or approved sheathing.

As would be expected, the fasteners, metal ties or other attachment devices must also be corrosion-resistant. The fasteners are to be spaced at no more than 24 inches (610 mm) vertically and horizontally, with at least four attachments for veneer panels exceeding 4 square feet (0.4 m²) in area. It is also important that some means of weather protection be provided for the exterior metal veneer, including the use of pressure-treated wood.

1405.12   Glass veneer. For safety reasons, the use of glass veneer as an exterior wall covering is highly regulated. Any single piece of veneer is limited to 6 square feet (0.56 m²) in area, unless located within 15 feet (4572 mm) of the sidewalk level or grade directly below, in which case it may be up to 10 square feet (0.93 m²) in area. The maximum height or width of any section is limited to 48 inches (1219 mm), with the thickness to be no less than 0.344 inch (8.7 mm).

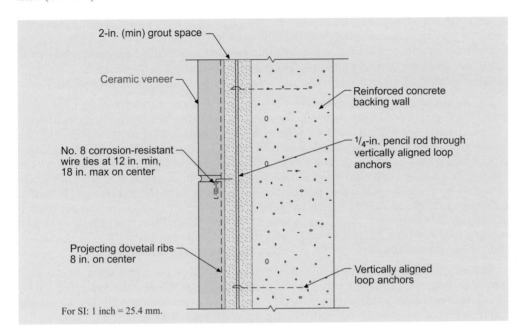

2-in. (min) grout space

Ceramic veneer

Reinforced concrete backing wall

$^1/_4$-in. pencil rod through vertically aligned loop anchors

No. 8 corrosion-resistant wire ties at 12 in. min, 18 in. max on center

Projecting dovetail ribs 8 in. on center

Vertically aligned loop anchors

For SI: 1 inch = 25.4 mm.

Figure 1405-4
Application of terra cotta or ceramic units

The application method requires at least one-half of the area of each veneer panel to be directly bonded to the backing with an approved mastic cement. Metal moldings shall be utilized for any veneer that extends to the sidewalk surface, with shelf angles required for additional support under certain conditions. At a height over 12 feet (3658 mm) or over show windows, additional support shall be provided by fasteners at either the four corners or the vertical or horizontal edges of the glass veneer. Design of the fasteners must allow for the independent support of the veneer. Exposed edges must also be adequately flashed to prevent the entrance of moisture between the backing and the glass veneer.

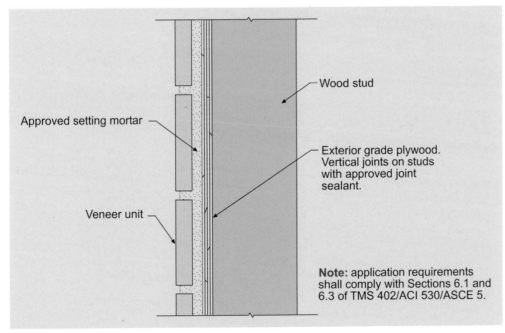

Wood stud

Approved setting mortar

Exterior grade plywood. Vertical joints on studs with approved joint sealant.

Veneer unit

Note: application requirements shall comply with Sections 6.1 and 6.3 of TMS 402/ACI 530/ASCE 5.

Figure 1405-5
Application of adhered masonry veneer

1405.13.2 **Window sills.** Historical data have shown that each year a considerable number of children fall from windows in residential buildings. It has been estimated that a sizable percentage of those falls occurred through windows with a low sill height. The minimum sill height of 24 inches (610 mm) was established to place the lowest point of the window opening above the center of gravity of the smallest children, thus reducing the number of falls. Therefore, the requirement is only applicable to those buildings where the hazard to young children is common—dwelling units classified as Group R-2 or R-3 occupancies. Sleeping units, such as those located in Group R-2 dormitories, are not regulated under these provisions.

The measurements on both the interior and exterior sides of the building are to be taken from the lowest portion of the clear window opening, providing for consistent application of the provisions. See Figure 1405-6. Where the lower window panel is inoperable, the measurements are to be taken to the lowest point of the lowest operable panel. Fixed glazing is permitted within 24 inches (610 mm) of the floor, as are openings that do no allow for the passage of a 4-inch (102 mm) diameter sphere. In addition, openings that are sufficiently protected by complying window guards are not regulated for the minimum sill height. Reference is made to ASTM F2006 for specifications regarding window fall prevention devices for nonemergency escape and rescue windows. Where the guard is to be used on a window required for emergency escape and rescue, it must comply with ASTM F 2090, addressing window fall devices with emergency egress release mechanisms.

The requirement for window sill height does not affect the application of the IBC provision addressing emergency escape and rescue windows. Where a window regulated by this section occurs in a sleeping room and is designated as the required emergency escape and rescue opening, it will have both a minimum and a maximum required sill height.

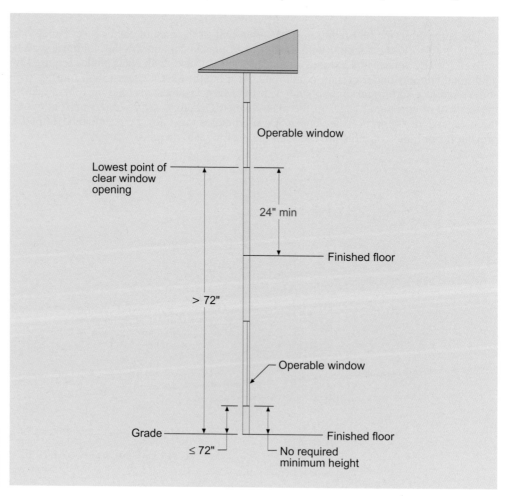

Figure 1405-6
Minimum window sill height

Vinyl siding. ASTM D 3679 establishes requirements and testing methods for extruded **1405.14**
single-wall siding manufactured from rigid polyvinyl chloride compound. It is important
that the building official determine if a particular siding material is code-complying by
referring to the packaging for information as to compliance with the material standard.

Siding regulated by this section is typically limited to installation on the exterior walls of
buildings of Type V construction located in areas where the wind speed, according to
Chapter 16, does not exceed 100 miles per hour (161 km/h) and the building height is
limited to 40 feet (12 192 mm) in Exposure C. More severe wind or height conditions would
require submittal data showing compliance with Chapter 16. However, most geographical
areas fall within the specified wind speed and exposure limit, and most Type V buildings are
under 40 feet (12 192 mm) in height. Although not specifically stated, the code intends that
the construction type be nonrated unless test data are provided to show that an exterior wall
utilizing vinyl siding has undergone fire testing to qualify as at least a 1-hour
fire-resistance-rated assembly. It is anticipated that vinyl siding will be used on wood-frame
construction, even though any type of construction could be classified as Type V
construction.

Vinyl siding must be installed over a wood-based sheathing material listed in Section
2304.6 and must satisfy the weather-resistive barrier requirements of Section 1403. This
section also sets forth specific requirements regarding the nailing of siding. Along with
these specific requirements, the section also states that "siding and accessories shall be
installed in accordance with approved manufacturer's instructions."

Fiber cement siding. A relatively new exterior wall covering material, fiber cement siding **1405.16**
is composed of fiber-reinforced cement. It is designed to be utilized as horizontal lap siding
or installed as panels.

Section 1406 *Combustible Materials on the Exterior Side of Exterior Walls*

Exterior walls of buildings of Types I, II, III and IV are typically required to be of
noncombustible construction. However, it is common for some limited combustible
elements to be installed on the exterior side of such exterior walls. Type V buildings are
permitted to utilize combustible materials for the entire wall construction. Where the
exterior walls of a building include combustible materials such as wall coverings, trim,
balconies and similar appendages, the materials must be in compliance with this section.
The only exception is for plastics, which are regulated by Chapter 26. The installation of
such materials in compliance with the code does not negatively impact the desired level of
fire safety.

The general requirements address the ignition resistance of the combustible wall
coverings, based upon the distance between the exterior wall and the lot line. Where the fire
separation distance is 5 feet (1524 mm) or less, the combustible wall covering must be
subjected to the test method of NFPA 268 and exhibit no sustained flaming. As the exterior
wall is moved farther and farther from the lot line, the tolerance to radiant heat energy need
not be as great, allowing the use of alternate materials. Table 1406.2.1.2 identifies the
tolerable level of incident radiant heat energy based on fire separation distance.

Section 1406.2.2 provides for the use of combustible wall coverings on otherwise
noncombustible exterior walls, provided the building is limited to 40 feet (12 192 mm) and
three stories in height. The amount of such materials is also limited where the fire separation
distance is 5 feet (1524 mm) or less. The reductions from the general requirements are
granted because of the restricted height and, in some cases, the limited amount of
combustible materials. At a height exceeding 40 feet (12 192 mm), the use of combustible
materials or supports is prohibited.

Balconies and similar projecting elements in buildings of Type I and II construction are to be constructed of noncombustible materials, unless the building is no more than three stories in height above grade plane. Under this condition, fire-retardant-treated wood may be used where the balcony or similar element is not used as a required egress path. Unless constructed of complying heavy-timber members, a combustible balcony or similar combustible projection must have a minimum fire-resistance rating equivalent to the required floor construction.

In Type III, IV and V buildings, balcony construction may be of any material permitted by the code, combustible or noncombustible. Where a fire-resistance rating is mandated by the code, it must be maintained at the projecting element unless sprinkler protection is provided or it is of Type IV construction.

In addition to these types of construction limitations, the aggregate length of all projections cannot exceed 50 percent of the building perimeter at each floor unless sprinkler protection is extended to the balcony areas. In all cases, the use of untreated wood is permitted for pickets, rails and similar guard elements when limited to a height of 42 inches (1067 mm).

Section 1407 *Metal Composite Materials*

Metal composite materials (MCM) consist of a thin, extruded plastic core encapsulated within metal facings. As the use of MCM continues to increase, it is important to provide detailed requirements addressing this unique building element. Many of the provisions are based on requirements from elsewhere in this chapter and Chapter 26. Although having some of the same characteristics as foam-plastic insulation, light-transmitting plastic, plastic veneer and combustible construction, metal composite materials used as exterior wall coverings are specifically regulated by this section.

Where installed on exterior walls of buildings, MCM systems are regulated for surface-burning characteristics. In buildings of other than Type V construction, the flame-spread index cannot exceed 75 and the smoke-developed index is limited to 450. In such buildings required to have noncombustible exterior walls, the use of a thermal barrier is necessary to separate the MCM from the interior of the building unless specifically approved by appropriate testing. Alternately, the installation of MCM to a limited height in compliance with Section 1407.11 modifies the general requirements.

Section 1408 *Exterior Insulation and Finish Systems (EIFS)*

Exterior Insulation and Finish Systems (EIFS) are nonload-bearing exterior wall coverings that are used extensively throughout North American, Europe and the Pacific Rim. The provisions of Section 1408 are primarily intended to reference the applicable ASTM standards that are specific to EIFS. Reference is made to E 2273 *Standard Test Method for Determining the Drainage Efficiency of Exterior Insulation and Finish Systems (EIFS) Clad Wall Assemblies*, E 2568 *Standard Specification for PB Exterior Insulation and Finish Systems (EIFS)*, and E 2570 *Standard Test Method for Evaluating Water-Resistive Barrier (WRB) Coatings Used Under Exterior Insulation and Finish Systems (EIFS) for EIFS with Drainage*. In addition to the several ASTM standards previously identified, reference is also made to various provisions found elsewhere in the IBC that are applicable to EIFS. Current ICC ES Acceptance Criteria further establish requirements for EIFS and related components, and numerous EIFS manufacturers hold evaluation reports to demonstrate code compliance.

KEY POINTS

- The interior of a building must be protected with a weather-resistant envelope, including wall coverings, flashing and drainage methods.

- The IBC regulates numerous veneer materials for exterior applications.

- The minimum sill height of operable exterior windows is regulated for dwelling units in Group R-2 and R-3 occupancies.

- A limited amount of combustible materials such as wall coverings, trim, balconies and similar appendages are permitted on exterior walls of Type I, II, III and IV buildings.

- The applicable referenced standards are identified for the installation of exterior insulation and finish systems (EIFS).

ROOF ASSEMBLIES AND ROOFTOP STRUCTURES

In addition to the requirements for roof assemblies and roof coverings, this chapter regulates roof insulation and rooftop structures. Rooftop structures include such elements as penthouses, tanks, cooling towers, spires, towers, domes and cupolas. Roofing materials and components are regulated for quality as well as installation.

The provisions in Chapter 15 for roof construction and roof covering are intended to provide a weather-protective barrier at the roof and, in most circumstances, to provide a fire-retardant barrier to prevent flaming combustible materials such as flying brands from nearby fires from penetrating the roof construction. The chapter is essentially prescriptive in nature and is based on decades of experience with the various traditional roof-covering materials. These prescriptive rules are very important to ensuring the satisfactory performance of the roof covering, even though the reason for a particular requirement may be lost. The provisions are based on an attempt to prevent observed past unsatisfactory performances of the various roofing materials and components.

Those measures that have been shown by experience to prevent past unsatisfactory performance generally are included in the manufacturer's instructions for application of the various roofing materials. In many cases, the manufacturer's instructions are incorporated in this code by reference. The code intends, then, that they be followed as if they were part of the code.

The overriding safety need of roofs is resistance to external fire factors. In this regard, the enforcement of this chapter is driven by Table 1505.1, as well as the appropriate standards for fire-retardant roof assemblies and roof coverings, including ASTM E 108, UL 790 and ASTM D 2898. Typically, a roof covering by itself cannot be a listed fire-retardant roof. Therefore, the regulations clearly separate assemblies and coverings to enforce construction of listed roof assemblies to the level that listed wall and floor/ceiling assemblies are regulated.

Section 1502 *Definitions*

As with other industries supplying specialty building products, the roof-covering industry has a language of its own. In order to properly understand and apply the provisions of the *International Building Code®* (IBC®), the unique terms employed in the IBC must be understood. The roofing industry publishes several publications containing excellent glossaries of the terms of their industry. Three of the more commonly used code terms are defined below.

PENTHOUSE. Regulated by Section 1509, penthouses are structures placed on the roofs of buildings to shelter mechanical equipment or vertical shaft openings. The definition clarifies that a penthouse is intended to be unoccupied. Structures such as tanks, towers, spires and domes are not considered penthouses, but are regulated to some degree by this chapter.

ROOF ASSEMBLY. A roof assembly may be either a single component serving as both the roof covering and the deck, or a combination of individual roof deck and roof-covering components used together to form a complete assembly. A roof assembly may selectively include the roof deck, vapor retarder, substrate, thermal barrier, insulation and roof covering.

ROOF COVERING. The roof covering is considered the covering placed upon a roof to provide the building with weather protection, fire retardancy or decoration.

Section 1503 *Weather Protection*

In all cases, a roof must be designed to provide protection from the elements. To ensure that the roof will adequately perform this function, it must be designed in accordance with this chapter. In addition, and just as important, the approved manufacturer's installation instructions must be adhered to.

This section requires flashing where the roof intersects vertical elements such as walls, chimneys, dormers, plumbing stacks, plumbing vents and other penetrations of the weather-protective barrier. Coping of parapet walls shall be accomplished with noncombustible, weather-proofed materials. Figures 1503-1 through 1503-4 depict some examples of roof flashing details at vertical surfaces.

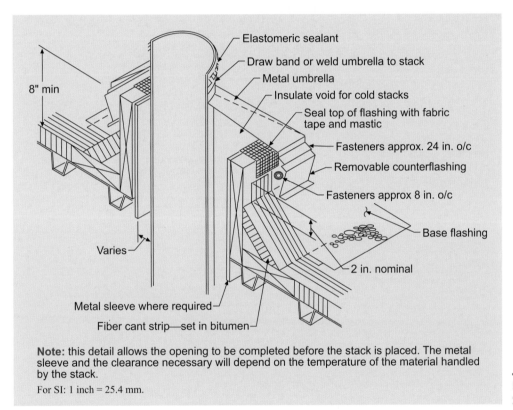

Note: this detail allows the opening to be completed before the stack is placed. The metal sleeve and the clearance necessary will depend on the temperature of the material handled by the stack.

For SI: 1 inch = 25.4 mm.

Figure 1503-1
Stack flashing

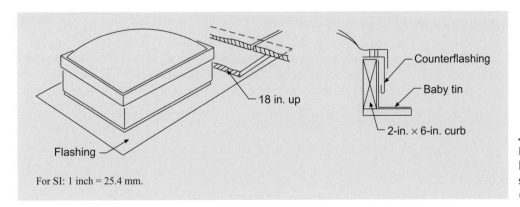

For SI: 1 inch = 25.4 mm.

Figure 1503-2
Flashing at skylight (wood roof)

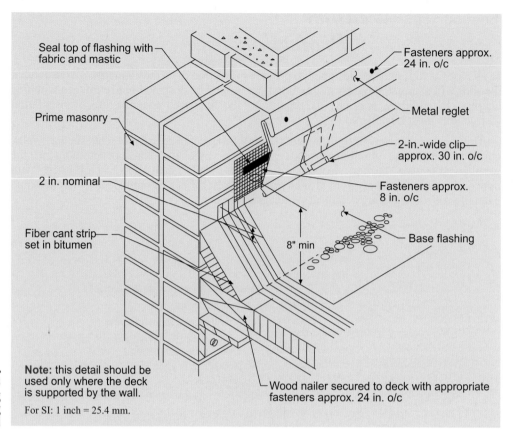

Seal top of flashing with fabric and mastic

Prime masonry

2 in. nominal

Fiber cant strip set in bitumen

Fasteners approx. 24 in. o/c

Metal reglet

2-in.-wide clip— approx. 30 in. o/c

Fasteners approx. 8 in. o/c

8" min

Base flashing

Wood nailer secured to deck with appropriate fasteners approx. 24 in. o/c

Note: this detail should be used only where the deck is supported by the wall.

For SI: 1 inch = 25.4 mm.

Figure 1503-3
Base flashing at bearing wall

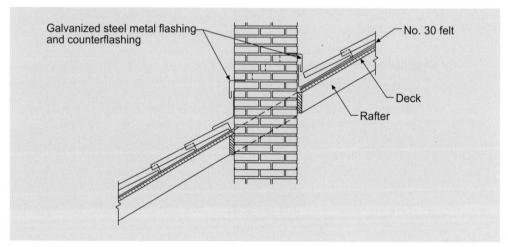

Galvanized steel metal flashing and counterflashing

No. 30 felt

Deck

Rafter

Figure 1503-4
Chimney flashing detail

Section 1504 *Performance Requirements*

Roof decks and roof coverings must be able to withstand the effects of nature in a satisfactory manner. This section of the code regulates the performance of a roof against three concerns: wind, weathering and impact. For wind resistance, roofs must comply with this section and Chapter 16. Low-slope roofs must demonstrate that they are resistant to both weathering and impact damage by complying with the appropriate standards.

The use of aggregate as a roof covering material and aggregate, gravel or stone as ballast is prohibited in specified locations in an effort to reduce property loss that is due to high winds. Field assessments of damage to buildings caused by high-wind events have shown that gravel or stone blown from the roofs of buildings has exacerbated damage to other buildings because of breakage of glass. The code prohibits the use of aggregate, gravel or stone on roofs of buildings in hurricane-prone regions. These regions are defined as areas along the United States Atlantic Ocean and Gulf of Mexico coasts where the basic wind speed exceeds 90 miles per hour (40 m/s) and also include the islands of Hawaii, Puerto Rico, Guam, Virgin Islands and American Samoa. Aggregate, gravel and stone roof covering materials are also prohibited on those buildings where the mean height exceeds that allowed by Table 1504.8 on the basis of the basic wind speed and exposure category. Under these conditions, there is a great enough potential for gravel stone, debris or other unsecured objects to become airborne and possibly break glass in buildings downwind.

Section 1505 *Fire Classification*

As a minimum, the IBC generally requires Class B or C roof coverings for most buildings. These are roof coverings that provide protection of the roof against moderate and light fire exposures, respectively. The various sizes of brands used for testing are shown in Figure 1505-1. These exposures are external and are generally created by fires in adjoining structures, wild fires (brush fires and forest fires, for example) and fire from the subject building that extends up the exterior and onto the top surface of the roof. Wild fires and some structural fires create flying and flaming brands that can ignite nonclassified roof coverings. With regard to clay tile roofing, which is defined in Section 1505.2 as a Class A roof assembly, it is of interest to note that the Spanish missionaries shipped clay roofing tile to North America to protect their mission buildings from fire caused by flaming arrows shot onto the roofs.

The roof assembly classifications required by the code, which are related primarily to type of construction, are delineated in Table 1505.1.

Section 1505 defines the following roof assemblies and roof coverings:

1. Class A roof assemblies. Roof assemblies recognized as Class A are effective against severe fire exposures. Roof coverings of brick, masonry, slate, clay or concrete roof tile, exposed concrete roof deck, and metal, ferrous or copper shingles or sheets are all considered Class A roof assemblies, as well as any roof assembly or roof covering tested and listed as Class A.

2. Class B roof assemblies. Class B roof assemblies are effective against moderate fire exposures and considered appropriate for all types of construction. There are no prescriptive roof assemblies or roof coverings considered as Class B, as they must be listed as such. Consistent with the universal acceptance of Class A assemblies, Class B roof assemblies are also permitted for use on all buildings.

3. Class C roof assemblies. Buildings not required to be of fire-resistant construction—Types IIB, IIIB and VB—are permitted to utilize Class C roof assemblies. Such assemblies are effective against light fire exposures.

4. Nonclassified roofing. Roof coverings that are considered nonclassified roof coverings are approved for use by the IBC on Group R-3 and U buildings where the roof is located at least 6 feet (1829 mm) from all lot lines. These roof coverings have been shown by experience to provide the necessary resistance to weather as intended by the code when the qualities of the materials comply with the appropriate requirements.

5. Fire-retardant-treated wood shingles and shakes. Extensive fire tests are mandated for wood shingles and shakes to ensure that they are compliant with the intent of the code. This section recognizes approved fire-retardant-treated shingles and shakes when classified in accordance with the testing.

6. Special purpose roofs. These roofs are either of wood shingles or wood shakes and are applied with a minimum $^5/_8$ inch (15.9 mm) Type X water-resistant gypsum backing board or gypsum sheathing panel. The intent of the provisions for special purpose roofs is to provide a roof covering that, although it may be ignited by flying brands, will not burn through to the interior of the building. Also, the special underlayment tends to prevent fires from the interior of the building from burning through to and igniting the roof covering, which helps prevent flying brands. Special purpose roofs are permitted in limited applications on buildings of Type IIB, IIIB and VB construction by Footnote c to Table 1505.1.

Because of the inconsistent use of the terms *roof assembly* and *roof covering* throughout Chapter 15, there is confusion as to the proper use of combustible materials at the roof. Based on Item 4 of Section 603.1, roof coverings that have an A, B or C classification are permitted in buildings of Type I or II construction. This does not include the structural deck materials, which must be of noncombustible construction or fire-retardant-treated wood in compliance with Item 1.3 of Section 603.1. It would, however, permit the use of wood structural panels or foam plastic insulation boards as a part of the classified roof covering where used in combination with a noncombustible wood deck. Where a Class A, B or C roofing assembly includes a combustible structural deck, other than fire-retardant-treated wood where permitted, it is limited to use on a building of Type III, IV or V construction.

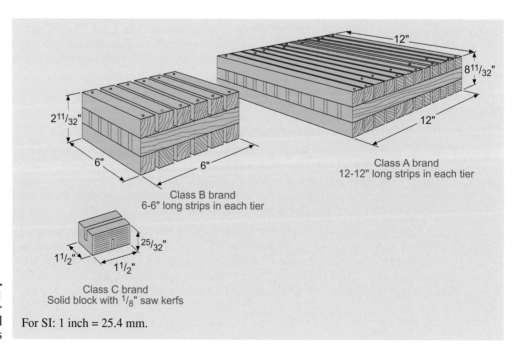

Figure 1505-1
Brands for Class A, B and C tests

Class B brand
6-6" long strips in each tier

Class A brand
12-12" long strips in each tier

Class C brand
Solid block with $^1/_8$" saw kerfs

For SI: 1 inch = 25.4 mm.

Section 1506 *Materials*

Certainly, roofing materials must comply with quality standards embodied in the IBC for Chapter 15. Furthermore, identification of the roofing materials is mandatory in order to verify that they comply with quality standards. In addition to bearing the manufacturer's label or identifying mark on the materials, roof-covering materials are required by the code to carry a label of an approved agency having a service for inspection of materials and finished products during manufacture.

Section 1507 *Requirements for Roof Coverings*

The following sections in the IBC contain the basic requirements for the installation of the more common roof-covering materials:

Section 1507.2 Asphalt shingles

Section 1507.3 Clay and concrete tile

Section 1507.4 Metal roof panels

Section 1507.8 Wood shingles

Section 1507.9 Wood shakes

Section 1507.10 Built-up roofing materials

Asphalt shingles. Figure 1507-1 shows a typical installation of asphalt shingles on roofs with a slope of 4:12 as required by the IBC. However, the IBC also permits application on a roof that has a slope as shallow as 2:12 when installed in accordance with Section 1507.2.8.

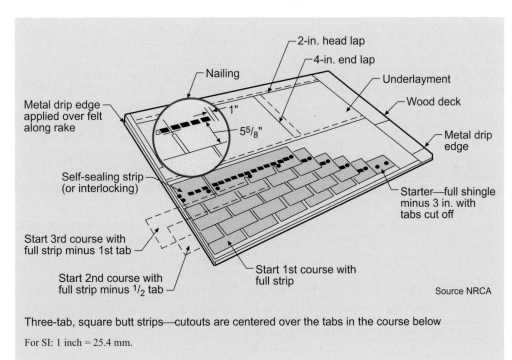

Three-tab, square butt strips—cutouts are centered over the tabs in the course below

For SI: 1 inch = 25.4 mm.

Figure 1507-1
Asphalt roofing shingles application high slope (4:12 minimum)

Where the roof slope is less than 4:12, water drainage from the roof is slowed down and has a tendency to back up under the roofing and cause leaks. Also, the effect of ice dams at eaves is more pronounced on low-slope roofs and, as a result, special precautions are necessary to ensure satisfactory performance of the roofing materials. Thus, the code requires that the underlayment be laid with two layers of shingled felt, which provides two thicknesses of the underlayment at any point. Figure 1507-2 shows the method of underlayment application, and Figure 1507-3 shows the method for shingle application for low-slope roofs. The code requires that underlayment of one layer of approved felt be provided under asphalt-shingle roof coverings with a slope of 4:12 or greater.

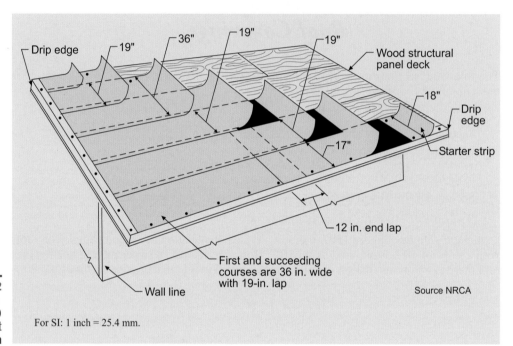

Figure 1507-2
Low slope (less than 4:12) underlayment application

For SI: 1 inch = 25.4 mm.

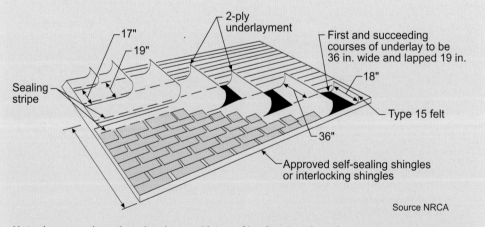

Figure 1507-3
Application of asphalt shingle slopes between 2:12 and 4:12

Note: In areas where there has been a history of ice forming along the eaves causing a backup of water, felt plies of underlayment should be cemented up from eaves far enough to overlie a point 24 in. inside the inside wall line of the building.

For SI: 1 inch = 25.4 mm.

Clay and concrete tile. Tile is among the oldest of roofing materials, and clay tile was used by the builders of ancient Greece and Egypt. Clay and concrete tile come in two generic configurations—roll tile and flat tile. Figures 1507-4 and 1507-5 give examples of some of the configurations for roofing tile. Either roll tile or flat tile may be interlocking, and Figure 1507-5 shows an example of interlocking tiles. Note that the ribs along the long edge of each tile are designed such that each adjacent tile has ribs that overlap and interlock with the adjoining tile. Table 1507.3.7 contains attachment requirements for these tiles.

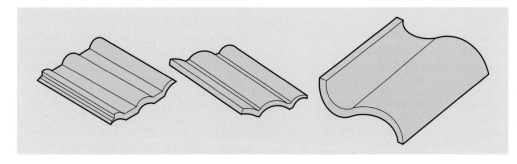

Figure 1507-4
Clay roll tile

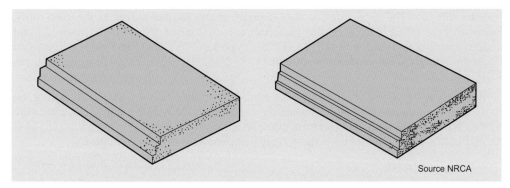

Source NRCA

Figure 1507-5
Concrete flat tile

Figure 1507-6 is an example of the application of roll tile and also provides details for ridge covering and the closure at a gable rake. Clay or concrete tile roofs may be installed over either a solid deck or spaced wood sheathing.

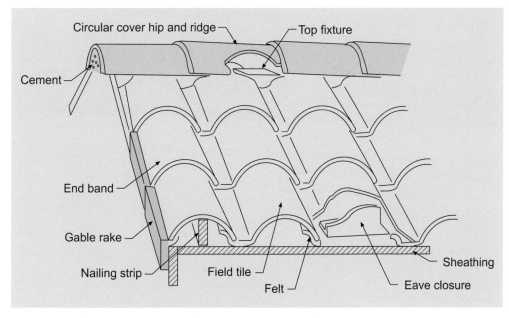

Figure 1507-6
Application of roll tile

Metal roof panels. Unless specifically designed to be applied to spaced supports, metal roof panel roof coverings must be applied to a solid or closely fitted deck. Depending upon the type of metal-roof system, the minimum slope of the roof deck varies between $\frac{1}{4}$:12 and 3:12. The attachment of metal roofing should be based upon the recommendations of the manufacturer. Where such recommendations are not provided, stainless-steel fasteners shall be used. Galvanized fasteners are permitted for galvanized roofs, and hard copper or copper-alloy fasteners shall be used for copper roofs. Table 1507.4.3(1) identifies the application rate or thickness for metal-sheet roof coverings installed over structural decking.

Metal roof shingles. The deck requirements for metal roof shingles are identical to those for metal roof panels. The minimum slope of the roof deck on which metal shingles are installed is to be 3:12. Provisions for underlayment and flashing are very similar in nature to those for other types of shingle applications.

Mineral-surfaced roll roofing. Limited to application on solidly sheathed roofs only, mineral-surfaced roll roofing shall not be applied on roof slopes less than 1:12. Conforming underlayment shall be provided and additional precautions must be taken in severe climate areas. Several publications provide detailed recommendations for the application of roll roofing.

Slate shingles. The application of slate shingles is also limited to solidly sheathed roofs. Because of the nature of these roofing materials, a deck slope of at least 4:12 is required, and underlayment complying with ASTM D 226, Type I or ASTM D 4869 must be installed below the shingles. The required minimum head lap for slate shingles is identified in Table 1507.7.5.

Wood shingles. Wood shingles may be applied on either solid or spaced sheathing, on roofs with a minimum slope of 3:12 when the exposure to the weather is in accordance with Table 1507.8.6, and with underlayment complying with ASTM D 226, Type I or ASTM D 4869. Figures 1507-7 and 1507-8 show typical applications of wood shingles.

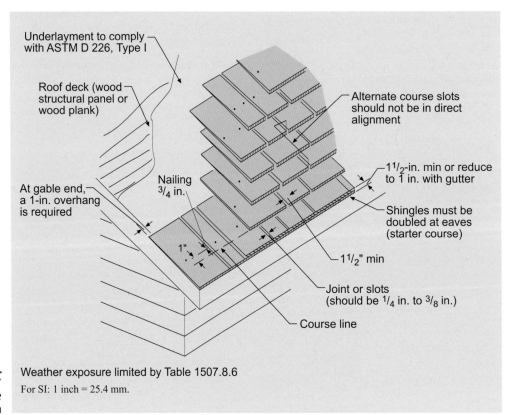

Weather exposure limited by Table 1507.8.6

For SI: 1 inch = 25.4 mm.

**Figure 1507-7
Wood shingle
application**

Figure 1507-8
Application of wood shingles at ridges

Wood shakes. Wood shakes are to be applied on roofs with a minimum slope of 4:12 on either spaced or solid sheathing. In addition to the required underlayment, felt interlayment is required to be shingled between the courses of wood shakes. As with other special requirements for shingle and tile roofs laid in areas of severe cold weather, extra precautions are required to prevent the backup of water on the roof and under the shingles with resulting leaks into the building. Figure 1507-9 illustrates an example of wood-shake application.

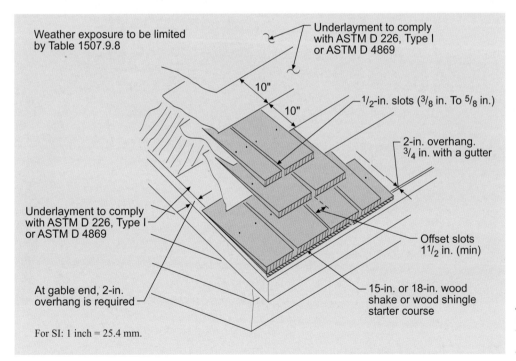

Figure 1507-9
Wood shake application

Both wood shingles and wood shakes are required by the code to have certain maximum exposures to the weather, and in the case of shingles, this exposure is dependent on the roof slope. These exposures for roofs with slopes of 4:12 or greater provide a minimum of three thicknesses of shingle with an overlap of $1\frac{1}{2}$ inches (38 mm) and have been shown through experience to be the maximum exposure that will provide a leak-free roof.

Other roof coverings. Sections 1507.10 through 1507.15 address those roof covering methods typically applicable to flat roofs. For most installations, a minimum roof slope of one vertical in 48 units horizontal ($^1/_4$:12) is specified. The appropriate material standards for each of the materials or systems are also identified. The following roof coverings are addressed:

1. Built-up roofs.

2. Modified bitumen roofing.

3. Thermoset single-ply roofing.

4. Thermoplastic single-ply roofing.

5. Sprayed polyurethane foam roofing.

6. Liquid-applied coatings.

Roof gardens and landscaped roofs. Where roofs are used for purposes in addition to weather protection, additional requirements may be applicable to address those concerns not typically encountered. As established in Section 1507.16, gardens and other landscaping installed on a roof are specifically addressed in regard to roof construction and structural integrity. For structural purposes, the Chapter 16 requirements for these types of special-purpose and landscaped roofs are cross-referenced. An example of a landscaped roof is shown in Figure 1507-10.

Figure 1507-10
Roof garden

Section 1508 *Roof Insulation*

Because of an increased level of energy consciousness, the use of roof insulation has become more and more prevalent, as it has distinct benefits not only in energy conservation but also in building occupant comfort. Insulation also provides a smooth uniform substrate for application of the roofing materials. The code requires that above-deck thermal insulation be covered by an approved covering and be in compliance with FM 4450 or UL 1256 when tested as an assembly. For foam plastic insulation used under roof coverings, see the commentary under Section 2603.

Section 1509 *Rooftop Structures*

Penthouses and other roof structures are regulated by the IBC as if they were appurtenances to the building rather than occupiable portions.

In fact, if a penthouse is used for any purpose other than shelter of mechanical equipment or shelter of vertical shaft openings, the code requires that it be considered an additional story of the building.

As intended by the IBC, roof structures are equipment shelters, equipment screens, platforms that support mechanical equipment, water-tank enclosures and other similar structures generally used to screen, support or shelter equipment on the roof of the building. This section also regulates towers and spires, which are addressed separately in Section 1509.5.

The IBC regulates penthouses, roof structures, tanks, towers and spires to prevent hazardous conditions that are due to internal and external fire concerns, structural inadequacy and to ensure their proper use as equipment shelters.

Penthouses. The code does not regulate the height of penthouses and roof structures on Type I buildings. However, for buildings of other construction types, the code limits the height of penthouses and roof structures to 18 feet (5486 mm) above the height of the roof, except where the penthouse is used to enclose a tank or elevator. In such cases, a maximum height of 28 feet (8534 mm) is permitted. As this section also limits the aggregate area of all penthouses and roof structures to one-third the area of the roof, the additional height permitted for penthouses and roof structures does not pose any significant fire- and life-safety hazard that is due to other restrictions that this section places on construction. See Figure 1509-1.

1509.2

Penthouses and other rooftop structures are also regulated by the provisions of Section 504.3, where the requirements are based on the effect of a rooftop structure on the allowable height permitted for the entire building. Where applicable, the provisions from both Chapters 5 and 15 are in effect, and where there is a conflict, the most restrictive condition will apply.

As the code has reduced requirements for construction of penthouses and roof structures, it is logical that their use should be limited as specified in this section. Thus, if other uses are made of penthouses or other roof structures, it also seems appropriate that they should be constructed as would be required for an additional story of the building. It is the intent of the code that a penthouse or roof structure not be used to create an additional story above that permitted by Section 503.

If a rooftop structure qualifies as a penthouse, the floor of the penthouse is only required to meet the roof provisions of Table 601. In addition, any fire-resistance-rated shaft that extends to or through the penthouse floor does not need to be protected at the floor line. Under both conditions, the floor of the penthouse is solely regulated as the building's roof. Regarding any required means of egress from the penthouse, the provisions for an occupied floor are not applicable, nor are the requirements of Section 1019.1 for an occupied roof. It would seem appropriate that the access provisions of the *International Mechanical Code*® (IMC®) for rooftop equipment also provide for adequate egress.

1509.2.4 **Type of construction.** The intent of the code is that penthouses or roof structures be constructed with the same materials and the same fire resistance as required for the main portion of the building. However, because of the nature of their use, the code does permit exceptions where the exterior walls of penthouses and roofs structures are at least 5 feet (1524 mm) [or in some cases 20 feet (6096 mm)] from the lot line. See Figure 1509-2.

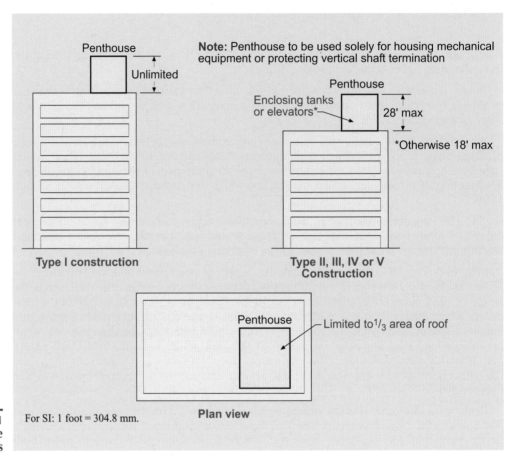

Note: Penthouse to be used solely for housing mechanical equipment or protecting vertical shaft termination

Penthouse — Unlimited

Type I construction

Penthouse
Enclosing tanks or elevators*
28' max
*Otherwise 18' max

Type II, III, IV or V Construction

Penthouse — Limited to ¹/₃ area of roof

Plan view

For SI: 1 foot = 304.8 mm.

Figure 1509-1
Penthouse limitations

In the case of unroofed mechanical equipment screens, fences and similar enclosures provided in Exception 4, the code permits combustible construction, provided the height of the structure is no more than 4 feet (1219 mm) above the adjacent roof surface. As this exception applies only to one-story buildings, the fire hazard is not significantly increased, as fire-fighting forces will have reasonable access to the roof from the exterior of the building.

1509.5 **Towers, spires, domes and cupolas.** The IBC intends that towers, spires, domes and cupolas be considered separately from penthouses and other roof structures. The towers contemplated in this section are towers such as radio and television antenna towers, church spires and other roof elements of similar nature that do not support or enclose any mechanical equipment and that are not occupied. As with penthouses and roof structures, the code intends to obtain construction and fire resistance consistent with that of the building to which they are attached. Under a variety of conditions, however, towers and similar elements are required to be constructed of noncombustible materials, regardless of the building construction.

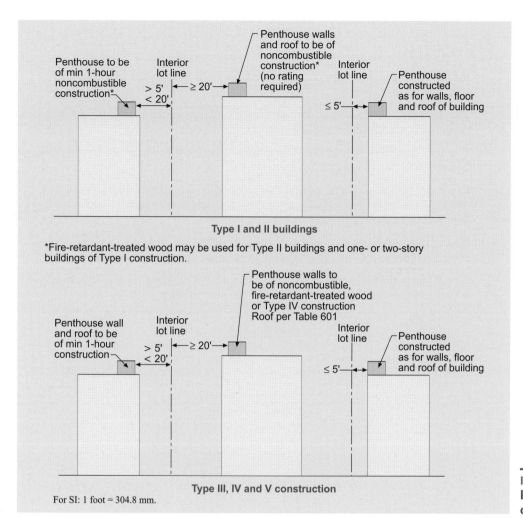

Type I and II buildings

*Fire-retardant-treated wood may be used for Type II buildings and one- or two-story buildings of Type I construction.

Type III, IV and V construction

For SI: 1 foot = 304.8 mm.

Figure 1509-2
Penthouse construction

Section 1510 *Reroofing*

This section addresses the problems associated with unregulated reroofing operations. The intent of these provisions is to ensure that when an existing building is reroofed, the existing roof is structurally sound and in a proper condition to receive the new roofing. With exceptions, the replacement of existing roof coverings is required where any one of the following conditions occur:

1. The existing roof or roof covering is water soaked or has deteriorated such that it is an inadequate base for additional roofing.

2. The existing roof covering consists of wood, slate, clay, cement or asbestos-cement tile.

3. There are currently two or more applications of roof covering on the existing roof.

KEY POINTS

- Roofs and roof coverings are addressed for both weather protection and fire retardancy.

- The performance of a roof for weather protection is regulated against three concerns—wind, weathering and impact.

- The IBC generally requires roof coverings that provide protection for the roof against moderate or light fire exposures.

- Roof assemblies and roof coverings are classified as either Class A, B or C roofing assemblies; nonclassified roofing; fire-retardant-treated wood shingles and shakes; or special-purpose roofs.

- Penthouses and other roof structures are regulated by the code as if they were appurtenances to the building rather than occupiable portions.

- Reroofing is limited to those buildings where the existing roof is structurally sound and in proper condition to receive the new roofing.

GLASS AND GLAZING

Chapter 24 regulates glass and glazing for essentially two important reasons:

1. To protect against breakage that is due to building distortion under lateral loads or that is due to wind loads applied perpendicularly to the surface of the glass.

2. To protect against breakage that is due to accidental impact by individuals adjacent to the glazing.

Glass, light-transmitting ceramic panels and light-transmitting plastic panels are regulated by Chapter 24 for their use in both vertical and sloped applications.

Section 2403 *General Requirements for Glass*

2403.1 **Identification.** Because glass is not manufactured at the building site and is usually incorporated into assemblies that are not manufactured at the building site, the code requires that the glass and glazing be identified as to type and thickness so that it is possible to determine compliance with this chapter as far as permitted areas of glass are concerned. The identifying information furnished on each light may be removable or permanent. Where approved by the building official, an affidavit furnished by the glazing contractor is acceptable in lieu of the manufacturer's mark located on each pane of glass or glazing material. The affidavit must certify that each light is glazed in accordance with approved construction documents, as well as the provisions of Chapter 24. Where an identification mark is utilized, it shall designate both the type and thickness of the material.

Unless used in a spandrel application, tempered glass is required to be permanently identified by the manufacturer. The identifying mark is to be acid etched, sandblasted, ceramic fired, embossed or of any other type that is unable to be removed without being destroyed. A removable paper marking provided by the manufacturer is permitted, but only for tempered spandrel glass.

2403.2 **Glass supports.** Glass firmly supported on all edges is considerably stronger than a glass light with one or more free edges. As a result, where the glass does not have firm support on all edges, the code requires that a design for the glass be submitted to the building official for approval. In this latter case, the design would be based on the number and location of any free edges.

2403.5 **Louvered windows or jalousies.** The requirements for louvered windows are based on there being no edge support on the longitudinal edges. Moreover, for safety purposes, the code requires that the exposed edge be smooth. For the same reason, wired glass, when used in jalousie or louvered windows, shall have no exposed wire projecting from the longitudinal edges. Where a louvered window complies with the provisions of this section, such an application is exempt from the requirements for safety glazing for use in hazardous locations.

Section 2404 *Wind, Snow and Dead Loads on Glass*

Exterior glass and glazing are subject to the same loads as the exterior cladding of the building; therefore, the code requires that glass and glazing in a vertical or near-vertical position should be designed for the same wind loads as specified in Section 1609 for cladding. For seismic considerations, ASCE 7 is referenced for the design of glass in glazed curtain walls, glazed storefronts and glazed partitions. This design will also include the increased pressures on local areas at discontinuities. As the slope of the glass increases, it may be necessary to address other loads as well, such as the dead load and any snow load.

Section 2405 *Sloped Glazing and Skylights*

By application, sloped glazing and skylights consist of glazing installed in roofs or walls that are on a slope of more than 15 degrees (0.26 rad) from the vertical. See Figure 2405-1. The provisions of the *International Building Code®* (IBC®) are intended to protect these glazed openings from flying fire brands and to provide adequate strength to carry the loads normally attributed to roofs. The provisions are also intended to protect the occupants of a building from the possibility of falling glazing materials.

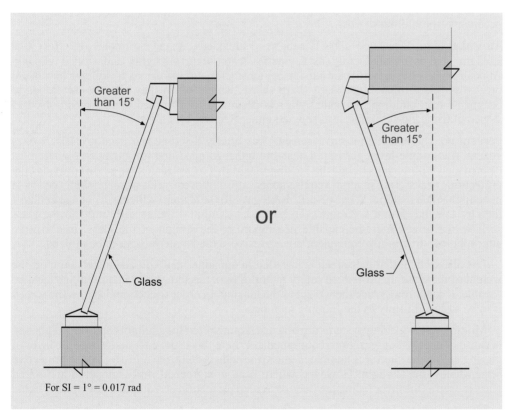

For SI = 1° = 0.017 rad

Figure 2405-1
Sloped glazing

Allowable glazing materials and limitations. Glazing materials and protective measures for sloped glazing and skylights are outlined in this section. The materials and their characteristics and limitations are as follows:

2405.2

Laminated glass. Laminated glass is usually furnished with an inner layer of polyvinyl butyral, which has a minimum thickness of 30 mil (0.76 mm). Such glass is highly resistant to impact and, as a result, requires no further protection below. When used within dwelling units of Group R-2, R-3 and R-4 occupancies without a protective screen, laminated glass is permitted to have a 15-mil (0.38 mm) polyvinyl butyral inner layer, provided each pane of glass is 16 square feet (1.5 m^2) or less in area, and the highest point of the glass is no more than 12 feet (3658 mm) above a walking surface or other accessible area.

Wired glass. Wired glass is resistant to impact and, when used as a single-layer glazing, requires no additional protection below.

Tempered glass. Tempered glass is glass that has been specifically heat-treated or chemically treated to provide high strength. When broken, the entire piece of glass immediately breaks into numerous small granular pieces. Because of its high strength and manner of breakage, tempered glass has been considered in the past to be a desirable glazing material for skylights without any protective screens. However, as a result of studies by the industry that show that tempered glass is subject to spontaneous breakage such that large chunks of glass may fall under this condition, the IBC requires screen protection below tempered glass.

Approved light-transmitting plastic materials. See discussion in this handbook under Section 2609.

Heat-strengthened glass. Heat-strengthened glass is glass that has been reheated to just below its melting point and then cooled. This process forms a compression on the outer surface and increases the strength of the glass. However, heat-strengthened glass has the unsatisfactory characteristic of breaking into shards, as does annealed glass. Thus, heat-strengthened glass requires screen protection below the skylight to protect the occupants from falling shards.

Annealed glass. Annealed glass is subject to breakage created by impact, has very low strength, and is unsatisfactory as a glazing material in skylights and sloped glazing. Annealed glass has a further unsatisfactory characteristic for use as a skylight because it breaks up under impact into large sharp shards which, when they fall, are hazardous to occupants of a building. Annealed glass is permitted to a limited degree where used as sloped glazing and skylights, as discussed under Section 2405.3.

2405.3   Screening. The use of protective screens was briefly discussed in Section 2405.2. As a general rule, single-layer glazing of heat-strengthened glass and fully tempered glass must be provided with screens below the glazing material. To ensure that the screen provides the protection necessary, several requirements are imposed. The screen shall be of a noncombustible material at least No.12 B&S gage (0.0808 inch) with mesh not larger than 1 inch by 1 inch (25 mm x 25 mm). To be located within 4 inches (102 mm) of the glass surface, the screen must be capable of supporting the weight of the glass. In corrosive atmospheres, structurally equivalent noncorrosive screen materials are to be utilized.

It is also critical that the screen is installed in a manner that will adequately support the weight of the glass. In utilizing a safety factor of two, the screen and its fastenings must be capable of supporting twice the weight of the glazing. In order to accomplish this, the screen is to be fastened firmly to the framing members.

Multiple-layer glazing systems are now used quite often for skylights and sloped glazing because of energy-conservation requirements. Where this is the case and where the layer of glazing facing the interior is laminated glass, the code permits the omission of the protective screen below the skylight. However, where heat-strengthened, fully tempered and wired glass is used as the layer facing the interior, screen protection is required below the skylight.

Five exceptions are provided to eliminate the need for protective screens in both monolithic and multilayer sloped-glazing systems. The first exception, shown in Figure 2405-2, permits fully tempered glass in near vertical wall sections, based on the low height plus the very low probability of breakage. Exception 4 allows the use of fully tempered glass as single glazing or as both panes in an insulating glass unit within individual dwelling units in Groups R-2, R-3 and R-4, provided each pane of the glass is limited in size and height above a walking surface.

Annealed glass may be used only (and is permitted without screening) under the following two circumstances:

1.   Where the accessible area below is permanently protected from falling glass.

2.   In greenhouses under the limitations outlined in Exception 3.

As previously discussed, screens are also not required below laminated glass with a 15-mil (0.38 mm) polyvinyl butyral inner layer within dwelling units of specific Group R occupancies. Such glazing is also limited in size and height above the walking surface.

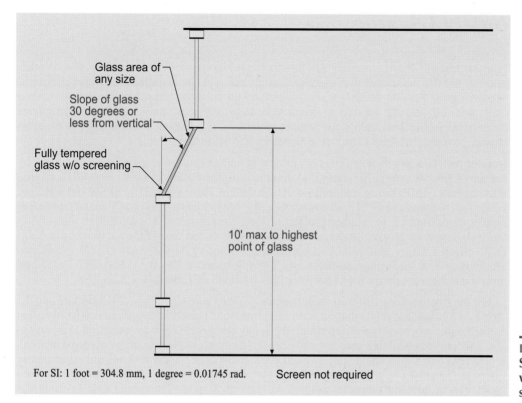

Glass area of any size

Slope of glass 30 degrees or less from vertical

Fully tempered glass w/o screening

10' max to highest point of glass

For SI: 1 foot = 304.8 mm, 1 degree = 0.01745 rad.   Screen not required

Figure 2405-2
Sloped glazing without screen

Framing. This is an omnibus section containing requirements related to: **2405.4**

1. Combustibility of materials, and

2. Leakage protection of the skylight juncture with the roof.

This section logically requires that skylight frames be constructed of noncombustible materials where erected on buildings for which the code requires noncombustible roof construction—that is, Type I and II construction. Where combustible roof construction is permitted by the code, combustible skylight frames are also permitted.

The provision requiring a curb at least 4 inches (102 mm) above the plane of the roof for mounting skylights is intended to provide a means for flashing between the skylight and the roofing to prevent leaks around the margin of the skylight. The 4-inch (102 mm) curb then provides a vertical surface up which the flashing can be extended and to which the counterflashing may be attached to cover the flashing.

Unit skylights. A unit skylight is considered a factory assembled, glazed roof unit   **2405.5** containing a single panel of glazing. Its purpose is simply the introduction of natural daylight through the roof while maintaining the roof assembly's weather resistant barrier. Unlike windows, the most critical uniform load on a skylight (positive or negative) will depend on the climate into which it is installed. In most cold climate areas where heavy snow loads are common, and design wind speeds are moderate, the positive load on the skylight from dead and snow loads is more critical than the negative load from wind uplift. The opposite is true in warm, coastal climates with high design wind speeds and little or no snow load. AAMA/WDMA/CSA 101/I.S.2/A440 *Specifications for Windows, Doors and Unit Skylights* establishes the performance requirements for skylights based on the desired performance grade rating. The minimum performance requirements include resistance to air leakage, water infiltration and the design load pressures prescribed by the *International Codes.*

Section 2406 *Safety Glazing*

2406.1 **Human impact loads.** In areas where it is likely that persons will impact glass or other glazing, the IBC mandates that specific glazing materials be installed. Such specific areas, as identified in Section 2406.4, are considered by the code as "hazardous locations." Unless exempted, all glazing located in hazardous locations must pass the test requirements established in Section 2406.2 for impact resistance. Additional criteria are placed upon plastic glazing, glass block, louvered windows and jalousies.

2406.2 **Impact test.** Glazing installed in areas subject to human impact as specifically identified by Section 2406.4 is generally required to comply with the CPSC 16 CFR, Part 1201, criteria for Category I or II glazing materials as established in Table 2406.2(1). SPSC 16 CFR, Part 1201, *Safety Standard for Architectural Glazing Materials,* is a federally mandated safety regulation of the U.S. Consumer Products Safety Commission. The exception also allows the installation of glazing materials that have been tested to a different standard, ANSI Z97.1 *Safety Glazing Materials Used in Buildings – Safety Performance Specifications and Methods of Test,* in limited applications. Glazing tested under the ANSI Z97.1 standard must meet the test criteria for Class A or B as set forth in Table 2406.2(2).

For the most part, the differences between the CPSC's 16 CFR Part 1201 standard and the ANSI Z97.1 standard relate to their scope and function. The CPSC standard is not only a test method and a procedure for determining the safety performance of architectural glazing, but also a federal standard that mandates where and when safety glazing materials must be used. It preempts any nonidentical state or local code. In contrast, ANSI Z97.1 is only a voluntary safety performance specification and test method. It does not indicate where and when safety glazing materials must be used.

Glazing in compliance with the appropriate test criteria of CPSC 16 CFR Part 1201 may be used in all hazardous locations. It is also acceptable to utilize safety glazing materials complying with ANSI Z97.1, but only to the extent of those applications other than storm doors, combination doors, entrance-exit doors, sliding patio doors and closet doors. See Figure 2406-1. In addition, such glazing is not permitted in doors and enclosures for hot tubs, whirlpools, saunas, steam rooms, bathtubs and showers. In all other areas subject to human impact (hazardous locations) as specified in Section 2406.4, required safety glazing is permitted to comply with the applicable requirements of either CPSC 16 CFR 1201 or ANSI Z97.1.

Glazing materials installed in those locations identified in Items 6 through 11 of Section 2406.4 must comply as either CPSC 16 CFR 1201 or ANSI Z97.1 safety glazing. The minimum category classification for glazing located as described in Items 6 and 7 is set forth both in Tables 2406.2(1) and 2406.2(2). Since there are no requirements in either table for glazing located per Items 8 through 11, it is assumed that Category I, Category II, Category A or Category B glazing can be used. The permitted use of ANSI A97.1 glazing in enclosures regulated by Item 5 of Section 2406.4 is limited to those portions of the building wall that are not considered as a part of the enclosure. Glazing in doors and those components determined to be part of the enclosure must comply with Table 2406.2(1). Proper application of Item 5 in Table 2406.2(2) is primarily based upon the determination of how the building wall is regulated as other than a part of the enclosure.

Table 2406.2(1)—Minimum Category Classification of Glazing Using CPSC 16 CFR 1201. Glazing tested in accordance with CPSC 16 CFR 1201 is classified as either Category I or II, based upon the specifics of the test. The Category II classification is more difficult to achieve and, thus, required in those areas where the highest degree of protection is required. Although Category II material is acceptable in all hazardous locations, Category I glazing may only be installed as relatively small lights in doors and areas adjacent to doors. Otherwise, only Category II glazing is acceptable. See Figure 2406-2.

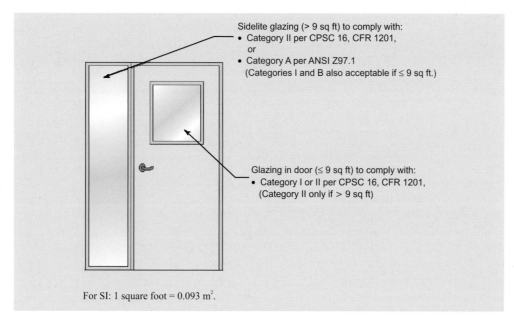

Sidelite glazing (> 9 sq ft) to comply with:
• Category II per CPSC 16, CFR 1201,
  or
• Category A per ANSI Z97.1
  (Categories I and B also acceptable if ≤ 9 sq ft.)

Glazing in door (≤ 9 sq ft) to comply with:
• Category I or II per CPSC 16, CFR 1201,
  (Category II only if > 9 sq ft)

For SI: 1 square foot = 0.093 m².

Figure 2406-1
Safety glazing classification

Table 2406.2(2) Minimum Category Classification of Glazing Using ANSI Z97.1. The minimum required category classification of glazing tested to ANSI Z97.1, established in Table 2406.2(2), is also based upon the glazing location and size of the lite. This approach is similar to that for determining the minimum required category classification of glazing tested to CPSC 16 CFR 1201 as set forth in Table 2406.2(1). ANSI Z97.1 addresses three separate impact categories or classes, based upon impact performance. ANSI Z97.1 Class A glazing materials are comparable to the CSCS Category II glazing materials and ANSI Z97.1 Class B glazing materials are comparable to the CPSC Category I glazing materials. Although there is also a Class C category recognized by ANSI Z97.1, applicable only for fire-resistant glazing materials, it is not viewed by the code as an acceptable safety glazing material. Only Class A and B ANSI Z97.1 glazing materials are recognized in the table. See Figure 2406-2.

Identification of safety glazing. The code requires the identification of safety glazing for the same reasons as for ordinary annealed glass not subject to human impact. However, in the case of safety glazing, the requirement carries more detail insofar as improperly installed annealed glass in areas of human impact can create a serious hazard. Therefore, it is doubly important that the glazing material be further identified to ensure that the proper glazing material is in place. Not only does proper marking assist in the inspection process, it could also help identify a location where safety glazing must be installed should future replacement be required. The code specifically requires the use of identification marks for glazing installed in hazardous locations.

2406.3

Each pane of safety glazing must be individually identified with a manufacturer's designation. It shall specify who applied the designation, the manufacturer or the installer, and the safety glazing standard with which the glazing complies, as well as the type and thickness of the glass or glazing material. Acceptable methods for the designation include acid-etching, sand blasting, ceramic firing or embossing. Additionally, the designation is permitted to be of a type that cannot be removed without being destroyed. This limitation ensures that a designation cannot be transferred to any other glazing materials. As another option, the building official may be willing to accept a certification or affidavit from the supplier and/or installer indicating that the appropriate glazing was provided and installed. Under such a situation, it is critical that the building official be completely satisfied that the appropriate material is utilized for the installation location. The acceptance of an affidavit or similar document is appropriate for all safety glazing materials other than tempered glass.

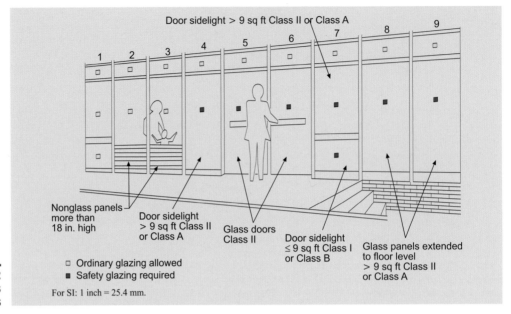

Door sidelight > 9 sq ft Class II or Class A

Nonglass panels more than 18 in. high

Door sidelight > 9 sq ft Class II or Class A

Glass doors Class II

Door sidelight ≤ 9 sq ft Class I or Class B

Glass panels extended to floor level > 9 sq ft Class II or Class A

□ Ordinary glazing allowed
■ Safety glazing required

For SI: 1 inch = 25.4 mm.

Figure 2406-2
Hazardous locations

The only variation to the general requirement for individual identification is for multipane glazed assemblies such as French doors. Where the individual panes do not exceed 1 square foot (0.0929 m^2) in exposed area, only one pane in the assembly is required to be marked as described above. However, all other panes in the assembly must still be identified with "CPSC 16 CFR, Part 1201." See Figure 2406-3. Where the multiple panes occur in other than a door or an enclosure regulated by Item 5 of Section 2406.4, glazing tested and labeled as "ANSI Z97.1" is also acceptable.

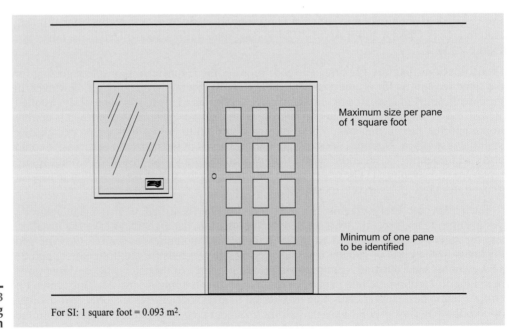

Maximum size per pane of 1 square foot

Minimum of one pane to be identified

For SI: 1 square foot = 0.093 m².

Figure 2406-3
Safety glazing identification

Hazardous locations. This section lists 11 specific hazardous locations where safety **2406.4** glazing is required. Some of these locations are shown in Figures 2406-5 through 2406-15. In addition to the hazardous locations shown in the various illustrations, safety glazing is also required for a number of other conditions, including fixed and sliding panels of sliding door assemblies, storm doors and glass railings.

Figure 2406-2 also illustrates several locations where safety glazing may or may not be required. To facilitate discussion, each panel has been numbered. Panels 1, 2, 3, 8 and 9 are addressed under Item 7 of Section 2406.4 Under this item, all four stated conditions must occur before safety glazing is required. These conditions are as follows:

1. The area of an individual pane must be greater than 9 square feet (0.84 m^2);

2. The bottom edge must be less than 18 inches (457 mm) above the floor;

3. The top edge must be more than 36 inches (914 mm) above the floor; and

4. One or more walking surfaces must be within 36 inches (914 mm), measured horizontally, of the glazed panel.

Panel 1 is not required to have safety glazing, because a protective bar has been installed in compliance with Exception 1 to Item 7, the requirements of which are illustrated in Figure 2406-4. Panels 2 and 3 do not require safety glazing, because their bottom edges are not less than 18 inches (457 mm) from the floor. If Panels 8 and 9 have a walking surface within 36 inches (914 mm) of the interior, safety glazing would be required. This would be true even though the bottom of the panel appears to be greater than 18 inches (457 mm) above the exterior walking surface, as the exterior condition would have no bearing on the determination. However, the exterior condition, because it is adjacent to a stairway, would be regulated by Item 10 and/or Item 11 for glazing adjacent to stairways. Therefore, multiple conditions mandate the need for safety glazing in Panels 8 and 9. Panels 4 and 7 require safety glazing because they are door sidelights. The exception mentioned above does not apply to panels adjacent to a door; therefore, though Panel 7 has been provided with a protective bar, safety glazing is still required.

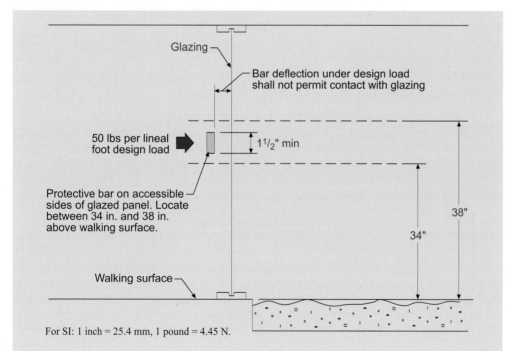

For SI: 1 inch = 25.4 mm, 1 pound = 4.45 N.

Figure 2406-4
Protective bar alternative

Figures 2406-5 and 2406-6 illustrate when safety glazing is required for panels adjacent to a door. This requirement applies to both fixed and operable panels. Where there is an intervening wall or permanent barrier as shown in Figure 2406-7, safety glazing would not be required. If the door serves only a shallow storage room or closet, adjacent glazing need not be safety glazing as depicted in Figure 2406-8. Figure 2406-9 illustrates an exception applicable only to one- and two-family dwellings and within dwelling units in Group R-2 uses.

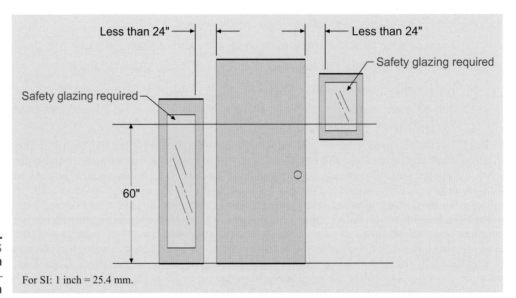

Figure 2406-5
Glass in sidelights— elevation

For SI: 1 inch = 25.4 mm.

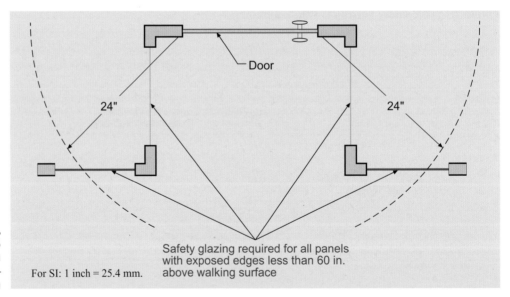

Figure 2406-6
Glass in sidelights— plan

For SI: 1 inch = 25.4 mm.

Safety glazing required for all panels with exposed edges less than 60 in. above walking surface

Panels 5 and 6 are glass doors, which require safety glazing based on the provisions of Item 1 or 4. All ingress and egress doors (except jalousies), unframed swinging doors and glazing in storm doors require safety glazing. See Figure 2406-10. There are several exceptions. If openings in a door will not pass a 3-inch-diameter (76 mm) sphere, the glazing is exempt, as are assemblies of leaded, faceted or carved glass used for decorative purposes. The latter exception not only applies to doors, but also to sidelights and other glazed panels covered by Items 6 and 7.

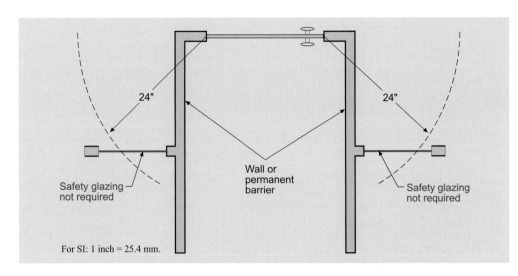

For SI: 1 inch = 25.4 mm.

Figure 2406-7
Barrier between glazing and door

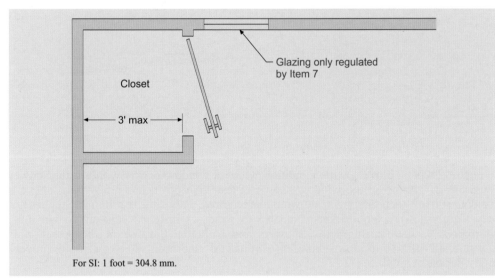

For SI: 1 foot = 304.8 mm.

Figure 2406-8
Glazing adjacent to closet door

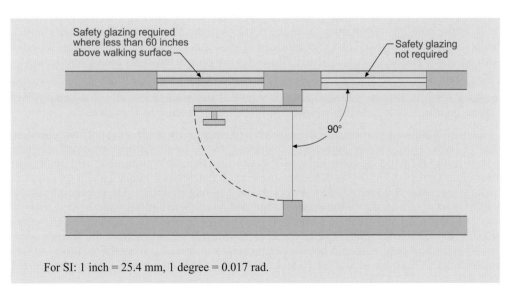

For SI: 1 inch = 25.4 mm, 1 degree = 0.017 rad.

Figure 2406-9
Applicable to dwelling units of Groups R-2 and R-3

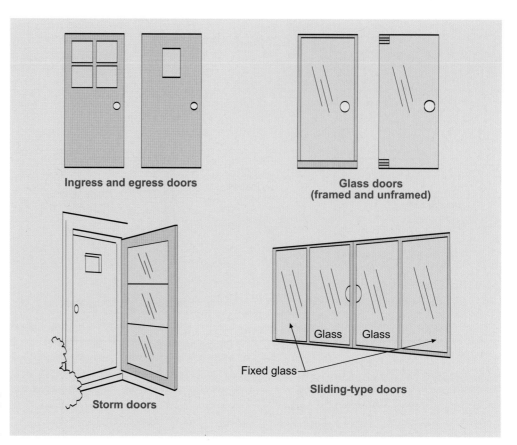

Ingress and egress doors

Glass doors
(framed and unframed)

Storm doors

Fixed glass

Glass   Glass

Sliding-type doors

Figure 2406-10
Glazing in
doors

Figure 2406-11 illustrates the condition where a window occurs within a tub/shower enclosure. If glazing in the window shown is less than 60 inches (1524 mm) above a standing surface, then safety glazing would be required. This same requirement applies not only to tub/shower combinations, but also to windows installed adjacent to hot tubs, whirlpools, saunas, steam rooms, showers and bathtubs. Because of the presence of moisture, all of these locations represent slip hazards and need safety glazing to prevent injury in case of a fall.

The language of Item 5 addresses two conditions. First, any glazing in the door, as well as the enclosure materials, of the referenced elements must be safety glazing. Second, where walls of the building construction surround portions of the tub, shower, etc., glazing in such walls located at a height of less than 60 inches (1524 mm) above the standing surface must also be safety glazing. Figure 2406-12 illustrates the proper locations for safety glazing.

Glass in railings, balusters panels and in-fill panels, regardless of height above the walking surface, require safety glazing. Because of the high probability that persons will impact guards, it is critical that an increased level of protection be provided.

Item 9 (regulating glazing in walls surrounding swimming pools and spas) and Items 10 and 11 (addressing glazing adjacent to stairways, ramps and landings) are relatively new provisions to the building code. Figure 2406-13 illustrates the requirements of Item 9. This provision applies to walls and fences used as a barrier for either indoor or outdoor swimming pools and spas. Before safety glazing is required, the glazed panels must be within 5 feet (1524 mm) of the water's edge and have their bottom edge be less than 60 inches (1524 mm) above the decking for a pool or spa. In reviewing the criteria for safety glazing, the horizontal *safety zone* of 5 feet (1524 mm) is measured from the water's edge. As a standard of care, it would be more appropriate to measure from the walking surface (pool deck). A distance of at least 10 feet (3048 mm) from the water's edge to the glazing would more appropriately provide the 60-inch (1524 mm) safety zone the code is anticipating.

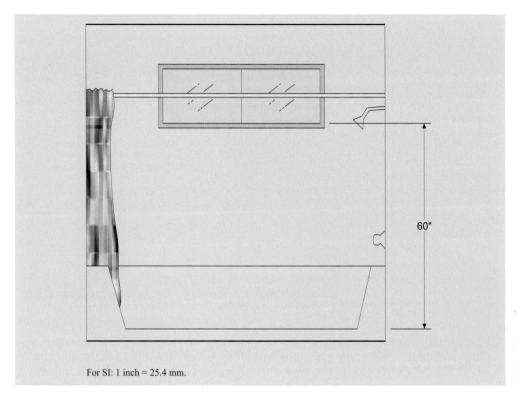

For SI: 1 inch = 25.4 mm.

Figure 2406-11
Glazing within a shower enclosure

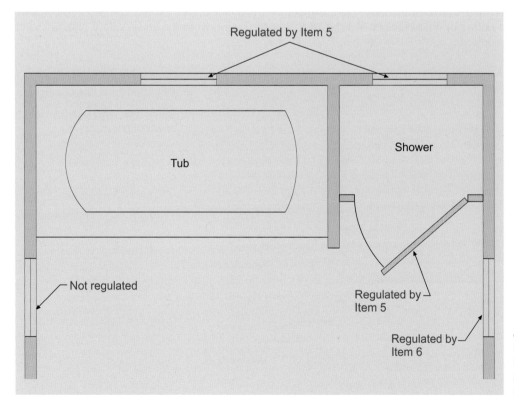

Regulated by Item 5

Tub

Shower

Not regulated

Regulated by Item 5

Regulated by Item 6

Figure 2406-12
Glazing in tub and shower enclosures

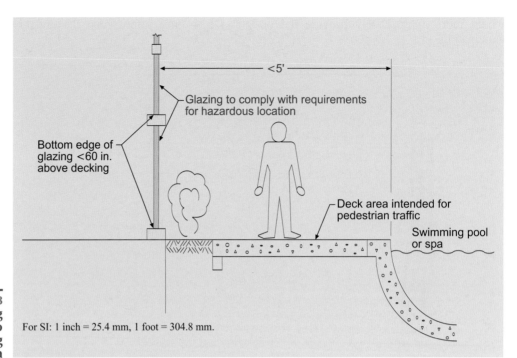

**Figure 2406-13
Glazing
adjacent to
swimming
pool or spa**

For SI: 1 inch = 25.4 mm, 1 foot = 304.8 mm.

Item 10 requires the installation of safety glazing for glazing adjacent to stairways, landings and ramps. Where the glazing is within 36 inches (914 mm) horizontally and 60 inches (1524 mm) vertically of the adjacent walking surface, safety glazing is mandated. See Figure 2406-14. In addition, Item 11 identifies a hazardous location to be within 5 feet (1524 mm) of the bottom tread of a stairway when the bottom edge of the glazing is less than 60 inches (1524 mm) above the nosing of the tread. See Figure 2406-15. Where protected by a railing or guard located at least 18 inches (457 mm) from the glass, safety glazing is not required.

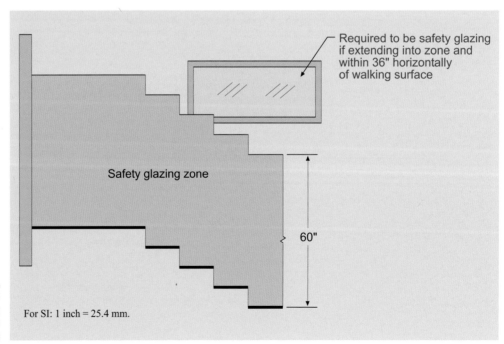

**Figure 2406-14
Glazing
adjacent to
stairways**

For SI: 1 inch = 25.4 mm.

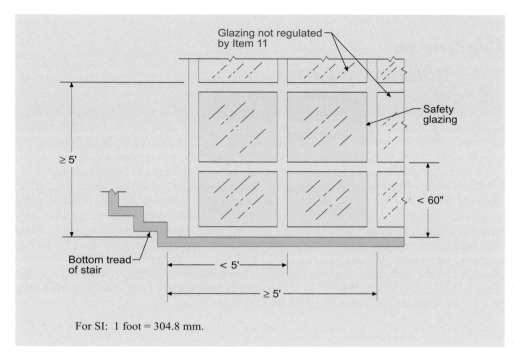

Glazing not regulated by Item 11

Safety glazing

≥ 5'

< 60"

Bottom tread of stair

< 5'

≥ 5'

For SI: 1 foot = 304.8 mm.

Figure 2406-15
Glazing adjacent to stairways

Fire department access panels. There may be conditions under which glass panels will be utilized for fire department access purposes. Certainly, such panels shall be of the type that will provide safe conditions under which fire department personnel can enter a building or a specific area of a building. For this reason, glass access panels must be of tempered glass. Where the glazed unit is an insulating panel, all panes shall be tempered. Because access must be provided completely through the panel, it is necessary that all panes be of this specific safety glazing material. **2406.5**

Section 2407 *Glass in Handrails and Guards*

The increased use of glass (usually tempered) in handrail assemblies and guard sections prompted the inclusion of these provisions in the IBC. These provisions provide uniform regulations identifying the specific types of safety glazing that may be used structurally. Fully tempered, laminated tempered, or laminated heat-strengthened glass are the only types considered by the code to be structurally adequate for this use so critical to life safety.

Only glazing conforming to the provisions of Section 2406.1.1 is permitted. This would limit glazed railing in-fill panels to materials that have passed the test requirements of CPSC 16 CFR 1201 or ANSI Z97.1. Regardless of the type of glazing, the minimum nominal thickness of the structural balustrade panels in railings shall be $^1/_4$ inch (6.4 mm). The panels and their supporting system must be designed to withstand the loads as specified in Section 1607.7, utilizing a safety factor of four. Not permitted to be installed without an attached handrail or guardrail, at least three glass balusters shall be used to support each handrail or guard section. The purpose for requiring at least three balusters is that should one fail, the remaining two balusters will continue to support the handrail or guard section. If another method is devised to provide continued support should a single baluster panel fail, such a method is acceptable.

Section 2408 *Glazing in Athletic Facilities*

Where glazing forms entire or partial wall sections, or is used as a door or as part of a door, in racquetball courts, squash courts, gymnasiums, basketball courts and similar athletic facilities subject to impact loading, it shall comply with Section 2408. In racquetball and squash courts, glass walls and glass doors must pass specific test criteria above and beyond those typically required of safety glazing materials. Such special test criteria are necessary to address those glazed areas where impact with the glass is not merely accidental, but rather expected because of the nature of the physical activities involved. Special conditions for compliance are also set forth for glazing subject to human impact in gymnasiums, basketball courts and other high-intensity activity areas where it is expected that contact with the glazing will occur more often, and with more force, than in most other hazardous locations addressed by the code. In such facilities, all glazing in hazardous locations identified by the code must meet the Category II requirement of CPSC 16 CFR 1201 or the Class A requirements of ANSI Z97.1. This would include glazing both in doors (only CPSC 16 CFR 1201 glazing permitted) and adjacent to doors.

KEY POINTS

- Glass and glazing must resist lateral loads in a manner consistent with other building components.

- Skylights and other sloped glazing are regulated as to the type of glazing material and the need for protective screening below the skylight.

- Safety glazing materials are to be installed in those areas subject to human impact, referred to as hazardous locations.

- The minimum required classification category of safety glazing materials is based on the size and location of the glazing material.

- Common safety glazing materials include tempered or laminated glass, as well as approved plastic.

- In order to verify compliance, the code specifically requires the use of identification marks for glazing installed in hazardous locations.

- Some of the most common locations identified as hazardous include those glazed areas in and around doors.

- Glazing adjacent to showers and bathtubs, where located in a position where impact is likely, must be safety glazed.

- Glass in handrails and guards is regulated for both structural adequacy and human impact.

- Special requirements are mandated for glazing subject to human impact in athletic facilities.

CHAPTER

25

GYPSUM BOARD AND PLASTER

> This chapter regulates the covering materials for walls and ceilings:
>
> 1. To provide weather protection for the exterior of the building.
>
> 2. To secure the material to the wall and ceiling framing so that it will remain in place during the expected life of the building.
>
> Where these materials are used or required for fire-resistance-rated construction, the code requires that they also comply with the provisions of Chapter 7.

Section 2501 *Scope*

This chapter of the *International Building Code*® (IBC®) covers the installation requirements for wall- and ceiling-covering materials, including their method of fastening and, in the case of plaster, the permitted materials for lath, plaster and aggregate.

Although plaster has many uses, including ornamental and decorative work, its use in the IBC is regulated purely as a wall- and ceiling-covering material.

The IBC regulates the installation of wall- and ceiling-covering materials as well as quality standards for the materials themselves. The primary wall- and ceiling-covering material in use today is gypsum wallboard; however, lath, plaster and wood paneling are sometimes utilized. As wood paneling is covered in Chapter 23, it follows that Chapter 25 only regulates gypsum wallboard, lath and plaster. However, in this section the code permits the installation of other wall- and ceiling-covering materials, provided the materials have been approved. On this basis, the manufacturer's recommendations and conditions of approval should be consulted.

Gypsum wallboard is a relatively new material for covering walls and ceilings. On the other hand, plaster is among the oldest of building materials still in use. The use of gypsum plaster dates back to about 4000 B.C., when the Egyptians applied it to the interior and exterior of the pyramids.

Section 2502 *Definitions*

The determination of whether or not a surface is considered interior or exterior is based on how it is viewed in relationship to the definition for weather-exposed surfaces. Any surface that can be considered weather-exposed under the definition in Section 2502 is considered an exterior surface. Surfaces other than weather-exposed surfaces are viewed as interior surfaces.

Surfaces of walls, ceilings, floors, roofs, soffits and similar elements, where exposed to the weather, are typically considered weather-exposed surfaces. There are three exceptions to the general criteria that would define such surfaces as interior where applying the code provisions. Those exterior conditions considered other than weather-exposed surfaces are illustrated in Figure 2502-1.

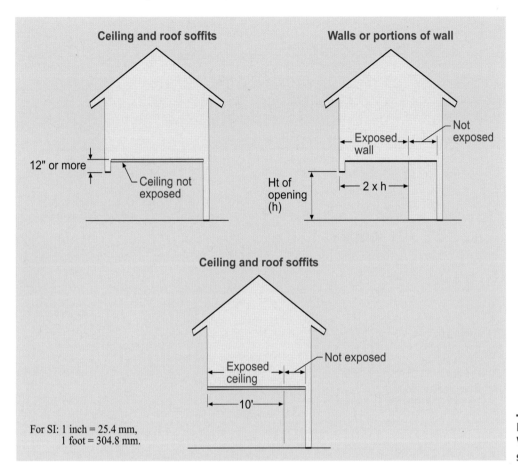

For SI: 1 inch = 25.4 mm,
1 foot = 304.8 mm.

**Figure 2502-1
Weather-exposed
surfaces**

Section 2504 *Vertical and Horizontal Assemblies*

The minimum dimensions required by this section for supports of lath or gypsum board are intended to prevent fastener failures and surface distortion that are due to warping of the wood supports. Where supporting lath or gypsum board, wood supports, stripping or furring is to be at least 2 inches (51 mm) nominal thickness in the least dimension. An exception permits a 1-inch by 2-inch (25 mm by 51 mm) wood furring strip to be installed over a solid backing.

Figure 2504-1 shows a solid plaster studless partition, permitted where constructed in compliance with the conditions set forth in Section 2504.1.2. As its name implies, it is a partition constructed solidly of plaster and lath without framing studs. Thus, the lath and plaster form the structural elements of the partition to resist lateral loads. These partitions are used as nonbearing partitions, but they must meet the horizontal load requirements specified in Section 1607.13.

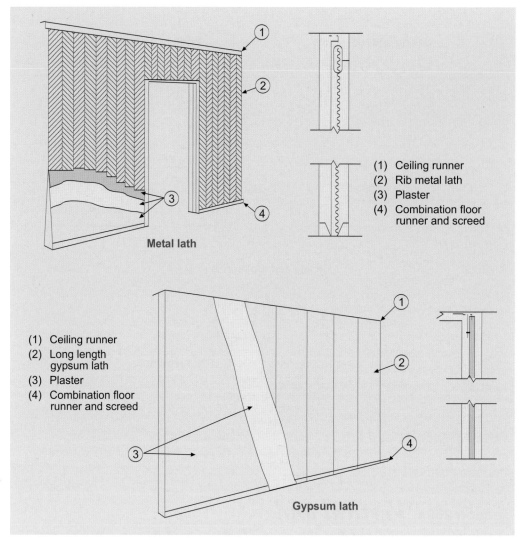

Metal lath

(1) Ceiling runner
(2) Rib metal lath
(3) Plaster
(4) Combination floor runner and screed

(1) Ceiling runner
(2) Long length gypsum lath
(3) Plaster
(4) Combination floor runner and screed

Gypsum lath

Figure 2504-1
Solid plaster studless partition

Section 2506 *Gypsum Board Materials*

To indicate compliance with the appropriate materials standards, gypsum board materials and related accessories shall be identified by the manufacturer's designation. Such standards, including those for gypsum sheathing, water-resistant gypsum backing board, nonload-bearing steel studs, steel screws and nails for gypsum board are referenced in Table 2506.2. It is also noted that any gypsum board materials utilized for fire-protection purposes must conform to the provisions found in Chapter 7.

Where acoustical and lay-in panel ceilings are provided, the metal-suspension systems for such ceilings are to conform with ASTM C 635 and Section 13.5.6 of ASCE 7 for installation in high seismic areas.

Section 2508 *Gypsum Construction*

This section addresses the installation of gypsum materials, primarily that of gypsum wallboard installed as wall and ceiling membranes. As gypsum board is a construction material utilized in almost every construction project, it is important that the application is in compliance with the IBC and the appropriate referenced standards.

General. The primary installation criteria for various gypsum materials are found in the referenced standards identified in Table 2508.1. The application of gypsum board varies based on gypsum board thickness, wall or ceiling installation, orientation of gypsum board to the framing members and maximum spacing of the framing members. Based on the specific conditions encountered, the maximum fastener spacing and size of fasteners is identified.   **2508.1**

Limitations. Because gypsum plaster and gypsum board are subject to deterioration from moisture, the code restricts their use to interior locations and weather-protected surfaces. The definition for weather-exposed surfaces is found in Section 2502 and illustrated in Figure 2502-1. Even where installed in interior locations, it is important that gypsum wallboard not be installed in areas of continuous high humidity or wet locations. The IBC further requires that interior gypsum board, gypsum plaster and gypsum lath shall not be installed until the installation has been weather-protected.   **2508.2**

Single-ply application. The application of gypsum wallboard is specified in this chapter for locations where fire-resistance-rated construction is not provided or for construction where diaphragm (shear wall) action is not required. Chapter 7 and fire-test reports will establish the means of fastening and supporting the ends and edges of gypsum wallboard for fire-resistance-rated assemblies. Table 2508.1 provides the installation standards for various types of gypsum construction.   **2508.3**

The code requires that the fit of gypsum wallboard sheets be such that the edges and ends are in moderate contact. However, wider gaps are permitted in concealed spaces where fire-resistance-rated construction or diaphragm action is not required. This requirement is based primarily on appearance. Therefore, where the wallboard application is concealed, it is not objectionable to have wider gaps than those resulting from moderate contact. However, where the wallboard surface is exposed as it normally is, moderate contact is required so that there will be no objectionable cracking when the joint between the sheets is finished.

Unless the wallboard is considered a shear-resisting element or an element of a fire-resistance-rated assembly, fasteners may be omitted at certain locations. It should be emphasized that where a fire-test report or other installation standard indicates that fasteners are required on supports or edges, the fastening pattern may not be modified. Otherwise, those fasteners located at the top and bottom plates of vertical assemblies are permitted to be omitted. In addition, fasteners need not be provided at the edges and ends of horizontal assemblies perpendicular to supports and at the wall line. Note that fasteners are to be applied in a manner in which the face paper is not fractured by the fastener head. The intent of the requirement is to provide a tight fastening but not damage the gypsum board to the extent its nail-holding power may be affected. Proper construction procedure for the nailing of gypsum wallboard panels is to use a drywall hammer that has a crowned head and use wallboard nails that have concave heads. Figure 2508-1 illustrates the case where a drywall hammer is used and the drywall nail is not overdriven. The intent is to create a dimple in the plaster board with no projection of the nail head above the plaster board.

Joint treatment. Although as a general rule the IBC requires joint and fastener treatment for fire-resistance-rated assemblies, the code exempts in Exception 1 those locations where the wallboard is to receive a decorative finish or any other similar application, which is considered to be equivalent to the joint treatment. Also, joint treatment is not required where joints occur over wood framing members, or where square-edge or tongue-and-groove edge   **2508.4**

gypsum board is used. In addition, Exception 5 indicates that joint treatment is not required for assemblies tested without joint treatment. In general, joint treatment does not materially increase the fire rating, and many partitions have passed the fire test without joint treatment. As indicated earlier in this section, joint treatment is primarily used for aesthetic reasons. One further exception addresses the condition where a multilayer system is constructed. Where two or more layers of gypsum board are utilized in the assembly, joint and fastener treatment is not required where the joints of adjacent layers are offset from each other.

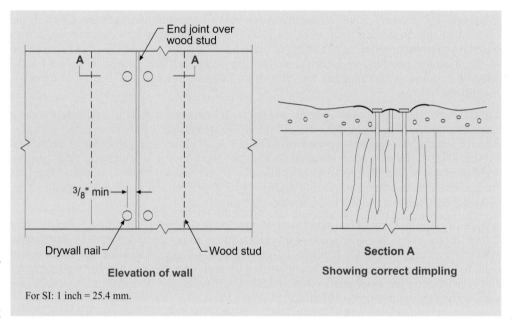

Figure 2508-1
Gypsum
wallboard
nailing

Elevation of wall

Section A
Showing correct dimpling

For SI: 1 inch = 25.4 mm.

Section 2509 *Gypsum Board in*
Showers and Water Closets

Special consideration is given to toilet and bathing areas subject to some degree of exposure to water or high humidity. In shower areas, it is mandated that complying glass mat water-resistant gypsum backing panels, discrete nonasbestos fiber-cement interior substrate sheets or nonasbestos fiber-mat reinforced cement substrate sheets be used as a base for wall tile and wall panels, as well as a base for ceiling panels. In tub areas, such backers are required as a base for wall tile. Where wall tile is installed over a gypsum board base on water closet compartment walls, it is mandated that the tile be installed over a base of water-resistant gypsum backing board. The use of glass mat water-resistant gypsum backing panels, discrete nonasbestos fiber-cement interior substrate sheets or nonasbestos fiber-mat reinforced cement substrate sheets is also permitted in such locations.

Two locations are identified in Section 2509.3 as areas where water-resistant gypsum backing board is not ever to be used. In shower or bathtub compartments, water-resistant gypsum backing board shall not be installed over a vapor retarder. In addition, such materials are prohibited in locations subject to direct water exposure or continuous high humidity, such as in saunas, steam rooms, gang showers or indoor pools. An additional provision permits the use of water-resistant gypsum backers on ceilings, but only under limited conditions. Because of the potential for sagging, the backer boards must be supported at close intervals. One-half-inch thick (12.7 mm) water-resistant gypsum backing board is prohibited on ceilings where the framing members are spaced in excess of 12 inches

(305 mm) on center. Such gypsum board, when $^5/_8$-inch (15.9 mm) in thickness, is permitted only where ceiling members are spaced a maximum of 16 inches (406 mm) on center.

Except for tub, showers and water closet compartment walls and ceilings of shower areas, regular gypsum board may be used as a backing for tile or wall panels. An example would be a backsplash at the rear of a kitchen counter. Even if this area is to be covered with wall panels or tile, regular gypsum board is permitted as the backing material.

Section 2511 *Interior Plaster*

Multicoat plastering has been the standard in the western world for over 100 years. It is generally the consensus of the industry that, particularly where plaster application is by hand, multicoat work is necessary for control of plaster thickness and density. Most of the materials used for plaster densify under hand application because of the pressure applied to the trowels, and it is believed that this change in density is more controllable and will be of a more uniform nature where the plaster is applied in thin, successive layers. For these reasons, the IBC requires three-coat plastering over metal or wire lath and two-coat work applied over other plaster bases approved for use by the code. Reducing the requirement for plaster bases other than metal or wire lath is based on the rigidity of the plaster base itself. More rigid plaster bases are not as susceptible to variations in thickness and flatness of the surface. In fact, it may be considered that the first coat applied in three-coat work on a flexible base, such as wire lath, is used to stiffen that base to provide the rigidity necessary to attain uniform thickness and surface flatness.

Fiber insulation board does not have the qualities necessary for a good performing plaster base. It absorbs excessive moisture from the plaster mix, creating problems of workmanship, and it does not have the stability and rigidity required for a proper functioning plaster base. Also, in the colder and damper climates, the fiberboard insulation retains the moisture absorbed from the plaster for a relatively longer period of time than other bases, causing premature failure of the plaster. For these reasons, the code prohibits its use as a plaster base.

Because portland cement plaster does not bond properly to gypsum plaster bases, the code prohibits its use over gypsum plaster bases. However, the code permits exterior plaster to be applied on horizontal surfaces, soffits, etc., over gypsum lath and gypsum board when used as a backing for metal lath.

Plaster grounds are utilized to establish the thickness of plaster and usually are wood or metal strips attached to the plaster base. The intent is that plaster grounds are used as a guide for the straightedge in determining the thickness. In many cases, door and window frames are used as plaster grounds.

In plaster work, a base coat is any coat beneath the finish coat. This is true whether the plaster is of two-coat or three-coat application. In three-coat work, the first coat is usually referred to in the trade as the *scratch* coat. It is usually applied over flexible bases, such as metal or wire fabric lath, and is intended to stiffen the base and provide a mechanical bond to the base. Also, as its name implies, the first coat is scratched with a scarifying tool, which provides horizontal ridges or scratches that are intended to provide mechanical keys for the application of the second coat (or brown coat). The brown coat usually constitutes the major bulk of the plaster and, consequently, materially affects the membrane strength. As a result, proportioning and workability are critical, and the mix should have high plasticity for proper application. The term *brown coat* is utilized by the trade to differentiate the relative color of the second coat to the finish coat, which is usually much lighter in color and is sometimes white, depending on the constituents.

The base coats in plaster work provide the strength for the plaster membrane but generally do not provide a proper surface texture for a finished surface. Therefore, a thin, almost veneer, coat of plaster is applied to the base coats as a finish coat. The finish coat may be applied in such a manner as to provide an ornamental or decorative finish, or it may be applied as a smooth surface to act as a flat base over which paint and wallpaper may be applied.

Section 2512 *Exterior Plaster*

2512.1 **General.** Portland cement plaster is the only material approved by the code for exterior plaster. Gypsum plaster deteriorates under conditions of weather and moisture, which are prevalent on the exterior surfaces of buildings. For this reason, Section 2512.3 states that gypsum plaster cannot be used on exterior surfaces. Exterior portland cement plaster is required by the code to be applied in not less than three coats when applied over metal lath, wire-fabric lath or gypsum board backing for the same reasons as discussed for interior plaster. When the portland cement plaster is applied over other approved plaster bases, the code requires only two-coat work. The code permits plaster work that is completely concealed to be of only two coats, provided the total thickness is that required by ASTM C 926, insofar as the finish code of plaster is to provide a surface for exterior finishes (such as paint) and to provide an aesthetic appearance. Thus, where the plaster surface is to be completely concealed, it is not necessary to provide a finish coat.

The code requires that the exterior plaster be installed to completely cover, but not extend below, the lath and paper on wood or metal-studded exterior wall construction supported by a nongrade concrete floor slab. This requirement, combined with the requirement in Section 2512.1.2 for a weep screed, is intended to prevent the entrapment of free moisture and the subsequent channeling of the moisture to the interior of the building. This requirement is depicted in Figure 2512-1.

2512.1.2 **Weep screeds.** Water can penetrate exterior plaster walls for a variety of reasons. Once it penetrates the plaster, the water will run down the exterior face of the water-resistive barrier until it reaches the sill plate or mudsill. At this point, the water will seek exit from the wall, and if the exterior plaster is not applied to allow the water to escape, it will exit through the inside of the wall and leak into the building. Thus, the IBC requires a weep screed that, when constructed as shown in Figure 2512-2, will permit the escape of the water to the exterior of the building. In addition, where weep screeds are not provided for plaster exterior walls constructed in cold-climate areas, it is possible that the trapped moisture will freeze and cause a premature failure of the exterior plaster. The water-resistive barrier required by the code must lap the weep screed's vertical flange. Although this section does not specify the amount of overlap, at least 2 inches (51 mm) should be adequate in keeping with the typical weather-resistive barrier lap requirements.

2512.2 **Plasticity agents.** Admixtures such as plasticizers should not be added to portland cement or blended cement unless approved by the building official. Some admixtures can have deleterious effects that more than offset the desired improvement in plasticity. It is preferable that plasticizers be added during the manufacture of the cement in order to ensure product uniformity and proper proportions. When plastic cement is used, the code does not permit any further additions of plasticizers as it is assumed that the amount added during the manufacturing process is adequate and is the maximum permitted. Hydrated lime and lime putty are time-tested plasticizers used with portland cement plaster, and their use is permitted by the code in the amounts set forth in ASTM C 926.

2512.4 **Cement plaster.** Portland cement plaster is affected by freezing in the same manner as portland cement mortar or portland cement concrete. When portland cement plaster is applied during freezing weather, it loses a high proportion of its strength and, therefore, does not meet the intent of the code. In addition to protecting the plaster coats from freezing for at least 24 hours after set has occurred, application of the plaster should only be done when the ambient temperature is higher than 40°F (4°C). Plaster may be applied in colder temperatures where provisions are made to keep the cement plaster work above 40°F (4°C) during application and for at least 48 hours thereafter.

It is also important that the plaster not be applied to frozen bases or those covered with frost, which will not only weaken the bond of the plaster to its base but will also freeze the layer of plaster adjacent to the frozen base. In those cases where portland cement plaster is mixed with frozen ingredients or applied to a frozen base, it loses a high percentage of its strength.

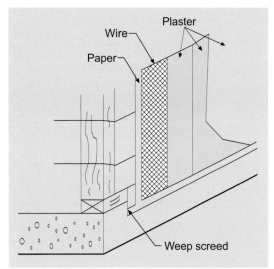

Figure 2512-1
Termination of exterior plaster at on-grade concrete floor for stud walls

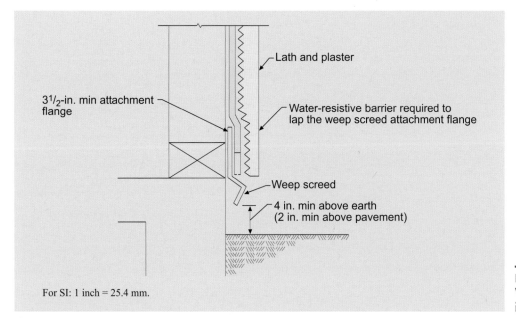

For SI: 1 inch = 25.4 mm.

Figure 2512-2
Weep screed installation

KEY POINTS

- Provisions for wall and ceiling coverings are expanded when used on weather-exposed surfaces.

- Because gypsum plaster, gypsum lath and gypsum board are subject to deterioration from moisture, the code limits their use to interior locations and weather-protected surfaces.

- When used as a base for tile or wall panels, or shower compartment walls, the code intends that glass mat water-resistant gypsum backing panels, discrete nonasbestos fiber-cement interior substrate sheets or nonasbestos fiber-mat reinforced cement substrate sheets be used.

- For exterior plaster walls, the IBC requires a weep screed that will permit the escape of water to the exterior of the building.

PLASTIC

This chapter covers several topics, all related to the use and installation of various types of plastic materials. Included are foam plastic insulation, light-transmitting plastics, plastic veneer and interior plastic trim.

Section 2602 *Definitions*

Several terms are defined in Chapter 26 in order to identify the materials to be regulated.

FOAM PLASTIC INSULATION. Considered to be an expanded plastic produced through use of a foaming agent, this reduced-density plastic contains voids consisting of open or closed cells distributed throughout the plastic, providing for thermal insulation or acoustic control. The density of the material is to be less than 20 pounds per cubic foot (320 kg/m^3).

LIGHT-TRANSMITTING PLASTIC WALL AND ROOF PANELS. Plastic materials that are fastened to structural members, or to structural panels or sheathing, and that are used to transmit light at the exterior walls or the roof.

Section 2603 *Foam Plastic Insulation*

During the early 1970s, the Federal Trade Commission (FTC) investigated claims made by some manufacturers in the plastics industry of "slow-burning" or "nonburning" as related to foam-plastic insulation materials. With assistance from the former National Bureau of Standards, now called the National Institute of Standards and Technology, the FTC concluded that these claims were erroneous because of improper testing. Because of the earlier criticisms aimed at the claims, the code changes that were finally adopted into the codes were, of necessity, somewhat conservative.

The provisions were developed by the Society of the Plastics Industry, Inc. (SPI), after numerous meetings, hearings and seminars relating to the hazardous characteristics of the materials. During this time, SPI funded an extensive program of research that reviewed the then-current test procedures, with a goal of establishing new test procedures where necessary to properly reflect the hazards of the material as it would actually be used in buildings.

The code provisions developed as a result of the extensive research were centered on two basic concepts:

1. An index limitation of the flame spread and smoke developed to 75 and 450, respectively.

2. Separation of the foam plastic insulation from the interior of the building by an approved thermal barrier. The adequacy of the thermal barrier is related to the time during which the thermal barrier is expected to remain in place under fire conditions.

2603.2 Labeling and identification. In addition to the flame-spread and smoke-developed criteria, the code also requires that the containers of foam plastic and foam plastic ingredients be labeled by an approved agency to show that the material is compliant. There are many foam plastic products on the market that do not comply with the code and that were not intended for use in construction. The labeling requirement is intended to prevent the misapplication of products not designed for this use.

Surface-burning characteristics. It is important that any foam plastic insulation or foam **2603.3** plastic core material found in manufactured assemblies be limited in flame spread and smoke development. In this section, the code limits such foam plastic materials to a flame-spread index of 75 and a smoke-developed index of 450 where tested at the maximum intended thickness of use. Various exceptions are provided for interior trim, cold-storage buildings and similar facilities, interior signs in covered mall buildings, listed roof assemblies and special approvals.

Thermal barrier. Because of the potential hazards involved, foam plastic must typically be **2603.4** separated from the interior of a building by an approved thermal barrier. Gypsum wallboard at least $^1/_2$ inch (12.7 mm) in thickness satisfies this requirement, as does any equivalent thermal barrier material complying with the criteria of this section. It must be demonstrated by approved testing that the thermal barrier will remain in place for the required 15-minute time period.

When the following conditions are met, the thermal barrier is not required:

Masonry or concrete construction. See Figure 2603-1. When foam plastics are encapsulated within concrete or masonry walls, or floor or roof systems, the code does not require a thermal barrier as long as the foam plastic is covered by a minimum of 1 inch (25 mm) thickness of the masonry or concrete.

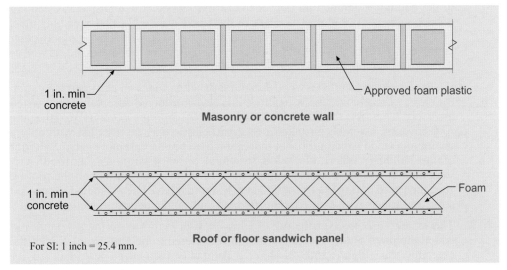

1 in. min concrete

Approved foam plastic

Masonry or concrete wall

1 in. min concrete

Foam

Roof or floor sandwich panel

For SI: 1 inch = 25.4 mm.

Figure 2603-1
Encapsulated foam plastic

Cooler and freezer walls. Cold storage uses provide a unique use for foam plastic insulation in that thicknesses are generally required to be greater than 4 inches (102 mm) for proper thermal insulation, although 4 inches (102 mm) is about the maximum that can be tested. The code, in all other cases, places limits on the thickness of the foam plastic insulation to that which was tested. However, because of the nature of the use in cold-storage facilities, the ignition hazards are not great. Therefore, the code permits foam plastic insulation in greater thicknesses, up to a maximum of 10 inches (25 mm), even though tested in a thickness of 4 inches (102 mm). The intent is that the foam plastic will be provided with a complying protective thermal barrier. In the case of interior rooms within a building, the foam plastic is required to be protected on both sides with a complying thermal barrier.

Provisions are included to permit cooler and freezer walls without a thermal barrier, provided the foam plastic has a flame-spread rating of 25 or less, has minimum allowable flash and self-ignition temperatures of 600°F and 800°F (316°C and 427°C), respectively, and the foam is protected by 0.032-inch-thick (0.8 mm) aluminum or 0.0160-inch-thick (0.4 mm) steel. The cooler or freezer and the portion of the building where the cooler or freezer is located must be sprinklered in this case. Again, the code presumes that with a low-hazard use, such as a cold storage and freezer box, the metal covering will prevent the actual impingement of any flames on the foam plastic, and the sprinkler system will provide the cooling necessary to maintain proper low temperatures to prevent ignition of the foam plastic.

Walk-in coolers. Where freestanding coolers and freezers have an aggregate floor area not exceeding 400 square feet (37 m²), the code contains an exception that, in effect, provides for no thermal barrier and no sprinkler protection as long as the foam plastics comply with the general provisions of this section. The foam plastic must be covered by an aluminum or steel facing of appropriate thickness. If the foam plastic material is over 4 inches (102 mm) in thickness, a complying thermal barrier must enclose the material.

Exterior walls—one-story buildings. For one-story buildings, metal-clad sandwich panels with foam plastic cores with thicknesses up to 4 inches (102 mm) are permitted to be installed without a thermal barrier, provided the metal cladding complies with the provisions outlined in the code and, furthermore, the building is protected with automatic fire sprinklers. In this case, the code assumes that the protection and cooling effect provided by automatic sprinklers is a reasonable alternative to the thermal barrier.

Roofing. This item covers two different cases involving roof coverings or roof assemblies:

1. The first case involves nonclassified roof assemblies or roof coverings. As there are generally no test standards for these prescriptive assemblies, the code provides that they may be applied over foam plastic when the foam is separated from the interior of the building by minimum 0.47-inch (12 mm) wood structural-panel sheathing bonded with exterior glue. The edges of the wood structural panel sheathing must be supported by blocking or be of tongue-and-groove construction or of any other approved type of edge support. In this case, the thermal barrier is waived, as well as the smoke-developed index.

    Based on the fact that a wood structural panel provides an adequate separation for ordinary roof-covering assemblies, it is also considered acceptable for a tested assembly. Thus, it is the intent of the *International Building Code*® (IBC®) that any roof covering assembly installed over foam plastic may be installed with only a complying wood structural-panel separation between the assembly and the interior of the building. Where the wood structural panel separation is utilized, it is important to recognize that the joints must be protected even though the roofing specimen used during the fire-retardancy test might have been installed over wood structural panels with abutted joints without any supplemental protection.

2. The second case involves the use of Class A, B or C roof covering assemblies in which the foam plastic insulation is also considered to be an integral part of the assembly. See Figure 2603-2. Here, a nationally recognized test standard for insulated roof decks is to be utilized. The test standards for insulated roof-deck construction are adequately conservative so that assemblies passing either of the two test standards are considered to meet the intent of the code without any limit on flame spread or smoke development. Furthermore, no thermal barrier is required.

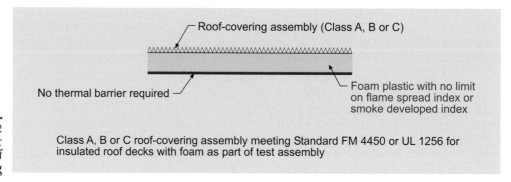

Figure 2603-2
Foam plastic used with roof covering

It should be noted that most insulated metal decks that are listed require that the deck be nonperforated—essentially a nonacoustical deck. Acoustical decks are commonly proposed in gymnasiums and auditoriums for sound control purposes and would therefore require a thermal barrier unless specifically listed under UL 1256 or FM 4450.

Attics and crawl spaces. See Figure 2603-3. This item describes specific methods used to protect foam plastics located within attics and crawl spaces (in lieu of a complying thermal barrier) where entry is provided only for service of utilities. The phrase "where entry is provided only for service of utilities" is intended to restrict these reduced requirements for a thermal barrier to those unused areas where there are no heat-producing appliances. In addition, drop lights or portable service lights are often utilized when serving equipment in such concealed spaces, and such lighting devices pose an ignition threat to the foam plastic. Thus, the reduced provisions are intended to provide a barrier whose only purpose is to prevent the direct impingement of flame on the foam plastic.

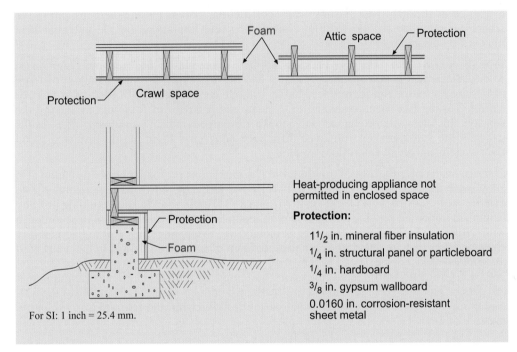

For SI: 1 inch = 25.4 mm.

Heat-producing appliance not permitted in enclosed space

Protection:

1 1/2 in. mineral fiber insulation

1/4 in. structural panel or particleboard

1/4 in. hardboard

3/8 in. gypsum wallboard

0.0160 in. corrosion-resistant sheet metal

Figure 2603-3
Foam plastic, attic and crawl spaces

The reduced level of protection is also applicable in those situations where the service of utilities is not an issue. If there are no utilities within the attic space or crawl space that require service, the minimum described degree of separation between the foam plastic and the enclosed space must still be provided. Where the attic or crawl space provides a suitable area that exists for a purpose other than the access to utilities, such as storage, a thermal barrier complying with Section 2603.4 is required.

Doors not required to have a fire-protection rating. Pivoted or side-hinged doors not required to have a fire protection rating are permitted to be installed without the thermal barrier, provided the door facings are of sheet metal of the thicknesses prescribed in this section. The rationale behind the waiver of the thermal barrier is that the foam plastic is completely encapsulated within the sheet-metal facings, and the quantity of foam plastic in protected doors is quite small.

Exterior doors in buildings of Groups R-2 and R-3. In specific residential occupancies, exterior doors to individual dwelling units are permitted to have a foam-plastic core. This allowance applies where the doors do not require a fire-resistance rating and the foam is covered with wood or other approved materials.

Garage doors. Garage doors, other than those in garages accessory to dwelling units, that contain foam plastic are allowed, provided the door does not require a fire-resistance rating and is faced with materials prescribed by this section. If the garage door containing the foam plastic does not have an aluminum, steel or wood facing of the minimum thickness prescribed, the door must be tested in accordance with DASMA 107 *Room Fire Test Standard for Garage Doors Using Foam Plastic Insulation.* This provision is intended to regulate the commercial applications of overhead, sectional and tilt-up types of doors.

Siding backer board. Where it is desired to insulate exterior walls under exterior siding, the code permits foam plastic to be used as a backer board for the siding, provided the insulation has a potential heat of not more than 2,000 Btu per square foot (22.7 MJ/m^2). The thermal barrier is not required under these circumstances as long as the siding backer board has a minimum thickness of $^1/_2$ inch (12.7 mm) and is separated from the interior of the building by not less than 2 inches (51 mm) of mineral-fiber insulation or the equivalent.

The code also permits the siding backer board without a thermal barrier when the siding is applied as residing over existing wall construction. This is reasonable considering the separation provided by the existing construction and limitations on the potential heat imposed by the code.

Type V construction. The use of spray-applied foam plastic has become common in wood-frame construction for the sill plates and headers. Such a limited amount of foam plastic insulation is considered acceptable without the protection afforded by a thermal barrier. Testing has been conducted to evaluate the behavior of foam plastic having the density, thickness, flame spread and smoke developed indices stipulated. The results indicated no substantial performance difference between a foam plastic insulated wood floor system and an all wood floor system.

2603.5 **Exterior walls of buildings of any height.** The provisions for foam plastic insulation also allow such material in the exterior walls of buildings required to have noncombustible exterior wall construction (Types I, II, III and IV). Applicable to such buildings of any height, an important provision of this section requires that the wall be tested in accordance with NFPA 285. This test provides a method of evaluating the wall's flammability characteristics that are due to the combustible foam plastic materials within the wall. Wall assemblies need not be tested where they can comply with the provisions of Section 2603.4.1.4 for fully-sprinklered, one-story buildings. Section 2603.5.1 also requires that test data be provided to show that if a fire-resistance rating is required, the rating of the wall containing the foam maintains the required rating. Moreover, the foam-plastic insulation must:

1. Be separated from the interior of the building with a thermal barrier meeting the provisions of Section 2603.4.

2. Not have a potential heat content exceeding that of the foam plastic insulation contained in the wall assembly as tested.

3. Have a maximum flame-spread index of 25 and smoke-developed index of 450. Exterior coatings and facings, tested individually, must also comply with these flame-spread and smoke-development limitations.

4. Be labeled by an approved agency.

5. Comply with the ignition limitations imposed by Section 2603.5.7.

2603.6 **Roofing.** As previously addressed, complying foam-plastic insulation may be utilized as a portion of a roof covering assembly, provided the assembly with the foam plastic insulation has been tested in accordance with ASTM E 108 or UL 790, and has been listed as a Class A, B or C roofing assembly.

2603.9 **Specific approval.** In this section, the code provides for those cases where foam plastic products and protective coverings do not comply with the specific requirements of Sections 2603.4 through 2603.7. The specific approvals are based on testing that is related to the actual end use of the products. The code refers to a number of test standards for determining specific approvals and, in addition, there are others that utilize some variation of the room test and are designed for testing exterior wall applications.

Section 2604 *Interior Finish and Trim*

The provisions of Chapter 8 for wall and ceiling finishes are applicable to those plastic materials installed as trim or interior finishes. In addition, foam plastics used as interior finish and trim must be in compliance with the provisions of this section, as well as the flame-spread index requirements of Chapter 8. By limiting the density, thickness, wall and ceiling area, as well as flame spread of foam plastic materials, the hazard level created by the exposure of foam plastics is low.

Section 2605 *Plastic Veneer*

Because it is a combustible material, plastic veneer used in the interior of a building is required by the code to comply with the interior finish requirements of Chapter 8.

Where plastic veneer is used on the exterior of a building, the code requires that the veneer be of approved plastic materials as defined in Section 2602. This places severe restrictions on the combustibility and smoke development of the plastic materials. Because plastic materials are combustible, the code limits their attachment on any exterior wall to a height no greater than 50 feet (15 240 mm) above grade. Furthermore, the IBC limits the area of plastic veneer to 300 square feet (27.9 m^2) in any one section and requires each section to be separated vertically by a minimum of 4 feet (1219 mm). The 4-foot (1219 mm) separation helps control the rapid vertical spread of fire. The code anticipates that local fire-fighting forces can effectively fight a fire that involves plastic veneer up to a height of about 50 feet (15 240 mm). Also, if the plastic veneer involves too large an area, it is conceivable that a fire could overtax local fire-fighting forces. The exception applies to Type VB buildings where the walls are not required to have a fire-resistance rating. In this case, the plastic materials do not present a greatly different hazard than the unprotected wood construction.

Plastic siding used on the exterior of a building is regulated separately from plastic veneer. The provisions of Section 2605 are not appropriate for exterior plastic siding, the requirements for exterior wall coverings established in Sections 1404 and 1405 are to be applied.

Section 2606 *Light-transmitting Plastics*

It is the intent of this section and Sections 2607 through 2611 to regulate the use of light-transmitting plastics—those plastics used in the building envelope or with interior lighting to transmit light to the interior of the building. Light-transmitting plastics are regulated because they are combustible materials. The unregulated use of combustible materials in the roof structure and for the exterior walls can possibly defeat the intent of the provisions of the code relating to types of construction. Thus, these six sections regulate these materials so that they do not materially affect the other requirements of the code regarding types of construction.

Any use of light-transmitting plastic materials must be approved by the building official and be based on technical data submitted to substantiate their use. As a basis of this approval, the building official should refer to Section 2602.1 for the definition of "Plastic, approved." The definition refers to the criteria of Section 2606.4 for the combustibility classifications of approved plastic materials, determined to be either Class CC1 or CC2 in accordance with ASTM D 635.

Materials of light-transmitting plastic, such as lenses, panels, grids or baffles, located below independent light sources are thought of as creating a light-diffusing system. Light-diffusing systems are specifically regulated in Section 2606.7. Regulated as to occupancy, location, support, installation and size, light-diffusing systems pose potential hazards that are due to their combustibility.

Section 2607 *Light-transmitting Plastic Wall Panels*

Exterior wall panels are regulated for the same reason as plastic glazing in openings. However, because exterior wall panels are sheet materials, they generally constitute larger unbroken areas than plastic glazing for openings do; as a result, their burn-rate characteristics are more critical than those for plastic glazing in openings addressed in Section 2608.

Section 2608 *Light-transmitting Plastic Glazing*

Because plastic glazing materials are combustible, their use is limited to openings not required to be fire protected. In the case of building construction other than Type VB, their use is further restricted. The glazing of openings not required to be fire protected in Type VB construction is essentially unlimited as to the area, height, percentage and separation requirements applicable to the individual glazed openings.

For plastic-glazed openings in buildings other than Type VB, restrictions are placed on the area, height, percentage and separation requirements for the individual glazed openings because plastic glazing materials are combustible. In other types of construction, unprotected combustible materials must be limited in accordance with their real extent and separation. Because of the combustibility of plastic glazing, the code requires flame barriers at each floor level for nonsprinklered multistory buildings to prevent the transmission of flame from one story to another by way of combustible openings.

As with other provisions of the code limiting the height of combustible materials above grade, this section also limits the height of plastic materials above grade to 75 feet (22 860 mm) unless the building is sprinklered throughout.

Section 2609 *Light-transmitting Plastic Roof Panels*

Plastic panels are regulated on the basis of three conditions, of which only one needs to be met in order to utilize light-transmitting plastic panels in roofs of all occupancies other than Groups H, I-2 and I-3. Light-transmitting plastic roof panels may be installed in buildings equipped throughout with an automatic sprinkler system, in buildings where the roof construction is not required to have a fire-resistance rating or where the roof panels meet the requirements for roof coverings in accordance with Chapter 15.

Because plastic roof panels constitute unprotected openings in the roof, the code requires that they be separated from each other by 4 feet (1219 mm) horizontally. The minimum 4-foot (1219 mm) separation is not mandated for fully-sprinklered buildings, nor is it required in buildings housing low-hazard occupancies as limited by Exception 2 or 3 to Section 2609.4. Furthermore, their location on the roof is regulated based upon the building's location in respect to lot lines. Roof panels shall be located at least 6 feet (1829 mm) from exterior walls that are located in a manner to require protected wall openings.

Because Class CC1 plastics have a slower burn rate than Class CC2 plastics, the code limits Class CC2 plastics to smaller areas than allowed for Class CC1 materials. The area limitations may be doubled based upon the installation of a sprinkler system, whereas plastic roof panels in low-hazard occupancy buildings, greenhouses and patio covers are not limited in area under specific conditions. As with exterior wall panels, the actual numbers relating to area, height and separation requirements must be somewhat arbitrary but are reasonable code limits determined by a consensus of knowledgeable experts.

Section 2610 *Light-transmitting Plastic Skylight Glazing*

In this section, the requirements for skylights are more detailed than those for roof panels in Section 2609 because there is no limit on the type of construction or fire-protection requirements for the roof assembly. Furthermore, skylights have unique requirements, such as those for flashing and resistance to burning brands. Also, as plastic-glazed skylights provide an unprotected combustible assembly in the roof, limitations must be placed on the area, percentage and separation of each unit. Each unit's location on the roof relative to property lines is regulated in a manner consistent with that for plastic roof panels.

Two of the primary concerns of plastic-glazed skylights are related to flashing at the intersection with the roof and their ability to resist the effects of flying, burning brands. Therefore, with one exception, the code requires that they be mounted on a curb at least 4 inches (102 mm) above the plane of the roof so that proper flashing may be accomplished. The exception involves skylights on roofs that have a minimum slope of 3 units vertical in 12 units horizontal (25 percent slope) and applies only to Group R-3 occupancies and on buildings having unclassified roof coverings. This slope should provide adequate roof drainage to accommodate skylights. The slope requirements for flat or corrugated plastic-glazed skylights and the rise requirement for dome-shaped skylights are based on the skylights' ability to shed flying brands. However, when the glazing material in the skylights can pass the Class B burning brands test specified in ASTM E 108 or UL 790, there is no limitation on slope, either of flat or corrugated glazed skylights, or on rise in the case of dome-shaped skylights.

The requirement for the protection of edges of plastic-glazed skylights or domes is to prevent the rapid spread of fire along the roof, as the edges of the plastic glazing material ignite more readily than the interior portions. Under those conditions where unclassified roof coverings are permitted, the metal or noncombustible edge material is not required.

As with roof panels, the various limitations on area, percentage and separation of skylights are somewhat arbitrary and, as with roof panels, are based on a consensus among knowledgeable experts on what is reasonable.

KEY POINTS

- Foam plastic is regulated for flame spread and smoke development.

- Separation with a thermal barrier must be provided between foam plastic insulation and the interior of the building.

- Containers of foam plastic and foam plastic ingredients must be labeled to prevent the misapplication of products not designed for their use.

- Foam plastics used in several applications, such as masonry or concrete construction, cooler and freezer walls, roofing, attics and crawl spaces, and doors not required to have a fire-protection rating, may be installed without a thermal barrier under specified conditions.

- When properly tested, foam plastic insulation is permitted in exterior walls of buildings required to have noncombustible exterior wall construction.

- Foam plastic used as interior finish and trim is acceptable where the density, thickness, wall and ceiling coverage, and flame spread of the foam plastic materials is limited.

- Light-transmitting plastics are regulated in part because they are combustible materials.

ELECTRICAL

Section 2702 Emergency and Standby
Power Systems

All electrical components, equipment and systems are to be designed and constructed in accordance with the provisions of NFPA: *National Electrical Code* (NEC). The *International Building Code®* (IBC®) specifically references the NEC for its technical provisions. The only electrical issues addressed within this chapter of the IBC are those emergency and standby power systems that are required by other provisions of the code. It should be noted that these provisions are maintained through the code change process of the *International Fire Code®* (IFC®).

Section 2702 *Emergency and Standby Power Systems*

NFPA Standards 70, 110 and 111 regulate the installation of emergency and standby power systems. This section identifies 20 situations where such systems must be in place. Stationary engine generators, where used to provide emergency and standby power, must comply with the requirements of UL 2200. This UL standard provides a benchmark for the evaluation of the safety and reliability of such generators.

Emergency power shall be provided as follows for:

1. Emergency voice/alarm communication systems in Group A occupancies with an occupant load of 1,000 or more.

2. All required exit signs, other than those that are self-luminous.

3. Means of egress illumination in occupancies where two or more means of egress are required.

4. Semiconductor fabrication facilities per Section 415.8.10.

5. Occupancies with highly toxic or toxic materials per IFC Section 3704.2.2.8.

6. Power-operated sliding doors or power-operated locks for swinging doors in Group I-3 occupancies, unless a remote mechanical operating release is provided.

Standby power is required under the following conditions for:

1. Smoke-control systems.

2. Elevators and platform lifts that are a portion of accessible means of egress.

3. Horizontal sliding doors utilized as a component of a means of egress.

4. Auxiliary-inflation systems in membrane structures exceeding 1,500 square feet (140 m^2) in area.

5. Occupancies where Class I and unclassified detonable organic peroxides are stored.

6. Emergency voice/alarm communication systems in covered mall buildings greater than 50,000 square feet (4645 m^2) in floor area.

7. Pressurization equipment, mechanical equipment, lighting, elevator-operator equipment, fire-alarm systems and smoke-detection systems in airport traffic-control towers more than 65 feet (19 812 mm) in height.

8. For operation of one or more elevators in a building.

9. Mechanical vestibule and stair shaft ventilation systems and automatic fire detection systems for smokeproof enclosures.

Both emergency power and standby power shall be provided in the following situations for:

1. High-rise buildings (with exceptions), defined as those structures having occupied floors more than 75 feet (22 860 mm) above the lowest level of fire-department vehicle access.

2. Underground buildings (with exceptions), defined as those building spaces having a floor level used for human occupancy more than 30 feet (9144 mm) below the lowest level of exit discharge.

Either emergency power or standby power shall be provided, in Group H occupancies where inside storage, dispensing or use of hazardous materials occurs, where the following systems are required:

Mechanical ventilation, treatment, temperature control, alarm, detection or other electrically operated systems as identified in Section 414.5.4.

MECHANICAL SYSTEMS

This chapter merely references the *International Mechanical Code*® (IMC®), *International Fuel Gas Code*® (IFGC®) and Chapter 21 of the *International Building Code*® (IBC®) for various requirements relating to the construction, installation and maintenance of mechanical equipment and systems. The appropriate code shall be utilized to address heating; air conditioning; refrigeration; mechanical and natural ventilation; plenums; and factory-built chimneys, fireplaces and barbecues. Reference is made to the IMC and Chapter 21 of the IBC for the regulation of masonry chimneys, fireplaces and barbecues.

29

PLUMBING SYSTEMS

Section 2902 Minimum Plumbing Facilities
Key Points

> The intent of Chapter 29 is to reference the *International Plumbing Code®* (IPC®) for the construction, installation and maintenance of plumbing systems and equipment, and the *International Private Sewage Disposal Code®* (IPSDC®) for the regulation of private sewage disposal systems. In addition, the provisions of this chapter provide for the determination of plumbing fixture counts based on occupancy classification and occupant loads.
>
> The provisions of Chapter 29 are maintained through the code change process of the IPC.

Section 2902 *Minimum Plumbing Facilities*

2902.1 Minimum number of fixtures. Section 2902.1 establishes the minimum number of plumbing fixtures that must be provided for various occupancies based upon Table 2902.1. The table is based upon the distinct occupancy classifications identified in Chapter 3 and the corresponding occupant loads calculated for the building. Those fixtures required in most occupancies include water closets, lavatories, drinking fountains and service sinks. In addition, bathtubs or showers, automatic clothes-washer connections and kitchen sinks are mandated in some residential occupancies. It is assumed at least one fixture of each type as required by Table 2902.1 will be provided in a building designed for human occupancy.

When determining the proper occupant load to be utilized in calculating plumbing fixture count, there is no specific methodology referenced. However, because the only provisions in the IBC addressing occupant load calculation are found in Chapter 10, it is typically assumed that the approach established in Section 1004 for the means of egress should also be used for plumbing fixture count. The basis for most occupant load determinations, other than areas with fixed seating, is the density factor established in Table 1004.1.1. An occupant load is determined by dividing the floor area under consideration by the appropriate density factor. It should be noted that the building official also has the authority as established in the exception to Section 1004.1.1 to base the occupant load on the actual number of occupants anticipated, rather than the calculated number. Although the use of this exception is typically inappropriate for egress and fire safety purposes, it is commonly applied for plumbing fixture count. The occupant load utilized for egress and fire protection requirements is purposely conservative by most counts because of the life safety concerns. The occupant load to be used in calculating the minimum plumbing fixture count should be based on more of a convenience concern, recognizing the need to satisfy any sanitation issues. Therefore, the occupant load used in the plumbing fixture count could differ from that used as the basis for the design of the means of egress system. The building official should rely on all available information that will assist in the appropriate determination of occupant load for fixture count purposes.

2902.1.1 Fixture calculations. Once the appropriate occupant load is determined, the minimum required number of fixtures is calculated by using Table 2902.1. The provisions of Section 2902.1.1 address the method in which the fixtures are to be distributed between the sexes.

For the determination of required plumbing fixtures, it is to be assumed that 50 percent of the occupants are male, 50 percent are female. The resulting occupant loads are then used when applying the table, with fixtures calculated individually for each of the sexes. Where the required number of fixtures contains a fraction, an additional fixture is required. See Application Examples 2902-1 and 2902-2. Note that the provisions of Section 419.2 of the IPC allow urinals as a substitution for water closets on a 1 to 1 basis, provided that, in assembly and educational occupancies, urinals account for no more than 67 percent of the required fixtures. For example, if nine water closets are required in a men's toilet room in a large nightclub, it is acceptable to provide six urinals and only three water closets. Where the occupancy classification is other than Group A or Group E, only one-half of the required number of water closets may be substituted with urinals.

The determination of the minimum plumbing fixture count becomes a bit more complex where the building contains a number of different occupancies. If toilet room facilities are provided independently for each of the occupancies in the building, the basic method of calculating the number of fixtures should be satisfactory. However, if common toilet facilities are designed to serve the occupants from multiple occupancy groups, a different approach is more appropriately warranted. In such a determination, the number of required fixtures would be calculated for each occupancy, then without rounding, added together to arrive at the minimum fixtures that must be provided for each of the sexes. An example of this methodology is shown in Application Example 2902-3. A similar approach could be utilized when substituting urinals for water closets.

Application Example 2902-1

GIVEN: An exhibition hall classified as a Group A-3 Occupancy. The hall's occupant load is determined to be 8680.

DETERMINE: The minimum required number of (a) water closets for the male occupants, (b) water closets for the female occupants, (c) lavatories for the male occupants and (d) lavatories for the female occupants.

1. Assume occupants as 50% male, 50% female per Section 2902.1.1:

    4340 males

    4340 females

2. (a) $\frac{4340}{125} = 34.72 = 35$ water closets* for males

    * IPC Section 419.2 would allow 12 or more water closets, with the remainder urinals, to make up the 35 fixtures

    (b) $\frac{4340}{65} = 66.77 = 67$ water closets for females

3. $\frac{8680}{200} = 43.4 = 44$ lavatories total

    (c) 50% of 44 = 22 lavatories for males

    (d) 50% of 44 = 22 lavatories for females

MINIMUM REQUIRED PLUMBING FIXTURES

Application Example 2902-2

GIVEN: A manufacturing facility classified as an F-1 Occupancy. The occupant load is determined to be 684.

DETERMINE: The minimum required number of water closets and lavatories.

1. Assuming a 50:50 split, assign 342 male and 342 female occupants.

2. $\frac{342}{100} = 3.42 = 4$ water closets minimum for each sex

    4 water closets* for males

    4 water closets for females

Same calculation for lavatories, a minimum of 4 for each sex.

    *If urinals are substituted for water closets, a minimum of 2 water closets must be provided.

MINIMUM REQUIRED PLUMBING FIXTURES

2902.2 **Separate facilities.** In most buildings where plumbing fixtures are required, separate facilities must be available for each sex. Simply, at minimum, one women's toilet room and one men's toilet room must be provided. There are conditions, however, where only a single toilet room is mandated. Separate-sex facilities need not be provided within dwelling units and sleeping units. In addition, common facilities are permitted where the number of people (both customers and employees) does not exceed 15. In mercantile occupancies with an occupant load of 50 or less, such as a small retail sales tenant, a single toilet room is also permitted.

2902.3 **Required public toilet facilities.** Only specific occupancies and uses are required to have public toilet facilities for use by customers, patrons and visitors of the building. Restaurants, nightclubs, places of assembly, business uses, mercantile occupancies and similar buildings and tenant spaces intended for public use must be provided with customer toilet facilities located within one story above or below the area under consideration, and with a path of travel not to exceed 500 feet (152 400 mm). Those uses where public use is not expected, such as warehouses, factories and similar buildings, only require employee toilet facilities.

Application Example 2902-3

GIVEN: A mixed-occupancy building containing a Group M with an occupant load of 368, a Group B with an occupant load of 56 and a Group S-1 with an occupant load of 78. All of the plumbing fixtures will be located at a single toilet room location.

DETERMINE: The minimum number of water closets that would be required in each toilet room.

SOLUTION:

Group M: 184/500 @ 1 per 500 occupants
Group B: 28/25 @ 1 per 25 for first 50 occupants
Group S-1: 39/100 @ 1 per 100 occupants

184/500 + 28/25 + 59/100 =
0.37 + 1.12 + 0.39 = 1.88 =

Minimum of two water closets required in each toilet room

MINIMUM REQUIRED PLUMBING FIXTURES

KEY POINTS

- The minimum required number of plumbing fixtures is based upon the use of the building and the anticipated number of occupants.

- Except for a limited number of situations, separate toilet facilities are required for each sex.

- Except in specific Group A occupancies, the required water closets, lavatories, showers and bathtubs are to be distributed equally between the sexes.

CHAPTER

30

ELEVATORS AND CONVEYING SYSTEMS

Elevators and other types of conveying systems are regulated under the provisions of this chapter. For the most part, the American Society of Mechanical Engineers (ASME) standards are utilized to address the specifics of elevator safety. ICC A117.1 must also be referenced for all elevators required to be accessible by Chapter 11 of the *International Building Code®* (IBC®).

Section 3002 *Hoistway Enclosures*

3002.1   **Hoistway enclosure protection.** This section is essentially a cross reference to Section 708 of the code, which contains the specific requirements for the enclosure of shafts in buildings. Elevator shafts are to be fire-resistance-rated enclosures, unless exempted by one of the exceptions of Section 708.2. If required to be fire-resistance rated, the shaft enclosure must have a 1-hour or 2-hour rating based on the number of stories connected by the shaft enclosure, as well as the required fire-resistance rating of the floor construction. Opening protectives for hoistway enclosures are also regulated by Chapter 7.

3002.2   **Number of elevator cars in hoistway.** These provisions were extracted from the elevator code insofar as they are more appropriate as building code requirements. The basis for limiting the number of cars in a single hoistway is to provide a reasonable level of assurance that a multilevel building served by several elevators would not have all of its elevator cars located in the same hoistway. This could result in a single emergency disabling all elevators within the building. For example, if all elevator cars were allowed to be located in the same hoistway, smoke that entered the enclosure during a fire would render all elevator cars unusable. The code provisions will increase the chance that some elevators within a major building would remain operational during a fire or other emergency. See Figure 3002-1.

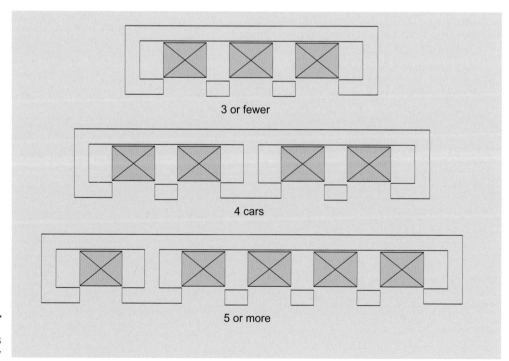

**Figure 3002-1
Elevator cars
in a hoistway**

**Figure 3002-2
Emergency
signs**

Emergency signs. In order to alert occupants to the fact that an elevator is not to be used for **3002.3**
egress purposes during a fire incident, this section mandates the placement of a sign
adjacent to each elevator call station on each floor of the building. The standardized
pictorial sign (an example is illustrated in Figure 3002-2) advises occupants to use the exit
stairways rather than the elevators in case of a fire. The emergency sign does not need to be
installed at those elevators complying as an accessible means of egress per Section 1007.4
or at occupant self-evacuation elevators as described in Section 3008.

Elevator car to accommodate ambulance stretcher. Where elevators are provided in **3002.4**
buildings of four stories or more in height, this section of the code requires that at least one
elevator serving all floors accommodate an ambulance stretcher. The ability to transport an
individual on a stretcher in an elevator in a multistory building is a basic life-safety
consideration. Immediate identification of elevators that accommodate stretchers is
necessary so that emergency-services personnel can quickly respond to emergency
conditions. For this reason, an identifying symbol as shown in Figure 3002-3 shall be placed
inside on both sides of the hoistway door frame.

Minimum elevator car size
requirements have been established to
ensure that the typical ambulance stretcher
can be accommodated. The minimum
24-inch by 84-inch (610 mm by 2134 mm)
size is further described to address
stretchers with rounded or chamfered
corners with a minimum 5-inch (127 mm)
radius. See Figure 3002-4. By specifically
identifying the minimum size
requirements, flexibility is provided to the
elevator industry in its efforts to provide
appropriately-sized cars. It is also
beneficial to stretcher manufacturers by
providing direction to aid in the
standardization of their products.

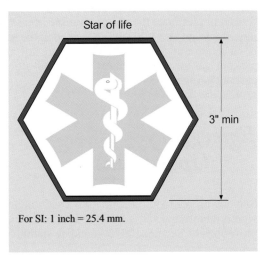

Star of life

3" min

For SI: 1 inch = 25.4 mm.

**Figure 3002-3
Sign denoting
accommodation
of ambulance
stretcher**

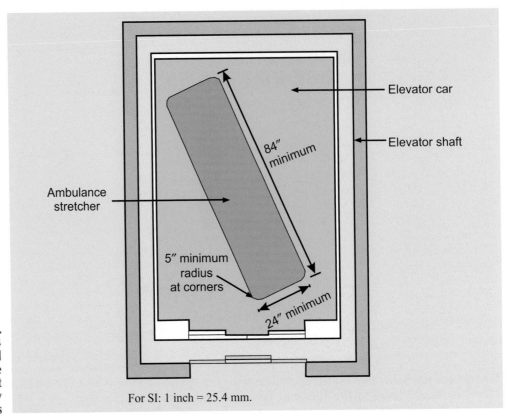

Figure 3002-4
Elevator used for fire department emergency access

For SI: 1 inch = 25.4 mm.

3002.5   **Emergency doors.** The provisions of ASME A17.1 regulate the installation of an emergency door when required by this section. It is necessary that an emergency door be provided in the blind portion of a hoistway or the blank face of an exterior wall where an elevator is installed in a single blind hoistway or on the outside of a building.

3002.6   **Prohibited doors.** The prohibition against installing additional doors at the point of access to an elevator car is to minimize the possibility of a person becoming trapped between doors, or the possibility of the elevator car being rendered unusable because of blocked access at a particular floor where such an additional door has been closed and locked. The only condition under which such doors are permitted occurs where the doors are readily openable from the car side without the use of a key, a tool, or any special effort or knowledge.

3002.7   **Common enclosure with stairway.** To better ensure that a single fire incident will not restrict or eliminate the use of multiple access or egress components within a building, it is important that elevators not be located in the same shaft enclosure as a stairway. Since there is no requirement for a shaft enclosure for an elevator or exit stairway serving an open parking garage, the isolation of the elevator from the stairway is not required in open parking garages.

Section 3003   *Emergency Operations*

Standby power requirements in this section are applicable for those buildings or structures where standby power is furnished to, or required to, operate an elevator. For example, the provisions of Section 1007.4 mandate standby power for any elevator used as an accessible means of egress. The requirement that standby power be manually transferable to all elevators in each bank is necessary to improve their reliability during emergency conditions. As an example, the elevator to which standby power is connected may be in a hoistway that is unusable because of smoke contamination. In this case, the transferability requirement provides for transferring the emergency power from the elevator in the affected hoistway to an elevator in a hoistway in the same bank that is usable.

Section 3004 *Hoistway Venting*

Hoistway ventilation is intended to remove any smoke that may have entered the hoistway, which would render the cars within the hoistway useless. In addition, unvented smoke can eventually spill out of the hoistway into the upper floors of the building. Inadequate top ventilation of the elevator shafts at the MGM Grand Hotel was one cause cited for smoke spreading into upper-story corridors. A portion of the required vent area must always be open so that venting would be immediately available in case of need. However, there is an allowance for maintaining all of the vent openings in a closed position, provided they all open upon actuation of any elevator lobby smoke detector, upon power failure, or upon activation of a manual override control.

Section 3006 *Machine Rooms*

Modern elevator control equipment is solid-state and is extremely sensitive to temperature. For this reason, provisions have been developed to keep smoke and heat out of the machine room by requiring that the room be provided with an independent ventilation or air-conditioning system.

Section 3007 *Fire Service Access Elevator*

To facilitate the rapid deployment of fire fighters, Section 403.6.1 contains provisions for a fire service access elevator in high-rise buildings that have at least one floor level more than 120 feet (36 576 mm) above the lowest level of fire department vehicle access. Usable by fire fighters and other emergency responders, the specific requirements for the elevator are set forth in Section 3007. See Figure 3007-1.

A fire service access elevator has a number of key features that will allow fire fighters to use the elevator for safely accessing an area of a building that may be involved in fire or for facilitating rescue of building occupants. The elevator is required to be protected by a shaft enclosure that complies with Section 708. For a building four or more stories in height, Section 708.4 requires the shaft have a minimum 2-hour fire-resistance rating.

An elevator lobby is mandated as a transitional element between the fire service access elevator and the remainder of the floor. Because the elevator lobby will be the location that fire fighters will use as the point of departure to the floor or area of fire origin, it must be constructed to limit the entrance of fire or smoke. The lobby must be enclosed by a smoke barrier with a minimum fire-resistance rating of 1 hour and the lobby door is required to have a minimum fire-resistance rating of 45 minutes. The lobby for the fire service access elevator must be designed such that it has direct access to an exit enclosure. By providing a direct connection between the elevator lobby and a stair enclosure, efficient access to and from other floors is increased. In addition the required standpipe hose valves located within the exit enclosure will be directly accessible from the fire service access elevator lobby, providing a protected location for fire fighters to prepare for manual fire-fighting operations and to assess interior fire and smoke behavior near the area of fire origin.

A unique requirement for fire service access elevators is that they must be designed so their status can be continuously monitored in the fire command center. The elevator is to be monitored by a standard emergency services interface meeting the requirements of NFPA 72, *National Fire Alarm Code*. The requirements stipulate that such an interface must be designed and arranged in accordance with the requirements of the organization that will use the device. In the case of a fire service access elevator, the fire department or fire service provider will need to be involved in the arrangement of interface.

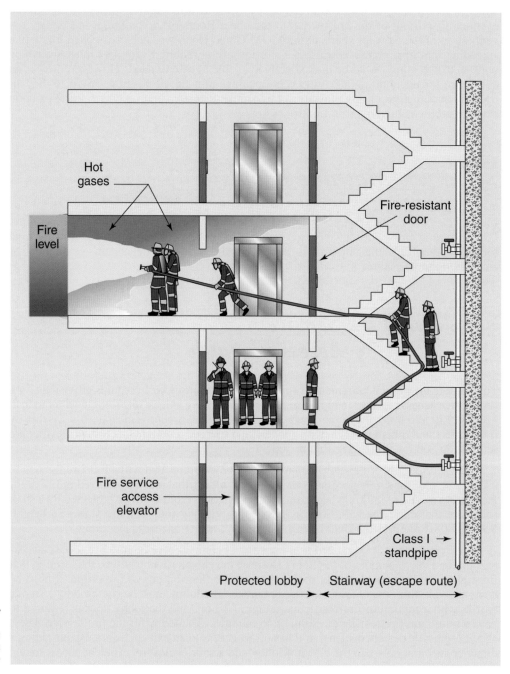

Figure 3007-1
Fire service access elevator

A fire service access elevator requires normal utility power and connection to a standby power system. Provisions further stipulate that the transfer switch for the standby power system operate within 60 seconds of utility power failure (Type 60), the power source is designed to operate for at least 2 hours under its design load (Class 2) and meet the requirements of NFPA 110 for Level 1 service. Level 1 service, as defined by NFPA 110, indicates that the standby power system is used in a building where the loss of electrical power could result in the death or serious injury to one or more occupants. Loads that must be connected to the standby power system include the elevator, its machine room ventilation and cooling equipment, and equipment provided to maintain the elevator controller within its temperature limits. An additional requirement for the electrical power system serving the fire service access elevator is the protection of wiring or cables. Electrical conductors that provide normal or standby power to the fire service access elevator must be protected located by a shaft or similar enclosure having a minimum 1-hour fire-resistance rating or circuit integrity cable having an equivalent fire resistance must be utilized.

Section 3008 *Occupant Evacuation Elevators*

Under the conditions of Section 3008, public-use passenger elevators are allowed to be utilized for the self-evacuation of occupants in high-rise buildings. Although such elevators may be used by building occupants during building evacuation, they are not intended to replace any means of egress facilities as required by Chapter 10. The only permitted reduction in required egress facilities due to the presence of complying occupant evacuation elevators is the elimination of the extra exit stairway mandated by Section 403.5.2 for high-rise buildings over 420 feet (128 m) in height. Under no conditions is the installation of such elevators required. The allowance for occupant evacuation elevators, although voluntary, provides tools for the architect to consider when designing tall buildings.

Where elevators are to be used for self-evacuation purposes, all passenger elevators in the building must be in compliance with the special provisions of Section 3008. The use of such elevators for occupant self-evacuation is limited to when the elevator is in the normal operating mode prior to Phase I Emergency Recall Operation per ASTM A17.1 and the building's fire safety and evacuation plan.

The occupant evacuation elevators must open directly into an elevator lobby that conforms to special requirements addressing access, enclosure, size and signage. Each elevator lobby must be provided with a status indicator arranged to display the following information as applicable:

- Illuminated green light and the message "Elevators available for occupant evacuation"
- Illuminated red light and the message "Elevators out of service, use exit stairs"
- No illumination or message when the elevators are in normal service operation

Additional provisions address the design and installation of a two-way communications system and instructions on the use of the system. The elevators must be continuously monitored at the fire command center or an approved central control point. A variety of other conditions are placed upon the design and installation of the occupant evacuation elevators to help ensure their reliability under emergency conditions.

KEY POINTS

- Elevator shafts and elevator machine rooms are typically required to be fire-resistance-rated enclosures, with the rating of either 1 hour or 2 hours based on the number of stories in the building and required fire-resistance rating of the building's floor construction.

- Additional doors are prohibited at the point of access to an elevator car unless they are readily openable from the car side without the use of a key, a tool or any special effort or special knowledge.

- Hoistway ventilation is intended to remove any smoke that may have entered the hoistway, which would render the cars within the hoistway useless.

- Fire service access elevators help facilitate the rapid deployment of fire fighters in applicable high-rise buildings.

- Although not intended to replace any required means of egress facilities, public-use passenger elevators are allowed to be utilized for the self-evacuation of occupants in high-rise buildings.

SPECIAL CONSTRUCTION

Those special types of elements or structures that are not conveniently addressed in other portions of the *International Building Code*® (IBC®) are found in this chapter. By special construction, the code is referring to membrane structures, pedestrian walkways, tunnels, awnings, canopies, marquees and similar building features that are unregulated elsewhere.

Section 3102 *Membrane Structures*

3102.1 **General.** Because membrane structures have several unique characteristics that set them apart from other buildings, they are regulated in Chapter 31 under the provisions for special construction. The regulations cover all such structures, including air-supported, air-inflated, cable-supported and frame-supported membrane structures. The intent of the provisions is that, except for the unique features of membrane structures, they otherwise comply with the code as far as occupancy requirements, allowable area and other regulations are concerned. Membrane structures are limited to one story in height insofar as there is insufficient experience to justify multilevel structures enclosed with a membrane.

The membrane structures regulated by the IBC are deemed to be permanent in nature, erected for a period of at least 180 days. Membrane structures in place for shorter periods of time, such as temporary tents, are to be regulated by the *International Fire Code*® (IFC®). Where a membrane structure is erected as a part of a permanent structure, such as a covering for a building, balcony or deck, it must comply with the provisions of Section 3102 for any time period.

Because of the limited hazards present in structures not used for human occupancy, such as water-storage facilities, water clarifiers, sewage-treatment plants and greenhouses, only a few provisions are applicable where membrane structures cover these types of facilities. Limitations on the membrane and interior liner material, as well as the structural design, are the only criteria in the IBC that apply to membrane structures covering facilities not typically used for human occupancy.

3102.3 **Type of construction.** In general, membrane structures are considered to be of Type V construction, except where the membrane structure is shown to be noncombustible. In this case, the membrane structure should be classified as Type IIB construction. Membrane structures supported by heavy-timber framing members are to be considered Type IV construction. The code permits the use of nonflame-resistant plastic material for the membrane of a greenhouse structure that is not available to the general public.

3102.6 **Mixed construction.** This section permits the use of a noncombustible membrane on a structure that would otherwise comply as Type IA, IB or IIA where the membrane is used exclusively as a roof or skylight and is located at least 20 feet (6096 mm) above any floor, balcony or gallery. This exception is similar to Footnote b of Table 601. This exception will permit nonrated noncombustible membranes to be constructed as roof systems for sports stadiums and similar buildings as well as for atriums. In other types of construction under the same conditions, the membrane need only be flame-resistant.

3102.8 **Inflation systems.** Where membrane structures are air-supported or air-inflated, this section addresses the regulations for equipment, standby power and support. The primary inflation system shall consist of one or more blowers, designed in such a manner that overpressurization is prevented. Air-supported or air-inflated structures exceeding 1,500 square feet (140 m²) in floor area must also be provided with an ancillary inflation system. This backup system, connected to an approved standby power-generating system, shall operate automatically to maintain the inflation of the structure if the primary system fails.

Additional support for the membrane must also be provided where covering structures having occupant loads of more than 50 and where covering swimming pools.

Section 3104 *Pedestrian Walkways and Tunnels*

This section regulates connecting elements between buildings, such as tunnels or pedestrian walkways, that are utilized for occupant circulation. The provisions of this section are only applicable to such tunnels and walkways designed primarily as circulation elements, typically for weather-protection purposes. A covered walkway or bridge connecting two buildings is the most common example of a pedestrian walkway. These elements may be located below, at or above grade. When in compliance with this section, pedestrian walkways and tunnels are not considered to contribute to the floor area or height of the connected buildings. In addition, those buildings connected by the pedestrian walkway or tunnel are permitted to be considered separate structures. These allowances establish the primary reason for the use of this section. Where multiple structures are attached, they would generally be considered by the code to be a single building and regulated as such. These provisions not only allow each building to be regulated independently, but also limit the requirements of the tunnel or pedestrian walkway to those provisions of this section. A common use of pedestrian walkways is the connection of buildings that are on separate lots, including situations where the buildings are on opposite sides of a public way. Often referenced as *skyways* in northern climates, pedestrian walkways allow for a method to connect buildings across lot lines without the normally mandated fire-resistance-rated exterior walls and opening prohibition associated with a fire separation distance of zero. The pedestrian walkway is treated as a *nonbuilding* and therefore is not regulated where it crosses the lot line. Instead, this approach allows for a level of fire protection at the connection between the buildings and the pedestrian walkway.

It is important to understand that this section is voluntary in application, and only utilized where the design professional chooses not to consider the walkway and connected buildings a single structure. There is always the option of regulating the entire structure as a single building, in which case this section would have no application. See Figure 3104-1.

This section establishes specific requirements for the protection of walls and openings between the connected buildings and the pedestrian walkway or tunnel, specifies minimum and maximum widths, and limits the length of exit access travel within a pedestrian walkway or tunnel. Section 3104.3 further requires that pedestrian walkways be constructed of noncombustible materials or of fire-retardant-treated wood unless all connected buildings are of combustible construction.

As previously mentioned, where pedestrian walkways and tunnels are designed and constructed in accordance with the provisions of this section, the code intends that they not be considered part of the connected buildings. Furthermore, they need not be considered in the determination of the allowable area or height of either of the connected buildings. In effect, this section is an exception to the general construction requirements, and its use is optional. Otherwise, the designer would

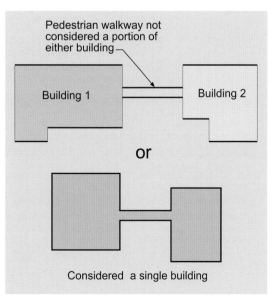

Figure 3104-1
Pedestrian walkways

consider the connected buildings to be a single building on the same lot, including the pedestrian walkway portion.

Section 3105 *Awnings and Canopies*

The definition of awning in Section 202 describes it as an architectural projection, comprising a lightweight, rigid skeleton structure over which a covering is attached, that provides weather protection, identity or decoration, and is completely supported by the building to which it is attached. Retractable awnings are also regulated by this chapter as well as Chapter 16; therefore, a definition is provided. Awnings are regulated for structural and construction features as well as cover materials.

Section 3106 *Marquees*

A marquee is defined in Section 202 as a permanent roof structure attached to and supported by a building and projecting over public property. Thus, a marquee is different from an awning, based on its permanence. The classic example of a marquee is the theater marquee or the entrance marquee at a hotel. The restrictions placed on the size, projection and clearances for marquees are intended to prevent interference with:

1. The free movement of pedestrians.
2. Trucks and other tall vehicles using the public street.
3. The fire department in its fire-fighting operations at a building.
4. Utilities.

Thus, the projection, clearance, length, height, construction and location of the marquee are regulated to meet the intent of the code. Figure 3106-1 depicts the permissible projection and clearances required for marquees.

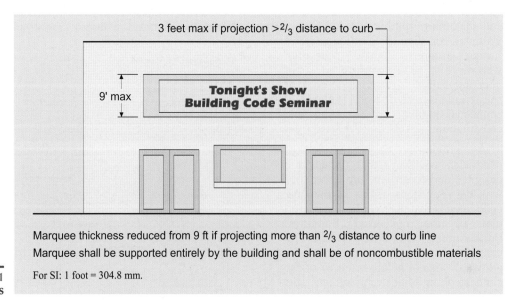

3 feet max if projection $>^2/_3$ distance to curb

9' max

Tonight's Show Building Code Seminar

Marquee thickness reduced from 9 ft if projecting more than $^2/_3$ distance to curb line
Marquee shall be supported entirely by the building and shall be of noncombustible materials

For SI: 1 foot = 304.8 mm.

Figure 3106-1
Marquees

Section 3109 *Swimming Pool Enclosures and Safety Devices*

This section addresses barriers for swimming pools, spas and hot tubs. Regulations are provided for both public and residential swimming pools; however, the primary criteria apply to residential facilities. Because of the supervision provided at public pools, the requirements for public swimming pools are much less extensive.

Public swimming pools. A fence, barrier or similar enclosure method is required around all public swimming pools. No less than 4 feet (1290 mm) in height, the fence must be equipped with self-closing and self-latching gates. Openings in the fence are also regulated so that the passage of a 4-inch (102 mm) sphere is not possible.   **3109.3**

Residential swimming pools. Unless a swimming pool is provided with a power safety cover or a spa is provided with a safety cover complying with the provisions of ASTM F 1346, a residential swimming pool, spa or hot tub must be enclosed in accordance with this section. The code addresses a number of different barrier designs, each intended to limit access to the pool area. See Figure 3109-1. Where a wall of a dwelling serves as a portion of the barrier, additional requirements are also in place to regulate unsupervised access to the pool or spa.   **3109.4**

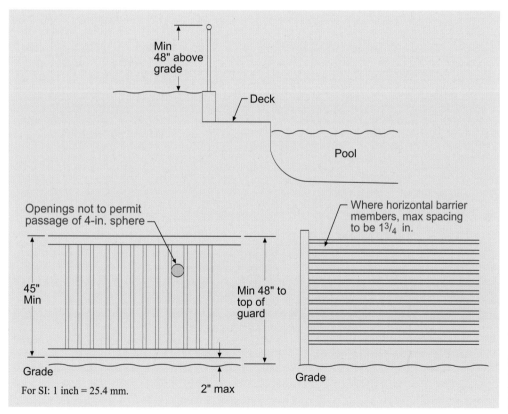

Figure 3109-1 Swimming pool enclosure barriers

KEY POINTS

- Membrane structures include those buildings that are air-supported, air-inflated, and cable- or frame-supported.

- Noncombustible membrane structures are classified as Type IIB construction, whereas combustible membrane structures are considered Type VB.

- Air-supported or air-inflated membrane structures are regulated for equipment, standby power and support.

- Complying pedestrian walkways and tunnels are not considered to contribute to the floor area or height of the connected buildings.

- Awnings are regulated for structural and construction features, as well as canopy materials.

- Marquees are restricted by size, projection and clearances.

- Unless provided with an approved safety cover, a swimming pool or spa must be enclosed by a fence, barrier or similar enclosure method.

ENCROACHMENTS INTO THE PUBLIC RIGHT-OF-WAY

By using the language "encroachments into the public right-of-way," the code is referring to the projections of appendages and other elements from the building that are permitted to project beyond the lot line of the building site and into the public right-of-way. The intent of the code is that projections from a building into the public right-of-way shall not interfere with its free use by vehicular and pedestrian traffic, and shall not interfere with public services such as fire protection and utilities.

Section 3201 *General*

The *International Building Code®* (IBC®) requires that any construction that projects into or across the public right-of-way be constructed as required by the code for buildings on private property. Furthermore, the intent is that the provisions of this chapter are not considered so as to permit a violation of other laws or ordinances regulating the use and occupancy of public property. Many jurisdictions have ordinances regulating the use and occupancy of public property that go beyond the provisions of this chapter, many of which may be more restrictive and not permit the projections permitted by the IBC. Where there are no other ordinances that prohibit such construction, it is the intent of this chapter to permit the construction of the connecting structure between buildings either over or under the public way.

These types of permanent uses of the public way are often permitted by the jurisdiction under a special license issued by the public works agency. Where these uses create no obstruction to the normal use of the public way, the jurisdiction, by authorizing connecting structures between buildings on either side of the public way, can derive revenue from a licensing agreement, which can be of assistance in maintaining the public right-of-way.

The code prohibits roof drainage water from a building to flow over a public walking surface, as well as any water collected from an awning, canopy or marquee, and condensate from mechanical equipment. The intent is not necessarily to prevent drainage water from flowing over public property in general, but rather to prevent drainage water from flowing over a sidewalk or pedestrian walkway that is between the building and the public street or thoroughfare.

There are at least two problems that arise when drainage water from a building or structure is allowed to flow over a public sidewalk:

1. Under proper conditions of light and temperature, algae will form where the water flows across the sidewalk and create a hazardous, slippery walking surface.

2. During heavy rain storms, the velocity and force of the water emitting from the drain can create hazardous walking conditions for pedestrians.

Therefore, the usual procedure is to carry the drain lines from the roof or other building elements inside the building through the wall of the building and under the sidewalk through a curb opening into the gutter.

Section 3202 *Encroachments*

The projection of any structure or appendage is generally regulated in regard to its relationship to grade. Those encroachments found below grade, such as structural supports, vaults and areaways, are governed in one fashion. Those encroachments located above grade, but below 8 feet (2438 mm) in height, are more severely controlled because of their immediate location adjacent to public areas. Although projections that are located 8 feet

(2438 mm) or more above grade tend to have fewer implications, such encroachments are also regulated.

Encroachments below grade. Footings, foundations, piles and similar structural elements **3202.1** that support the building and are erected below grade are not permitted to project beyond the lot lines. There is an allowance for footings that support exterior walls, provided the projection does not extend more than 12 inches (305 mm) beyond the street lot line, and that the footings or supports are located at least 8 feet (2438 mm) below grade. In this exception, the code assumes that there will be no utilities or other public facilities below this depth adjacent to the building and, therefore, permits footings to project a limited amount where below the 8-foot (2438 mm) depth. It should be noted that this encroachment is permitted for street walls only and does not apply to interior lot lines. See Figure 3202-1.

Where vaults, basements and other enclosed spaces are located below grade in the public right-of-way, they shall be further regulated by the ordinances and conditions of the applicable governing authority. In many cases, the space below the sidewalk or other area adjacent to the building in the public right-of-way is not used for public purposes such as utilities, sewers and storm drains (except for catch basins); therefore, where local ordinances do not prohibit the use of space beneath the sidewalk or other public area, the code permits its use by adjoining property owners. Usually the basement or other enclosed space of an adjoining building is extended beneath the public right-of-way and is used for any of the uses permitted by the code. This is a revocable permission, as the jurisdiction may find at a later date that there is a public need for the public use of the space beneath grade.

Encroachments above grade and below 8 feet in height. Building components that may **3202.2** arbitrarily project into the public right-of-way pose a considerable hazard to pedestrians or other users of the right-of-way. Therefore, doors and windows are prohibited from opening into such an area. However, certain features of the building that are permanent in nature are permitted to encroach into the public right-of-way to a limited degree. Where steps are permitted by other provisions of the IBC, they may project into the right-of-way up to 12 inches (305 mm), provided they are protected by guards at least 3 feet (914 mm) high or

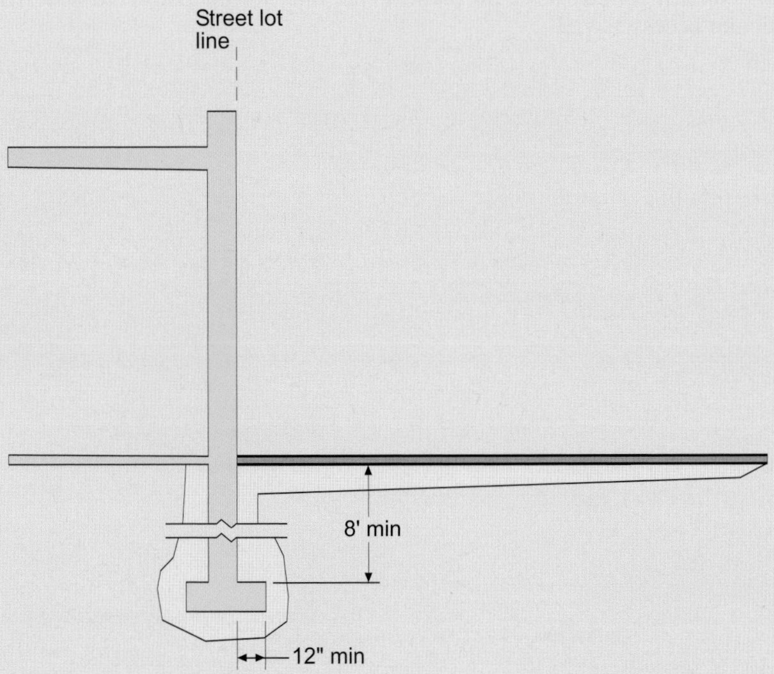

For SI: 1 inch = 25.4 mm, 1 foot = 304.8 mm.

Figure 3202-1
Encroachments below grade

similar protective elements such as columns or pilasters. The intent of the provision is that the encroachment be small enough so as not to obstruct pedestrian or vehicular traffic and yet provide some protection for occupants of the building to look each way before proceeding into the right-of-way.

Encroachments into the public right-of-way for certain architectural features are also permitted, but again only a limited projection is allowed. For columns or pilasters, the maximum projection is 12 inches (305 mm). Other architectural features such as lintels, sills and pediments are limited to a 4-inch (102 mm) projection. Awnings, including valances, may only extend into the public right-of-way when the vertical clearance from the right-of-way to the lowest part of the awning is at least 7 feet (2134 mm).

3202.3 Encroachment 8 feet or more above grade. The code permits the projection of awnings, canopies, marquees, signs, balconies and similar architectural features at a height of 8 feet (2438 mm) or more in accordance with Figures 3202-2 and 3202-3. The intent of the code is that no projection should be permitted near the level of the public right-of-way and up to 8 feet (2438 mm) in height so that the free passage of pedestrians along the sidewalk or other walking surface will not be inhibited. At the 8-foot (2438 mm) height and above, projections are permitted as long as they do not interfere with public utilities. It is generally assumed that utility lines for telephones or power will not occupy this zone except for the service entrances to the buildings. There are jurisdictions that have high-voltage power lines running along the sidewalk, and the regulations of the agency that regulates the power companies generally require certain clearances from these lines. Therefore, in addition to the requirements shown in Figures 3202-2 and 3202-3, power-line clearances should also be checked, and the requirements of the *National Electrical Code*® should be reviewed when it is adopted by the jurisdiction. Where such awnings, canopies, balconies and similar building elements are located 15 feet (4572 mm) or more above grade, their encroachment is unlimited.

It is not uncommon to find a pedestrian walkway or a similar circulation element to be constructed over a public right-of-way. Such a condition is permitted subject to the approval of the local authority having jurisdiction. In no case, however, should the vertical clearance between the right-of-way and the lowest part of the walkway be less than 15 feet (4572 mm). A clearance of a least 15 feet (4572 mm) allows not only for unobstructed pedestrian travel but for vehicular access as well.

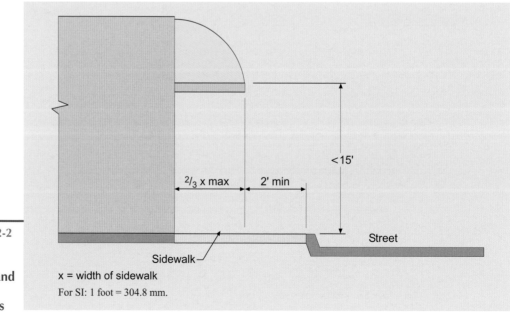

Figure 3202-2
Awning, canopy, marquee and sign projections

x = width of sidewalk
For SI: 1 foot = 304.8 mm.

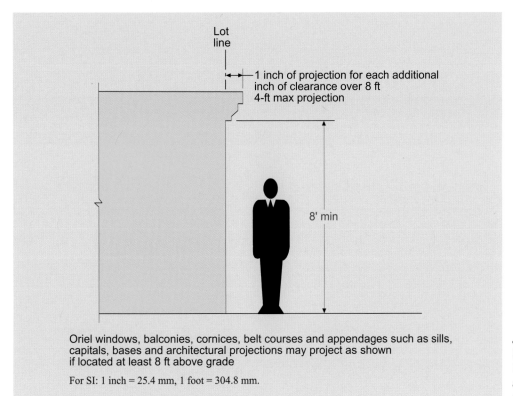

Lot line

1 inch of projection for each additional inch of clearance over 8 ft
4-ft max projection

8' min

Oriel windows, balconies, cornices, belt courses and appendages such as sills, capitals, bases and architectural projections may project as shown if located at least 8 ft above grade

For SI: 1 inch = 25.4 mm, 1 foot = 304.8 mm.

Figure 3202-3
Balcony and appendages projections

33

SAFEGUARDS DURING CONSTRUCTION

This chapter deals with safety practices during the construction process, as well as the protection of property, both public and private, adjacent to the construction site. Included in the provisions are requirements addressing site work; protection of pedestrians; and temporary use of streets, alleys and public property. Special issues relating to fire extinguishers, exits, standpipes and automatic sprinkler systems are also addressed.

Section 3302 *Construction Safeguards*

Where an existing building continues to be occupied during the process of remodeling or constructing an addition, it is very important that the life- and fire-safety features of the building continue to be in place. Required exits shall be maintained, the strength of structural elements shall not be diminished, fire-protection devices are to remain in working condition and sanitary facilities shall be fully functional. These features must be maintained at all times during any remodeling, alterations, repairs or additions to the building under consideration. Obviously, there will be times when the specific protective feature is the element or device being altered or repaired. In this case, an equivalent method shall be provided to safeguard the occupants.

Section 3303 *Demolition*

As the work of demolishing a building is subject to so many variations and so many different hazards, the *International Building Code*® (IBC®) authorizes the building official to require the submission of plans and a complete schedule of the demolition. Under certain circumstances, pedestrian protection may be required during the demolition process. Such protection shall be provided in compliance with other provisions of Chapter 33. For some multilevel buildings, and for certain types of demolition operations, it may also be necessary to temporarily close the street. Once the structure has been demolished, any resulting excavation is to be filled consistent with the existing grade, or in any other manner in conformance with the ordinances of the jurisdiction.

Section 3304 *Site Work*

The provisions in Section 3304.1 for excavations and fills are intended to apply to the specific area where the building or structure will be located. This section is not intended to address massive grading on a site. For those requirements, one would turn to Section 1803, as well as any local grading ordinance that might be in effect. To prevent decay and to eliminate an avenue of entrance for termites and other insects, the code requires that the area of the site occupied by the building be free of all stumps, roots, and any loose or casual lumber. Typically applicable in the construction of wood-framed structures, the requirements also address wood forms that have been used in placing concrete, as well as any loose or casual wood that might be in direct contact with the ground under the building.

The IBC limits the slopes for permanent fills or cut slopes for permanent excavations for structures to two units horizontal in one unit vertical (50-percent slope). Although steeper cut slopes are permitted where substantiating data are submitted, the limitation on filled slopes of two units horizontal in one unit vertical (50-percent slope) is an absolute limitation, and filled slopes may not be steeper. Although cut slopes into natural soils may be excavated with a slope steeper than 50 percent, depending on the nature of the soil materials and the density, the code reasons that fill slopes must not be steeper than 50 percent to cover unprecedented circumstances such as heavy rains creating overly saturated soils or, in the cases of seismic activity, vibration of the soils during an earthquake causing failures of steep-filled slopes.

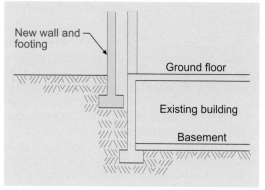

Figure 3304-1
Surcharge load on adjacent building

Understandably, the code requires that fill or surcharge loads should not be placed adjacent to an existing building or structure unless the existing building or structure is structurally capable of resisting the additional loads caused by the fills or surcharge. See Figure 3304-1. Alternatively, the existing building can be strengthened in order to resist the additional loading.

The code requires that existing footings or foundations that may be affected by an excavation be adequately underpinned or otherwise protected. This concerns excavations on the same lot and under the control of the same individual who has control of the existing buildings. For obvious reasons, it is the intent of the code that excavations in close proximity to an existing footing shall not be made unless proper protective measures are taken for the existing building. These measures may involve underpinning of the existing foundations, or shoring and bracing of the excavations, so that the existing building foundations will not settle or lose lateral support.

Where the excavation is for a new building foundation and the new footing is at an elevation below, but within reasonably close proximity to, an existing foundation, underpinning is the usual procedure to protect the existing foundation. If the existing footing is for a structure that creates a horizontal thrust, the means of providing lateral support may take the form of a buttressed retaining wall designed to resist the lateral thrust of the existing foundation.

Where buildings are to be supported by fills, the code requires that the fills be placed in accordance with accepted engineering practice. Thus, fills utilized for the support of buildings must be designed by a geotechnical engineer (soils engineer) utilizing the principles of geotechnical engineering so as to provide a proper and adequate foundation for the structure above, and one that will limit settlements to tolerable levels. In order to verify the adequacy of the geotechnical design of the fills, the IBC permits the building official to ask for a soil investigation report outlining the geotechnical design of the fill materials, as well as a report of the satisfactory placement of the fill.

Section 3306 *Protection of Pedestrians*

Both Section 3306 and Chapter 32 regulate the use of public streets and sidewalks. Section 3306 provides those general regulations and criteria for the temporary use of public property, which are generally found to apply to most jurisdictions.

As the nature and philosophy of each jurisdiction varies, so do the regulations each promulgates regarding the use of public property and, in particular, the use of streets and sidewalks. Adjacent property owners have rights of access to their property. The public street also provides access for public services such as fire and police protection, street sweeping and maintenance, street lighting, trash pickup, and other services provided by the jurisdiction or its contractors. Utilities serving adjacent property also use the public street for access.

Pedestrians must be protected from the potential hazards that exist during construction or demolition operations adjacent to the public way. The type of protection depends on the type of operation being conducted. For example, an excavation directly adjacent to a pedestrian path would necessitate a minimum of a construction railing along the side facing the excavation. It is also possible that a barrier would be required. Where the pedestrian path extends into the public street, a construction railing is required on the street side of the walkway to protect the pedestrians from vehicular traffic. Table 3306.1 provides the criteria for determining whether or not additional protection is required, depending on the height and proximity of the construction operations to the pedestrian walkway. The level of protection mandated by the table is shown in Figure 3306-1. Depending on these various parameters, the protection required will vary from none or merely a construction railing, to a barrier or covered walkway.

3306.2 **Walkways.** A public jurisdiction provides sidewalks and streets for the free passage of vehicular and pedestrian traffic. In the case of pedestrian traffic, the usual procedure is to provide a sidewalk on each side of the street. Therefore, when construction or demolition operations are conducted on property adjacent to the sidewalk, the code requires a walkway at least 4 feet (1219 mm) wide be maintained in front of the building site for pedestrian use. However, in those cases where pedestrian traffic is unusually light, the jurisdiction may authorize the fencing and closing of the sidewalk to pedestrian use. The walking surface, in addition to being durable, must be in compliance with the accessibility provisions of Chapter 11. The walkway shall also be designed to support all imposed loads, with a minimum design live load of 150 pounds per square foot (7.2 kN/m^2).

3306.3 **Directional barricades.** Where the temporary walkway used for pedestrian travel around a construction or demolition site extends into the street, it is critical that a sufficient barrier be provided to direct vehicular traffic away from the walkway. The construction of the barrier is to be large enough to ensure that motorists will easily and quickly identify the revised traffic path.

3306.4 **Construction railings.** The code only requires an open-type guardrail 3 feet, 6 inches (1067 mm) in height where a construction railing is required. The intent of the railing is to direct pedestrians as they travel adjacent to construction areas, as well as to provide a very limited degree of protection. Where access must be further regulated, such as adjacent to an excavation, additional provisions must be applied. It is incumbent that the jurisdiction adequately protect pedestrians from the hazards associated with construction or demolition work.

3306.5 **Barriers.** The intent of the code is that a barrier, where required, be solid and sturdy enough to prevent impacts from construction operations from penetrating the barrier and injuring passing pedestrians. Because of this intent, the code also requires that the barrier be at least 8 feet (2438 mm) in height and extend along the entire length of the building site. Although the IBC requires openings in the barrier to be protected by doors that are normally kept closed, the code does not intend to prevent the use of small viewports at eye level so that passing pedestrians may stop and view the construction operations if they wish.

To provide a reasonable level of protection from the construction operations, barriers are to be designed to resist loads as required in Chapter 16. As an option, specific construction procedures are outlined in Section 3306.6. The procedures address the size and spacing of studs as well as the size and span limitations for wood structural panels used in the barrier.

3306.7 **Covered walkways.** A covered walkway is required for the same reasons as a protective barrier—to prevent falling objects from endangering passing pedestrians. Therefore, the code requires that the design be such that the roof and supporting structure is capable of preventing falling objects from breaking through. With a minimum required clear height of 8 feet (2438 mm) measured from the floor surface to the canopy overhead, the covered walkway must be adequately illuminated at all times. Pedestrians using walkways adjacent to building sites where construction or demolition operations are taking place have every right to safe passage and to be protected from falling debris and other hazards of construction operations. Furthermore, they should not be subjected to the indignities of dirt, water and other foreign material sifting through leaks in the canopy.

Because a covered walkway is intended to protect pedestrians against construction operations, including demolition, the canopy structure should be structurally designed to serve that purpose. Thus, the code provides two means for the design of the canopy:

1. *A structural design* to withstand the actual loads to which it will be subjected, provided the design live load is not less than 150 pounds per square foot (7.18 kN/m^2).

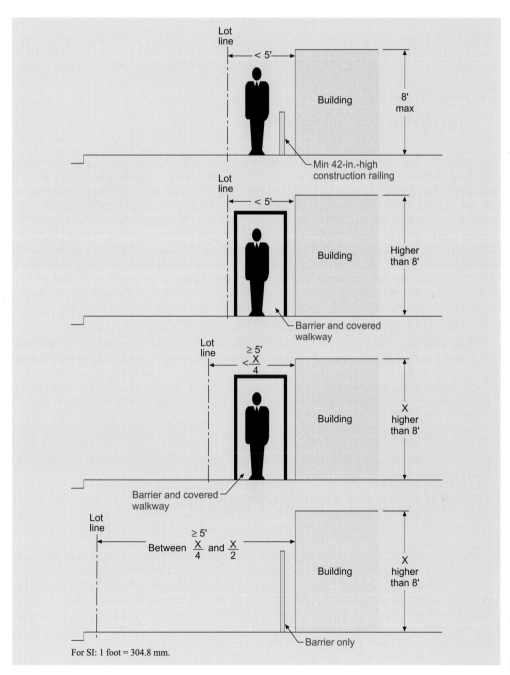

For SI: 1 foot = 304.8 mm.

Figure 3306-1
Pedestrian protection

2. *A prescriptive design* in accordance with this section of the code where the covered walkway will not be subjected to a live load greater than 150 pounds per square foot (7.18 kN/m²).

The exception provides for a permissive design based on a live load of 75 pounds per square foot (3.56 kN/m²) for the covered walkway where the construction operation is limited to the erection of a new, small, light-frame building not more than two stories above grade plane.

3306.9 **Adjacent to excavations.** Where a site excavation occurs within 5 feet (1524 mm) of the street lot line, it must be enclosed with a barrier not less than 6 feet (1829 mm) in height. The barrier must be constructed to resist the wind pressures specified in Chapter 16, but need not meet the general barrier requirements of Section 3306.6. The building official also has the authority to require a barrier around excavations in those cases where the excavation is more than 5 feet (1524 mm) from the street lot line. Because of the potential liability involved, it is probable that most excavations of considerable size or hazard would be enclosed by a complying barrier.

Section 3307 *Protection of Adjoining Property*

It is critically important that during construction, remodeling or demolition work, any adjoining public or private property be protected from damage. Essentially, all portions of the neighboring structure must be protected from damage, including footings, foundations, exterior or party walls, chimneys, skylights, roofs and other building elements. In addition, water run-off must be controlled to prevent erosion. Although there are no specific requirements laid out in the IBC for the level of protection required, the intent of the protection is based on common-law precepts, such as that of lateral support where it comes to footings and foundations. Over the years, common law and, more recently, statute law have established the requirements for lateral support. That is, the owner of a piece of real property shall be entitled to the lateral support of the property by adjacent property. Responsible practice would also dictate a satisfactory level of protection for other portions of the adjacent site and building during the construction process.

Where excavation work is to be performed, the person responsible for such work shall notify the owners of any adjoining buildings. The written notice, to be delivered no less than 10 days prior to the initiation of the excavation work, shall indicate that excavation work is to take place and that protection of the adjoining buildings should be considered. Appropriate action should be taken where access to an adjoining site or building is desired in order to provide the proper protective measures.

Section 3308 *Temporary Use of Streets, Alleys and Public Property*

It is important that access to essential facilities be maintained where construction materials and equipment are temporarily stored on public property. Obstructions shall not block access to such fire protection features as standpipes, fire hydrants or alarm boxes, as well as any catch basins or manholes that might be present. In regard to maintaining safe and effective vehicular traffic in the area where construction materials and equipment are stored, such materials or equipment must be located at least 20 feet (6096 mm) from the street intersection, or otherwise located where sight lines are not obstructed.

Section 3309 *Fire Extinguishers*

It is quite common for combustible debris and waste materials to accumulate in and around a building under construction. Therefore, at least one approved portable fire extinguisher sized for at least ordinary hazard and complying with Section 906 of the *International Fire Code* (IFC) is to be provided at every stairway on all floor levels where combustible material has accumulated. In addition to those extinguishers required within the building, every storage shed and construction shed shall also contain an approved fire extinguisher. Additional extinguishers may be required by the building official where any special or unique hazards exist, such as the presence of flammable or combustible liquids.

Section 3310 *Exits*

Once a building under construction reaches a considerable height, deemed to be 50 feet (15 240 mm) or four stories by the IBC, there is a need to provide at least one stairway that is available and usable for egress purposes. The temporary stairway shall be adequately illuminated during those times where there are occupants who may need to utilize the stairway. A temporary stairway need not be provided where there is at least one permanent stairway that is maintained and usable as the construction progresses. Where an existing building exceeding 50 feet (15 240 mm) in height is undergoing an alteration, at least one lighted stairway must continue to be available.

For those buildings required to have a permanent standpipe system per Section 905.3.1,

Section 3311 *Standpipes*

during construction operations a standpipe system must also be installed before the construction height exceeds 40 feet (12 192 mm) above the lowest level of fire department access. As construction continues to proceed upward, the standpipes shall also be extended in a timely manner. A standpipe shall always be available within one floor of the highest point of construction having secured decking or flooring. During construction operations, the amount of combustible materials from concrete forms, scaffolding, plastic and canvas tarpaulins, and other materials are prevalent not only throughout the building, but throughout the construction site itself. Thus, in many cases, the standpipe system provides the only source of water for fire-fighting operations.

A minimum of one standpipe is required, with hose connections to be located in accessible areas adjacent to usable stairs. The standpipes may be either temporary or permanent, but in all cases shall meet the minimum requirements of Section 905 for capacity, outlets and materials. The water supply may also be either temporary or permanent, as long as it is available at the first sign of combustible material accumulation.

KEY POINTS

- Where an existing building continues to be occupied during the process of remodeling or constructing an addition, it is very important that the life- and fire-safety features of the building continue to be in place.

- Once a structure has been demolished, any resulting excavation is to be filled, consistent with the existing grade.

- The code requires that fills or surcharge loads not be placed adjacent to an existing building unless the existing building is structurally capable of resisting the additional loads caused by the fills or surcharge.

- Pedestrians must be protected from the potential hazards that exist during construction or demolition operations adjacent to the public way.

- Depending on various conditions, required pedestrian protection varies from none, to merely a construction railing, to a barrier, to a covered walkway.

- During construction, remodeling or demolition work, any adjoining public or private property must be protected from damage.

EXISTING STRUCTURES

This chapter covers all aspects of existing buildings, including their occupancy and maintenance, as well as additions, alterations or repairs to existing buildings. In addition, accessibility provisions for existing structures are found in Section 3411.

A building in existence when a new code edition is adopted has the right to have its existing use or occupancy continued, provided all devices or safeguards have been maintained in conformance with the code edition under which they were installed. Should the owner subsequently want to change the use or occupancy, Section 3408 provides the guidelines for doing so. Additional provisions are contained in Section 3412, which are primarily intended to provide alternatives that will maintain or increase the current degree of public safety, and are to be used in those situations where full compliance with the *International Building Code*® (IBC®) is not possible.

In existing buildings, there are times where full and strict compliance with the minimum accessibility requirements for new construction cannot be achieved. There is little chance the alteration can be made to be fully accessible, because of existing structural conditions that require removal or alteration of a load-bearing member that is an essential part of the structural frame, or because of the fact that there are other existing physical or site restraints that limit or prohibit a fully accessible structure. The code refers to such a situation as *technically infeasible*. In those situations where full accessibility is deemed to be technically infeasible, exceptions are provided allowing for a reduced level of access.

Unless modified by jurisdictional amendment, the use of this chapter for the regulation of existing buildings is optional. Another code published by the International Code Council, the *International Existing Building Code*® (IEBC®), provides an alternative approach to the regulation of existing buildings that undergo repair, alterations, additions or a change of occupancy. The IEBC, referenced in Section 3401.5 of the IBC, is deemed to be an acceptable alternative to the provisions of Chapter 34. The primary intent of the IEBC is to encourage the use and reuse of existing buildings by providing flexibility to permit the use of alternative approaches to achieve compliance with the minimum requirements. The value of the IEBC extends beyond its potential as an alternative compliance tool. It can be a valuable resource to assist in the application of IBC Chapter 34, particularly in regard to buildings that undergo a change in use or occupancy.

Section 3404 *Alterations*

The provisions in this section are, for the most part, simple and direct. Any alteration work must be in full compliance with the code requirements for new construction. Therefore, all construction performed in the alteration to an existing building must comply with all of the appropriate provisions of this code. The IBC also intends that the alteration work will not cause the existing building to be in violation of the code. For example, any new alterations should not remove or block existing exits. On the other hand, those portions of an existing building that are not involved or affected in the alteration work are not required to comply with the current code provisions. Simply, the portions of the building under construction are regulated as for a new building, the portions that are not part of the work are exempted.

The basic requirement of Section 3404.3 is that any structural alteration or repair made to an existing building cannot increase the force in any structural element by more than 5 percent, unless the new force levels remain in compliance with the IBC for new structures. The strength of any existing structural elements cannot be lessened to a level below that required by the code for new structures.

Section 3408 *Change of Occupancy*

Because each occupancy group contemplates a different level of hazard, the IBC intends that when a use is changed, and particularly when the change in use increases the level of hazard, the building must be brought to conform to the requirements of the code for the proposed occupancy.

The building official is granted broad authority in determining whether or not the new or proposed use is less hazardous than the existing use. Therefore, each change of occupancy must be analyzed based upon the hazards consequent to the new use. In addition, the fire- and life-safety and structural features of the building must be evaluated to determine if they are as required by the code for the new use. It is the intent of the code that the change of occupancy shall not be made unless the building is made to comply with the requirements of the code that are related to the proposed use. Thus, the structural requirements of the code must also be met where they are more restrictive for the new use.

Wide authority is also granted to the building official in using judgment to determine which code requirements will be exacted. The building official should develop the rules that are to be followed upon a change of occupancy so that consistent enforcement will result. If it is determined that the proposed or new use is less hazardous than the existing use, the building official may approve a change in occupancy without requiring that the building conform to all the requirements of the code for the new use. Given this wide authority, a building official may require that the building comply with only those requirements for the new use that are important to fire and life safety. Under these circumstances, the IBC permits the building official to choose not to require those changes that are deemed unnecessary as far as fire and life safety are concerned, based on the character of the new use.

A practical application. In closing, let us briefly discuss a change of occupancy determination that building officials face occasionally—the conversion of a single-family dwelling to an office use. Districts that were formally residential in nature have seen the expansion of neighboring commercial areas; now someone wants to save an attractive old Victorian residence and convert it into offices. To what extent does the building need to be modified to meet the intent of this section? To determine this, the building official must evaluate the fire- and life-safety features and structural features of the building to determine if existing conditions satisfy the code, and determine to what extent the new use poses a greater or a lesser hazard than the existing use. For example, the adequacy of the floor system would need to be reviewed because a dwelling allows for a lower design live load than does an office. Thus, floor system structural capacity is an important factor to consider. Another item that always arises with such conversions is what to do about exterior wall and opening protection. Many of the dwellings that are converted are historical in nature or, at the very least, are a part of the community's cultural fabric. As a result, there is an understandable desire to maintain their present appearance. In such cases, some jurisdictions permit the use of exterior fire sprinklers to serve as exterior protection. In other jurisdictions, a comparison is made between the fire- and life-safety hazards of the two occupancies. Such things as fire loading, alertness of the building occupants, building size, and similar considerations may be used by the building official in determining if the converted use poses a greater or lesser hazard than the former one.

Section 3411 *Accessibility for Existing Buildings*

This section addresses the level of accessibility required when maintenance or a change of occupancy occurs, or where additions or alterations are made to existing buildings. In general, the provisions contained in this section are less detailed than those for new buildings and place a greater reliance on determinations made by the building official.

Asking a number of questions may be helpful in determining compliance with Section 3411.6 on alterations:

1. What is the current level of accessibility?
2. What is the scope of the alteration?
3. How will the alteration affect the current level of accessibility?
4. Does the alteration increase the accessibility or use of the building?
5. If it is technically infeasible, what is the maximum extent the alteration can be made to meet accessibility requirements?

A number of building elements are addressed in this section because of the importance of their features. To maintain a certain degree of accessibility to these elements, modifications to the standard provisions lessen the degree of alteration required. However, alterations must occur at these minimum levels, regardless of their perceived technical infeasibility. The following elements are addressed in the scoping provisions of Section 3411.8 as they relate to alterations of a building:

1. Entrances
2. Elevators and platform lifts
3. Stairs, ramps and escalators
4. Performance areas
5. Dwelling or sleeping units
6. Jury boxes and witness stands
5. Toilet rooms
6. Dressing, fitting and locker rooms
7. Fuel dispensers
8. Thresholds

Historic buildings and facilities that undergo alterations or a change of occupancy are regulated for accessibility by Section 3411.9. The IBC describes a historic building as one that is listed or is eligible for listing in the National Register of Historic Places, or designated as historic under an appropriate state or local law. The provisions address key building elements such as accessible routes, entrances and toilet facilities. Where the historic significance of the building will be threatened or destroyed through full accessibility compliance, modifications for these key elements are permitted.

Section 3412 *Compliance Alternatives*

In the conservation and rehabilitation of existing building stock, it is often difficult to fully comply with the provisions of Chapters 2 through 33 of the IBC. This section is designed to provide an alternate means for meeting the intent and purpose of the code, which is safeguarding the public's safety, health and general welfare. Unless an unsafe condition exists as determined by the building official, compliance with this section shall be considered acceptable. The compliance alternatives addressed in this section apply to all occupancies other than Group H or I.

Under the method presented in this section, the three fundamental categories of fire safety, means of egress and general safety must be evaluated. These categories are further broken down into more specific issues. Fire safety is considered to be structural fire-resistance, automatic fire detection, fire alarm and fire suppression system features. Means of egress includes the configuration, characteristics and support features of exiting systems. The parameters of fire safety and means of egress are included within the general safety category.

In the evaluation process, the following issues are addressed:

1. Building height and area

2. Compartmentation

3. Tenant and dwelling unit separations

4. Corridor walls

5. Vertical openings

6. Heating, ventilating and air-conditioning systems

7. Automatic fire detection and fire alarm systems

8. Smoke control

9. Means of egress capacity and number

10. Travel distance and dead ends

11. Elevator control

12. Means of egress emergency lighting

13. Mixed occupancies and specific occupancy areas

14. Sprinklers and standpipes

15. Incidental accessory occupancies

After all of the building safety parameters have been evaluated, the values of the parameters are tabulated based on the three major categories. The score of each category, based on the tabulations, is compared with the mandatory safety score for each category. The result is quite simple—it either passes or fails.

KEY POINTS

- A building in existence when a new code edition is adopted has the right to have its existing use or occupancy continued, provided all devices and safeguards have been maintained in conformance with the code edition under which they were installed.

- The IEBC is deemed to be an acceptable alternative to the provisions of Chapter 34.

- Existing buildings may be altered, repaired or modified without complying with all provisions of the code as long as the new work complies.

- The IBC intends that when a use is changed, and particularly when the change in use increases the level of hazard, the building must be brought to conform to the requirements of the code for the proposed occupancy.

- The accessibility provisions for existing buildings tend to be less detailed than those for new buildings.

- Through the use of compliance alternatives, an alternate means is available for meeting the intent and purpose of the code.

Accessible and Usable
Buildings and Facilities

2003

CHAPTER

35

REFERENCED STANDARDS

The *International Building Code* (IBC) is for the most part a performance-based code. Thus, the IBC contains numerous references to standards that are intended to assist in the application of the code. Where standards are referenced in the body of the IBC, they are considered a part of the code. Thus, when a jurisdiction adopts the IBC as its building code, it also adopts the standards identified in this chapter.

Standards can be divided into several different categories—structural engineering standards, materials standards, installation standards and testing standards. Developed by a consensus process, the standards referenced in the IBC are all identified in this chapter. Organized by the promulgating agency of the standard, the chapter provides the standard number, standard title, and the code section or sections where the standard is referenced in the IBC.

Where one of the listed standards is referenced in the IBC, only the applicable portions of the standard relating to the specific code provision are in force. For example, Chapter 11 references ICC A117.1 for the technical requirements relating to accessibility. The provisions of ICC A117.1 address accessible telephones; however, Chapter 11 does not require telephones to be accessible. Therefore, that portion of the standard is not considered part of the IBC, and accessible telephones are not required. Accessible telephones are regulated in Appendix E, so if the jurisdiction adopts the appendix chapter, the technical provisions of ICC A117.1 for telephones will be in force. As a reminder, Section 102.4 also indicates that where the provisions of the code and a standard are in conflict, the provisions of the code apply.

APPENDICES

The appendix chapters to the *International Building Code®* (IBC®) contain subjects that have been determined to be an optional part of the code rather than mandatory, with each jurisdiction adopting all, parts of or none of the appendix chapters—depending on its needs for enforcement in any given area. The provisions of IBC Section 101.2.1 indicate that the requirements contained in the appendices are only applicable where specifically adopted by the jurisdiction. It is important that each jurisdiction review the appendix chapters in detail prior to their adoption to ensure their appropriateness.

Appendix A *Employee Qualifications*

The provisions of this appendix are intended to assist the jurisdiction in qualifying individuals to be employed in key roles in the Department of Building Safety. Employee qualifications are provided for the position of building official as well as those for chief inspectors, inspectors and plans examiners. An overview of the qualifications is shown in Figure A-1.

In addition to the education and experience criteria, an important consideration is the professional qualification obtained through certification. A comprehensive certification program is available through the International Code Council (ICC) that recognizes individuals for their knowledge relating to code enforcement.

Position	Experience			Certification
	Total years	Supervisory years	Type	
Building official	10	5	Architect, engineer, inspector, contractor, superintendent of construction	Certified building official
Chief inspector				Certified inspector for appropriate trade: building, plumbing, mechanical, electrical, combination
Inspector	5	–	Types listed above plus foreman or mechanic in charge of construction	
Plans examiner				Certified plans examiner for appropriate trade

Figure A-1

Appendix B *Board of Appeals*

This appendix expands on the provisions of Section 112 relating to the board of appeals. Issues dealing with the filing of an appeal, the board membership, meeting notices and board decisions are also addressed.

This appendix specifies that the board consist of five individuals, representing various disciplines or professions. In addition, two alternate members should be appointed to serve in the absence or disqualification of a regular member. All individuals are to be registered design professionals or contractors, qualified by registration or experience to rule on technical matters that may come before the board. The following disciplines are identified in Section B101.2.2:

1. Architecture or building construction

2. Structural engineering

3. Plumbing and mechanical engineering or contracting

4. Electrical engineering or contracting

5. Fire-protection engineering or contracting

It is important that all hearings before the board are considered open meetings and are available to all interested parties. The appellant, the appellant's representative or counsel, the building official and all other persons who have an affected interest in the decision shall be permitted to address the board. In order to overturn or modify the decision of the building official, a minimum two-thirds vote of the board is required. Unless acceptable to the appellant, all five members must be present for the board to act on an appeal, with at least four concurring votes necessary to modify or reverse the building official's decision.

Appendix C *Group U Agricultural Buildings*

The provisions of Appendix C were developed to address the needs of those jurisdictions (primarily unincorporated county territory) whose primary development is agricultural. In these cases, agricultural property usually consists of large tracts of land on which agricultural buildings are placed, usually with large open spaces and with essentially no congestion. Therefore, the provisions for agricultural buildings classify the structures as Group U occupancies and include barns, shade structures, grain silos, stables and horticultural structures, as well as buildings used for livestock and poultry shelters, equipment and machinery storage, and milking operations.

Because of the generally large open spaces that usually surround the buildings and the relatively low occupant load, the limitations imposed on construction, height, area, mixed uses and exiting are generally more liberal than the requirements in the body of the code for Group F or S occupancies that would generally otherwise apply.

Appendix D *Fire Districts*

Maintained by the *International Fire Code*® (IFC®) and its code change committee, this appendix is available for adoption by jurisdictions wishing to establish fire districts. The use of fire districts provides a method to address fire hazards that are created by a variety of conditions, with the primary concerns based on occupancy and structure density. The provisions apply to new buildings built within the fire district, as well as those buildings undergoing alterations or a change of occupancy.

It is necessary to first establish the territory that is to be included within the fire district. The code identifies three basic types of areas that are of importance in the regulation of fire districts. These include adjoining blocks, buffer zones and developed blocks. The specifics of each of these areas are illustrated in Figure D–1.

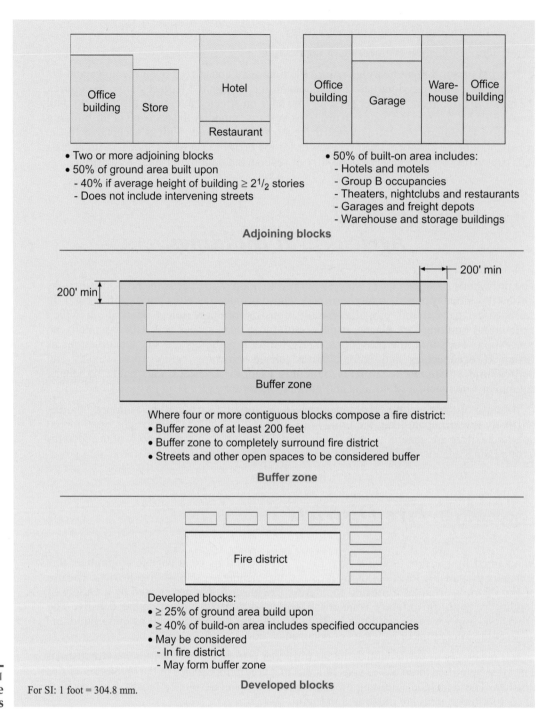

• Two or more adjoining blocks
• 50% of ground area built upon
  - 40% if average height of building ≥ 2¹/₂ stories
  - Does not include intervening streets

• 50% of built-on area includes:
  - Hotels and motels
  - Group B occupancies
  - Theaters, nightclubs and restaurants
  - Garages and freight depots
  - Warehouse and storage buildings

Adjoining blocks

200' min

200' min

Buffer zone

Where four or more contiguous blocks compose a fire district:
• Buffer zone of at least 200 feet
• Buffer zone to completely surround fire district
• Streets and other open spaces to be considered buffer

Buffer zone

Fire district

Developed blocks:
• ≥ 25% of ground area build upon
• ≥ 40% of build-on area includes specified occupancies
• May be considered
  - In fire district
  - May form buffer zone

Developed blocks

Figure D-1
**Fire
districts**

For SI: 1 foot = 304.8 mm.

One of the primary concepts of fire districts is that buildings of Type V construction are not permitted to be built, nor are Group H occupancies allowed. Roof coverings must be either Class A or B roof assemblies, and the primary building elements, such as floors, roofs, walls and supporting members, must be at least 1-hour fire-resistance-rated construction unless one of five exceptions is applicable. Other building elements regulated within a fire district include the fire resistance of exterior walls, architectural trim, plastic signs and veneers, canopies, and roof structures.

Those buildings already existing when a fire district is established may not be increased in height or area unless in compliance with the code for new buildings. Any new construction must also be of a type permitted within the fire district. Alterations may not increase the level of fire hazard in the building, nor may the occupancy be changed to a classification that is not permitted in the fire district. Any buildings located partially in the fire district are to be regulated as if they are in the district, provided at least 50 percent of the structure lies within the district or the building extends more than 10 feet (3048 mm) inside the fire district's boundaries.

Section D105 identifies a number of types of uses and structures permitted within a fire district that might otherwise be excluded. The listed uses are all relatively minor in nature and do not pose a significant hazard to the fire-safety level that is mandated. Such uses include small private garages and sheds, fences, tanks and towers, small greenhouses and wood decks. This section also permits a limited amount of alteration on dwellings of Type V construction.

Appendix E *Supplementary Accessibility Requirements*

This appendix addresses the design and construction of those accessible facilities not typically addressed by a building code. Although it is important that all reasonable efforts are made to make buildings accessible and usable, the body of the code in Chapter 11 contains only those requirements that directly relate to structures. Therefore, additional criteria are provided in this appendix to expand on the other accessibility features of the built environment.

Many of the facilities regulated by this chapter involve furnishings or equipment. In Section E103.1, the code requires that an accessible route be provided to a raised platform used as a head table or speaker's lectern. Section E104.2 regulates transient lodging features such as accessible beds and access to such beds, while communications features in Group I-3 occupancies, and transient lodging facilities are addressed in Section E104.3.

Water coolers, portable toilet and bathing rooms, laundry equipment, vending machines, mailboxes, automatic-teller machines, and fare machines are other possible features of a building that, through the regulation of this section, can be made more usable for individuals with physical disabilities. Telephones are fully addressed in Section E106, including provisions for wheelchair access, volume controls and TTYs. Section E107.2 mandates that where permanent signage designates the use or description of a room or area, tactile identification is also required. Directional and informational signs, as well as other special types of signage, are also addressed.

Three of the remaining sections of this appendix regulate specific types of uses or buildings. Bus stops and bus shelters must be designed and constructed in a manner that makes them accessible. Fixed transportations facilities, such as stations for rapid rail, light rail, commuter rail, high-speed rail and other fixed-guideway systems, must selectively have station entrances, signage, fare machines, platforms, TTYs, track crossings, public-address systems and clocks that are accessible or usable. Some of these same features are regulated for airports as well.

Appendix F *Rodentproofing*

In an effort to reduce the possibility of rodents entering a building, this appendix sets forth construction methods to seal those potential entry points. The provisions not only apply to habitable and occupiable rooms but also to any spaces containing feed, food or foodstuffs. The obvious intent is to prevent unsanitary conditions and the potential spread of disease that may follow.

All openings in the foundation walls are to be covered or sealed in a prescribed manner to prevent the passage of rodents. Doors and windows are regulated when located adjacent to ground level. It is also the intent of this appendix that an apron or similar protective barrier be installed where the foundation wall is not continuous. The intent of this appendix is an attempt to eliminate all potential avenues for rodent entry that occur around the exterior of a building.

Appendix G *Flood-resistant Construction*

Most jurisdictions in the United States have specific areas that are subject to flood conditions. This appendix is designed to reduce those losses, both public and private, that occur because of flooding. Administrative procedures and land-use limitations are set forth, intending to meet or exceed the regulations of the National Flood Insurance Program (NFIP).

The building-sciences provisions for flood-resistant design and construction are located in Section 1612. In conjunction with the provisions of this appendix, the regulations are consistent with the NFIP regulations.

Appendix H *Signs*

The design and installation of outdoor signs is regulated by this appendix. Signs can come in many shapes and sizes, and are used for many purposes. This appendix classifies signs based on their location such as ground signs, roof signs, wall signs, projecting signs, marquee signs and portable signs. The types of signs, including internally illuminated signs, combustible signs and animated devices, are also addressed.

This appendix identifies the areas of concern when signs are placed upon structures. It is important that any exit signs, fire escapes or egress openings remain unobstructed. Required natural ventilation openings must also remain available. Signs must be able to withstand all imposed loads, including any wind or seismic loads that may be encountered. The combustibility of signs is also regulated, with specific provisions for plastic materials.

Appendix I *Patio Covers*

Patio covers are one-story structures not exceeding 12 feet (3657 mm) in height. Enclosure walls may have any configuration, provided the open area of the longer wall and one additional wall is equal to at least 65 percent of the area below a minimum of 6 feet, 8 inches (2032 mm) of each wall, measured from the floor. Openings may be enclosed with insect screening, translucent or transparent plastic not more than $1/_8$ inch (3.2 mm) in thickness, or glass.

Patio covers may be detached or attached to dwelling units. Patio covers shall be used only for recreational, outdoor living purposes and not as carports, garages, storage rooms or habitable rooms.

Exterior openings required for light or ventilation may open into a patio structure conforming to this section. Where emergency egress or rescue openings from sleeping rooms lead to a patio structure, the structure shall be unenclosed. Where an exit from the dwelling unit passes through the patio structure, the structure shall be unenclosed or exits shall be provided in conformance with Chapter 10.

Patio covers shall be designed and constructed to sustain the applicable snow loads or all dead loads plus a vertical live load of 10 pounds per square foot (0.48 kN/m^2), whichever is greater. The minimum wind and seismic loads shall also be considered in the design.

A patio cover may be supported on concrete slab on grade without footings, provided the slab is not less than 3$^1/_2$ inches (89 mm) thick and further provided that the columns do not support live and dead loads in excess of 750 pounds (3.34 kN) per column.

Appendix J *Grading*

Not every jurisdiction is located in an area where the topography of the terrain requires grading operations on private property. In those areas where developers need to grade private property, Appendix J provides appropriate administrative and technical regulations to assure the jurisdiction of reasonable safety against slope failure, landslides and other soil failure hazards.

Appendix K *Administrative Provisions*

Appendix K is provided to allow those communities who adopt NFPA 70, the *National Electrical Code*® (NEC) to include administrative provisions that will assist in their implementation and enforcement. These provisions assist in the administration of the NEC by providing administrative language that correlates with that of the *International Codes*. In addition, the provisions established in Section K111 address technical issues that are additions or modifications to the requirements of the NEC.

INDEX

C

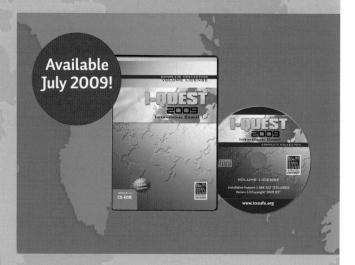

Don't Miss Out On Valuable ICC Membership Benefits. Join ICC Today!

Join the largest and most respected building code and safety organization. As an official member of the International Code Council®, these great ICC® benefits are at your fingertips.

EXCLUSIVE MEMBER DISCOUNTS

ICC members enjoy exclusive discounts on codes, technical publications, seminars, plan reviews, educational materials, videos, and other products and services.

TECHNICAL SUPPORT

ICC members get expert code support services, opinions, and technical assistance from experienced engineers and architects, backed by the world's leading repository of code publications.

FREE CODE—LATEST EDITION

Most new individual members receive a free code from the latest edition of the International Codes®. New corporate and governmental members receive one set of major International Codes (Building, Residential, Fire, Fuel Gas, Mechanical, Plumbing, Private Sewage Disposal).

FREE CODE MONOGRAPHS

Code monographs and other materials on proposed International Code revisions are provided free to ICC members upon request.

PROFESSIONAL DEVELOPMENT

Receive Member Discounts for on-site training, institutes, symposiums, audio virtual seminars, and on-line training! ICC delivers educational programs that enable members to transition to the I-Codes®, interpret and enforce codes, perform plan reviews, design and build safe structures, and perform administrative functions more effectively and with greater efficiency. Members also enjoy special educational offerings that provide a forum to learn about and discuss current and emerging issues that affect the building industry.

ENHANCE YOUR CAREER

ICC keeps you current on the latest building codes, methods, and materials. Our conferences, job postings, and educational programs can also help you advance your career.

CODE NEWS

ICC members have the inside track for code news and industry updates via e-mails, newsletters, conferences, chapter meetings, networking, and the ICC website (www.iccsafe.org). Obtain code opinions, reports, adoption updates, and more. Without exception, ICC is your number one source for the very latest code and safety standards information.

MEMBER RECOGNITION

Improve your standing and prestige among your peers. ICC member cards, wall certificates, and logo decals identify your commitment to the community and to the safety of people worldwide.

ICC NETWORKING

Take advantage of exciting new opportunities to network with colleagues, future employers, potential business partners, industry experts, and more than 50,000 ICC members. ICC also has over 300 chapters across North America and around the globe to help you stay informed on local events, to consult with other professionals, and to enhance your reputation in the local community.

JOIN NOW! 1-888-422-7233, x33804 | www.iccsafe.org/membership

INTERNATIONAL CODE COUNCIL®

People Helping People Build a Safer World™

09-01530

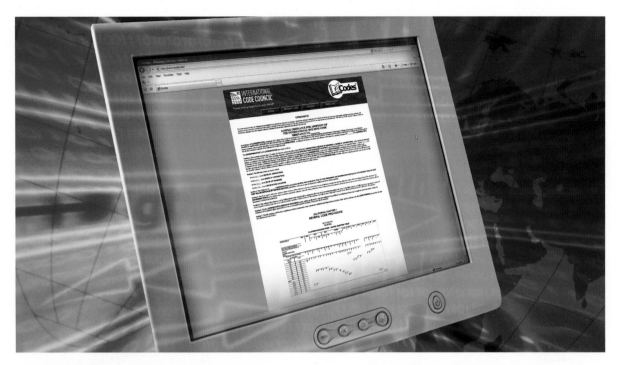 

Most Widely Accepted and Trusted

Innovative Building Products

Make sure they are up to code with ICC-ES Evaluation Reports

The ICC-ES Solution

ICC Evaluation Service® (ICC-ES®), a subsidiary of ICC®, was created to assist code officials and industry professionals in verifying that new and innovative building products meet code requirements. This is done through a comprehensive evaluation process that results in the publication of ICC-ES Evaluation Reports for those products that comply with requirements in the code or acceptance critera. Today, more code officials prefer using ICC-ES Evaluation Reports over any other resource to verify products comply with codes.

FREE Access to ICC-ES Evaluation Reports!

ICC EVALUATION SERVICE

Most Widely Accepted and Trusted

ICC-ES Evaluation Report | **ESR-4802**

Issued March 1, 2008
This report is subject to re-examination in one year.

www.icc-es.org | 1-800-423-6587 | (562) 699-0543 | *A Subsidiary of the International Code Council®*

DIVISION: 07—THERMAL AND MOISTURE PROTECTION
Section: 07410—Metal Roof and Wall Panels

REPORT HOLDER:

ACME CUSTOM-BILT PANELS
52380 FLOWER STREET
CHICO, MONTANA 43820
(808) 664-1512
www.custombiltpanels.com

EVALUATION SUBJECT:

CUSTOM-BILT STANDING SEAM METAL ROOF PANELS: CB-150

1.0 EVALUATION SCOPE

Compliance with the following codes:
- 2006 *International Building Code®* (IBC)
- 2006 *International Residential Code®* (IRC)

Properties evaluated:
- Weather resistance
- Fire classification
- Wind uplift resistance

2.0 USES

Custom-Bilt Standing Seam Metal Roof Panels are steel panels complying with IBC Section 1507.4 and IRC Section R905.10. The panels are recognized for use as Class A roof coverings when installed in accordance with this report.

3.0 DESCRIPTION

3.1 Roofing Panels:

Custom-Bilt standing seam roof panels are fabricated in steel and are available in the CB-150 and SL-1750 profiles. The panels are roll-formed at the jobsite to provide the standing seams between panels. See Figures 1 and 3 for panel profiles. The standing seam roof panels are roll-formed from minimum No. 24 gage [0.024 inch thick (0.61 mm)] cold-formed sheet steel. The steel conforms to ASTM A 792, with an aluminum-zinc alloy coating designation of AZ50.

3.2 Decking:

Solid or closely fitted decking must be minimum 15/32-inch-thick (11.9 mm) wood structural panel or lumber sheathing, complying with IBC Section 2304.7.2 or IRC Section R803, as applicable.

4.0 INSTALLATION

4.1 General:

Installation of the Custom-Bilt Standing Seam Roof Panels must be in accordance with this report, Section 1507.4 of the IBC or Section R905.10 of the IRC, and the manufacturer's

published installation instructions. The manufacturer's installation instructions must be available at the jobsite at all times during installation. The roof panels must be installed on solid or closely fitted decking, as specified in Section 3.2. Accessories such as gutters, drip angles, fascias, ridge caps, window or gable trim, valley and hip flashings, etc., are fabricated to suit each job condition. Details must be submitted to the code official for each installation.

4.2 Roof Panel Installation:

4.2.1 CB-150: The CB-150 roof panels are installed on roof shaving a minimum slope of 2:12 (17 percent). The roof panels are installed over the optional underlayment and secured to the sheathing with the panel clip. The clips are located at each panel rib side lap spaced 6 inches (152 mm) from all ends and at a maximum of 4 feet (1.22 m) on center along the length of the rib, and fastened with a minimum of two No. 10 by 1-inch pan head corrosion-resistant screws. The panel ribs are mechanically seamed twice, each pass at 90 degrees, resulting in a double-locking fold.

4.3 Fire Classification:

The steel panels are considered Class A roof coverings in accordance with the exception to IBC Section 1505.2 and IRC Section R902.1.

4.4 Wind Uplift Resistance:

The systems described in Section 3.0 and installed in accordance with Sections 4.1 and 4.2 have an allowable wind uplift resistance of 45 pounds per square foot (2.15 kPa).

5.0 CONDITIONS OF USE

The standing seam metal roof panels described in this report comply with, or are suitable alternatives to what is specified in, those codes listed in Section 1.0 of this report, subject to the following conditions:

5.1 Installation must comply with this report, the applicable code, and the manufacturer's published installation instructions. If there is a conflict between this report and the manufacturer's published installation instructions, this report governs.

5.2 The required design wind loads must be determined for each project. Wind uplift pressure on any roof area must not exceed 45 pounds per square foot (2.15 kPa).

6.0 EVIDENCE SUBMITTED

Data in accordance with the ICC-ES Acceptance Criteria for Metal Roof Coverings (AC166), dated October 2007.

7.0 IDENTIFICTION

Each standing seam metal roof panel is identified with a label bearing the product name, the material type and gage, the Acme Custom-Bilt Panels name and address, and the evaluation report number (ESR-4802).

ICC-ES Evaluation Reports are not to be construed as representing aesthetics or any other attributes not specifically addressed, nor are they to be construed as an endorsement of the subject of the report or a recommendation for its use. There is no warranty by ICC Evaluation Service, Inc., express or implied, as to any finding or other matter in this report, or as to any product covered by the report.

© 2008 Copyright

 ANSI

Page 1 of 1

William Gregory
Building and Plumbing Inspector
Town of Yorktown, New York

"We've been using ICC-ES Evaluation Reports as a basis of product approval since 2002. I would recommend them to any jurisdiction building department, particularly in light of the many new products that regularly move into the market. It's good to have a group like ICC-ES evaluating these products with a consistent and reliable methodology that we can trust."

Becky Baker, CBO
Director/Building Official
Jefferson County, Colorado

"The ICC-ES Evaluation Reports are designed with the end user in mind to help determine if building products comply with code. The reports are easily accessible, and the information is in a format that is useable by plans examiners and inspectors as well as design professionals and contractors."

VIEW ICC-ES EVALUATION REPORTS ONLINE!
www.icc-es.org

09-02246